S0-BNQ-327

HARFORD COMMUNITY COLLEGE LIBRARY
BEL AIR, MARYLAND 21014

SOLUTIONS MANUAL
TO ACCOMPANY

BRADY / HOLUM

FUNDAMENTALS OF CHEMISTRY

Third Edition

PREPARED BY

Paul L. Gaus

The College of Wooster

John Wiley & Sons

New York Chichester Brisbane Toronto Singapore

66711

Copyright © 1988 by John Wiley & Sons, Inc.

All rights reserved.

Reproduction or translation of any part of this work beyond that permitted by Section 107 or 108 of the 1976 United States Copyright Act without the permission of the copyright owner is unlawful. Requests for permission or further information should be addressed to the Permissions Department, John Wiley & Sons, Inc.

ISBN 0-471-62808-5

Printed in the United States of America

10987654321

QD31.2
.B69
c.20

Contents

CHAPTER ONE

PRACTICE EXERCISES

1. (a) 1-Ni and 2-Cl
   (b) 1-Fe, 1-S, and 4-O
   (c) 3-Ca, 2-P, and 8-O
   (d) 1-Cu, 1-S, 9-O, and 10-H

2. 1-Mg, 4-H, 2-O, and 2-Cl
   $Mg(OH)_2(s) + 2HCl(aq) \rightarrow MgCl_2(aq) + 2H_2O(\ell)$

3. $Mg(OH)_2(s) + 2HCl(aq) \rightarrow MgCl_2(aq) + 2H_2O(\ell)$

4. (a) Volume = Length(m) × Width(m) × Height(m) = m^3
   (b) Speed = Distance(m)/Time(s) = m/s

5. (a) nanometer (b) centimeter (c) kilometer (d) picometer
   (e) millimeter

6. °C = 5/9(°F - 32) = 5/9(86 °F - 32) = 30 °C
   °F = 9/5(°C) + 32 = 9/5(-17.8 °C) + 32 = -0.0400 °F

7. °C = 5/9(°F - 32) = 5/9(90 °F - 32) = 32 °C
   = 5/9(85 °F - 32) = 29 °C

8. °C = K - 273
   For 300 K, °C = 300 - 273 = 27
   For 315 K, °C = 315 - 273 = 42
   The temperature increase is 42 - 27 = 15 °C.

9. (a) 4.8 × 392 = 1881.6, which should be rounded off to 1900 so as to have two significant figures.
   (b) 7.255 ÷ 81.334 = 0.08920
   (c) 0.2983 + 1.52 = 1.82
   (d) 14.5403 - 0.022 = 14.518

10. (a) $64.25 \text{ in} \times \frac{1 \text{ ft}}{12 \text{ in.}} \times \frac{1 \text{ yd}}{3 \text{ ft}} = 1.785 \text{ yd}$

    (b) $64.25 \text{ in} = \frac{1 \text{ ft}}{12 \text{ in.}} \times \frac{1 \text{ mi}}{5280 \text{ ft}} = 0.001014 \text{ mi}$

11. (a) 2.3×10^4 (b) 2.17×10^7 (c) 1.5×10^{-3}

    (d) 2.7×10^{-5}

12. (a) 2,700 (b) 35,000,000,000,000,000,000,000,000,000
(c) 0.000000000002

13. (a) $3.00\ \text{yd} \times \frac{36\ \text{in.}}{1\ \text{yd}} = 108\ \text{in.}$

(b) $1.25\ \text{km} \times \frac{1000\ \text{m}}{\text{km}} \times \frac{100\ \text{cm}}{\text{m}} = 1.25 \times 10^5\ \text{cm}$

(c) $3.27\ \text{mm} \times \frac{1\ \text{cm}}{10\ \text{mm}} \times \frac{1\ \text{in.}}{2.540\ \text{cm}} \times \frac{1\ \text{ft}}{12\ \text{in.}} = 0.0107\ \text{ft}$

(d) $\frac{1\ \text{gal}}{20.2\ \text{mi}} \times \frac{4\ \text{qts}}{1\ \text{gal}} \times \frac{1\ \text{L}}{1.057\ \text{qts}} \times \frac{0.6215\ \text{mi}}{1\ \text{km}} = 0.116\ \text{L/km}$

14. $°\text{F} = \frac{9\ °\text{F}}{5\ °\text{C}}(30\ °\text{C}) + 32°\text{F} = 86\ °\text{F}$

15. $\text{density} = \text{mass/volume} = \frac{3.92\ \text{g}}{1.45\ \text{mL}} = 2.70\ \text{g/mL}$

16. $2.86\ \text{g silver} \times \frac{1\ \text{cm}^3}{10.5\ \text{g}} = 0.272\ \text{cm}^3$

$16.3\ \text{cm}^3\ \text{silver} \times \frac{10.5\ \text{g}}{1\ \text{cm}^3} = 171\ \text{g silver}$

17. $\text{sp. gr.} = \frac{2.70\ \text{g/ml}}{1.00\ \text{g/ml}} = 2.70$

$\frac{2.70\ \text{g}}{1.00\ \text{mL}} \times \frac{1.00\ \text{lb}}{453.6\ \text{g}} \times \frac{1.00\ \text{mL}}{1.00\ \text{cm}^3} \times \frac{(2.54\ \text{cm})^3}{1.00\ \text{in.}^3} \times \frac{(12.0\ \text{in.})^3}{1.00\ \text{ft}^3} = 169\ \text{lb/ft}^3$

18.

$\text{sp.gr.} = d_{\text{ethyl acetate}}/d_{\text{water}} = 0.902$

$d_{\text{ethyl acetate}} = d_{\text{water}} \times 0.902 = 1.00\ \text{g/mL} \times 0.902 = 0.902\ \text{g/mL}$

$d_{\text{ethyl acetate}} = d_{\text{water}} \times 0.902 = 8.34\ \text{lb/gal} \times 0.902 = 7.52\ \text{lb/gal}$

Chapter 1

REVIEW EXERCISES

General

1.1 Student answer.

1.2 The science of chemistry concerns the composition of substances and the way in which their composition relates to their properties.

1.3 Student answer.

1.4 Student answer.

Chemical Reactions

1.5 A chemical reaction is a process that transforms substances.

1.6 A chemical reaction has taken place because the substances that were originally present have been transformed into new substances having different characteristic properties.

1.7 Yes, a chemical reaction has occurred, because baking soda has reacted to give a new substance, carbon dioxide.

Matter and Energy

1.8 Matter is tangible stuff, and it characteristically has mass and occupies space. Energy is the capacity or potential to do work. All of the items in the question are examples of matter.

1.9 The mass that an object has does not depend on gravity, whereas its weight does depend on the particular force of gravity at the object's location.

1.10 (a) Kinetic energy is energy of motion.
(b) Potential energy is energy that is stored in a system or an object. Kinetic energy is determined by an object's mass and speed.

1.11 $KE = 1/2mv^2$, where m = mass and v = speed (velocity) of the object in motion.

1.12 Heat is a form of energy, and temperature is a measure of the amount (intensity) of heat energy in an object. Heat flows spontaneously from an object of higher temperature to one of lower temperature.

1.13 The carbon dioxide and water vapor have less potential energy, because the reaction of oxygen with gasoline (combustion) releases potential energy from the latter. The energy appears in the form of heat.

1.14 Mechanical energy is kinetic energy.

Properties of Matter

1.15 Physical properties are characteristic properties possessed by substances. A physical property can be measured and used to characterize a pure substance without reference to other substances. A chemical property of a substance governs the reactivity of the substance either with other substances or to give other substances. Thus a physical property characterizes the way a substance exists and a chemical property governs the way in which a substance reacts.

1.16 An extensive property depends on the extent or size of the sample that is at hand, whereas an intensive property is one that would be possessed by a substance regardless of the size of the sample that was being considered.

1.17 Student answer.

1.18 (a) Volume is an extensive property, and cannot be used to distinguish substances.
(b) Either liquid could occupy either volume, and it would not be possible to distinguish the two substances unless their intensive characteristics were known.

1.19 Intensive properties that could be used to distinguish water from gasoline are density, odor, color, etc. Flammability would be a chemical property that could distinguish the two.

Elements, Compounds and Mixtures

1.20 (a) An element is a pure substance that cannot be decomposed into something simpler.
(b) A compound is a pure substance that is composed of elements in some fixed and characteristic proportion.
(c) Mixtures result from combinations of pure substances in varying proportions.
(d) A homogeneous mixture has one phase.
(e) A heterogeneous mixture has more than one phase.
(f) A phase is a region of a mixture that has properties that are different from other regions of the mixture.
(g) A solution is a homogeneous mixture.
(h) A physical change is one that does not alter a substance's chemical

identity.

1.21 A chemical change is needed to separate a compound substance into the elements that compose it.

1.22 These are physical changes.

Chemical Symbols

1.23 (a) Cl (b) S (c) Fe (d) Ag (e) Na (f) P (g) I (h) Cu
(i) Hg (j) Ca

1.24 (a) potassium (e) manganese (h) aluminum
(b) zinc (f) magnesium (i) carbon
(c) silicon (g) nickel (j) nitrogen
(d) tin

Chemical Formulas

1.25 The smallest particle that is representative of a particular element is the atom of that element. A molecule is a representative unit that is made up of two or more atoms linked together.

1.26 (a) 1 (b) 2 (c) 8 (d) 4 (e) 8 (f) 10

1.27 (a) 3 Na, 1 P, 4 O (d) 3 Fe, 2 As, 8 O
(b) 1 Ca, 4 H, 2 P, 8 O (e) 1 Cu, 2 N, 6 O
(c) 4 C, 10 H (f) 1 Mg, 1 S, 11 O, 14 H

1.28 3 Ca, 5 Mg, 8 Si, 24 O, 2 H

1.29 (a) Zn and S (b) Mg and N (c) Ca and P (d) C and Cl

1.30 Hydrogen, H_2; Nitrogen, N_2; Oxygen, O_2; Fluorine, F_2; Chlorine, Cl_2; Bromine, Br_2; Iodine, I_2.

Chapter 1

Chemical Equations

1.31 (a) Magnesium reacts with oxygen to give (yield) magnesium oxide.
(b) The reactants are Mg and O_2.
(c) The product is MgO.

1.32 coefficients

1.33 (a) 6 (b) 3 (c) 27

1.34 (a) 16 (b) 36 (c) 50

1.35

$$2C_8H_{18}(\ell) + 25O_2(g) \rightarrow 16CO_2(g) + 18H_2O(\ell)$$

Scientific Method

1.36 The steps of the scientific method are: 1. observations (performing scientific experiments), 2. hypothesis (formulation of explanations and theories), and 3. testing of explanations by further observations).

1.37 It provides conditions necessary to carry out experiments under controlled settings so that the results are reproducible.

1.38 (a) data - empirical facts
(b) hypotheses - tentative explanations
(c) law - generalization based on the results of experiments and observation
(d) theory - tested explanation

Metric and SI Units

1.39 In the metric system, changes from one size of unit to another simply involve moving the decimal place.

1.40 Student answer. Some common answers are food packages of all sorts, soft drink containers, and the like.

1.41 Le System International d'Unites, or the International System of Units.

1.42 Mass

1.43 kg m^2/s^2

1.44 (a) 10^{-2} (b) 10^{-3} (c) 10^{3} (d) 10^{-6} (e) 10^{-9}

(f) 10^{-12} (g) 10^{6}

1.45 (a) c (b) m (c) k (d) μ (e) n (f) p (g) M

1.46 (a) centimeters or millimeters (b) millimeters (c) grams

1.47 Weight is the force with which a given object is attracted to the earth by gravity. Mass is the amount of matter in an object.

1.48 The mass of an object is measured by comparing its weight to that of objects with a known mass, using a balance. A scale measures only an object's weight.

1.49 (a) $1 \text{ cm} \times \frac{1 \text{ m}}{100 \text{ cm}} = 0.01 \text{ m}$

(b) $1 \text{ km} \times \frac{1000 \text{ m}}{1 \text{ km}} = 1000 \text{ m}$

(c) $1 \text{ m} \times \frac{1 \text{ pm}}{10^{-12} \text{ m}} = 1 \times 10^{12} \text{ pm}$

(d) $1 \text{ dm} \times \frac{1 \text{ m}}{10 \text{ dm}} = 0.1 \text{ m}$

(e) $1 \text{ g} \times \frac{1 \text{ kg}}{1000 \text{ g}} = 1 \times 10^{-3} \text{ kg}$

(f) $1 \text{ mg} \times \frac{1 \text{ g}}{1000 \text{ mg}} = 1 \times 10^{-3} \text{ g}$

1.50 (a) $1 \text{ nm} \times \frac{10^{-9} \text{ m}}{1 \text{ nm}} = 10^{-9} \text{ m}$

(b) $1 \text{ μg} \times \frac{10^{-6} \text{ g}}{1 \text{ μg}} = 10^{-6} \text{ g}$

(c) $1 \text{ kg} \times \frac{1000 \text{ g}}{1 \text{ kg}} = 1000 \text{ g}$

(d) $1 \text{ Mg} \times \frac{10^{6} \text{ g}}{1 \text{ Mg}} = 10^{6} \text{ g}$

(e) $1 \text{ mg} \times \frac{1 \text{ g}}{1000 \text{ mg}} = 10^{-3} \text{ g}$

(f) $1 \text{ dg} \times \frac{1 \text{ g}}{10 \text{ dg}} = 0.1 \text{ g}$

Temperature

1.51 The boiling point and the freezing point of water serve as reference temperatures. Water boils at 100 °C and freezes at 0 °C.

1.52 (a) °F = 9/5(°C) + 32 = 9/5(24) + 32 = 75 °F, when rounded to the proper number (two) of significant figures.

(b) °F = 9/5(10) + 32 = 50 °F

(c) °C = 5/9(°F - 32) = 5/9(41 - 32) = 5 °C

(d) °C = 5/9(50 - 32) = 10 °C

(e) K = °C + 273 = 30 + 273 = 303 K

(f) K = (-10) + 273 = 263 K

1.53 (a) °C = 5/9(°F - 32) = 5/9(85 - 32) = 29 °C, when rounded to the proper number (two) of significant figures.

(b) °C = 5/9(°F - 32) = 5/9[(-5) - 32] = -2×10^1, when rounded to the proper number (one) of significant figures.

(c) °F = 9/5(°C) + 32 = 9/5(-40) + 32 = -40 °F

(d) °C = K - 273 = 215 - 273 = -58 °C

(e) °C = 315 - 273 = 42 °C

(f) K = °C + 273 = 25 + 273 = 298 K

1.54 °F = 9/5(°C) + 32 = 9/5(37.13) + 32 = 98.83 °F

1.55 °C = 5/9(°F - 32) = 5/9[(-96 - 32)] = -71 °C

1.56 °C = K - 273 = 4 - 273 = -269 °C
°F = 9/5(-269) + 32 = -452 °F

1.57 °C = K - 273 = 5800 - 273 = 5527 °C

Significant Figures

1.58 The digits that are significant figures in a quantity are those that are known (measured) with certainty, plus the last digit, which contains some uncertainty.

1.59 The accuracy of a measured value is the closeness of the measured value to the true value of the quantity. The precision of a number of repeated measurements of the same quantity is the closeness of the measurements to one another.

1.60 (a) three (b) four (c) three (d) two (e) four (f) one

1.61 (a) three (b) five (c) one (d) four (e) six (f) four

1.62 (a) 6.9 (b) 83.14 (c) 0.006 (d) 22.5 (e) 946.5

1.63 (a) 2.06 (b) 4.02 (c) 8×10^4 (d) 0.01216 (e) 6×10^{-4}

Scientific Notation

1.64 (a) 2.45×10^{2} (b) 3.10×10^{4} (c) 2.87×10^{-3}
(d) 4.50×10^{7} (e) 4.00×10^{-8} (f) 3.24×10^{5}

1.65 (a) 3.389×10^{3} (b) 2.5×10^{-5} (c) 8.13×10^{7}
(d) 2.25×10^{-2} (e) 2.33×10^{0} (f) 1.83×10^{4}

1.66 (a) 2100 (b) 0.000335 (c) 3,800,000
(d) 0.00000000046 (e) 0.346 (f) 85,000

1.67 (a) 0.000427 (b) 71,100,000 (c) 0.0000335 (d) 0.00285
(e) 50,000 (f) 1,720,000

1.68 (a) 2.0×10^{4} (b) 8.0×10^{7} (c) 1.0×10^{3}
(d) 2.4×10^{5} (e) 2.0×10^{18}

1.69 (a) 4.0×10^{-2} (b) 2.0×10^{5} (c) 5.0×10^{39}
(d) 1.1×10^{5} (e) 2.55×10^{-2}

Factor-Label Method

1.70 A conversion factor is a fraction constructed from a valid relationship between quantities that have different units, for example from the relationship 12 in = 1 ft. The equation 1 yd = 2 ft is not a correct relationship between the the units yards and feet, so we cannot use it to construct a valid conversion factor. The relationship 1 cm = 1000 m is not correct; the conversion factor between cm and m must be constructed from the relationship 1 cm = 0.1 m.

1.71 Multiply 250 s by the conversion factor: 1 hr/3600 s. Multiply 3.84 hr by the conversion factor: 3600 s/hr.

1.72 Round off to four significant figures. The number 4.165 has four significant figures. The values for the conversion factors are exact quantities, and hence may be expressed with any desirable number of significant figures. It is the number 5.165, therefore, that requires the use of only four significant figures in the answer.

Unit Conversions by the Factor-Label Method

1.73

(a) $10.0 \text{ cm} \times \frac{1 \text{ m}}{100 \text{ cm}} \times \frac{1 \text{ km}}{1000 \text{ m}} = 1 \times 10^{-4} \text{ km}$

(b) $5.3 \text{ g} \times \frac{1 \text{ mg}}{10^{-3} \text{ g}} = 5.3 \times 10^{3} \text{ mg}$

(c) $5.3 \text{ mg} \times \frac{10^{-3} \text{ g}}{1 \text{ mg}} \times \frac{1 \text{ kg}}{1000 \text{ g}} = 5.3 \times 10^{-6} \text{ kg}$

(d) $37.5 \text{ mL} \times \frac{10^{-3} \text{ L}}{1 \text{ mL}} = 3.75 \times 10^{-2} \text{ L}$

(e) $0.125 \text{ L} \times \frac{1 \text{ mL}}{10^{-3} \text{ L}} = 125 \text{ mL}$

(f) $342 \text{ nm} \times \frac{10^{-9} \text{ m}}{1 \text{ nm}} \times \frac{1 \text{ mm}}{10^{-3} \text{ m}} = 3.42 \times 10^{-4} \text{ mm}$

1.74

(a) $1.83 \text{ nm} \times \frac{10^{-9} \text{ m}}{1 \text{ nm}} \times \frac{100 \text{ cm}}{1 \text{ m}} = 1.83 \times 10^{-7} \text{ cm}$

(b) $3.55 \times \frac{1 \text{ mg}}{10^{-3} \text{ g}} = 3.55 \times 10^{3} \text{ mg}$

(c) $8.44 \text{ km} \times \frac{1000 \text{ m}}{1 \text{ km}} \times \frac{100 \text{ cm}}{1 \text{ m}} = 8.44 \times 10^{5} \text{ cm}$

(d) $33 \text{ m} \times \frac{1000 \text{ mm}}{1 \text{ m}} = 3.3 \times 10^{4} \text{ mm}$

(e) $0.55 \text{ dm} \times \frac{0.1 \text{ m}}{1 \text{ dm}} \times \frac{1 \text{ km}}{1000 \text{ m}} = 5.5 \times 10^{-5} \text{ km}$

(f) $5.38 \text{ kg} \times \frac{1000 \text{ g}}{1 \text{ kg}} \times \frac{1 \text{ mg}}{10^{-3} \text{ g}} = 5.38 \times 10^{7} \text{ mg}$

1.75

(a) $36 \text{ in.} \times \frac{2.540 \text{ cm}}{1 \text{ in.}} = 91 \text{ cm}$

(b) $5.0 \text{ lb} \times \frac{1 \text{ kg}}{2.205 \text{ lb}} = 2.3 \text{ kg}$

(c) $3.0 \text{ qt} \times \frac{946.4 \text{ mL}}{1 \text{ qt}} = 2.8 \times 10^3 \text{ mL}$

(d) $8 \text{ oz} \times \frac{29.6 \text{ mL}}{1 \text{ oz}} = 237 \text{ mL}$

(e) $55 \frac{\text{mi}}{\text{hr}} \times \frac{1.609 \text{ km}}{1 \text{ mi}} = 88 \text{ km/hr}$

(f) $50.0 \text{ mi} \times \frac{1.609 \text{ km}}{1 \text{ mi}} = 80.5 \text{ km}$

1.76 (a) $250 \text{ mL} \times \frac{1 \text{ qt}}{946.4 \text{ mL}} = 0.26 \text{ qt}$

(b) $2.0 \text{ ft} \times \frac{1 \text{ yd}}{3 \text{ ft}} \times \frac{0.9114 \text{ m}}{1 \text{ yd}} = 0.61 \text{ m}$

(c) $1.33 \text{ kg} \times \frac{2.205 \text{ lb}}{1 \text{ kg}} = 2.93 \text{ lb}$

(d) $1.75 \text{ L} \times \frac{1000 \text{ ml}}{1 \text{ L}} \times \frac{1 \text{ fluid oz}}{29.6 \text{ mL}} = 59.1 \text{ fluid oz}$

(e) $75 \frac{\text{km}}{\text{hr}} \times \frac{0.6215 \text{ mi}}{1 \text{ km}} = 47 \text{ mi/hr}$

(f) $80.0 \text{ km} \times \frac{0.6215 \text{ mi}}{1 \text{ km}} = 49.7 \text{ mi}$

1.77 $12 \text{ fluid oz} \times \frac{29.6 \text{ mL}}{1 \text{ fluid oz}} = 3.6 \times 10^2 \text{ mL}$

1.78 $2 \text{ L} \times \frac{1000 \text{ mL}}{1 \text{ L}} \times \frac{1 \text{ fluid oz}}{29.6 \text{ mL}} = 7 \times 10^1 \text{ fluid oz}$

1.79 $1000 \text{ kg} \times \frac{2.205 \text{ lb}}{1 \text{ kg}} = 2.205 \times 10^3 \text{ lb}$

1.80 $2240 \text{ lb} \times \frac{1 \text{ kg}}{2.205 \text{ lb}} \times \frac{1 \text{ metric ton}}{1000 \text{ kg}} = 1.016 \text{ metric ton}$

1.81 $5 \text{ ft} \times \frac{12 \text{ in.}}{1 \text{ ft}} + 8 \text{ in.} = 68 \text{ in.}$

$68 \text{ in.} \times \frac{2.54 \text{ cm}}{1 \text{ in.}} = 1.7 \times 10^2 \text{ cm}$

1.82 $74.3 \text{ kg} \times \dfrac{2.205 \text{ lb}}{1 \text{ kg}} = 164 \text{ lb}$

1.83 (a) $8.0 \text{ yd}^2 \times \dfrac{0.9144 \text{ m}}{1 \text{ yd}} \times \dfrac{0.9144 \text{ m}}{1 \text{ yd}} = 6.7 \text{ m}^2$

(b) $3.4 \text{ in.}^2 \times \dfrac{(2.540 \text{ cm})^2}{1 \text{ in.}^2} = 22 \text{ cm}^2$

(c) $1.5 \text{ ft}^3 \times \left(\dfrac{12 \text{ in.}}{1 \text{ ft}}\right)^3 \times \left(\dfrac{1 \text{ m}}{39.37 \text{ in.}}\right)^3 \times \left(\dfrac{1 \text{ dm}}{0.1 \text{ m}}\right)^3 \times \dfrac{1 \text{ L}}{1 \text{ dm}^3} = 42 \text{ L}$

1.84 (a) $85 \text{ cm}^2 \times \dfrac{1 \text{ in.}}{2.540 \text{ cm}} \times \dfrac{1 \text{ in.}}{2.540 \text{ cm}} = 13 \text{ in.}^2$

(b) $3.3 \text{ m}^3 \times \left(\dfrac{39.37 \text{ in.}}{1 \text{ m}}\right)^3 \times \left(\dfrac{1 \text{ ft}}{12 \text{ in.}}\right)^3 = 12 \times 10^1 \text{ ft}^3$

(c) $144 \text{ in.}^2 \times \dfrac{1 \text{ m}}{39.37 \text{ in.}} \times \dfrac{1 \text{ m}}{39.37 \text{ in.}} = 9.29 \times 10^{-2} \text{ m}^2$

<u>More Practice with the Factor-Label Method</u>

1.85

$$124 \text{ francs} \times \frac{1 \text{ head cabbage}}{31 \text{ francs}} \times \frac{3 \text{ cans potatoes}}{1 \text{ head cabbage}} \times \frac{17 \text{ francs}}{1 \text{ can potatoes}} =$$

204 francs spent on potatoes

1.86

$$2155 \frac{\text{ft}}{\text{s}} \times \frac{60 \text{ s}}{1 \text{ min}} \times \frac{60 \text{ min}}{1 \text{ hr}} \times \frac{1 \text{ yd}}{3 \text{ ft}} \times \frac{0.9144 \text{ m}}{1 \text{ yd}} \times \frac{1 \text{ km}}{1000 \text{ m}} = 2365 \text{ km/hr}$$

1.87

$$1 \text{ yr} \times \frac{365 \text{ days}}{1 \text{ yr}} \times \frac{24 \text{ hr}}{1 \text{ day}} \times \frac{60 \text{ min}}{1 \text{ hr}} \times \frac{60 \text{ s}}{1 \text{ min}} \times \frac{3.0 \times 10^8 \text{ m}}{1 \text{ s}} \times$$

$$\times \frac{1 \text{ km}}{1 \times 10^3 \text{ m}} \times \frac{0.6215 \text{ mi}}{1 \text{ km}} = 5.9 \times 10^{12} \text{ mi}$$

1.88 55 mi/hr × 1 km/0.6215 mi = 88 km/hr

1.89 Each revolution moves the point by a distance equal to one circumference, that is $\pi d = 12\pi$ in. Therefore:

$$\frac{33.3 \text{ revs}}{1 \text{ min}} \times \frac{12(3.142) \text{ in.}}{1 \text{ rev}} \times \frac{1 \text{ ft}}{12 \text{ in.}} \times \frac{1 \text{ mi}}{5280 \text{ ft}} \times \frac{60 \text{ min}}{1 \text{ hr}} = 1.2 \text{ mi/hr}$$

1.90 The required round-trip distance is 478,000 mi.

$$4.78 \times 10^5 \text{ mi} \times \frac{1 \text{ km}}{0.6215 \text{ mi}} \times \frac{1000 \text{ m}}{1 \text{ km}} \times \frac{1 \text{ s}}{3.0 \times 10^8 \text{ m}} = 2.6 \text{ s}$$

Density and Specific Gravity

1.91 $25 \text{ gal} \times \frac{3.786 \text{ L}}{1 \text{ gallon}} \times \frac{1000 \text{ mL}}{1 \text{ L}} = 9.47 \times 10^4 \text{ mL}$

$9.47 \times 10^4 \text{ mL} \times 0.65 \frac{\text{g}}{\text{mL}} = 6.16 \times 10^4 \text{ g}$

$6.16 \times 10^4 \text{ g} \times \frac{1 \text{ kg}}{1000 \text{ g}} = 61.6 \text{ kg}$, which should be rounded to the value 62 kg.

$62 \text{ kg} \times 2.205 \text{ lb/kg} = 1.4 \times 10^2 \text{ lb}$

1.92 $\text{density} = \frac{\text{mass}}{\text{volume}} = \frac{25.3 \text{ g}}{31.7 \text{ mL}} = 0.798 \text{ g/mL}$

1.93 $10.0 \text{ g} \times \frac{1 \text{ mL}}{0.791 \text{ g}} = 12.6 \text{ mL}$

1.94 $21.335 \text{ g } H_2O \times \frac{1 \text{ mL}}{0.99704 \text{ g } H_2O} = 21.398 \text{ mL}$

1.95 $\text{density} = \frac{(62.00 - 27.35) \text{ g}}{(18.3 - 15.0) \text{ mL}} = 11 \text{ g/mL}$

1.96 (a) The volume of the pycnometer is found from the mass of the water that it contained:

$\text{volume} = \frac{(36.842 - 27.314) \text{ g}}{0.99704 \text{ g/mL}} = 9.556 \text{ mL}$

(b) The above volume is then used to determine the density of chloroform:

$\text{density} = \frac{(41.428 - 27.314) \text{ g}}{9.556 \text{ mL}} = 1.477 \text{ g/mL}$

1.97 The specific gravity of ethyl ether is the ratio of its density to that of water, if both densities are expressed with the same units:

$$\text{sp. gr.(ethyl ether)} = \frac{0.715\ \text{g/mL}}{1.00\ \text{g/mL}} = 0.715$$

1.98 As in problem 1.97:

$$\text{sp. gr.} = \frac{8.65\ \text{lb/gal}}{8.34\ \text{lb/gal}} = 1.04$$

1.99 The product of the specific gravity of trichloroethylene and the density of water is equal to the density of trichloroethylene:

$$\text{density} = 1.47 \times 0.998203 = 1.47\ \text{g/mL}$$

1.100 The specific gravity of gold (19.3) is the ratio between its density (in units lb/ft^3) and the density of water (also expressed in the units lb/ft^3). This means that the density of gold may be calculated by multiplying the specific gravity of gold by the density of water:

$$\text{density of gold} = 19.3 \times 62.4\ \text{lb/ft}^3 = 1.20 \times 10^3\ \text{lb/ft}^3$$

Hence one cubic foot of gold weighs 1.20×10^3 lb.

1.101 Because mass and volume are extensive properties, their individual values cannot be used to distinguish the liquids in question. The density however, being equal to the ratio of the mass to the volume of a sample, is an intensive property. As such, density can be used to identify and distinguish the sample of gasoline from the sample of water.

CHAPTER TWO

PRACTICE EXERCISES

1. The atomic number (94) is equal to the number of protons, and the mass number (94 + 146 = 240) is equal to the number of protons and neutrons:

$$^{240}_{94}Pu$$

2. The mass number minus the atomic number is equal to the number of neutrons: 4 - 2 = 2.

3. 2.24845 × 12 amu = 26.9814 amu

4. Copper is 63.546 ÷ 12 = 5.2955 times as heavy as carbon.

5. (a) For hydrogen we have (1.0079 ÷ 12) × 20 = 1.6798 amu.
   (b) For iron: (55.847 ÷ 12) × 20 = 93.078.
   (c) For sulfur: (32.06 ÷ 12) × 20 = 53.43.

6. (a) K, Ar, Al (b) Cl (c) Ba (d) Ne (e) Li (f) Ce

7. (a) NaF (b) Na_2O (c) MgF_2 (d) Al_2S_3

8. (a) $CrCl_3$ and Cr_2O_3 (b) CuCl, $CuCl_2$, Cu_2O and CuO

9. (a) Na_2CO_3 (b) $(NH_4)_2SO_4$ (c) $KC_2H_3O_2$

   (d) $Sr(NO_3)_2$ (e) $Fe(C_2H_3O_2)_3$

10. (a) potassium sulfide (b) magnesium phosphide
    (c) nickel(II) chloride (d) iron(III) oxide

11. (a) Al_2S_3 (b) SrF_2 (c) TiO_2 (d) $CrBr_2$

12. phosphorus trichloride, sulfur dioxide, dichlorine heptaoxide

13. hydrofluoric acid, and hydrobromic acid

14. sodium arsenate

15. $NaHSO_3$

Chapter 2

REVIEW EXERCISES

Laws of Chemical Combination and Dalton's Theory

2.1 Law of Conservation of Mass: Mass is neither created nor destroyed (gained nor lost) in a chemical reaction, i.e. mass is conserved. Law of Definite Proportions: In a given chemical compound, the elements are always found to be combined in the same proportion by mass. Law of Multiple Proportions: When two elements form more than one compound, the different masses of one element that are found to combine with a fixed mass of the other element in the various compounds are in the ratio of small whole numbers.

2.2 These are given on page 48 of the text. The essential ideas are the following: There are atoms; they are indestructible; all atoms of the same element have identical masses; the atoms of the different elements have different masses; chemical changes bring atoms together in definite proportions.

2.3 Atoms are indestructible.

2.4 Elements combine in definite ratios, as atoms. This guarantees that elements combine in definite mass ratios, assuming that atoms are indestructible.

2.5 Atoms can be broken, but not in chemical reactions. Also, not all of the atoms in a sample of a single element have identical masses, because of the occurrence of isotopes. An element in fact has an average atomic mass, and in chemical reactions elements behave as if they each had a single (average) mass.

2.6 Conservation of mass derives from the postulate that atoms are not destroyed in normal chemical reactions. The Law of Definite Proportions derives from the notion that compound substances are always composed of the same types and numbers of atoms of the various elements in the compound.

2.7 This is the Law of Definite Proportions, which guarantees that a single pure substance is always composed of the same ratio of masses of the elements that compose it.

2.8 An authentic sample of laughing gas must have a mass ratio of nitrogen/oxygen of 1.75 to 1.00. The only possibility in this list is item (c), which has the ratio of mass of nitrogen to mass of oxygen of 8.84/5.05 = 1.75.

2.9 The amount of oxygen per gram of nitrogen in NO_2 should be exactly twice that in NO, as required by the formulas of the two substances.

2.10 (a) This ratio should be 4/2 = 2/1, as required by the formulas of the two compounds.
(b) Twice 0.597, or 1.19 g.

Atoms and Isotopes

2.11 See Table 2.1 of the text, page 51.

2.12 Nearly all of the mass of an atom is in its nucleus, because this is the portion of the atom where the protons and neutrons are located.

2.13 atomic mass unit; μ

2.14 neutrons and protons

2.15 The atomic number is equal to the number of protons in the nucleus of the atom, and the mass number is the sum of the neutronsand protons.

2.16 The atomic number, which is numerically equal to the number of electrons on a neutral atom, is a better indication of the chemistry of an element, because it is the electrons which often govern the reactivity of an element.

2.17 The isotopes of an element have identical atomic numbers (identical numbers of protons and electrons), but differing numbers of neutrons.

2.18 Certain isotopes of different elements may have the same mass number, because the two different elements may coincidentally have the same total number of protons and neutrons.

2.19 (a) mass number (b) atomic number (c) electrical charge
(d) number of such atoms

2.20 (a) ${}^{131}_{53}I$ (b) ${}^{90}_{38}Sr$ (c) ${}^{137}_{55}Ce$ (d) ${}^{18}_{9}F$

2.21

	neutrons	protons	electrons
(a)	138	88	88
(b)	8	6	6
(c)	124	82	82
(d)	12	11	11

2.22

	neutrons	protons	electrons
(a)	82	55	55
(b)	78	53	53
(c)	146	92	92
(d)	118	79	79

Atomic Weights

2.23 This is carbon-12, which has a mass of 12 amu by definition:

$$^{12}_{6}C$$

2.24 $12 \times 1.6605665 \times 10^{-24}$ g = $1.9926798 \times 10^{-23}$

2.25 Since we know that the formula is CH_4, we know that one fourth of the

total mass due to the hydrogen atoms constitutes the mass that may be compared with that of carbon. Hence we have 0.33597 g ÷ 4 = 0.083993 g assigned to hydrogen and 1.000 grams assigned to the amount of C-12 in the compound. Then it is necessary to realize that the ratio 1.000 ÷ 12 for carbon is equal to the ratio 0.083993 ÷ X, where X equals the relative atomic mass of hydrogen:

8.3391 X / X

$$\frac{1.000}{12} = \frac{0.83993}{X} \qquad \therefore X = 1.008$$

2.26 Regardless of the definition, the ratio of the mass of hydrogen to that of oxygen would be the same. If C-12 were assigned a mass of 24 (twice its accepted value), then hydrogen would also have a mass twice its current value, or 2.0158 amu.

Molecules, Ions and Compounds

2.27 (a) atom-Na, ion-Na^+ (b) Yes (c) Yes (d) Yes (e) No

2.28 An atom that gains an electron acquires a negative charge.

2.29

	number of electrons	number of protons	number of neutrons
(a)	18	17	18
(b)	23	26	30
(c)	27	29	35
(d)	18	19	20

2.30 (a) An ionic compound is formed by the transfer of electrons, and is accompanied by the formation of ions of opposite charge.
(b) Molecular compounds arise from the sharing of electrons between atoms, rather than from the complete transfer of electrons as in (a).

The Periodic Table

2.31 Mendeleev observed that when the elements were arranged in order of increasing atomic weight, there was a periodic recurrence of properties of the elements.

2.32 Strontium and calcium are in the same Group of the periodic table, so they are expected to have similar chemical properties. Strontium should therefore form compounds that are similar to those of calcium, including the sorts of compounds found in bone.

2.33 Cadmium is in the same periodic table Group as zinc, but silver is not. Therefore, cadmium would be expected to have properties similar to those of zinc, whereas silver would not.

2.34 Silver and gold are in the same periodic table Group as copper, so they might well be expected to occur together in nature, because of their similar properties and tendencies to form similar compounds.

2.35 For selenium, the following predictions are reasonable:
atomic weight = 79.8, melting point = 281 °C, boiling point = 717 °C, formula of the oxide = XO_2, melting point of the oxide = 330 °C, density = 4.16 g/cm^3.

2.36 When the chemical elements are arranged in order of increasing atomic weight, there is a periodic recurrence of their properties.

2.37 A period in the periodic table is a horizontal row of elements. A group is one of the vertical columns of the periodic table.

2.38 Although all of the elements had not yet been discovered, Mendeleev left spaces for the ones that he predicted would eventually be discovered.

2.39 These occur between cobalt and nickel, thorium and protactinium, uranium and neptunium.

2.40 When the chemical elements are arranged in the order of their increasing atomic number, there is a periodic recurrence of their properties.

2.41 This is shown on page 60 of the text.

2.42 (a) The representative elements are designated, using the traditional US system as Groups IIIA through VIIA and 0, as shown in Figure 2.5 on page 59 of the text.

(b) The representative elements are designated, using the IUPAC notation as Groups 13 through 18, as shown in Figure 2.5 on page 59 of the text.

2.43 (a) Group IA = Group 1.
(b) Group VIIA = Group 17
(c) Group IIIB = Group 3
(d) Group IB = Group 11
(e) Group IVA = Group 14

2.44 This would be highly unlikely because it would require the element to have a fractional atomic number.

2.45 Li

2.46 I

2.47 W

2.48 Xe

2.49 Sm

2.50 Pu

2.51 Mg

Physical Properties of Metals, Nonmetals and Metalloids

2.52 Luster, electrical conductivity, thermal conductivity, ductility, and malleability are the characteristic properties of metals.

2.53 Mercury is used in thermometers because it is a liquid and tungsten is used in light bulbs because it has such a high melting point.

2.54 ductility

2.55 malleability

2.56 copper and gold

2.57 the noble gases: He, Ne, Ar, Kr, Xe, and Rn.

2.58 $H_2(g)$, $N_2(g)$, $O_2(g)$, $F_2(g)$, $Cl_2(g)$, $Br_2(\ell)$, and $I_2(s)$

2.59 mercury (a metal) and $Br_2(\ell)$

2.60 They are semiconductors.

2.61 This is shown in Figure 2.7 on page 61 of the text.

2.62 Metals that are used to make jewelry are those that do not corrode: silver, gold, and platinum. Iron would be useless for jewelry, because it is susceptible to rusting.

2.63 Metals react with nonmetals.

2.64 An ionic compound contains ions (e.g. Na^+ and Cl^-) which are held together by the electrostatic attraction among oppositely charged ions in an extended geometrical array. Individual molecules cannot be found in ionic compounds because such substances are extended geometrical arrays of ions in the solid.

2.65 Binary compounds such as CCl_4 contain two elements only. A

diatomic substance is composed of molecules having two atoms. In the latter, the two atoms may or may not be the same (e.g. HCl and Cl_2).

2.66 A post-transition metal is one that occurs after a row of transition elements. Examples are gallium (Ga), indium (In) and thallium (Tl).

2.67 A cation is an ion with a positive charge, whereas an anion is an ion with a negative charge.

Formulas, Ions and Ionic Compounds

2.68 (a) K^+ (b) Br^- (c) Mg^{2+} (d) S^{2-} (e) Al^{3+}

2.69 (a) Ba^{2+} (b) O^{2-} (c) F^- (d) Sr^{2+} (e) Rb^+

2.70 The formulas for ionic compounds are always electrically neutral, overall, and they are empirical formulas. The cation (positive ion) is written first.

2.71 (a) NaBr (b) KI (c) BaO (d) $MgBr_2$ (e) BaF_2

2.72 The incorrect ones are a, d, and e.

2.73 (a) $Ca(s) + Cl_2(g) \rightarrow CaCl_2(s)$

(b) $2Mg(s) + O_2(g) \rightarrow 2MgO(s)$

(c) $4Al(s) + 3O_2(g) \rightarrow 2Al_2O_3(s)$

(d) $S(s) + 2Na(s) \rightarrow Na_2S(s)$

or $S_8(s) + 16Na(s) \rightarrow 8Na_2S(s)$

2.74 (a) CN^- (b) NH_4^+ (c) NO_3^- (d) SO_3^{2-} (e) ClO_3^-

2.75 (a) OCl^- (b) HSO_4^- (c) PO_4^{3-} (d) $H_2PO_4^-$ (e) MnO_4^-

2.76 (a) dichromate ion (b) hydroxide ion (c) acetate ion (d) carbonate ion (e) perchlorate ion

2.77 (a) KNO_3 (b) $Ca(C_2H_3O_2)_2$ (c) NH_4Cl (d) $Fe_2(CO_3)_3$

(e) $Mg_3(PO_4)_2$

2.78 (a) $Zn(OH)_2$ (b) Ag_2CrO_4 (c) $BaSO_3$ (d) Rb_2SO_4

(e) $LiHCO_3$

2.79 Ti^{4+}

2.80 (a) PbO and PbO_2 (b) SnO and SnO_2 (c) MnO and Mn_2O_3

(d) FeO and Fe_2O_3 (e) Cu_2O and CuO

2.81 (a) $CdCl_2$ (b) AgCl (c) $ZnCl_2$ (d) $NiCl_2$

Nonmetal-Nonmetal Compounds

2.82 Nonmetals combine with metals and other nonmetals.

2.83 The atoms in nonmetal-nonmetal compounds are held to each other by the sharing of electrons between the atoms. In compounds of a nonmetal with a metal, ions are formed, and the ions are held together in the solid by the electrostatic attractions between oppositely charged ions.

2.84 Metals are found to combine only with the nonmetals, whereas the nonmetals are also found in combination with other nonmetal elements.

2.85 The noble gases, which are fairly unreactive, monatomic elements.

2.86 (a) CH_4 (b) NH_3 (c) H_2Te (d) HI

2.87 HAt

2.88 SnH_4

2.89 Bi_2O_3 and Bi_2O_5

Chapter 2

Properties of Ionic and Molecular Compounds

2.90 They are normally hard solids. They have high melting points. Although they are nonconductors in the solid state, they are good conductors of electricity when either molten or in aqueous solution.

2.91 In the molten state, the ions of an ionic compound are free to move about, which is required for electrical conduction.

2.92 The ions become separated and move about in solution freely, allowing electrical conduction.

2.93 In molecular substances, the atoms are held together by the sharing of electrons between adjacent atoms. In ionic compounds, ions are formed by transfer of electrons between atoms; the resulting ions of opposite charge then attract one another and form a solid. See also the answer to question 2.83.

2.94 The ionic compounds have regular arrangements of ions in a three dimensional array. When there is slight movement of one layer of ions with respect to another, ions of like charge find themselves adjacent to one another. The resulting repulsion is destabilizing, and the solid crystal splits (cleaves) along the layer line.

2.95 The ionic bond (the attraction of ions of opposite charge within a crystalline ionic compound) is very strong, and the melting of the solid requires this bond to be disrupted, allowing the individual ions to move freely.

2.96 Napthalene is a molecular compound; an ionic one would have a higher melting point.

2.97 Ionic compounds are hard, and they have high melting points. Ionic compounds are always, therefore, solids. Molecular compounds tend to be soft, and they can have low melting points. Furthermore, molecular compounds can be solids, liquids or gases.

Acids and Bases

2.98 An acid is a substance that produces the H_3O^+ ion in water solution. A base is a substance that produces the OH^- ion in water solution. A neutralization reaction is the reaction of an acid with a base, the most fundamental being:

$$H_3O^+ + OH^- \rightarrow 2H_2O$$

Otherwise the reaction of an acid with a base also produces a salt. A salt is the product (other than water) of the reaction of an acid with a base.

2.99 the hydronium ion

2.100 The H_3O^+ ion is simply a proton, H^+, with a water molecule attached.

2.101 $Mg(OH)_2 + 2HCl \rightarrow MgCl_2 + 2H_2O$

2.102 $Al(OH)_3 + 3HCl \rightarrow AlCl_3 + 3H_2O$

2.103 (a) $HNO_2 + KOH \rightarrow KNO_2 + H_2O$

(b) $2HCl + Ca(OH)_2 \rightarrow CaCl_2 + 2H_2O$

(c) $H_2SO_4 + 2NaOH \rightarrow Na_2SO_4 + 2H_2O$

(d) $3HClO_4 + Al(OH)_3 \rightarrow Al(ClO_4)_3 + 3H_2O$

(e) $2H_3PO_4 + 3Ba(OH)_2 \rightarrow Ba_3(PO_4)_2 + 6H_2O$

2.104 Monoprotic acid - an acid that can provide only one H^+ ion per molecule of acid. Polyprotic acid - an acid that can provide more than one H^+ ion per molecule of acid. Diprotic acid - an acid that can provide precisely two H^+ ions per molecule of acid.

2.105 An acidic anhydride reacts with water to give an acid. Acidic anhydrides are also usually nonmetal oxides. A basic anhydride reacts with water to give a base. Basic anhydrides are also often metal oxides.

2.106 Na_2HPO_4 and NaH_2PO_4

2.107 X is a nonmetal because it behaves as an acidic anhydride. It is likely from Group V of the periodic table, because it has the oxidation state 3+ in its oxide.

Naming Compounds

2.108 (a) calcium sulfide (c) aluminum bromide (e) sodium phosphide
(b) sodium fluoride (d) magnesium carbide (f) lithium nitride

2.109 (a) chromic chloride; chromium(III) chloride
(b) manganic oxide; manganese(III) oxide
(c) cupric oxide; copper(II) oxide
(d) mercurous chloride; mercury(I) chloride
(e) stannic oxide; tin(IV) oxide
(f) plumbous sulfide; lead(II) sulfide

2.110 (a) silicon dioxide
(b) chlorine trifluoride
(c) xenon tetrafluoride
(d) disulfur dichloride
(e) tetraphosphorus decaoxide
(f) dinitrogen pentaoxide

2.111 periodic acid

2.112 (a) sodium nitrite
(b) potassium phosphate
(c) potassium permanganate
(d) ammonium acetate
(e) barium sulfate
(f) ferric carbonate; iron(III) carbonate

2.113 H_2CrO_4

2.114

(a) Na_2HPO_4	(e) $Ni(CN)_2$	(i) Al_2Cl_6
(b) Li_2Se	(f) Fe_2O_3	(j) As_4O_{10}
(c) NaH	(g) SnS_2	(k) $Mg(OH)_2$
(d) $Cr(C_2H_3O_2)_3$	(h) SbF_5	(l) $Cu(HSO_4)_2$

2.115

(a) $(NH_4)_2S$	(e) Fe_2O_3	(i) SiF_4
(b) $Cr_2(SO_4)_3$	(f) $Ca(BrO_3)_2$	(j) BCl_3
(c) MoS_2	(g) $Hg(C_2H_3O_2)_2$	(k) SnS
(d) $SnCl_4$	(h) $Ba(HSO_3)_2$	(l) Ca_3P_2

2.116 (a) hypochlorous acid and sodium hypochlorite
(b) iodous acid and sodium iodite
(c) bromic acid and sodium bromate
(d) perchloric acid and sodium perchlorate

CHAPTER THREE

PRACTICE EXERCISES

1. The conversion factor that allows this calculation is the ratio of the mass of chlorine to that of sodium:

$$\frac{2.059 \text{ g Cl}}{1.335 \text{ g Na}} = \frac{1.542 \text{ g Cl}}{1 \text{ g Na}}$$

The Law of Definite Proportions guarantees that this same ratio of Cl to Na will be found in every authentic sample of sodium chloride. Therefore, we may write:

$$2.366 \text{ g Cl} \times \frac{1 \text{ g Na}}{1.542 \text{ g Cl}} = 1.534 \text{ g Na}$$

2. (a) 1Na + 1Cl =
(1 × 23.0) + (1 × 35.5) = 58.5

(b) 12C + 22H + 11O =
(12 × 12.0) + (22 × 1.01) + (11 × 16.0) = 324.0

(c) 1Ca + 6C + 10H + 4O =
(1 × 40.1) + (6 × 12.0) + (10 × 1.01) + (4 × 16.0) = 186.2

(d) 2N + 8H + 1Fe + 2S + 8O =
(2 × 14.0) + (8 × 1.01) + (1 × 55.8) + (2 × 32.1) + (8 × 16.0)
= 284.1

3.

$$250 \text{ g } H_2O \times \frac{1 \text{ mol } H_2O}{18.0 \text{ g } H_2O} \times \frac{6.02 \times 10^{23} \text{ molecules}}{1 \text{ mole}} = 8.36 \times 10^{24} \text{ molecules}$$

4.

$$0.22 \text{ mol O} \times \frac{2 \text{ mol Fe}}{3 \text{ mol O}} = 0.15 \text{ mol Fe}$$

5.

$$0.50 \text{ mol } P_4O_{10} \times \frac{4 \text{ mol P}}{1 \text{ mol } P_4O_{10}} = 2.0 \text{ mol P}$$

$$0.50 \text{ mol } P_4O_{10} \times \frac{10 \text{ mol O}}{1 \text{ mol } P_4O_{10}} = 5.0 \text{ mol O}$$

6.

(a)
$$0.586 \text{ mol } H_2O \times \frac{18.0 \text{ g } H_2O}{1 \text{ mol } H_2O} = 10.5 \text{ g } H_2O$$

(b)
$$0.586 \text{ g } C_6H_{12}O_6 \times \frac{180 \text{ g } C_6H_{12}O_6}{1 \text{ mol } C_6H_{12}O_6} = 105 \text{ g } C_6H_{12}O_6$$

(c)
$$0.586 \text{ mol Fe} \times \frac{55.8 \text{ g Fe}}{1 \text{ mol Fe}} = 32.7 \text{ g Fe}$$

(d)
$$0.586 \text{ mol } CH_4 \times \frac{16.0 \text{ g } CH_4}{1 \text{ mol } CH_4} = 9.38 \text{ g } CH_4$$

7.

(a)
$$100.0 \text{ g } NH_3 \times \frac{1 \text{ mol } NH_3}{17.03 \text{ g } NH_3} = 5.87 \text{ mol } NH_3$$

(b)
$$100.0 \text{ g } C_{27}H_{46}O \times \frac{1 \text{ mol } C_{27}H_{46}O}{386.6 \text{ g } C_{27}H_{46}O} = 0.2587 \text{ mol } C_{27}H_{46}O$$

(c)
$$100.0 \text{ g Au} \times \frac{1 \text{ mol Au}}{197.0 \text{ g Au}} = 0.5076 \text{ mol Au}$$

(d)
$$100.0 \text{ g } C_2H_6O \times \frac{1 \text{ mol } C_2H_6O}{46.07 \text{ g } C_2H_6O} = 2.171 \text{ mol } C_2H_6O$$

8. This is because the formula weight of ammonia (NH_3 = 17.03 grams per mole) is smaller than that of $C_{27}H_{46}O$ (386.6 grams per mole). One mole of the latter weighs more than a mole of ammonia, although a mole of each substance contains the same number of molecules.

9.

$$\frac{40.4 \text{ g}}{12.04 \times 10^{23} \text{ atoms}} \times \frac{6.02 \times 10^{23} \text{ atoms}}{1 \text{ mol}} = \frac{20.2 \text{ g}}{\text{mol}}$$

10. The % by weight of an element in a compound is given by the formula:

$$\frac{\text{weight of the element}}{\text{given weight of the compound}} \times 100$$

(a) $\% \text{ C} = \frac{0.1945 \text{ g}}{0.4620 \text{ g}} \times 100 = 42.10$

(b) $\% \text{ H} = \frac{0.02977 \text{ g}}{0.4620 \text{ g}} \times 100 = 6.444$

(c) $\% \text{ O} = \frac{0.2377 \text{ g}}{0.4620 \text{ g}} \times 100 = 51.45$

Note also that the sum of these values (42.10 + 6.44 + 52.45) is equal to 100 %, as it should be if no other elements are present in the substance.

11. The formula weight for C_4H_{10} is given by the following:

$$4\text{C} + 10\text{H} = (4 \times 12.01) + (10 \times 1.009) = 58.12 \text{ g/mole}$$

The % by weight of an element in a compound is given by the formula:

$$\frac{\text{weight of the element in 1 mole of the compound}}{\text{formula weight of the compound}} \times 100$$

(a) $\% \text{ C in } C_4H_{10} = \frac{48.04 \text{ g C}}{58.12 \text{ g } C_4H_{10}} \times 100 = 82.66$

(b) $\% \text{ H in } C_4H_{10} = \frac{10.08 \text{ g H}}{58.12 \text{ g } C_4H_{10}} \times 100 = 17.34$

Note that the sum of these two values equals 100 %, as it should since this compound has only C and H in it.

12. The formula weight of $Cr(CO)_2$ is given by:

$$1\text{Cr} + 2\text{C} + 2\text{O} = (1 \times 52.00) + (2 \times 12.01) + (2 \times 16.00) = 108.02$$

Similarly the formula weight of $Cr(CO)_6$ is given by:

$$1\text{Cr} + 6\text{C} + 6\text{O} = (1 \times 52.00) + (6 \times 12.01) + (6 \times 16.00) = 220.06$$

With this information it is possible to solve for the expected % by weight of Cr, C, and O for each of the two possible compositions:

(a) for $Cr(CO)_2$:

$$\% \text{ Cr} = \frac{52.00 \text{ g Cr}}{108.02 \ Cr(CO)_2} \times 100 = 48.14$$

$$\% \text{ C} = \frac{24.02 \text{ g C}}{108.02 \text{ g Cr(CO)}_2} \times 100 = 22.24$$

$$\% \text{ O} = \frac{32.00 \text{ g O}}{108.02 \text{ g Cr(CO)}_2} \times 100 = 29.62$$

(b) for $Cr(CO)_6$:

$$\% \text{ Cr} = \frac{52.00 \text{ g Cr}}{220.06 \text{ g Cr(CO)}_6} \times 100 = 23.63$$

$$\% \text{ C} = \frac{72.06 \text{ g C}}{220.06 \text{ g Cr(CO)}_6} \times 100 = 32.75$$

$$\% \text{ O} = \frac{96.00 \text{ g O}}{220.06 \text{ g Cr(CO)}_6} \times 100 = 43.62$$

The data are clearly more consistent with the formula $Cr(CO)_6$.

13. We begin by converting the composition, which has been given in terms of the number of grams of each element, to a corresponding composition in terms of moles of each element:

$$2.549 \text{ g Fe} \times \frac{1 \text{ mol Fe}}{55.85 \text{ g Fe}} = 4.564 \times 10^{-2} \text{ mol Fe}$$

$$1.947 \text{ g O} \times \frac{1 \text{ mol O}}{16 \text{ g O}} = 1.217 \times 10^{-1} \text{ mol O}$$

$$0.9424 \text{ g P} \times \frac{1 \text{ mol P}}{30.97 \text{ g P}} = 3.043 \times 10^{-2} \text{ mol P}$$

Next the relative number of moles of each element is determined by dividing the number of moles of each by the smallest of the various numbers, in this case 3.043×10^{-2} moles:

$4.564 \times 10^{-2} \div 3.043 \times 10^{-2} = 1.500$ moles of Fe relative to moles of P

$1.217 \times 10^{-1} \div 3.043 \times 10^{-2} = 3.999$ moles of O relative to moles of P

$3.043 \times 10^{-2} \div 3.043 \times 10^{-2} = 1.000$ moles of P

Although these relative mole amounts are numerically correct, by convention we represent empirical formulas using integers. Thus each of the above relative mole amounts is multiplied by the number (in this case 2) that will transform each relative mole amount into an integer. The resulting formula is $Fe_3P_2O_8$.

14. It is convenient to convert the % by weight data into the number of grams of each element that are found in 100 grams of the compound, because these

weights are given by the % values directly: 84.98 g Hg and 15.02 g Cl. Next we determine the number of moles of each element that are represented by these weights:

$$84.98 \text{ g Hg} \times \frac{1 \text{ mol Hg}}{200.6 \text{ g Hg}} = 0.4236 \text{ mol Hg}$$

$$15.02 \text{ g Cl} \times \frac{1 \text{ mol Cl}}{35.45 \text{ g Cl}} = 0.4237 \text{ mol Cl}$$

The relative numbers of moles of Hg and Cl are properly expressed as integers, which in this case is the simple ratio of 1 to 1. The empirical formula is thus HgCl.

15. It is necessary to use the combustion analysis to determine the original number of grams of Ba and of S in the sample, and then to determine the number of grams of O by difference.

(a) The number of grams of Ba in the original sample are determined as follows:

$$0.5771 \text{ g BaO} \times \frac{1 \text{ mol BaO}}{153.4 \text{ g BaO}} \times \frac{1 \text{ mol Ba}}{1 \text{ mol BaO}} \times \frac{137.3 \text{ g Ba}}{1 \text{ mol Ba}}$$

$= 0.5165$ g Ba in the original compound, as well as in the product.

(b) The number of moles of S in the original sample are determined as follows:

$$0.3012 \text{ g } SO_3 \times \frac{1 \text{ mol } SO_3}{80.07 \text{ g } SO_3} \times \frac{1 \text{ mol S}}{1 \text{ mol } SO_3} \times \frac{32.06 \text{ g S}}{1 \text{ mol S}}$$

$= 0.1206$ g S in the original compound, as well as in the SO that is produced.

(c) The mass of O may now be determined by difference, because all of the remaining mass of the original compound is due only to O:

$$\text{mass of O} = 0.8778 - (0.5165 + 0.1206) = 0.2407 \text{ g}$$

The calculation is continued by converting the mass amounts of each element in the compound into mole amounts:

$$0.5165 \text{ g Ba} \times \frac{1 \text{ mol Ba}}{137.3 \text{ g Ba}} = 3.762 \times 10^{-3} \text{ mol Ba}$$

$$0.1206 \text{ g S} \times \frac{1 \text{ mol S}}{32.06 \text{ g S}} = 3.762 \times 10^{-3} \text{ mol S}$$

$$0.2407 \text{ g O} \times \frac{1 \text{ mol O}}{16.0 \text{ g O}} = 1.504 \times 10^{-2} \text{ mol O}$$

The relative number of moles of each element is thus given by:

$3.762 \times 10^{-3} \div 3.762 \times 10^{-3} = 1.000$ mol Ba

$3.762 \times 10^{-3} \div 3.762 \times 10^{-3} = 1.000$ mol S

$1.504 \times 10^{-2} \div 3.762 \times 10^{-3} = 3.999$ mol O

The empirical formula is $BaSO_4$, and the name of the substance is barium sulfate.

16. The molecular formula will either be the same as the empirical formula, or it will be a simple integer multiple of the empirical formula. The unit weight of HgCl is 236.0, which is half the known value for molecular weight. This means that the molecular weight must be twice that of the unit weight, and it means that the molecular formula must be twice HgCl, namely Hg_2Cl_2.

17. First it is necessary to determine the empirical formula by using the % composition data. In 100 grams of the compound there are:

$$87.42 \text{ g N} \times \frac{1 \text{ mol N}}{14.01 \text{ g N}} = 6.240 \text{ mol N}$$

$$12.58 \text{ g H} \times \frac{1 \text{ mol H}}{1.008 \text{ g H}} = 12.48 \text{ mol H}$$

Next we determine the relative mole proportions of the two elements in the compound:

$6.240 \div 6.240 = 1.000$ mol N
$12.48 \div 6.240 = 2.000$ mol H

The empirical formula is therefore NH_2, and the weight of the empirical unit is 1N + 2H = 16.03. The molecular mass divided by the empirical mass gives us the factor by which the empirical formula is to be multiplied in order to obtain the molecular formula:

$$32.1 \div 16.03 = 2.00$$

Hence the molecular formula is twice the empirical formula, namely N_2H_4.

Chapter 3

REVIEW EXERCISES

Chemical Calculations - Avogadro's Number and Moles

3.1 A conversion factor may be constructed by using the masses of aluminum and carbon that are required in forming the carbide:

$$\frac{18.74 \text{ g Al}}{6.26 \text{ g C}}$$

The Law of Definite Proportions allows us to use the same mass ratio for other samples of aluminum carbide:

$$8.240 \text{ g C} \times \frac{18.74 \text{ g Al}}{6.26 \text{ g C}} = 24.7 \text{ g Al}$$

The factor-label method as applied in the above calculation assures us that the conversion factor is constructed properly.

3.2 If the two compounds are identical, then the Law of Definite Proportions guarantees that they will have the same ratio of masses of elements. Hence we can use a conversion factor as follows:

$$10.00 \text{ g silicon} \times \frac{8.348 \text{ g chlorine}}{1.652 \text{ g silicon}} = 50.53 \text{ g chlorine}$$

The factor-label method shows that units cancel properly in the above calculation.

3.3 The term formula weight may apply to any substance that is composed of atoms, namely to elements, isotopes of elements and compounds of any variety.

3.4 (a) Na_2CO_3 = 2Na + 1C + 3O
= (2 × 23.0) + (12.0) + (3 × 16.0)
= 106.0

(b) $(NH_4)_2B_4O_7$ = 2N + 8H + 4B + 7O
= (2 × 14.0) + (8 × 1.01) + (4 × 10.8) + (7 × 16.0)
= 191.2

(c) $Ca(C_6H_{12}NSO_3)_2$ = Ca + 12C + 24H + 2N + 2S + 6O
= 40.1 + (12 × 12.0) + (24 × 1.01) + (2 × 14.0) + (2 × 32.1) + (6 × 16.0)
= 396.5

3.5 (a) $Ca(NO_3)_2$ = Ca + 2N + 6O
= 40.1 + (2 × 14.0) + (6 × 16.0)
= 164.1

(b) $Mg_3(PO_4)_2$ = 3Mg + 2P + 8O
= (3 × 24.3) + (2 × 31.0) + (8 × 16.0)
= 262.9

(c) $C_6H_8O_6$ = 6C + 8H + 6O
= (6 × 12.0) + (8 × 1.01) + (6 × 16.0)
= 176.1

3.6 A mole of water has more mass than a molecule of water, since there are 6.02×10^{23} molecules in a mole.

3.7 A mole of a molecular compound contains Avogadro's number (6.02×10^{23}) of molecules. The single molecular unit is the molecule.

3.8 The molar mass of a substance is its formula weight expressed in the units grams per mole (g/mol). The two values are numerically equal, but the molar mass is the number of grams that contain a mole of the material.

3.9 grams per mole, g/mol

3.10

$$39.1 \text{ g K} \times \frac{1 \text{ mol K}}{39.1 \text{ g K}} \times \frac{6.02 \times 10^{23} \text{ atoms}}{1 \text{ mol}} = 6.02 \times 10^{23} \text{ atoms of K}$$

which is Avogadro's number of atoms.

3.11

(a) $$12.04 \times 10^{23} \text{ atoms Mg} \times \frac{1 \text{ mol}}{6.02 \times 10^{23} \text{ atoms}} = 1.999 \text{ mol Mg}$$

(b) $$1.999 \text{ mol Mg} \times \frac{24.31 \text{ g Mg}}{1 \text{ mol Mg}} = 48.60 \text{ g Mg}$$

3.12

(a) $$0.100 \text{ mol } Na_2CO_3 \times \frac{106 \text{ g } Na_2CO_3}{1 \text{ mol } Na_2CO_3} = 10.6 \text{ g } Na_2CO_3$$

(b) $$0.100 \text{ mol } (NH_4)_2B_4O_7 \times \frac{191 \text{ g } (NH_4)_2B_4O_7}{1 \text{ mol } (NH_4)_2B_4O_7} = 19.1 \text{ g } (NH_4)_2B_4O_7$$

(c) $$0.100 \text{ mol } Ca(C_6H_{12}NSO_3)_2 \times \frac{397 \text{ g } Ca(C_6H_{12}NSO_3)_2}{1 \text{ mol } Ca(C_6H_{12}NSO_3)_2}$$

$$= 39.7\ \text{g Ca}(C_6H_{12}NSO_3)_2$$

3.13

(a)

$$0.250\ \text{mol Ca}(NO_3)_2 \times \frac{164\ \text{g Ca}(NO_3)_2}{1\ \text{mol Ca}(NO_3)_2} = 41.0\ \text{g Ca}(NO_3)_2$$

(b)

$$0.250\ \text{mol } Mg_3(PO_4)_2 \times \frac{263\ \text{g } Mg_3(PO_4)_2}{1\ \text{mol } Mg_3(PO_4)_2} = 65.8\ \text{g } Mg_3(PO_4)_2$$

$$0.250\ \text{g } C_6H_8O_6 \times \frac{176\ \text{g } C_6H_8O_6}{1\ \text{mol } C_6H_8O_6} = 44.0\ \text{g } C_6H_8O_6$$

3.14

$$9.00\ \text{g } H_2O \times \frac{1\ \text{mol } H_2O}{18.0\ \text{g } H_2O} = 0.500\ \text{mol } H_2O$$

3.15

$$1.70\ \text{g NaNO}_3 \times \frac{1\ \text{mol NaNO}_3}{85.0\ \text{g NaNO}_3} = 2.00 \times 10^{-2}\ \text{mol NaNO}_3$$

3.16

$$35.8\ \text{mol } (NH_4)_2SO_4 \times \frac{132\ \text{g } (NH_4)_2SO_4}{1\ \text{mol } (NH_4)_2SO_4} = 473 \times 10^1\ \text{g } (NH_4)_2SO_4$$

3.17

$$0.500\ \text{mol } C_{12}H_{22}O_{11} \times \frac{342\ \text{g } C_{12}H_{22}O_{11}}{1\ \text{mol } C_{12}H_{22}O_{11}} = 171\ \text{g } C_{12}H_{22}O_{11}$$

3.18

$$840\ \text{g ZnCl}_2 \times \frac{1\ \text{mol ZnCl}_2}{136\ \text{g ZnCl}_2} = 6.18\ \text{mol ZnCl}_2$$

3.19

$$1.00\ \text{kg } Tl_2SO_4 \times \frac{1000\ \text{g}}{1\ \text{kg}} \times \frac{1\ \text{mol } Tl_2SO_4}{505\ \text{g } Tl_2SO_4} = 1.98\ \text{mol } Tl_2SO_4$$

3.20

$$3.02 \times 10^{23} \text{ atoms B} \times \frac{1 \text{ atom N}}{1 \text{ atom B}} = 3.02 \times 10^{23} \text{ atoms N}$$

$$3.02 \times 10^{23} \text{ atoms N} \times \frac{1 \text{ mol N}}{6.02 \times 10^{23} \text{ atoms N}} \times \frac{14.0 \text{ g N}}{1 \text{ mol N}} = 7.02 \text{ g N}$$

3.21

$$12.04 \times 10^{23} \text{ atoms C} \times \frac{1 \text{ atom Ca}}{2 \text{ atoms C}} = 6.02 \times 10^{23} \text{ atoms Ca}$$

$$6.02 \times 10^{23} \text{ atoms Ca} \times \frac{1 \text{ mol Ca}}{6.02 \times 10^{23} \text{ atoms Ca}} \times \frac{40.1 \text{ g Ca}}{1 \text{ mol Ca}} = 40.1 \text{ g Ca}$$

3.22

$$0.500 \text{ mol } Ca(IO_3)_2 \times \frac{2 \text{ mol I}}{1 \text{ mol } Ca(IO_3)_2} \times \frac{6.02 \times 10^{23} \text{ atoms}}{1 \text{ mol}} = 6.02 \times 10^{23} \text{ atoms of I}$$

$$0.500 \text{ mol } Ca(IO_3)_2 \times \frac{390 \text{ g } Ca(IO_3)_2}{1 \text{ mol } Ca(IO_3)_2} = 195 \text{ g } Ca(IO_3)_2$$

3.23

$$0.750 \text{ mol } (NH_4)_2CO_3 \times \frac{2 \text{ mol N}}{1 \text{ mol } (NH_4)_2CO_3} \times \frac{6.02 \times 10^{23} \text{ atoms}}{1 \text{ mol}}$$

$$= 9.03 \times 10^{23} \text{ atoms N}$$

$$0.750 \text{ mol } (NH_4)_2CO_3 \times \frac{96.1 \text{ g } (NH_4)_2CO_3}{1 \text{ mol } (NH_4)_2CO_3} = 72.1 \text{ g } (NH_4)_2CO_3$$

3.24

$$0.665 \text{ mol } NH_4NO_3 \times \frac{2 \text{ mol N}}{1 \text{ mol } NH_4NO_3} \times \frac{6.02 \times 10^{23} \text{ atoms}}{\text{mol}} = 8.01 \times 10^{23} \text{ N atoms}$$

$$0.665 \text{ mol } NH_4NO_3 \times \frac{80.1 \text{ g } NH_4NO_3}{1 \text{ mol } NH_4NO_3} = 53.3 \text{ g } NH_4NO_3$$

3.25

$$6.26 \text{ mol} \times \frac{132 \text{ g}}{\text{mol}} = 826 \text{ g } (NH_4)_2HPO_4$$

3.26

$$4.34 \text{ mol} \times \frac{234 \text{ g}}{\text{mol}} = 1.02 \times 10^3 \text{ g } Ca(H_2PO_4)_2$$

3.27

$$100 \text{ lb} \times \frac{454 \text{ g}}{1 \text{ lb}} \times \frac{1 \text{ mol } (NH_4)_2HPO_4}{132 \text{ g } (NH_4)_2HPO_4} = 344 \text{ mol } (NH_4)_2HPO_4$$

3.28

$$525 \text{ lb} \times \frac{454 \text{ g}}{1 \text{ lb}} \times \frac{1 \text{ mol } Ca(H_2PO_4)_2}{234 \text{ g } Ca(H_2PO_4)_2} = 1.02 \times 10^3 \text{ mol}$$

3.29

$$4.65 \text{ mol } NaBO_3 \times \frac{81.8 \text{ g } NaBO_3}{\text{mol}} = 380 \text{ g } NaBO_3$$

3.30

$$0.568 \text{ mol } BaSO_4 \times \frac{233 \text{ g } BaSO_4}{\text{mol}} = 132 \text{ g } BaSO_4$$

3.31

$$13 \text{ oz NaOH} \times \frac{28.4 \text{ g}}{1 \text{ oz}} \times \frac{1 \text{ mol NaOH}}{40.0 \text{ g NaOH}} = 9.23 \text{ mol NaOH}$$

$$6.0 \text{ lb tallow} \times \frac{454 \text{ g}}{1 \text{ lb}} \times \frac{1 \text{ mol tallow}}{891 \text{ g tallow}} = 3.06 \text{ mol tallow}$$

$$\text{mole ratio} = \frac{9.23 \text{ mol NaOH}}{3.06 \text{ mol tallow}} = 3.02$$

3.32

(a) For KNO_3:

$$1.50 \text{ oz} \times \frac{28.4 \text{ g}}{1 \text{ oz}} = 42.6 \text{ g } KNO_3$$

$$42.6 \text{ g} \times \frac{1 \text{ mol}}{101 \text{ g}} = 0.422 \text{ mol } KNO_3$$

(b) For $CaSO_4$:

$$1.00 \text{ oz} \times \frac{28.4 \text{ g}}{1 \text{ oz}} = 28.4 \text{ g } CaSO_4$$

$$28.4 \text{ g} \times \frac{1 \text{ mol}}{136 \text{ g}} = 0.209 \text{ g } CaSO_4$$

(c) For $MgSO_4$:

$$0.750 \text{ oz} \times \frac{28.4 \text{ g}}{1 \text{ oz}} = 21.3 \text{ g } MgSO_4$$

$$21.3 \text{ g} \times \frac{1 \text{ mol}}{120 \text{ g}} = 0.178 \text{ mole } MgSO_4$$

(d) For $CaHPO_4$:

$$0.500 \text{ oz} \times \frac{28.4 \text{ g}}{1 \text{ oz}} = 14.2 \text{ g } CaHPO_4$$

$$14.2 \text{ g} \times \frac{1 \text{ mol}}{136 \text{ g}} = 0.104 \text{ mol } CaHPO_4$$

(e) For $(NH_4)_2SO_4$:

$$0.250 \text{ oz} \times \frac{28.4 \text{ g}}{1 \text{ oz}} = 7.10 \text{ g } (NH_4)_2SO_4$$

$$7.10 \text{ g} \times \frac{1 \text{ mol}}{132 \text{ g}} = 0.0538 \text{ mol } (NH_4)_2SO_4$$

3.33

$$\frac{32 \times 10^9 \text{ lb } NH_3}{1 \text{ year}} \times \frac{454 \text{ g}}{1 \text{ lb}} \times \frac{1 \text{ mol } NH_3}{17.0 \text{ g } NH_3} = 8.5 \times 10^{11} \text{ mol } NH_3 \text{ /year}$$

3.34

$$40 \times 10^6 \text{ tons } H_2SO_4 \times \frac{2000 \text{ lb}}{1 \text{ ton}} \times \frac{454 \text{ g}}{1 \text{ lb}} \times \frac{1 \text{ mol } H_2SO_4}{98.1 \text{ g } H_2SO_4} = 3.7 \times 10^{11} \text{ mol } H_2SO_4$$

Percent Composition

3.35 (a) For NH_3, the molar mass (1N + 3H) equals 17.0 g/mole, and the % by weight N is given by:

$$\frac{14.0 \text{ g N}}{17.0 \text{ g } NH_3} \times 100 = 82.4 \text{ \% N}$$

(b) For $CO(NH_2)_2$, the molar mass (2N + 4H + 1C + 1O) equals 60.0 g/mole, and the % by weight N is given by:

$$\frac{28.0 \text{ g N}}{60.0 \text{ g } CO(NH_2)_2} \times 100 = 46.7 \text{ \% N}$$

3.36 (a) For $NH_4H_2PO_4$, the molar mass (1N + 6H + 1P + 4O) equals 115.1 g/mole, and the % by weight values are:

$$\frac{14.0 \text{ g N}}{115 \text{ g } NH_4H_2PO_4} \times 100 = 12.2 \text{ \% N}$$

$$\frac{31.0 \text{ g P}}{115 \text{ g } NH_4H_2PO_4} \times 100 = 27.0 \text{ \% P}$$

(b) For $(NH_4)_2HPO_4$, the molar mass (2N + 9H + 1P + 4O) equals 132 g/mole, and the % by weight values are:

$$\frac{28.0 \text{ g N}}{132 \text{ g } (NH_4)_2HPO_4} \times 100 = 21.2 \text{ \% N}$$

$$\frac{31.0 \text{ g P}}{132 \text{ g } (NH_4)_2HPO_4} \times 100 = 23.5 \text{ \% P}$$

3.37 For $C_{17}H_{25}N$, the molar mass (17C + 25H + 1N) equals 243.38 g/mole, and the three theoretical values for % by weight are calculated as follows:

(a) Theoretical % by weight carbon in a pure sample:

$$\frac{204.2 \text{ g C}}{243.38 \text{ g } C_{17}H_{25}N} \times 100 = 83.901 \text{ \% C}$$

(b) Theoretical % by weight hydrogen in a pure sample:

$$\frac{25.20 \text{ g H}}{243.38 \text{ g } C_{17}H_{25}N} \times 100 = 10.35 \text{ \% H}$$

(c) Theoretical % by weight nitrogen in a pure sample:

$$\frac{14.01 \text{ g N}}{243.38 \text{ g } C_{17}H_{25}N} \times 100 = 5.756 \text{ \% N}$$

These data are consistent with the experimental values cited in the problem.

3.38 For $C_{20}H_{25}N_3O$, the molar mass (20C + 25H + 3N + O) equals 323.43 g/mole, and the theoretical values for % by weight are calculated as follows:

(a) Theoretical % by weight carbon:

$$\frac{240.2 \text{ g C}}{323.43 \text{ g } C_{20}H_{25}N_3O} \times 100 = 74.27 \text{ \% C}$$

(b) Theoretical % by weight H:

$$\frac{25.20 \text{ g H}}{323.43 \text{ g } C_{20}H_{25}N_3O} \times 100 = 7.792 \text{ \% H}$$

(c) Theoretical % by weight N:

$$\frac{42.03 \text{ g N}}{323.43 \text{ g } C_{20}H_{25}N_3O} \times 100 = 13.00 \text{ \% N}$$

(d) Theoretical % by weight O:

$$\frac{16.00 \text{ g O}}{323.43 \text{ g } C_{20}H_{25}N_3O} \times 100 = 4.947 \text{ \% O}$$

Empirical and Molecular Formulas

3.39 We need the % by weight composition to determine the empirical formula.

3.40 The empirical formula gives only the ratio of numbers of atoms, expressed as the smallest whole numbers that are possible. The molecular formula is generally some integer multiple of the empirical formula, and it gives the number of atoms that compose one molecule of the substance.

3.41 The molecular formula is some integer multiple of the empirical formula. This means that we can divide the molecular formula by the largest possible whole number that gives an integer ratio among the atoms in the empirical formula.

(a) C_1H_1 (b) CH_2O (c) C_4H_9 (d) $N_2H_4O_3$, or NH_4NO_3

3.42 We need the molar mass in order to determine the integer that, when multiplied by the formula weight, will give the molecular weight.

3.43 The value of % by weight for each element in a compound substance represents the mass of that element that is present in 100 g of the compound. These % by weight values in grams can be converted into the number of moles of each element that are present in the same 100 g of the compound. The empirical formula is then given by the lowest whole number ratio of the moles of the various elements that are present in the compound.

3.44 The lowest whole number ratio among the atoms in a compound substance gives the empirical formula only. This ratio in general will have to be multiplied by some whole number in order to obtain the molecular formula. It is only the molecular formula that will always be consistent with the molecular weight. In certain instances, the empirical formula and the molecular formula may be identical.

3.45 We begin by realizing that the mass of oxygen in the compound may be determined by difference:

$$0.9872 \text{ g total} - (0.1220 \text{ g Na} + 0.5255 \text{ g Tc}) = 0.3397 \text{ g O}$$

Next we can convert each mass of an element into the corresponding number of moles of that element as follows:

(a) for sodium:

$$0.1220 \text{ g} \times \frac{1 \text{ mol}}{22.99 \text{ g}} = 5.307 \times 10^{-3} \text{ mol Na}$$

(b) for technetium:

$$0.5255 \text{ g} \times \frac{1 \text{ mol}}{98 \text{ g}} = 5.4 \times 10^{-3} \text{ mol Tc}$$

(c) for oxygen:

$$0.3397 \text{ g} \times \frac{1 \text{ mol}}{16.00 \text{ g}} = 2.123 \times 10^{-2} \text{ mol O}$$

Next we divide each of these numbers of moles by the smallest of the three numbers, in order to obtain the simplest mole ratio among the three elements in the compound:

for Na, $5.307 \times 10^{-3} \div 5.307 \times 10^{-3} = 1.000$

for Tc, $5.4 \times 10^{-3} \div 5.307 \times 10^{-3} = 1.0$

for O, $2.123 \times 10^{-2} \div 5.307 \times 10^{-3} = 4.000$

These relative mole amounts give us the empirical formula directly, since the relative mole amounts are each obviously the smallest possible whole numbers: $NaTcO_4$.

3.46 First the mass of oxygen in the compound can be determined by subtraction: 0.8162 g total - (0.1936 g S + 0.2361 g K) = 0.3865 g O
Next, the mole amounts of the various elements are determined:

(a) for potassium:

$$0.2361 \text{ g} \times \frac{1 \text{ mol}}{39.10 \text{ g}} = 6.038 \times 10^{-3} \text{ mol K}$$

(b) for sulfur:

$$0.1936 \text{ g} \times \frac{1 \text{ mol}}{32.06 \text{ g}} = 6.039 \times 10^{-3} \text{ mol S}$$

(c) for oxygen:

$$0.3865 \text{ g} \times \frac{1 \text{ mol}}{16.00 \text{ g}} = 2.416 \times 10^{-2} \text{ mol O}$$

Next determine the relative number of moles for each element by dividing the above mole amounts by that which is the smallest:

for K, $6.038 \times 10^{-3} \div 6.038 \times 10^{-3} = 1.000$

for S, $6.039 \times 10^{-3} \div 6.038 \times 10^{-3} = 1.000$

for O, $2.416 \times 10^{-2} \div 6.038 \times 10^{-3} = 4.001$

Since these relative mole amounts are already obviously whole numbers, within experimental error, then the empirical formula is given by the above numbers directly: KSO_4. The weight of this empirical unit (135)

must be some whole number multiple of the true formula weight. In fact it is exactly half the value for the known formula weight:

$$270/135 = 2$$

We can now deduce that the molecular formula is twice the empirical formula, namely $K_2S_2O_8$.

3.47 First the mass of oxygen in the compound is determined by difference: 0.8138 g total - (0.1927 + 0.02590 + 0.1124 + 0.1491) = 0.3337 g O
Next, as in problem 3.45 and 3.46, we convert from grams of an element to moles of that element:

(a) for carbon:

$$0.1927\ \text{g} \times \frac{1\ \text{mol}}{12.01\ \text{g}} = 1.605 \times 10^{-2}\ \text{mol C}$$

(b) for hydrogen:

$$0.02590\ \text{g} \times \frac{1\ \text{mol}}{1.008\ \text{g}} = 2.569 \times 10^{-2}\ \text{mol H}$$

(c) for nitrogen:

$$0.1124\ \text{g} \times \frac{1\ \text{mol}}{14.01\ \text{g}} = 8.023 \times 10^{-3}\ \text{mol N}$$

(d) for phosphorus:

$$0.1491\ \text{g} \times \frac{1\ \text{mol}}{30.97\ \text{g}} = 4.814 \times 10^{-3}\ \text{mol P}$$

(e) for oxygen:

$$0.3337\ \text{g} \times \frac{1\ \text{mol}}{16.00\ \text{g}} = 2.086 \times 10^{-2}\ \text{mol O}$$

Next convert these mole amounts to relative values by dividing each by the smallest of the four:

for C, $1.605 \times 10^{-2} \div 4.814 \times 10^{-3} = 3.334$

for H, $2.569 \times 10^{-2} \div 4.814 \times 10^{-3} = 5.337$

for N, $8.023 \times 10^{-3} \div 4.814 \times 10^{-3} = 1.667$

for P, $4.814 \times 10^{-3} \div 4.814 \times 10^{-3} = 1.000$

for O, $2.086 \times 10^{-2} \div 4.814 \times 10^{-3} = 4.333$

Although the above relative mole amounts represent numerically correct values, by convention we express the empirical formula using the smallest possible whole numbers. If all of the above relative mole amounts are multiplied by the number 3, then we obtain integers for use in an empirical formula, which becomes $C_{10}H_{16}N_5P_3O_{13}$. The formula weight of the empirical unit is 507 g/mole, the sum of all of the atomic weights composing the unit. Since the formula weight of the empirical unit is equal to the molar mass, the molecular weight is also 507 g/mole, and the molecular formula and the empirical formula are the same.

3.48 First convert the number of grams of each element to the number of moles of each that are present in the compound:

(a) for arsenic:

$$0.4774 \text{ g As} \times \frac{1 \text{ mol}}{74.92 \text{ g}} = 6.372 \times 10^{-3} \text{ mol As}$$

(b) for sulfur, the number of grams is first found by difference:
0.6817 g total - 0.4774 g As = 0.2043 g S

$$0.2043 \text{ S} \times \frac{1 \text{ mol}}{32.06 \text{ g S}} = 6.372 \times 10^{-3} \text{ mol S}$$

Next, as in problems 3.45 through 3.47, the relative mole amounts are determined:

for As, $6.372 \times 10^{-3} \div 6.372 \times 10^{-3} = 1.000$

for S, $6.372 \times 10^{-3} \div 6.372 \times 10^{-3} = 1.000$

Hence the empirical formula is As_1S_1. The formula weight from the empirical formula is the sum of the atomic weights of As and S, 107 g/mole. The molar mass or true formula weight is 428. The empirical formula is thus a whole number multiple of the molecular formula:

$$428 \div 107 = 4$$

and the molecular formula is four times the empirical formula, namely, As_4S_4.

3.49 Proceeding as in problems 3.45 through 3.48:

For carbon,

$$0.5555 \text{ g} \times \frac{1 \text{ mol}}{12.01 \text{ g}} = 4.625 \times 10^{-2} \text{ mol C}$$

For hydrogen,

0.6481 g total - 0.5555 g C = 0.0926 g H

$$0.0926 \text{ g} \times \frac{1 \text{ mol}}{1.008 \text{ g}} = 9.19 \times 10^{-2} \text{ mol H}$$

The moles of each element relative to those of carbon (the smaller of the two) are then:

$4.625 \times 10^{-2} \div 4.625 \times 10^{-2} = 1.000$ relative moles of C

$9.19 \times 10^{-2} \div 4.625 \times 10^{-2} = 1.99$ relative moles of H and the empirical formula is C_1H_2. The formula weight from the empirical formula is thus (1C + 2H) = 14 g/mole. The factor by which the empirical formula is to be multiplied in arriving at the molecular formula is determined by the ratio:

$$57 \div 14 = 4$$

and the molecular formula is four times the empirical formula, or C_4H_8.

3.50 Proceeding as in problems 3.45 through 3.49:

(a) for carbon,

$$0.1570 \text{ g} \times \frac{1 \text{ mol}}{12.01 \text{ g}} = 1.307 \times 10^{-2} \text{ mol C}$$

(b) for hydrogen,

$$0.01317 \text{ g} \times \frac{1 \text{ mol}}{1.008 \text{ g}} = 1.307 \times 10^{-2} \text{ mol H}$$

(c) for nitrogen,

$$0.1832 \text{ g} \times \frac{1 \text{ mol}}{14.01 \text{ g}} = 1.308 \times 10^{-2} \text{ mol N}$$

(d) for oxygen, which constitutes the remaining mass of the sample,

$$0.2093 \text{ g} \times \frac{1 \text{ mol}}{16.00 \text{ g}} = 1.308 \times 10^{-2} \text{ mol O}$$

Next the relative mole amounts of each element are determined by dividing each mole amount by the smallest mole amount:

for C, $1.307 \times 10^{-2} \div 1.307 \times 10^{-2} = 1.000$

for H, $1.307 \times 10^{-2} \div 1.307 \times 10^{-2} = 1.000$

for N, $1.308 \times 10^{-2} \div 1.307 \times 10^{-2} = 1.000$

for O, $1.308 \times 10^{-2} \div 1.307 \times 10^{-2} = 1.000$

and the empirical formula is seen to be CHNO, which has a formula weight of 43. It can be seen that the number 43 must be multiplied by the integer 3 in order to obtain the molar mass (3 × 43 = 129), and this means that the empirical formula should similarly be multiplied by 3 in order to arrive at the molecular formula, $C_3H_3N_3O_3$.

3.51 The empirical formula is obtained by the same procedure used in problems 3.45 through 3.50:

(a) for hydrogen,

$$0.0220 \text{ g} \times \frac{1 \text{ mole}}{1.008 \text{ g}} = 2.183 \times 10^{-2} \text{ mol H}$$

(b) for phosphorus,

$$0.3374 \text{ g} \times \frac{1 \text{ mol}}{30.97 \text{ g}} = 1.089 \times 10^{-2} \text{ mol P}$$

(c) for oxygen,

$$0.5227 \text{ g} \times \frac{1 \text{ mol}}{16.00 \text{ g}} = 3.267 \times 10^{-2} \text{ mol O}$$

The relative mole amounts are next determined:

for H, $2.183 \times 10^{-2} \div 1.089 \times 10^{-2} = 2.005$

for P, $1.089 \times 10^{-2} \div 1.089 \times 10^{-2} = 1.000$

for O, $3.267 \times 10^{-2} \div 1.089 \times 10^{-2} = 3.000$

and the empirical formula is seen to be H_2PO_3, having a formula weight of

81 g/mole. Since this is half the molecular weight, the empirical formula is to be multiplied by two in obtaining the molecular formula, $H_4P_2O_6$.

3.52 If we assume that we have a sample weighing 100 g, then the % by weight values conveniently are transformed directly into mass values to be used in arriving at an empirical formula, exactly as in problems 3.45 through 3.51:

(a) for strontium,

$$47.70 \text{ g} \times \frac{1 \text{ mol}}{87.62 \text{ g}} = 0.5444 \text{ mol Sr}$$

(b) for sulfur,

$$17.46 \text{ g} \times \frac{1 \text{ mol}}{32.06 \text{ g}} = 0.5446 \text{ mol S}$$

(c) for oxygen,

$$34.84 \text{ g} \times \frac{1 \text{ mol}}{16.00 \text{ g}} = 2.178 \text{ mol O}$$

Next we determine relative mole amounts:

for Sr, $0.5444 \div 0.5444 = 1.000$

for S, $0.5446 \div 0.5444 = 1.000$

for O, $2.178 \div 0.5444 = 4.001$

from which we deduce that the empirical formula is $SrSO_4$. Since the formula weight from the empirical formula (184) is equal to the known molecular weight, the empirical formula and the molecular formulas are the same.

3.53 We proceed as in problem 3.52:

(a) for mercury,

$$77.26 \text{ g} \times \frac{1 \text{ mol}}{200.6 \text{ g}} = 0.385 \text{ mol Hg}$$

(b) for carbon,

$$9.25 \text{ g} \times \frac{1 \text{ mol}}{12.01 \text{ g}} = 0.770 \text{ mol C}$$

(c) for hydrogen,

$$1.17 \text{ g} \times \frac{1 \text{ mol}}{1.008 \text{ g}} = 1.16 \text{ mol H}$$

(d) for oxygen,

$$12.32 \text{ g} \times \frac{1 \text{ mol}}{16.00 \text{ g}} = 0.7700 \text{ mol O}$$

The relative mole amounts are determined as follows:

for Hg, $0.3851 \div 0.3851 = 1.000$

for C, $0.770 \div 0.3851 = 2.00$

for H, $1.16 \div 0.3851 = 3.01$

for O, $0.7700 \div 0.3851 = 1.999$

and the empirical formula is $HgC_2H_3O_2$. The empirical formula weight is 260 g/mole, which must be multiplied by 2 in order to obtain the molecular weight. This means that the molecular formula is twice the empirical formula, or $Hg_2C_4H_6O_4$.

3.54 We again assume a 100 g sample so that the known values for % by weight become mass values directly, making the calculation convenient, and the empirical formula is determined using these mass values, as in problems 3.45 through 3.53:

(a) for carbon,

$$75.42 \text{ g} \times \frac{1 \text{ mol}}{12.01 \text{ g}} = 6.280 \text{ mol C}$$

(b) for hydrogen,

$$6.63 \text{ g} \times \frac{1 \text{ mol}}{1.008 \text{ g}} = 6.58 \text{ mol H}$$

(c) for nitrogen,

$$8.38 \text{ g} \times \frac{1 \text{ mol}}{14.01 \text{ g}} = 0.598 \text{ mol N}$$

(d) for oxygen,

$$9.57 \text{ g} \times \frac{1 \text{ mol}}{16.00 \text{ g}} = 0.598 \text{ mol O}$$

The relative mole amounts become:

for C, $6.280 \div 0.598 = 10.5$

for H, $6.58 \div 0.598 = 11.0$

for N, $0.598 \div 0.598 = 1.00$

for O, $0.598 \div 0.598 = 1.00$

and the empirical formula is $C_{21}H_{22}N_2O_2$. The formula weight from the empirical formula is 334 g/mole, which is equal to the molecular weight. Hence the molecular formula is the same as the empirical formula.

3.55 This problem is solved in the same fashion as 3.52 through 3.54:

(a) for carbon,

$$40.00 \text{ g} \times \frac{1 \text{ mol}}{12.01 \text{ g}} = 3.331 \text{ mol C}$$

(b) for hydrogen,

$$6.71 \text{ g} \times \frac{1 \text{ mol}}{1.008 \text{ g}} = 6.66 \text{ mol H}$$

(c) for oxygen,

$$53.29 \text{ g} \times \frac{1 \text{ mol}}{16.00 \text{ g}} = 3.331 \text{ mol O}$$

Next, relative mole amounts are determined:

for C, $3.331 \div 3.331 = 1.000$

for H, 6.66 ÷ 3.331 = 2.00

for O, 3.331 ÷ 3.331 = 1.000

and the empirical formula is CH_2O. The empirical formula weight is to be multiplied by 3 in order to obtain the molecular weight (90), so the molecular formula is three times the empirical formula, $C_3H_6O_3$.

3.56 (a)

$$0.4681 \text{ g Na} \times \frac{1 \text{ mol Na}}{22.99 \text{ g Na}} = 2.036 \times 10^{-2} \text{ mol Na}$$

$$0.3258 \text{ g O} \times \frac{1 \text{ mol O}}{16.00 \text{ g O}} = 2.036 \times 10^{-2} \text{ mol O}$$

Since these mole amounts are clearly in a ratio of 1 to 1, the empirical formula is NaO.

(b)

$$1.000 \text{ g O} \times \frac{0.4681 \text{ g Na}}{0.3258 \text{ g O}} = 1.437 \text{ g Na}$$

or, as an alternate method of calculation,

$$1.000 \text{ g O} \times \frac{1 \text{ mol O}}{16.00 \text{ g O}} \times \frac{1 \text{ mol Na}}{1 \text{ mol O}} \times \frac{22.99 \text{ g Na}}{1 \text{ mol Na}} = 1.437 \text{ g Na}$$

(c) Any integer multiple or any whole number fraction of the value 1.437 g will be a possible amount of Na to combine with 1.000 g of O to make a new compound of sodium and oxygen.

(d)

$$1.000 \text{ g O} \times \frac{1.145 \text{ g Na}}{0.3983 \text{ g O}} = 2.875 \text{ g Na}$$

Notice that in this different compound of sodium and oxygen, the mass ratio of the two elements is different from that found in part (b) to this question, as is guaranteed by the Law of Definite Proportions.

(e)

$$\frac{1.437 \text{ g Na}/1.000 \text{ g O in sodium peroxide}}{2.875 \text{ g Na}/1.000 \text{ g O in sodium oxide}} = \frac{1}{2}$$

Since this gives a ratio of small whole numbers, it is entirely consistent with the Law of Multiple Proportions.

(f)

$$1.145 \text{ g Na} \times \frac{1 \text{ mol}}{22.99 \text{ g}} = 4.980 \times 10^{-2} \text{ mol Na}$$

$$0.3983 \text{ g O} \times \frac{1 \text{ mol}}{16.00 \text{ g}} = 2.489 \times 10^{-2} \text{ mol O}$$

4.980×10^{-2} mole ÷ 2.489×10^{-2} mole = 2.000 relative moles of Na

2.489×10^{-2} ÷ 2.489×10^{-2} = 1.000 relative moles of O

and the empirical formula is seen to be Na_2O.

3.57 (a) First determine the number of moles of each element that are in the compound:

for potassium:

$$0.5634 \times \frac{1 \text{ mol}}{39.10 \text{ g}} = 1.441 \times 10^{-2} \text{ mol K}$$

for oxygen:

$$0.4611 \text{ g} \times \frac{1 \text{ mol}}{16.00 \text{ g}} = 2.882 \times 10^{-2} \text{ mol O}$$

Next determine the relative number of moles of each element:

for K, 1.441×10^{-2} mole ÷ 1.441×10^{-2} mole = 1.000

for O, 2.882×10^{-2} mole ÷ 1.441×10^{-2} mole = 2.000

and the empirical formula is KO_2.

(b)

$$1.000 \text{ g K} \times \frac{0.4611 \text{ g O}}{0.5634 \text{ g K}} = 0.8184 \text{ g O}$$

(c) Any whole number multiple of or integer fraction of the number 0.8184 will be possible.

(d) Proceeding as in part (a) for this new compound we have:

for potassium,

$$0.8298 \text{ g} \times \frac{1 \text{ mol}}{39.10 \text{ g}} = 2.122 \times 10^{-2} \text{ mol K}$$

for oxygen,

$$0.1698 \text{ g} \times \frac{1 \text{ mol}}{16.00 \text{ g}} = 1.061 \times 10^{-2} \text{ mol O}$$

and the relative mole amounts are:

for K, 2.122×10^{-2} ÷ 1.061×10^{-2} = 2.000

for O, $1.061 \times 10^{-2} \div 1.061 \times 10^{-2} = 1.000$

giving an empirical formula of K_2O.

(e)

$$1.000 \text{ g K} \times \frac{0.1698 \text{ g O}}{0.8298 \text{ g K}} = 0.2046 \text{ g O}$$

(f)

$$\frac{0.8184 \text{ g O/1.000 g K in potassium superoxide}}{0.2046 \text{ g O/1.000 gK in potassium oxide}} = \frac{4}{1}$$

which illustrates the Law of Multiple Proportions.

3.58 This sort of combustion analysis takes advantage of the fact that the entire amount of carbon in the original sample appears among the products as carbon that is in CO_2. Hence the mass of carbon in the original sample must be equal to the mass of carbon that is found in the CO_2.

$$20.08 \times 10^{-3} \text{ g } CO_2 \times \frac{12.01 \text{ g C}}{44.01 \text{ g } CO_2} = 5.480 \times 10^{-3} \text{ g C}$$

Similarly, the entire mass of hydrogen that was present in the original sample ends up in the products as H_2O:

$$5.023 \times 10^{-3} \text{ g } H_2O \times \frac{2.016 \text{ g H}}{18.02 \text{ g } H_2O} = 5.620 \times 10^{-4} \text{ g H}$$

Next the empirical formula may be determined:

$$5.480 \times 10^{-3} \text{ g C} \times \frac{1 \text{ mol C}}{12.01 \text{ g C}} = 4.563 \times 10^{-4} \text{ mol C}$$

$$5.620 \times 10^{-4} \text{ g H} \times \frac{1 \text{ mol H}}{1.008 \text{ g H}} = 5.575 \times 10^{-4} \text{ mol H}$$

The mass of oxygen is determined by subtracting the mass due to C and H from the total mass:

6.853 mg total - (5.480 mg C + 0.5620 mg H) = 0.811 mg O

$$0.811 \times 10^{-3} \text{ g O} \times \frac{1 \text{ mol O}}{16.00 \text{ g O}} = 5.07 \times 10^{-5} \text{ mol O}$$

The relative mole amounts are:

for C, $4.563 \times 10^{-4} \div 5.07 \times 10^{-5} = 9.00$

for H, $5.575 \times 10^{-4} \div 5.07 \times 10^{-5} = 11.0$

for O, $5.07 \times 10^{-5} \div 5.07 \times 10^{-5} = 1.00$

and the empirical formula is $C_9H_{11}O$, for which the formula weight is 135 g/mole. Since the molar mass is twice the formula weight of the empirical formula, the molecular formula must be obtained by multiplying the empirical formula by 2, giving $C_{18}H_{22}O_2$.

3.59 We proceed as in problem 3.58.

The mass of carbon in the original sample is:

$$17.536 \times 10^{-3}\ g\ CO_2 \times \frac{12.01\ g\ C}{44.01\ g\ CO_2} = 4.785 \times 10^{-3}\ g\ C$$

The mass of H in the original sample is:

$$5.850 \times 10^{-3}\ g\ H_2O \times \frac{2.016\ g\ H}{18.02\ g\ H_2O} = 6.545 \times 10^{-4}\ g\ H$$

The mass of oxygen is determined by difference:

5.676 mg total - (4.785 mg C + 0.6545 mg H) = 0.237 mg O

The mole amounts of C, H, and O in the original compound are:

$$4.785 \times 10^{-3}\ g\ C \times \frac{1\ mol}{12.01\ g} = 3.984 \times 10^{-4}\ mol\ C$$

$$6.545 \times 10^{-4}\ g\ H \times \frac{1\ mol\ H}{1.008\ g\ H} = 6.493 \times 10^{-4}\ mol\ H$$

$$0.237 \times 10^{-3}\ g\ O \times \frac{1\ mol}{16.00\ g} = 1.48 \times 10^{-5}\ mol\ O$$

Next we determine relative mole amounts for the three elements.

for C, $3.984 \times 10^{-4} \div 1.48 \times 10^{-5} = 26.9$

for H, $6.493 \times 10^{-4} \div 1.48 \times 10^{-5} = 43.9$

for O, $1.48 \times 10^{-5} \div 1.48 \times 10^{-5} = 1.00$

and the empirical formula is $C_{27}H_{44}O$. The formula weight of the empirical unit is 384 g/mole, within experimental error of the known molar mass. This means that the molecular formula is the same as the empirical formula.

3.60

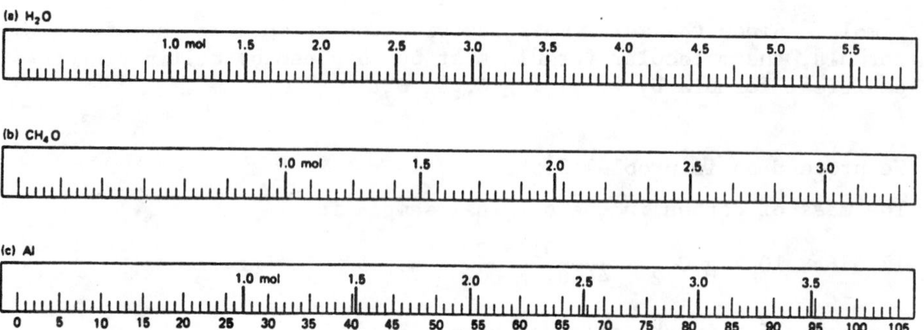
(a) H_2O
1.0 mol
1.5
2.0
2.5
3.0
3.5
4.0
4.5
5.0
5.5
(b) CH_4O
1.0 mol
1.5
2.0
2.5
3.0
(c) Al
1.0 mol
1.5
2.0
2.5
3.0
3.5
0 5 10 15 20 25 30 35 40 45 50 55 60 65 70 75 80 85 90 95 100 105

CHAPTER FOUR

PRACTICE EXERCISES

1. (a) $4P + 5O_2 \rightarrow P_4O_{10}$ (b) $N_2 + 3H_2 \rightarrow 2NH_3$

2. (a) $2Mg + O_2 \rightarrow 2MgO$

   (b) $CH_4 + 2Cl_2 \rightarrow 4HCl + CCl_4$

   (c) $2NO + O_2 \rightarrow 2NO_2$

   (d) $2NaOH + H_2SO_4 \rightarrow Na_2SO_4 + 2H_2O$

   (e) $CH_4 + 2O_2 \rightarrow CO_2 + 2H_2O$

   (f) $2C_2H_6 + 7O_2 \rightarrow 4CO_2 + 6H_2O$

   (g) $2Al(OH)_3 + 3H_2SO_4 \rightarrow Al_2(SO_4)_3 + 6H_2O$

3.

$$18 \text{ mol CO} \times \frac{2 \text{ mol Fe}}{3 \text{ mol CO}} = 12 \text{ mol Fe}$$

4.

$$6 \text{ mol } O_2 \times \frac{4 \text{ mol } H_2O}{3 \text{ mol } O_2} = 8 \text{ mol } H_2O$$

5. (a)

$$\frac{3 \text{ mol } Cl_2}{2 \text{ mol Fe}} \quad \text{and} \quad \frac{2 \text{ mol Fe}}{3 \text{ mol } Cl_2}$$

$$\frac{2 \text{ mol Fe}}{2 \text{ mol } FeCl_3} \quad \text{and} \quad \frac{2 \text{ mol } FeCl_3}{2 \text{ mol Fe}}$$

$$\frac{3 \text{ mol } Cl_2}{2 \text{ mol } FeCl_3} \quad \text{and} \quad \frac{2 \text{ mol } FeCl_3}{3 \text{ mol } Cl_2}$$

(b)

$$24 \text{ mol } Cl_2 \times \frac{2 \text{ mol } FeCl_3}{3 \text{ mol } Cl_2} = 16 \text{ mol } FeCl_3$$

(c)

$$24 \text{ mol } Cl_2 \times \frac{2 \text{ mol Fe}}{3 \text{ mol } Cl_2} = 16 \text{ mol Fe}$$

(d)

$$0.500 \text{ mol Fe} \times \frac{3 \text{ mol } Cl_2}{2 \text{ mol Fe}} = 0.750 \text{ mol } Cl_2$$

$$0.500 \text{ mol Fe} \times \frac{2 \text{ mol } FeCl_3}{2 \text{ mol Fe}} = 0.500 \text{ mol } FeCl_3$$

6.

$$16.5 \text{ g } Fe_2O_3 \times \frac{1 \text{ mol } Fe_2O_3}{160 \text{ g } Fe_2O_3} = 0.103 \text{ mol } Fe_2O_3$$

$$0.103 \text{ mol } Fe_2O_3 \times \frac{2 \text{ mol Fe}}{1 \text{ mol } Fe_2O_3} = 0.206 \text{ mol Fe}$$

$$0.206 \text{ mol Fe} \times \frac{55.8 \text{ g Fe}}{1 \text{ mol Fe}} = 11.5 \text{ g Fe}$$

7.

$$35.8 \text{ g } I_2 \times \frac{1 \text{ mol } I_2}{254 \text{ g } I_2} = 0.141 \text{ mol } I_2$$

$$0.141 \text{ g } I_2 \times \frac{1 \text{ mol } K_2Cr_2O_7}{3 \text{ mol } I_2} = 0.0470 \text{ mol } K_2Cr_2O_7$$

$$0.0470 \text{ mol } K_2Cr_2O_7 \times \frac{294 \text{ g } K_2Cr_2O_7}{1 \text{ mol } K_2Cr_2O_7} = 13.8 \text{ g } K_2Cr_2O_7$$

8.

$$28.6 \text{ g ICl} \times \frac{1 \text{ mol ICl}}{162 \text{ g ICl}} = 0.177 \text{ mol ICl}$$

$$0.177 \text{ mol ICl} \times \frac{2 \text{ mol } I_2}{5 \text{ mol ICl}} = 0.0708 \text{ mol } I_2$$

$$0.0708 \text{ mol } I_2 \times \frac{254 \text{ g } I_2}{1 \text{ mol } I_2} = 18.0 \text{ g } I_2$$

9.

$$11.4 \text{ g } KNO_2 \times \frac{1 \text{ mol } KNO_2}{85.1 \text{ g } KNO_2} = 0.134 \text{ mol } KNO_2$$

$$0.134 \text{ mol } KNO_2 \times \frac{2 \text{ mol } KMnO_4}{5 \text{ mol } KNO_2} = 0.0536 \text{ mol } KMnO_4$$

$$0.0536 \text{ mol } KMnO_4 \times \frac{158 \text{ g } KMnO_4}{1 \text{ mol } KMnO_4} = 8.47 \text{ g } KMnO_4$$

10.

$$63.4 \text{ g HF} \times \frac{1 \text{ mol HF}}{20.0 \text{ g HF}} = 3.17 \text{ mol HF}$$

$$3.17 \text{ mol HF} \times \frac{1 \text{ mol } SiF_4}{4 \text{ mol HF}} = 0.793 \text{ mol } SiF_4$$

$$0.793 \text{ mol } SiF_4 \times \frac{104 \text{ g } SiF_4}{1 \text{ mol } SiF_4} = 82.5 \text{ g } SiF_4$$

11.

(a) First calculate the mass of HI that is required:

$$10.4 \text{ g } CuSO_4 \times \frac{1 \text{ mol } CuSO_4}{160 \text{ g } CuSO_4} = 0.0650 \text{ mol } CuSO_4$$

$$0.0650 \text{ mol } CuSO_4 \times \frac{4 \text{ mol HI}}{2 \text{ mol } CuSO_4} = 0.130 \text{ mol HI}$$

$$0.130 \text{ mol HI} \times \frac{128 \text{ g HI}}{1 \text{ mol HI}} = 16.6 \text{ g HI needed.}$$

Thus the total mass of reactants is 16.6 + 10.4 = 27.0 g.

(b) Next calculate the mass of the products that is expected:

for CuI,

$$0.0650 \text{ mol } CuSO_4 \times \frac{2 \text{ mol CuI}}{2 \text{ mol } CuSO_4} \times \frac{190 \text{ g CuI}}{1 \text{ mol CuI}} = 12.4 \text{ g CuI}$$

for H_2SO_4,

$$0.0650 \text{ mol } CuSO_4 \times \frac{2 \text{ mol } H_2SO_4}{2 \text{ mol } CuSO_4} \times \frac{98.1 \text{ g } H_2SO_4}{1 \text{ mol } H_2SO_4} = 6.38 \text{ g } H_2SO_4$$

for I_2,

$$0.0650 \text{ mol } CuSO_4 \times \frac{1 \text{ mol } I_2}{2 \text{ mol } CuSO_4} \times \frac{254 \text{ g } I_2}{1 \text{ mol } I_2} = 8.26 \text{ g } I_2$$

The total mass of products is 12.4 + 6.38 + 8.26 = 27.0 g

The total mass of products should be equal to the total mass of the reactants, according to the Law of Conservation of Mass.

12.

$$11.4 \text{ g } KNO_2 \times \frac{1 \text{ mol } KNO_2}{85.1 \text{ g } KNO_2} \times \frac{2 \text{ mol } KMnO_4}{5 \text{ mol } KNO_2} \times \frac{158 \text{ g } KMnO_4}{1 \text{ mol } KMnO_4} = 8.47 \text{ g } KMnO_4$$

13.

$$63.4 \text{ g HF} \times \frac{1 \text{ mol HF}}{20.0 \text{ g HF}} \times \frac{1 \text{ mol } SiF_4}{4 \text{ mol HF}} \times \frac{104 \text{ g } SiF_4}{1 \text{ mol } SiF_4} = 82.4 \text{ g } SiF_4$$

14.

$$16.5 \text{ g } Fe_2O_3 \times \frac{1 \text{ mol } Fe_2O_3}{160 \text{ g } Fe_2O_3} \times \frac{2 \text{ mol Fe}}{1 \text{ mol } Fe_2O_3} \times \frac{55.8 \text{ g Fe}}{1 \text{ mol Fe}} = 11.5 \text{ g Fe}$$

$$35.8 \text{ g } I_2 \times \frac{1 \text{ mol } I_2}{254 \text{ g } I_2} \times \frac{1 \text{ mol } K_2Cr_2O_7}{3 \text{ mol } I_2} \times \frac{294 \text{ g } K_2Cr_2O_7}{1 \text{ mol } K_2Cr_2O_7} = 13.8 \text{ g } K_2Cr_2O_7$$

15.

$$28.6 \text{ g ICl} \times \frac{1 \text{ mol ICl}}{162 \text{ g ICl}} \times \frac{2 \text{ mol } I_2}{5 \text{ mol ICl}} \times \frac{254 \text{ g } I_2}{1 \text{ mol } I_2} = 17.9 \text{ g } I_2$$

16. First determine how much CO is needed to react completely, according to the equation, with the given amount (0.300 mole) of Fe_2O_3:

$$0.300 \text{ mol } Fe_2O_3 \times \frac{3 \text{ mol CO}}{1 \text{ mol } Fe_2O_3} = 0.900 \text{ mol CO are required.}$$

Since 1.20 mol of CO are available, it is present in excess, and Fe_2O_3 must be the limiting reagent. It will be used up first, and then the reaction will stop, some CO being left over. The theoretical yield of Fe must, therefore, be based on the amount of Fe_2O_3 that is available:

$$0.300 \text{ mol } Fe_2O_3 \times \frac{2 \text{ mol Fe}}{1 \text{ mol } Fe_2O_3} = 0.600 \text{ mol of Fe may be expected.}$$

17. The balanced equation is:

$$2Na + Cl_2 \rightarrow 2NaCl$$

$$10.0 \text{ g Na} \times \frac{1 \text{ mol Na}}{23.0 \text{ g Na}} = 0.435 \text{ mol Na are available.}$$

$$20.0 \text{ g } Cl_2 \times \frac{1 \text{ mol } Cl_2}{70.9 \text{ g } Cl_2} = 0.282 \text{ mol } Cl_2 \text{ are available.}$$

If all of the Na were to react, then

$$0.435 \text{ mol Na} \times \frac{1 \text{ mol } Cl_2}{2 \text{ mol Na}} = 0.218 \text{ mol } Cl_2 \text{ would be needed.}$$

The Chlorine is present in excess, and the calculation must be based on the limiting reagent, Na.

18. The theoretical yield is calculated in the following manner:

$$1.75 \text{ kg } SO_2 \times \frac{1 \text{ mol } SO_2}{64.1 \times 10^{-3} \text{ kg } SO_2} \times \frac{2 \text{ mol } SO_3}{2 \text{ mol } SO_2} \times \frac{80.1 \times 10^{-3} \text{ kg } SO_3}{1 \text{ mol } SO_3}$$

$$= 2.19 \text{ kg } SO_3$$

The % yield is based on the actual yield in the following way:

$$\% \text{ yield} = \frac{\text{actual yield}}{\text{theoretical yield}} \times 100$$

and in this question, the calculation is:

$$\frac{1.72 \text{ kg } SO_3}{2.19 \text{ kg } SO_3} \times 100 = 78.5 \%$$

19. Molarity (M) = mol solute ÷ L solution
Using the factor-label method we have:

$$0.00100 \text{ mol glucose} \times \frac{1000 \text{ mL solution}}{0.200 \text{ mol glucose}} = 5.00 \text{ mL}$$

An alternate, but equivalent, method is:

$$\frac{0.00100 \text{ mol glucose}}{0.200 \text{ mol/L}} = 5.00 \times 10^{-3} \text{ L of solution, or } 5.00 \text{ mL}$$

20. First calculate the number of moles of $NaHCO_3$ that are required, realizing that 250 mL equals 0.250 L:

$$0.250 \text{ L} \times \frac{0.200 \text{ mol}}{1 \text{ L}} = 0.0500 \text{ mol } NaHCO_3$$

Next convert to grams:

$$0.0500 \text{ mol } NaHCO_3 \times \frac{84.0 \text{ g } NaHCO_3}{1 \text{ mol } NaHCO_3} = 4.20 \text{ g } NaHCO_3$$

The solution should be prepared by dissolving 4.20 g of $NaHCO_3$ in enough water to make 250 mL of solution.

21. The desired volume (0.250 L) and the desired molarity (0.0500 mol/L) may be multiplied to determine the number of moles of glucose that are required:

$$0.0500 \text{ mol/L} \times 0.250 \text{ L} = 0.0125 \text{ mol glucose}$$

The number of grams are thus:

$$0.0125 \text{ mol} \times 180 \text{ g/mole} = 2.25 \text{ g}$$

The solution is to be prepared by dissolving 2.25 g of glucose in water, bringing the final volume of the solution to 250 mL.

22. Molarity may be expressed either as moles solute/L solution or as moles solute/1000 mL of solution:

$$18.6 \text{ mL } H_2SO_4 \times \frac{0.156 \text{ mol } H_2SO_4}{1000 \text{ mL}} \times \frac{2 \text{ mol KOH}}{1 \text{ mol } H_2SO_4} \times \frac{1000 \text{ mL KOH}}{0.337 \text{ mol KOH}} =$$

$$= 17.2 \text{ mL KOH}$$

23. As in practice exercise 22, but using the stepwise method of example 4.15:

First calculate the number of moles of Na_2CO_3 that are involved:

$$0.0242 \text{ L} \times \frac{0.284 \text{ mol}}{\text{L}} = 6.87 \times 10^{-3} \text{ mol } Na_2CO_3$$

Next calculate the number of moles of HCl that are required:

$$6.87 \times 10^{-3} \text{ mol } Na_2CO_3 \times \frac{2 \text{ mol HCl}}{1 \text{ mol } Na_2CO_3} = 1.37 \times 10^{-2} \text{ mol HCl}$$

Last, determine the volume of HCl solution that is needed in order to deliver this amount of HCl:

$$\frac{1.37 \times 10^{-2} \text{ mol HCl}}{0.224 \text{ mol/L}} = 0.0614 \text{ L of solution}$$

24. This exercise is worked as in example 4.16:

$V_i = ?$ $\quad$ $M_i = 0.50$ mol/L

$V_f = 500$ mL $\quad$ $M_f = 0.20$ mol/L

It is in general true that

$$V_i \times M_i = V_f \times M_f$$

and $\quad V_i = V_f \times M_f \div M_i$

Hence: $\quad V_i = (0.500 \text{ L}) \times (0.20 \text{ mol/L}) \div (0.50 \text{ mol/L})$
$\quad = 0.200$ L

Place 0.200 L of the NaOH solution having a molarity of 0.50 mol/L in a flask, and add sufficient water to bring the final volume of the diluted solution to 0.500 L.

25. As in practice exercise 24:

$$V_i \times M_i = V_f \times M_f$$

$$V_f = V_i \times M_i \div M_f$$

so $\quad V_f = (0.300 \text{ L}) \times (0.500 \text{ mol/L}) \div (0.200 \text{ mol/L})$

$$V_f = 0.750 \text{ L}$$

Thus enough water is to be added to the original solution to bring the final volume up to 750 mL, i.e. add 450 mL of water.

REVIEW EXERCISES

Balanced Equations

4.1 The coefficients indicate the mole proportions of reactants and products, which together with the formula weights of the reactants and products, are needed in order to calculate the masses of substances to be employed in chemical reactions or to calculate the masses of products to be expected from chemical reactions.

4.2 coefficients

4.3 $H_2O + SO_3 \rightarrow H_2SO_4$

Sulfur trioxide and water react in a 1 to 1 mole ratio to give 1 mole of sulfuric acid.

4.4 $4Fe + 3O_2 \rightarrow 2Fe_2O_3$

4.5 (a) $Ca(OH)_2 + 2HCl \rightarrow CaCl_2 + 2H_2O$

(b) $2AgNO_3 + CaCl_2 \rightarrow Ca(NO_3)_2 + 2AgCl$

(c) $2Fe_2O_3 + 3C \rightarrow 4Fe + 3CO_2$

(d) $2NaHCO_3 + H_2SO_4 \rightarrow Na_2SO_4 + 2H_2O + 2CO_2$

(e) $2C_4H_{10} + 13O_2 \rightarrow 8CO_2 + 10H_2O$

4.6 (a) $2SO_2 + O_2 \rightarrow 2SO_3$

(b) $P_4O_{10} + 6H_2O \rightarrow 4H_3PO_4$

(c) $Pb(NO_3)_2 + Na_2SO_4 \rightarrow PbSO_4 + 2NaNO_3$

(d) $Fe_2O_3 + 3H_2 \rightarrow 2Fe + 3H_2O$

(e) $2Al + 3H_2SO_4 \rightarrow Al_2(SO_4)_3 + 3H_2$

4.7 (a) $Mg(OH)_2 + 2HBr \rightarrow MgBr_2 + 2H_2O$

(b) $2HCl + Ca(OH)_2 \rightarrow CaCl_2 + 2H_2O$

(c) $Al_2O_3 + 3H_2SO_4 \rightarrow Al_2(SO_4)_3 + 3H_2O$

(d) $2KHCO_3 + H_3PO_4 \rightarrow K_2HPO_4 + 2H_2O + 2CO_2$

(e) $C_9H_{20} + 14O_2 \rightarrow 9CO_2 + 10H_2O$

4.8 (a) $CaO + 2HNO_3 \rightarrow Ca(NO_3)_2 + H_2O$

(b) $Na_2CO_3 + Mg(NO_3)_2 \rightarrow MgCO_3 + 2NaNO_3$

(c) $(NH_4)_3PO_4 + 3NaOH \rightarrow Na_3PO_4 + 3NH_3 + 3H_2O$

(d) $Mg(HCO_3)_2 + 2HCl \rightarrow MgCl_2 + 2H_2O + 2CO_2$

(e) $C_4H_{10}O + 6O_2 \rightarrow 4CO_2 + 5H_2O$

Stoichiometric Calculations

4.9 Samples that possess equal numbers of molecules also possess equal numbers of moles.

4.10 the limiting reagent

4.11 Their formula weights must be identical.

4.12 (a) Stoichiometry is the fact that the coefficients in the balanced equation are 1:1:1.
(b) The scale of the reaction is the amount of material that a chemist actually invests in the laboratory process, namely 0.10 mole of Mg and Cl_2.

4.13 $2H_2O_2 \rightarrow 2H_2O + O_2$

4.14 (a)

$$4 \text{ mol octane} \times \frac{25 \text{ mol } O_2}{2 \text{ mol octane}} = 50 \text{ mol } O_2$$

(b)

$$1 \text{ mol octane} \times \frac{16 \text{ mol } CO_2}{2 \text{ mol octane}} = 8 \text{ mol } CO_2$$

(c)

$$6 \text{ mol octane} \times \frac{18 \text{ mol } H_2O}{2 \text{ mol octane}} = 54 \text{ mol } H_2O$$

(d)

$$8 \text{ mol } CO_2 \times \frac{25 \text{ mol } O_2}{16 \text{ mol } CO_2} = 12.5 \text{ mol } O_2$$

$$8 \text{ mol } CO_2 \times \frac{2 \text{ mol octane}}{16 \text{ mol } CO_2} = 1 \text{ mol octane}$$

4.15 (a)

$$25 \text{ mol } C_2H_6O \times \frac{3 \text{ mol } O_2}{1 \text{ mol } C_2H_6O} = 75 \text{ mol } O_2$$

(b)

$$30 \text{ mol } O_2 \times \frac{1 \text{ mol } C_2H_6O}{3 \text{ mol } O_2} = 10 \text{ mol } C_2H_6O$$

$$30 \text{ mol } O_2 \times \frac{2 \text{ mol } CO_2}{3 \text{ mol } O_2} = 20 \text{ mol } CO_2$$

(c)

$$23 \text{ mol } CO_2 \times \frac{3 \text{ mol } O_2}{2 \text{ mol } CO_2} = 35 \text{ mol } O_2$$

(d)

$$41 \text{ mol } H_2O \times \frac{1 \text{ mol } C_2H_6O}{3 \text{ mol } H_2O} = 14 \text{ mol } C_2H_6O$$

$$41 \text{ mol } H_2O \times \frac{3 \text{ mol } O_2}{3 \text{ mol } H_2O} = 41 \text{ mol } O_2$$

$$41 \text{ mol } H_2O \times \frac{2 \text{ mol } CO_2}{3 \text{ mol } H_2O} = 27 \text{ mol } CO_2$$

4.16 (a)

$$2.46 \text{ g } H_2O \times \frac{1 \text{ mol } H_2O}{18.0 \text{ g } H_2O} = 1.37 \times 10^{-1} \text{ mol } H_2O$$

(b)

$$0.137 \text{ mol } H_2O \times \frac{2 \text{ mol } C_4H_{10}}{10 \text{ mol } H_2O} = 2.74 \times 10^{-2} \text{ mol } C_4H_{10}$$

(c)

$$2.74 \times 10^{-2} \text{ mol } C_4H_{10} \times \frac{58.1 \text{ g } C_4H_{10}}{1 \text{ mol } C_4H_{10}} = 1.59 \text{ g } C_4H_{10}$$

(d)

$$0.137 \text{ mol } H_2O \times \frac{13 \text{ mol } O_2}{10 \text{ mol } H_2O} = 0.178 \text{ mol } O_2$$

$$0.178 \text{ mol } O_2 \times \frac{32.0 \text{ g } O_2}{1 \text{ mol } O_2} = 5.70 \text{ g } O_2$$

4.17 (a)

$$25 \text{ mol } Fe_2O_3 \times \frac{2 \text{ mol Fe}}{1 \text{ mol } Fe_2O_3} = 50 \text{ mol Fe}$$

(b)

$$30 \text{ mol Fe} \times \frac{3 \text{ mol } H_2}{2 \text{ mol Fe}} = 45 \text{ mol } H_2$$

(c)

$$120 \text{ mol } H_2O \times \frac{1 \text{ mol } Fe_2O_3}{3 \text{ mol } H_2O} \times \frac{160 \text{ g } Fe_2O_3}{1 \text{ mol } Fe_2O_3} = 6.40 \times 10^3 \text{ g } Fe_2O_3$$

4.18

$$154 \text{ g } C_8H_{10} \times \frac{1 \text{ mol } C_8H_{10}}{106 \text{ g } C_8H_{10}} \times \frac{1 \text{ mol } C_8H_6O_4}{1 \text{ mol } C_8H_{10}} = 1.45 \text{ mol } C_8H_6O_4$$

$$1.45 \text{ mol } C_8H_6O_4 \times \frac{166 \text{ g } C_8H_6O_4}{1 \text{ mol } C_8H_6O_4} = 241 \text{ g } C_8H_6O_4$$

4.19 (a)

$$40.0 \text{ mol } C_6H_{10}O_4 \times \frac{5 \text{ mol } O_2}{2 \text{ mol } C_6H_{10}O_4} = 100 \text{ mol } O_2$$

(b)

$$164 \text{ g } C_6H_{12} \times \frac{1 \text{ mol } C_6H_{12}}{84.2 \text{ g } C_6H_{12}} \times \frac{2 \text{ mol } C_6H_{10}O_4}{2 \text{ mol } C_6H_{12}} = 1.95 \text{ mol } C_6H_{10}O_4$$

$$1.95 \text{ mol } C_6H_{10}O_4 \times \frac{146 \text{ g } C_6H_{10}O_4}{1 \text{ mol } C_6H_{10}O_4} = 285 \text{ g } C_6H_{10}O_4$$

4.20 (a)

$$100 \text{ mol NaCl} \times \frac{1 \text{ mol } Na_2CO_3}{2 \text{ mol NaCl}} = 50.0 \text{ mol } Na_2CO_3$$

(b)

$$546 \text{ g NaCl} \times \frac{1 \text{ mol NaCl}}{58.5 \text{ g NaCl}} \times \frac{1 \text{ mol } Na_2CO_3}{2 \text{ mol NaCl}} = 4.67 \text{ mol } Na_2CO_3$$

$$4.67 \text{ mol } Na_2CO_3 \times \frac{106 \text{ g } Na_2CO_3}{1 \text{ mole } Na_2CO_3} = 495 \text{ g } Na_2CO_3$$

4.21 The formula weight of phosphate rock is 384 g/mole, and 1 metric ton equals 1×10^6 g. We first calculate the number of moles of H_2SO_4 that are needed for the reaction of 125 metric tons (125×10^6 g) of phosphate rock:

(a)

$$125 \times 10^6 \text{ g rock} \times \frac{1 \text{ mol rock}}{384 \text{ g rock}} \times \frac{3 \text{ mol } H_2SO_4}{1 \text{ mol rock}} = 9.77 \times 10^5 \text{ mol } H_2SO_4$$

$$9.77 \times 10^5 \text{ mol } H_2SO_4 \times \frac{98.1 \text{ g } H_2SO_4}{1 \text{ mol } H_2SO_4} \times \frac{1 \text{ metric ton}}{1 \times 10^6 \text{ g}}$$

$$= 95.8 \text{ metric ton } H_2SO_4$$

(b)

$$125 \times 10^6 \text{ g rock} \times \frac{1 \text{ mol rock}}{384 \text{ g rock}} \times \frac{1 \text{ mole } Ca(H_2PO_4)_2}{1 \text{ mole rock}}$$

$$= 3.26 \times 10^5 \text{ mol } Ca(H_2PO_4)$$

$$3.26 \times 10^5 \text{ mol } Ca(H_2PO_4)_2 \times \frac{234 \text{ g}}{\text{mol}} \times \frac{1 \text{ metric ton}}{1 \times 10^6 \text{ g}}$$

$$= 76.3 \text{ metric ton } Ca(H_2PO_4)_2$$

$$125 \times 10^6 \text{ g rock} \times \frac{1 \text{ mol}}{384 \text{ g}} \times \frac{3 \text{ mol } Ca(SO_4)_2}{1 \text{ mol rock}}$$

$$= 9.77 \times 10^5 \text{ mole } Ca(SO_4)_2$$

$$9.77 \times 10^5 \text{ mole } Ca(SO_2)_4 \times \frac{136 \text{ g}}{\text{mol}} \times \frac{1 \text{ metric ton}}{1 \times 10^6 \text{ g}}$$

$$= 133 \text{ metric ton } Ca(SO_2)_4$$

4.22 (a)

$$324 \text{ g } Fe_2O_3 \times \frac{1 \text{ mol } Fe_2O_3}{160 \text{ g } Fe_2O_3} \times \frac{2 \text{ mol Fe}}{1 \text{ mol } Fe_2O_3} = 4.05 \text{ mol Fe}$$

$$4.05 \text{ mol Fe} \times \frac{55.8 \text{ g Fe}}{1 \text{ mol Fe}} = 226 \text{ g Fe}$$

(b)

$$\% \text{ yield} = \frac{198 \text{ g}}{226 \text{ g}} \times 100 = 87.6 \%$$

4.23 (a) First calculate the theoretical yield realizing that the stoichiometry is 1:1:

$$13{,}800 \text{ g C} \times \frac{1 \text{ mol}}{12.0 \text{ g}} \times \frac{1 \text{ mol } CH_4}{1 \text{ mol C}} \times \frac{16.0 \text{ g } CH_4}{1 \text{ mol } CH_4} = 1.84 \times 10^4 \text{ g } CH_4$$

Next calculate the % yield:

$$\frac{580 \text{ g}}{1.84 \times 10^4 \text{ g}} \times 100 = 3.15 \%$$

(b) Now only 44% of the starting mass is actually C, and only 44% of the theoretical yield is possible. The new theoretical yield thus becomes:

$$0.44 \times 1.84 \times 10^4 \text{ g} = 8.10 \times 10^3 \text{ g } CH_4$$

and the new % yield is:

$$\frac{580 \text{ g}}{8.10 \times 10^3 \text{ g}} \times 100 = 7.16 \% \text{ } CH_4$$

4.24 (a)

$$88.6 \text{ g } P_4O_{10} \times \frac{1 \text{ mol } P_4O_{10}}{284 \text{ g } P_4O_{10}} \times \frac{4 \text{ mol } H_3PO_4}{1 \text{ mol } P_4O_{10}} = 1.25 \text{ mol } H_3PO_4$$

$$1.25 \text{ mol } H_3PO_4 \times \frac{98.0 \text{ g } H_3PO_4}{1 \text{ mol } H_3PO_4} = 123 \text{ g } H_3PO_4$$

(b) Remember that 1 L equals 1000 mL

$$\frac{125 \text{ mol } H_3PO_4}{0.500 \text{ L}} = 2.50 \text{ M } H_3PO_4$$

(c)

$$18.41 \times 10^9 \text{ lb} \times \frac{453.6 \text{ g}}{\text{lb}} \times \frac{1 \text{ mol}}{98.00 \text{ g}} = 8.521 \times 10^{10} \text{ mol } H_3PO_4$$

$$8.521 \times 10^{10} \text{ mol } H_3PO_4 \times \frac{1 \text{ mol } P_4O_{10}}{4 \text{ mol } H_3PO_4} = 2.130 \times 10^{10} \text{ mol } P_4O_{10}$$

$$2.130 \times 10^{10} \text{ mole } P_4O_{10} \times \frac{283.88 \text{ g}}{\text{mol}} \times \frac{1 \text{ metric ton}}{1 \times 10^6 \text{ g}}$$

$$= 6.047 \times 10^6 \text{ metric ton } P_4O_{10}$$

4.25 (a)

$$400 \times 10^6 \text{ metric ton} \times \frac{2.5 \text{ g S}}{100 \text{ g coal}} = 1.0 \times 10^7 \text{ metric ton S}$$

(b) First convert the above amount of S to moles of S:

$$1.0 \times 10^7 \text{ metric ton S} \times \frac{1 \times 10^6 \text{ g}}{1 \text{ metric ton}} \times \frac{1 \text{ mol}}{32.06 \text{ g}}$$

$$= 3.1 \times 10^{11} \text{ mole S}$$

Next convert to moles of SO_2:

$$3.1 \times 10^{11} \text{ mol S} \times \frac{1 \text{ mol } SO_2}{1 \text{ mol S}} = 3.1 \times 10^{11} \text{ mole } SO_2$$

and then convert to mass of SO_2:

$$3.1 \times 10^{11} \text{ mol } SO_2 \times \frac{64.0 \text{ g}}{\text{mol}} \times \frac{1 \text{ metric ton}}{1 \times 10^6 \text{ g}} = 2.0 \times 10^7 \text{ metric ton of } SO_2$$

(c) This calculation starts with the number of moles of SO_2, and uses the stoichiometry of the second reaction:

$$3.1 \times 10^{11} \text{ mol } SO_2 \times \frac{1 \text{ mol } CaSO_3}{1 \text{ mol } SO_2} \times \frac{120 \text{ g}}{\text{mol}} = 3.7 \times 10^{13} \text{ g } CaSO_3$$

$$3.7 \times 10^{13} \text{ g} \times \frac{1 \text{ metric ton}}{1 \times 10^6 \text{ g}} = 3.7 \times 10^7 \text{ metric ton } CaSO_3$$

(d)

$$3.7 \times 10^7 \text{ metric ton} \times \frac{1 \text{ railroad car}}{100 \text{ metric ton}} = 3.7 \times 10^5 \text{ railroad cars}$$

4.26 We must compare 452 g actual yield to the theoretical yield for each possibility in order to determine which metal has been involved:

for copper sulfide:

$$543 \text{ g CuS} \times \frac{1 \text{ mol}}{95.6 \text{ g}} \times \frac{2 \text{ mol CuO}}{2 \text{ mol CuS}} \times \frac{79.5 \text{ g CuO}}{1 \text{ mol CuO}} = 452 \text{ g CuO}$$

for lead sulfide:

$$543 \text{ g PbS} \times \frac{1 \text{ mol}}{239 \text{ g}} \times \frac{2 \text{ mol PbO}}{2 \text{ mol PbS}} \times \frac{223 \text{ g PbO}}{1 \text{ mol PbO}} = 507 \text{ g PbO}$$

It is the theoretical yield for CuO that is consistent with the actual yield of the metal oxide, and we conclude that the original sulfide must have been CuS.

4.27 First calculate the mass of Hg that is to be prepared:

$$2.55 \text{ L} \times \frac{1000 \text{ mL}}{1 \text{ L}} \times \frac{13.6 \text{ g Hg}}{1 \text{ mL Hg}} = 3.47 \times 10^4 \text{ g Hg}$$

and the mass of HgS that this will require:

$$3.47 \times 10^4 \text{ g Hg} \times \frac{1 \text{ mol}}{201 \text{ g}} \times \frac{4 \text{ mol HgS}}{4 \text{ mol Hg}} \times \frac{233 \text{ g HgS}}{1 \text{ mol HgS}} = 4.02 \times 10^4 \text{ g HgS}$$

$= 40.2$ kg HgS

4.28 This sort of analysis relies on the fact that all hydrogen that was originally present in the compound that was combusted ends up in the products only as that in H_2O. Similarly, the only carbon-containing product is CO_2, which is transformed directly into $CaCO_3$. The calculation begins by determining the number of moles of each product and ends by determining the number of moles of C and H that must have been present in the original compound:

$$3.739 \times 10^{-2} \text{ g CaCO}_3 \times \frac{1 \text{ mol}}{100.1 \text{ g}} = 3.735 \times 10^{-4} \text{ mole CaCO}_3$$

$$3.735 \times 10^{-4} \text{ mol CaCO}_3 \times \frac{1 \text{ mol C}}{1 \text{ mol CaCO}_3} = 3.735 \times 10^{-4} \text{ mol C}$$

$$7.570 \times 10^{-3} \text{ g H}_2\text{O} \times \frac{1 \text{ mol}}{18.02 \text{ g H}_2\text{O}} = 4.201 \times 10^{-4} \text{ mol H}_2\text{O}$$

$$4.201 \times 10^{-4} \text{ mol H}_2\text{O} \times \frac{2 \text{ mol H}}{1 \text{ mol H}_2\text{O}} = 8.402 \times 10^{-4} \text{ mol H}$$

The relative moles of C and H are next determined:

for C, $3.735 \times 10^{-4}/3.735 \times 10^{-4} = 1.000$

for H, $8.402 \times 10^{-4}/3.735 \times 10^{-4} = 2.249$

These mole ratios are themselves clearly a ratio of approximately 1 to 2.25, which is numerically the same as a ratio of $4 \times (1/2.25) =$

4 to 9. The empirical formula is thus C_4H_9, and the formula weight of the empirical unit is 57 g/mole. Since the molar mass is twice the value of the empirical formula weight:

$$114 \div 57 = 2$$

we deduce that the molecular formula is twice the empirical formula, or C_8H_{18}.

Limiting Reagents and Percent Yields

4.29 (a) First determine the amount of Fe_2O_3 that would be required to react completely with the given amount of Al:

$$3 \text{ mol Al} \times \frac{1 \text{ mol } Fe_2O_3}{2 \text{ mol Al}} = 1.50 \text{ mol } Fe_2O_3$$

Since only 1.25 mol of Fe_2O_3 are supplied, it is the limiting reagent. This can be confirmed by calculating the amount of Al that would be required to react completely with all of the available Fe_2O_3:

$$1.25 \text{ mol } Fe_2O_3 \times \frac{2 \text{ mol Al}}{1 \text{ mol } Fe_2O_3} = 2.50 \text{ mol Al}$$

Since an excess (3.00 mol - 2.50 mol = 0.50 mol) of Al is present, Fe_2O_3 must be the limiting reagent, as determined above.

(b)

$$1.25 \text{ mol } Fe_2O_3 \times \frac{2 \text{ mol Fe}}{1 \text{ mol } Fe_2O_3} = 2.50 \text{ mol Fe, theoretical yield}$$

4.30 (a) First calculate the number of moles of water that are needed to react completely with the given amount of PCl_5:

$$0.600 \text{ mol } PCl_5 \times \frac{4 \text{ mol } H_2O}{1 \text{ mol } PCl_5} = 2.40 \text{ mol } H_2O$$

Since this is less than the amount of water that is supplied, the limiting reagent must be PCl_5. This can be confirmed by the following calculation:

$$4.80 \text{ mol } H_2O \times \frac{1 \text{ mol } PCl_5}{4 \text{ mol } H_2O} = 1.20 \text{ mol } PCl_5$$

which also demonstrates that the limiting reagent is PCl_5.

(b)

$$0.600 \text{ mol } PCl_5 \times \frac{1 \text{ mol } H_3PO_4}{1 \text{ mol } PCl_5} = 0.600 \text{ mol } H_3PO_4$$

$$0.600 \text{ mol } PCl_5 \times \frac{5 \text{ mol HCl}}{1 \text{ mol } PCl_5} = 3.00 \text{ mol HCl}$$

4.31 (a) If all of the Zn were to react, then the required amount of S would be:

$$25.0 \text{ g Zn} \times \frac{1 \text{ mol Zn}}{65.4 \text{ g Zn}} \times \frac{1 \text{ mol S}}{1 \text{ mol Zn}} \times \frac{32.1 \text{ g S}}{1 \text{ mol S}} = 12.3 \text{ g S}$$

Similarly, the amount of Zn that would be required to react with all of the available S would be:

$$30.0 \text{ g S} \times \frac{1 \text{ mol S}}{32.1 \text{ g S}} \times \frac{1 \text{ mol Zn}}{1 \text{ mol S}} \times \frac{65.4 \text{ g Zn}}{1 \text{ mol Zn}} = 61.2 \text{ g Zn}$$

Since only 25.0 g Zn are available, Zn is the limiting reagent.

(b)

$$25.0 \text{ g Zn} \times \frac{1 \text{ mol Zn}}{65.4 \text{ g Zn}} \times \frac{1 \text{ mol ZnS}}{1 \text{ mol Zn}} \times \frac{97.5 \text{ g ZnS}}{1 \text{ mol ZnS}} = 37.3 \text{ g ZnS}$$

(c) From part (a) we know that 12.3 g S will react completely with all of the 25.0 g Zn. The unreacted amount of S is therefore:

30.0 g S total - 12.3 g S reacted = 17.7 g S unused.

4.32 (a) $3AgNO_3 + FeCl_3 \rightarrow 3AgCl + Fe(NO_3)_3$

(b) Calculate the amount of $FeCl_3$ that are required to react completely with all of the available silver nitrate:

$$25.0 \text{ g } AgNO_3 \times \frac{1 \text{ mol } AgNO_3}{170 \text{ g } AgNO_3} \times \frac{1 \text{ mol } FeCl_3}{3 \text{ mol AgCl}} \times \frac{162 \text{ g } FeCl_3}{1 \text{ mol } FeCl_3}$$

$= 7.94$ g $FeCl_3$ required for complete reaction.

Since more than this minimum amount is available, $FeCl_3$ is present in excess, and $AgNO_3$ must be the limiting reagent.

(c) The calculation of yield must be based on the amount of

$AgNO_3$ that is available:

$$25.0 \text{ g } AgNO_3 \times \frac{1 \text{ mol}}{170 \text{ g}} \times \frac{1 \text{ mol AgCl}}{1 \text{ mol } AgNO_3} = 0.147 \text{ mol AgCl}$$

(d)

$$0.147 \text{ mol AgCl} \times \frac{143 \text{ g AgCl}}{1 \text{ mol AgCl}} = 21.0 \text{ g AgCl}$$

(e) From part (b) we know that only 7.94 g $FeCl_3$ will be used. Therefore, the amount left unused is:

$$45.0 \text{ g total} - 7.94 \text{ g used} = 37.1 \text{ g } FeCl_3$$

4.33 (a) We proceed to determine which reagent is present in a limiting quantity by calculating the reagent that is first depleted. The amount of SO_2 that would be required to react completely with all of the available $CaCO_3$ would be:

$$255 \text{ g } CaCO_3 \times \frac{1 \text{ mol}}{100 \text{ g}} \times \frac{1 \text{ mol } SO_2}{1 \text{ mol } CaCO_3} \times \frac{64.1 \text{ g } SO_2}{1 \text{ mol } SO_2} = 163 \text{ g } SO_2$$

Since we are given only 135 g SO_2, and since 163 g are needed to use up all of the available $CaCO_3$, then SO_2 must be the limiting reagent. This can be confirmed by calculating the amount of $CaCO_3$ that would be required to react completely with all of the available SO_2:

$$135 \text{ g } SO_2 \times \frac{1 \text{ mol}}{64.1 \text{ g}} \times \frac{1 \text{ mol } CaCO_3}{1 \text{ mol } SO_2} \times \frac{100 \text{ g } CaCO_3}{1 \text{ mol } CaCO_3} = 211 \text{ g } CaCO_3$$

If only 211 g of $CaCO_3$ are needed, when 255 g are supplied, then $CaCO_3$ is present in excess, and it cannot be the limiting reagent. The theoretical yield must be based on the amount of that reagent that is limiting:

$$135 \text{ g } SO_2 \times \frac{1 \text{ mol}}{64.1 \text{ g}} \times \frac{1 \text{ mol } CaSO_3}{1 \text{ mol } SO_2} \times \frac{120 \text{ g } CaSO_3}{1 \text{ mol } CaSO_3} = 253 \text{ g } CaSO_3$$

(b) % yield = 198 g / 253 g × 100 = 78.3 %

4.34 First calculate the quantity of benzene that would be needed to give 100 g of chlorobenzene based on a yield of 100 %:

$$100\ g\ C_6H_5Cl \times \frac{1\ mol}{113\ g} = 0.885\ mol\ C_6H_5Cl$$

$$0.885\ mol\ C_6H_5Cl \times \frac{1\ mol\ C_6H_6}{1\ mol\ C_6H_5Cl} \times \frac{78.1\ g\ C_6H_6}{1\ mol\ C_6H_6} = 69.1\ g\ C_6H_6$$

Next calculate the amount that would be needed if the reaction were only 65 % effective:

69.1 g ÷ 0.65 = 106 g C_6H_6 would be needed.

4.35 Proceeding as in problem 4.34:

$$10.0\ g\ KC_7H_5O_2 \times \frac{1\ mol}{160\ g} = 0.0625\ mol\ KC_7H_5O_2$$

$$0.0625\ mol\ KC_7H_5O_2 \times \frac{1\ mol\ C_7H_8}{1\ mol\ KC_7H_5O_2} \times \frac{92.1\ g\ C_7H_8}{1\ mol\ C_7H_8} = 5.76\ g\ C_7H_8$$

Since the above amount is that needed assuming 100 % yield, it must be recalculated now to give the amount needed for a reaction that is only 68 % efficient:

5.76 g ÷ 0.68 = 8.47 g C_7H_8 would be needed.

Solutions

4.36 Yes, saturated solutions of CuS or $PbSO_4$, for example, are very dilute.

4.37 Yes, ammonium chloride, calcium chloride, or sodium hydroxide solutions could be prepared to be very concentrated, although they might not at the same time be saturated.

4.38 Solubility refers to the concentration of a solution that is, at some specified temperature, saturated. Concentration describes the amount of solute to that of solvent regardless of whether or not the solution is saturated.

4.39 (a) Since this is a solution that will be saturated, we can take its

66711

concentration directly from Table 4.1: 4.3×10^{-3} g/100 g H_2O.

(b) This solution is saturated because more solid is provided than can dissolve in the given amount of water.
(c) dilute
(d) Since the solution is saturated, no more lead sulfate can dissolve, and the new material would simply join the rest of the undissolved lead sulfate solid that rests on the bottom of the container.

4.40 Either the alcohol or the water could be called the solvent. Normally one refers to the liquid that is present in the greater amount as the solvent.

Molarity

4.41 The mixture contains 0.500 mol of NaCl in each liter of solution.

4.42 No, it tells only the ratio of moles of solute to liter of solution.

4.43 No, it tells only the ratio of moles of solute to liter of solution.

4.44 This is because it is the solution, not just the solvent, that is dispensed when the mixture is used in the laboratory.

4.45 1 mol NaCl = 1 mole of sodium chloride
1 M NaCl = a solution of NaCl having a concentration of 1 molar, namely 1 mole of NaCl per liter of solution.

4.46 Molecule: the smallest possible unit of a molecular compound.
Mole: Avogadro's number of something, e.g. of molecules or atoms.
Molar concentration: the number of moles of solute that are dissolved in a liter of solution.

4.47 The concentration does not change, although some of the solution has been dispensed.

4.48 Fill the flask to the mark with pure water, place it in the constant temperature bath set at 20.0 °C, and wait until the temperature of the water in the flask has reached 20.0 °C. Make a final adjustment of the amount of water in the flask, using a long medicine dropper to add or remove water until the water in the flask is at the level of the mark on the flask that is supposed to represent 100 mL. Dry the exterior of the flask, with its stopper in place and weigh the flask, the stopper, and the water that is inside the flask. Next determine the weight of the dry flask and stopper. The difference between the two weights is the mass of water that the flask contained when filled to the mark. Last, use the density of water at 20.0 °C and the mass determined above to calculate the volume of water that the flask contained. This value should be compared with the volume etched on the label.

4.49 1000 mmol = 1 mol

4.50 A millimole equals 1×10^{-3} mol. A milliliter equals 1×10^{-3} L. Hence:

$$\frac{\text{millimoles}}{\text{milliliter}} = \frac{1 \times 10^{-3}\ \text{mol}}{1 \times 10^{-3}\ \text{L}} = \frac{\text{mol}}{\text{L}}$$

4.51 (a) $$\frac{0.100\ \text{mol NaCl}}{1\ \text{L}} \times 0.250\ \text{L} \times \frac{58.5\ \text{g NaCl}}{1\ \text{mol NaCl}} = 1.46\ \text{g NaCl}$$

(b) $$\frac{0.440\ \text{mol}\ C_6H_{12}O_6}{1\ \text{L}} \times 0.100\ \text{L} \times \frac{180\ \text{g}\ C_6H_{12}O_6}{1\ \text{mol}\ C_6H_{12}O_6} = 7.92\ \text{g}\ C_6H_{12}O_6$$

(c) $$\frac{0.500\ \text{mol}\ H_2SO_4}{1\ \text{L}} \times 0.500\ \text{L} \times \frac{98.1\ \text{g}\ H_2SO_4}{1\ \text{mol}\ H_2SO_4} = 24.5\ \text{g}\ H_2SO_4$$

4.52 (a) $$\frac{0.100\ \text{mol}\ Na_2SO_4}{1\ \text{L}} \times 0.250\ \text{L} \times \frac{142\ \text{g}\ Na_2SO_4}{1\ \text{mol}\ Na_2SO_4} = 3.55\ \text{g}\ Na_2SO_4$$

(b) $$\frac{0.250\ \text{mol}\ Na_2CO_3}{1\ \text{L}} \times 0.100\ \text{L} \times \frac{106\ \text{g}\ Na_2CO_3}{1\ \text{mol}\ Na_2CO_3} = 2.65\ \text{g}\ Na_2CO_3$$

(c) $$\frac{0.400\ \text{mol NaOH}}{1\ \text{L}} \times 0.500\ \text{L} \times \frac{40.0\ \text{g NaOH}}{1\ \text{mol NaOH}} = 8.00\ \text{g NaOH}$$

4.53 Assume water has a density of 1.00 g/mL:

$$\frac{1.00\ \text{g}\ H_2O}{1.00\ \text{g}\ H_2O} \times \frac{1000\ \text{mL}}{1\ \text{L}} \times \frac{1\ \text{mol}\ H_2O}{18.0\ \text{g}\ H_2O} = 55.6\ \text{M}\ H_2O$$

4.54 $$\frac{1.94\ \text{g}\ H_2SO_4}{1\ \text{mL}\ H_2SO_4} \times \frac{1000\ \text{mL}}{1\ \text{L}} \times \frac{1\ \text{mol}\ H_2SO_4}{98.1\ \text{g}\ H_2SO_4} = 19.8\ \text{M}\ H_2SO_4$$

4.55 $$\frac{26\ \text{g}\ NH_3}{0.100\ \text{L}} \times \frac{1\ \text{mol}\ NH_3}{17.0\ \text{g}\ NH_3} = 15\ \text{M}\ NH_3$$

4.56

$$\frac{1.199\ \text{g soln}}{1.000\ \text{mL soln}} \times \frac{1000\ \text{mL}}{1\ \text{L}} \times \frac{26.00\ \text{g NaCl}}{100.0\ \text{g soln}} \times \frac{1\ \text{mol NaCl}}{58.45\ \text{g NaCl}} = 5.333\ \text{M}$$

4.57 First calculate the mass of one liter of the solution, using its known density:

$$\frac{1.0077 \text{ g soln}}{1 \text{ mL soln}} \times \frac{1000 \text{ mL}}{1 \text{ L}} \times \frac{1007.7 \text{ g soln}}{1 \text{ L soln}}$$

Next calculate the fraction of this mass that is due to the acetic acid only:

$$\left[\frac{5.50 \text{ g } HC_2H_3O_2}{(5.50 \text{ g } HC_2H_3O_2 + 100.0 \text{ g } H_2O)}\right] = 0.0521$$

Hence the amount of the total weight of one liter of solution (1007.7 g) that is mass due to acetic acid is given by:

$1007.7 \text{ g} \times 0.0521 = 52.5 \text{ g } HC_2H_3O_2$ per liter of solution.

The molarity is thus:

$$\frac{52.5 \text{ g } HC_2H_3O_2}{1 \text{ L soln}} \times \frac{1 \text{ mol } HC_2H_3O_2}{60.1 \text{ g } HC_2H_3O_2} = 0.874 \text{ M } HC_2H_3O_2$$

Reactions in Solution

4.58 (a) Start by calculating the number of moles of $CaCO_3$ that are to be made:

$$12.0 \text{ g } CaCO_3 \times \frac{1 \text{ mol}}{100 \text{ g}} = 0.120 \text{ mol } CaCO_3$$

Next determine the volume of each solution that will be required:

$$0.120 \text{ mol } CaCO_3 \times \frac{1 \text{ mol } CaCO_3}{1 \text{ mol } CaCO_3} \times \frac{1 \text{ L}}{0.500 \text{ mol } CaCl_2} = 0.240 \text{ L } CaCl_2$$

$$0.120 \text{ mol } CaCO_3 \times \frac{1 \text{ mol } K_2CO_3}{1 \text{ mol } CaCO_3} \times \frac{1 \text{ L}}{0.750 \text{ mol } K_2CO_3} = 0.160 \text{ L } K_2CO_3$$

(b) $0.240 \text{ L} \div 0.92 = 0.261 \text{ L } CaCl_2$ solution

$0.160 \text{ L} \div 0.92 = 0.174 \text{ L } K_2CO_3$ solution

4.59 (a) Calculate the amount of sodium bicarbonate that will be required:

$$40 \text{ mL} \times \frac{6.0 \text{ mol } H_2SO_4}{1000 \text{ mL}} \times \frac{2 \text{ mol } NaHCO_3}{1 \text{ mol } H_2SO_4} = 0.48 \text{ mol } NaHCO_3$$

Next calculate the amount of sodium bicarbonate that is given:

$$25 \text{ g NaHCO}_3 \times \frac{1 \text{ mol NaHCO}_3}{84 \text{ g NaHCO}_3} = 0.30 \text{ mol NaHCO}_3$$

There is less than enough sodium bicarbonate to neutralize the acid.

(b) $$\frac{0.30 \text{ mol NaHCO}_3 \text{ given}}{0.48 \text{ mol NaHCO}_3 \text{ needed}} \times 100 = 63 \text{ \% neutralization}$$

4.60 First calculate the number of moles of Na_2CO_3 that have reacted:

$$\frac{0.144 \text{ mol HCl}}{1000 \text{ mL}} \times 35.4 \text{ mL} \times \frac{1 \text{ mol Na}_2\text{CO}_3}{2 \text{ mol HCl}} = 2.55 \times 10^{-3} \text{ mol Na}_2\text{CO}_3$$

This is also the number of moles of sodium carbonate that are present in the sample. Next convert to mass:

$$2.55 \times 10^{-3} \text{ mol Na}_2\text{CO}_3 \times \frac{106 \text{ g Na}_2\text{CO}_3}{1 \text{ mol Na}_2\text{CO}_3} = 0.270 \text{ g Na}_2\text{CO}_3$$

The percent purity is then given by:

$0.270 \text{ g} \div 0.321 \text{ g} \times 100 = 84.1 \text{ \%}$

4.61 Calculate the number of moles of sulfuric acid that have reacted:

$$\frac{0.1048 \text{ mol NaOH}}{1000 \text{ mL}} \times 33.48 \text{ mL} \times \frac{1 \text{ mol H}_2\text{SO}_4}{2 \text{ mol NaOH}} = 1.754 \times 10^{-3} \text{ mol H}_2\text{SO}_4$$

The molarity is therefore given by:

$$\frac{1.754 \times 10^{-3} \text{ mol H}_2\text{SO}_4}{0.01546 \text{ L}} = 0.1135 \text{ M H}_2\text{SO}_4$$

4.62 First calculate the number of moles of lead sulfate that have been produced:

$$1.13 \text{ g PbSO}_4 \times \frac{1 \text{ mol PbSO}_4}{303 \text{ g PbSO}_4} = 3.73 \times 10^{-3} \text{ mol PbSO}_4$$

Since the stoichiometry of the reaction is 1:1:1, the number of moles of $PbSO_4$ that are formed is also equal to the number of moles of the

two reactants that are used. The molarity of each reactant solution is thus given by:

$$\frac{0.00373 \text{ mol } Na_2SO_4}{0.0353 \text{ L } Na_2SO_4} = 0.106 \text{ M } Na_2SO_4 \text{ solution}$$

$$\frac{0.00373 \text{ mol } Pb(NO_3)_2}{0.0325 \text{ L } Pb(NO_3)_2} = 0.115 \text{ M } Pb(NO_3)_2 \text{ solution}$$

4.63 First calculate the number of moles of AgCl that have been formed:

$$0.696 \text{ g AgCl} \times \frac{1 \text{ mol AgCl}}{143 \text{ g AgCl}} = 4.87 \times 10^{-3} \text{ mol AgCL}$$

Next calculate the number of moles of $MgCl_2$ that have reacted:

$$4.87 \times 10^{-3} \text{ mol AgCl} \times \frac{1 \text{ mol } MgCl_2}{2 \text{ mol AgCl}} = 2.44 \times 10^{-3} \text{ mol } MgCl_2$$

Finally, calculate the molarity of the $MgCl_2$ solution:

2.44×10^{-3} moles ÷ 0.0195 L = 0.125 M

Similarly, the molarity of the silver nitrate is equal to:

4.87×10^{-3} mol ÷ 0.0258 L = 0.189 M

4.64 This problem can be solved by determining the formula weight of the product. First calculate the number of moles of HCl that have been consumed:

$$\frac{1.24 \text{ mol HCl}}{\text{L}} \times 0.0558 \text{ L} = 0.0692 \text{ mol HCl}$$

Next convert to the number of moles of MCl_2 that must have been formed:

$$0.0692 \text{ mol HCl} \times \frac{1 \text{ mol } MCl_2}{2 \text{ mol HCl}} = 0.0346 \text{ mol } MCl_2$$

Finally calculate the formula weight of the product MCl_2:

4.72 g ÷ 0.0346 mol = 136 g/mol

which is consistent only with a formula MCl_2 where M = Zn.

4.65 First calculate the number of moles of $AgNO_3$ that have reacted:

0.1184 mol/L × 0.02476 L = 0.002932 mol $AgNO_3$

Next calculate the number of moles of either $CaCl_2$ or $FeCl_3$ that would be needed to react with this much $AgNO_3$:

For $CaCl_2$:

$$2.932 \times 10^{-3} \text{ mol } AgNO_3 \times \frac{1 \text{ mol } CaCl_2}{2 \text{ mol } AgNO_3} = 1.466 \times 10^{-3} \text{ mol } CaCl_2$$

For $FeCl_3$:

$$2.932 \times 10^{-3} \text{ mol } AgNO_3 \times \frac{1 \text{ mol } FeCl_3}{3 \text{ mol } AgNO_3} = 9.773 \times 10^{-4} \text{ mol } FeCl_3$$

Next calculate the molarity of the solution that was reacted, assuming first that it was $CaCl_2$ and then that it was $FeCl_3$:

$$1.466 \times 10^{-3} \text{ mol } CaCl_2 \div 0.01435 \text{ L} = 0.1022 \text{ M } CaCl_2$$

or

$$9.773 \times 10^{-4} \text{ mol } FeCl_3 \div 0.01435 \text{ L} = 0.06810 \text{ M } FeCl_3$$

Next calculate the number of grams of each of these that would be present in the original volume (250 mL) of solution that was prepared:

$$\frac{0.1022 \text{ mol } CaCl_2}{1 \text{ L}} \times 0.250 \text{ L} \times \frac{111.0 \text{ g}}{1 \text{ mol } CaCl_2} = 2.836 \text{ g } CaCl_2$$

as compared to:

$$\frac{0.06810 \text{ mol } FeCl_3}{1 \text{ L}} \times 0.250 \text{ L} \times \frac{162.2 \text{ g}}{1 \text{ mol } FeCl_3} = 2.761 \text{ g } FeCl_3$$

The actual mass used to prepare the original 250 mL of solution is closer to that calculated for $CaCl_2$, and we conclude that the solid must have been $CaCl_2$, not $FeCl_3$.

4.66 (a) The volume of the block is:

$$40.0 \text{ cm} \times 40.0 \text{ cm} \times 100.0 \text{ cm} = 1.60 \times 10^5 \text{ cm}^3$$

and the mass of the block is:

$$1.60 \times 10^5 \text{ cm}^3 \times 10.5 \text{ g/cm}^3 = 1.68 \times 10^6 \text{ g}$$

The theoretical yield of $AgNO_3$ is:

$$1.68 \times 10^6 \text{ g Ag} \times \frac{1 \text{ mol Ag}}{108 \text{ g Ag}} \times \frac{1 \text{ mol AgNO}_3}{1 \text{ mol Ag}} \times \frac{170 \text{ g AgNO}_3}{1 \text{ mol AgNO}_3}$$

$$= 2.63 \times 10^3 \text{ g AgNO}_3 = 2.63 \times 10^6 \text{ kg AgNO}_3$$

(b)

$$1.68 \times 10^6 \text{ g Ag} \times \frac{1 \text{ mol Ag}}{108 \text{ g Ag}} \times \frac{4 \text{ mol HNO}_3}{3 \text{ mol Ag}} = 2.07 \times 10^4 \text{ mol HNO}_3$$

$$2.07 \times 10^4 \text{ mol} \times \frac{1 \text{ L}}{16.0 \text{ mol HNO}_3} = 1.29 \times 10^3 \text{ L}$$

4.67

$$\frac{0.100 \text{ L} \times 0.500 \text{ mol/L}}{16 \text{ mol/L}} = 0.0031 \text{ L} = 3.1 \text{ mL}$$

4.68

$$\frac{1.00 \text{ mol/L} \times 1.00 \text{ L}}{17.4 \text{ mol/L}} = 0.0575 \text{ L} = 57.5 \text{ mL}$$

4.69 (a) In general, $V_i \times M_i = V_f \times M_f$, and in this problem we have:

$$\frac{0.450 \text{ M} \times 0.150 \text{ L}}{0.100 \text{ M}} = 0.675 \text{ L} = 675 \text{ mL of solution, total}$$

The volume of water to be added is then:

675 mL - 150 mL = 525 mL

(b)
$$\frac{0.080 \text{ L} \times 0.500 \text{ M}}{0.140 \text{ L}} = 0.286 \text{ M}$$

4.70 (0.20 M × 0.250 L) ÷ 0.100 M = 0.50 L, the final volume. Therefore, add 250 mL of water to 250 mL of the original solution.

4.71 (0.15 M × 0.100 L) ÷ 0.250 L = 0.060 M

4.72 The molarity of the original solution was:

10 g NaOH/L ÷ 40 g/mol = 0.25 mol/L

In general,

$$V_i \times M_i = V_f \times M_f$$

and in this case, the initial molarity is 0.250 mol/L. Also, the final volume is equal to (0.500 L + X), where X equals the initial volume of the NaOH solution, whose value we are to determine. The above relationship thus becomes:

$(X) \times (0.250\ mol/L) = (0.500 + X)L \times 0.15\ mol/L$

Solving for X we have:

$0.25X = (0.500)(0.15) + 0.15X$

$0.10X = 0.075,$

so that X = 0.75 L, the original volume of the NaOH solution.

CHAPTER FIVE

PRACTICE EXERCISES

1. The amount of heat transferred into the water is:

$$\frac{4.18\ \text{J}}{\text{g °C}} \times 250\ \text{g} \times 5.0\ \text{°C} = 5.2 \times 10^3\ \text{J}$$

$$5.2 \times 10^3\ \text{J} \times \frac{1\ \text{cal}}{4.184\ \text{J}} = 1.2 \times 10^3\ \text{cal}$$

2. The amount of heat absorbed by the water is:

$$\frac{4.18\ \text{J}}{\text{g °C}} \times 335\ \text{g} \times 1.9\ \text{°C} = 2.7 \times 10^3\ \text{J} = 2.7\ \text{kJ}$$

$$2.7 \times 10^3\ \text{J} \times \frac{1\ \text{cal}}{4.184\ \text{J}} = 6.4 \times 10^2\ \text{cal} = 0.64\ \text{kcal}$$

3. q = specific heat × mass × temperature change
= 4.184 J/g °C × (175 g + 4.90 g) × (14.9 °C - 10.0 °C)

$= 3.7 \times 10^3$ J = 3.7 kJ of heat released by the process.

This should then be converted to a value representing kJ per mole of reactant, remembering that the sign of ΔH is to be negative, since the process releases heat energy to surroundings. The number of moles of sulfuric acid is:

$$4.90\ \text{g} \times \frac{1\ \text{mol}}{98.1\ \text{g}} = 0.0499\ \text{mol}\ H_2SO_4$$

and the enthalpy change in kJ/mole is given by:

$$\Delta H = \frac{-3.7\ \text{kJ}}{0.0499\ \text{mol}\ H_2SO_4} = -74\ \text{kJ/mol}$$

4. The heat released by the combustion is:

4.26 °C × 9.43 kJ/°C = 40.2 kJ

The number of moles of steric acid is given by:

$1.02\ \text{g} \div 285\ \text{g/mol} = 3.58 \times 10^{-3}$ mol

The enthalpy change is negative, because the process releases heat, and the numerical value is:

$$\Delta H = -\frac{40.2\ kJ}{3.58 \times 10^{-3}\ mol} = -1.12 \times 10^4\ kJ/mol$$

5. $H_2(g) + \frac{1}{2}O_2(g) \rightarrow H_2O(\ell) \quad \Delta H° = -258.9\ kJ$

6. This problem requires that we add the reverse of the second equation (remembering to change the sign of the associated ΔH value) to the first equation:

$$C_2H_4(g) + 3O_2(g) \rightarrow 2CO_2(g) + 2H_2O(\ell), \quad \Delta H° = -1411.1\ kJ$$

$$+\ 2CO_2(g) + 3H_2O(\ell) \rightarrow C_2H_5OH(\ell) + 3O_2(g), \quad \Delta H° = +1367.1\ kJ$$

which gives the following net equation and value for ΔH:

$$C_2H_4(g) + H_2O(\ell) \rightarrow C_2H_5OH(\ell), \quad \Delta H = -44.0\ kJ$$

7. (a) $\Delta H° = \text{sum}\ \Delta H_f°[\text{products}] - \text{sum}\ \Delta H_f°[\text{reactants}]$

$= 2\Delta H_f°[NO_2(g)] - \{2\Delta H_f°[NO(g)] + \Delta H_f°[O_2(g)]\}$

$= 2\ mol \times 33.8\ kJ/mol - [2\ mol \times 90.37\ kJ/mol + 1\ mol \times 0\ kJ/mol]$

$= -113\ kJ/mol$

(b) $\Delta H° = \{\Delta H_f°[H_2O(g)] + \Delta H_f°[NaCl(s)]\}$

$- \{\Delta H_f°[NaOH(s)] + \Delta H_f°[HCl(g)]\}$

$= [(-241.8\ kJ/mol) + (-411.0\ kJ/mol)]$

$- [(-426.8\ kJ/mol) + (-92.30\ kJ/mol)]$

$= -133.4\ kJ/mol$

8. From example 5.5 we know the heat of combustion of sucrose:

$$C_{12}H_{22}O_{11}(s) + 12O_2(g) \rightarrow 12CO_2(g) + 11H_2O(\ell), \quad \Delta H° = -5.65 \times 10^3\ kJ$$

Also, this same value could be obtained by the following calculation:

$\Delta H° = \{11\Delta H_f°[H_2O(\ell)] + 12\Delta H_f°[CO_2(g)]\}$

$- \{12\Delta H_f°[O_2(g)] + \Delta H_f°[C_{12}H_{22}O_{11}]\}$

$-5.65 \times 10^3\ kJ = [11\ mol \times (-285.9\ kJ/mol) + 12\ mol \times (-393.5\ kJ/mol)]$

$- \{12\ mol \times (0\ kJ/mol) + \Delta H_f°[\text{sucrose}]\}$

Solving for the heat of formation of sucrose, we have:

$\Delta H^\circ_f[C_{12}H_{22}O_{11}] = -2226$ kJ/mol

REVIEW EXERCISES

Energy Sources and Units

5.1 Chemical energy is the potential energy in substances, which changes into other forms of energy when substances undergo chemical reactions.

5.2 A source of energy is any process, or any substance that can be induced to undergo a change leading to a process, that releases energy in a form that is useful to humans.

5.3 The principal fossil fuels are coal, oil, and natural gas. The term fossil signifies that these fuels formed eons ago from the remains of living things.

5.4 $CO_2 + H_2O \rightarrow (CH_2O) + O_2$

5.5 $1\ J = 1\ kg\ m^2/s^2$. It is the SI unit of energy, and it equals 4.184 cal.

5.6 The calorie was originally defined to be the energy necessary to raise the temperature of 1 g of water from 14.5 °C to 15.5 °C.

5.7 1 J = 4.184 cal

5.8 1 kJ = 4.184 kcal

5.9

$$458\ \text{kcal} \times \frac{1000\ \text{cal}}{1\ \text{kcal}} \times \frac{4.184\ \text{J}}{1\ \text{cal}} = 1.92 \times 10^6\ \text{J}$$

$$1.92 \times 10^6\ \text{J} \times \frac{1\ \text{kJ}}{1000\ \text{J}} = 1.92 \times 10^3\ \text{kJ}$$

5.10

$$5225\ \text{J} \times \frac{1\ \text{cal}}{4.184\ \text{J}} \times \frac{1\ \text{kcal}}{1000\ \text{cal}} = 1.249\ \text{kcal}$$

5.11

$$1.00\ \text{cal} \times \frac{4.184\ \text{J}}{1\ \text{cal}} \times \frac{9.48 \times 10^{-4}\ \text{BTU}}{1.00\ \text{J}} = 3.97 \times 10^{-3}\ \text{BTU}$$

5.12

$$\frac{1.00 \text{ cal}}{3.97 \times 10^{-3} \text{ BTU}} \times \frac{1 \text{ kcal}}{1000 \text{ cal}} \times \frac{1.00 \times 10^{15} \text{ BTU}}{1.00 \text{ quad}} = 2.5 \times 10^{14} \text{ kcal/quad}$$

5.13 In general, kinetic energy is given by the following formula, where mass (m) is expressed in the units kg, and velocity (v) is expressed in units m/s:

$$KE = \tfrac{1}{2}mv^2$$

First it is necessary to convert mass in lb to kg units:

$$4000 \text{ lb} \times \frac{1 \text{ kg}}{2.205 \text{ lb}} = 1814 \text{ kg}$$

and to convert velocity to units of m/s:

$$\frac{55 \text{ mi}}{1.0 \text{ hr}} \times \frac{1 \text{ km}}{0.6215 \text{ mi}} \times \frac{1000 \text{ m}}{1 \text{ km}} \times \frac{1 \text{ hr}}{3600 \text{ s}} = 25 \text{ m/s}$$

Finally, kinetic energy is calculated:

$$KE = 0.5(1814 \text{ kg})(25 \text{ m/s})^2 = 5.75 \times 10^5 \text{ J} = 5.7 \times 10^2 \text{ kJ}$$

5.14 First convert 5 oz to mass in the units kg:

$$5.0 \text{ oz} \times \frac{28.35 \text{ g}}{1 \text{ oz}} \times \frac{1 \text{ kg}}{1000 \text{ g}} = 0.14 \text{ kg}$$

and then convert velocity in units mi/hr to the units m/s:

$$\frac{66 \text{ mi}}{1.0 \text{ hr}} \times \frac{1 \text{ km}}{0.6215 \text{ m}} \times \frac{1000 \text{ m}}{1 \text{ km}} \times \frac{1 \text{ hr}}{3600 \text{ s}} = 29 \text{ m/s}$$

Last, calculate kinetic energy as in problem 5.13:

$$KE = 0.5(0.14 \text{ kg})(29 \text{ m/s})^2 = 59 \text{ J} = 0.059 \text{ kJ}$$

Specific Heat and Heat Capacity

5.15 The energy depends directly on the specific heat, so the material with the higher specific heat requires the higher energy input for a given rise in temperature.

5.16 Low specific heat

5.17 The numerical values would not change because 1000 cal/1000 g is equal to the ratio 1 kcal/1 kg, and the magnitude of the kelvin is the same as that of the Celsius degree.

5.18 A substance such as water, having a high specific heat, can absorb more energy without much of a temperature change, compared to a substance with a low specific heat.

5.19

$$\frac{1.00\ \text{cal}}{\text{g}\ ^\circ\text{C}} \times 225\ \text{g}\ H_2O \times (25.0\ ^\circ\text{C} - 10.0\ ^\circ\text{C}) = 3.37 \times 10^3\ \text{cal}$$

$$3.37 \times 10^3\ \text{cal} \times \frac{1\ \text{kcal}}{1000\ \text{cal}} = 3.37\ \text{kcal}$$

5.20

$$\frac{1.00\ \text{cal}}{\text{g}\ ^\circ\text{C}} \times 1.0\ \text{kg}\ H_2O \times \frac{1000\ \text{g}}{1\ \text{kg}} \times (99\ ^\circ\text{C} - 25\ ^\circ\text{C}) \times \frac{1\ \text{kcal}}{1000\ \text{cal}}$$

$= 74$ kcal

5.21 (a)

$$1.0\ \text{lb tissue} \times \frac{454\ \text{g}}{1.0\ \text{lb}} \times \frac{85\ \text{g fat}}{100\ \text{g fat tissue}} \times \frac{9.0\ \text{kcal}}{1.0\ \text{g fat}}$$

$= 3.5 \times 10^3$ kcal

(b)

$$\frac{1.0\ \text{lb fat}}{3.5 \times 10^3\ \text{kcal}} \times \frac{500\ \text{kcal}}{\text{hr}} = 0.14\ \text{lb/hr}$$

$$\frac{8.0\ \text{mi/hr}}{0.14\ \text{lb/hr}} = 57\ \text{mi}$$

5.22

$$\frac{1.0\ \text{day}}{0.25\ \text{lb butter}} \times \frac{1.0\ \text{lb}}{454\ \text{g}} \times \frac{1\ \text{g butter}}{9.0\ \text{kcal}} \times \frac{3500\ \text{kcal}}{0.5\ \text{lb fat tissue}}$$

= 7 days per pound of fat tissue

5.23

$$0.4498\ \text{J/g}\ ^\circ\text{C} \times \frac{55.85\ \text{g Fe}}{1\ \text{mol Fe}} = 25.12\ \text{J/mol}\ ^\circ\text{C}$$

5.24

$$0.586\ \text{cal/g}\ ^\circ\text{C} \times \frac{4.184\ \text{J}}{1\ \text{cal}} \times \frac{46.1\ \text{g ethyl alcohol}}{1\ \text{mol ethyl alcohol}} = 113\ \text{J/mol}\ ^\circ\text{C}$$

Enthalpy

5.25 The system is that part of the universe that is under study, and as such, it is separated from the surroundings by real or imaginary boundaries.

5.26 The system is everything inside of and attached to the vat. The vat serves as an insulating boundary, and it allows the heat to flow from the reaction to only those things that are in contact with the water.

5.27 Ideally, the insulation is part of the surroundings, because its purpose is to contain the heat of reaction, allowing heat flow only to the water and things in contact with the water. The system under study is the water and the heat absorbing parts of the calorimeter.

5.28 The heat content is the enthalpy of a system. H_{final} refers to the total heat content (enthalpy) of the system at the conclusion of the change, and $H_{initial}$ refers to enthalpy before the change starts.

5.29 $\Delta H - H_{products} - H_{reactants}$

5.30 The total energy of the universe is constant; it cannot be created or destroyed, but only transferred or transformed.

5.31 As described in the answer to question 5.30, the enthalpy of the surroundings must decrease by 100 kJ, since the total energy must remain constant.

5.32 negative

5.33 A state function is a function whose value is independent of the path (route) that is taken to achieve it. A state function has a value that depends only on the difference between the final and initial states, not on the path taken to change from the initial to the final state.

Calorimetry

5.34 Calorimetry is the study of the quantities of heat that are involved in chemical or physical changes.

5.35 A bomb calorimeter is a device within which a reaction is allowed to occur at constant volume. The reactions are usually those that can be initiated by an igniting device such as an eletrical spark. The calorimeter is designed to allow the heat flow as a result of the reaction to be measured by a temperature change for the bomb, its water bath and the other devices as diagramed in Figure 5.5, on page 161 of the text.

5.36 calorie. The additional multiplication that is required is by mass.

5.37 The bomb calorimeter is designed to operate at constant volume. It is heat flow at constant pressure that is termed enthalpy. The latter sorts of heat flows are normally measured in open devices, such as "coffee cup" calorimeters.

5.38 q = specific heat × mass × temperature change

$= (1.00 \text{ cal/g °C}) \times (5.45 \times 10^3 \text{ g}) \times (60.30 \text{ °C} - 57.60 \text{ °C})$

$= 1.47 \times 10^4 \text{ cal} = 14.7 \text{ kcal}$

$14.7 \text{ kcal} \times 4.184 \text{ kJ/kcal} = 61.5 \text{ kJ}$

5.39 Keep in mind that the total mass must be considered in this calculation, and that both liquids, once mixed, undergo the same temperature increase:

$q = (1.00 \text{ cal/g °C}) \times (55.0 \text{ g} + 55.0 \text{ g}) \times (31.8 \text{ °C} - 23.5 \text{ °C})$

$= 9.1 \times 10^2 \text{ cal}$

$9.1 \times 10^2 \text{ cal} \times 4.184 \text{ J/cal} = 3.8 \times 10^3$ J of heat energy released.

Next determine the number of moles of reactant involved in the reaction:

$0.0550 \text{ L} \times 1.3 \text{ mol/L} = 0.072$ mol of acid and of base.

Thus the enthalpy change is:

$$\Delta H = -9.1 \times 10^2 \text{ cal} \times \frac{1 \text{ kcal}}{1000 \text{ cal}} \times \frac{1}{0.072 \text{ mol}} = -13 \text{ kcal/mol}$$

5.40

$16.44 \text{ kJ} = (18.56 \text{ kJ/°C}) \times (t_{final} - 23.518 \text{ °C})$

Solving for the final temperature we have:

$(t_{final} - 23.518 \text{ °C}) = 16.44 \text{ kJ} \div 18.56 \text{ kJ/mol}$

$t_{final} = 24.404 \text{ °C}$

5.41

$1.05 \times 10^1 \text{ kJ} = (26.6 \text{ kJ/°C}) \times (26.13 \text{ °C} - t_{initial})$

Solve for the initial temperature as follows:

$(26.13 \text{ °C} - t_{initial}) = 1.05 \times 10^1 \text{ kJ} \div 26.6 \text{ kJ/°C}$

$t_{initial} = 25.74 \text{ °C}$

5.42 The heat of neutralization is released to three "independent" components of the system, all of which undergo the same temperature increase: $\Delta T = 20.610\ °C - 16.784\ °C = 3.826\ °C$. Also, the total heat capacity of the system is the sum of the three heat capacities:

$$\text{heat capacity}_{HCl} + \text{heat capacity}_{NaOH} + \text{heat capacity}_{calorimeter}$$

$$= (4.031\ J/g\ °C \times 610.29\ g) + (4.046\ J/g\ °C \times 615.31\ g) + 77.99\ J/°C$$

$$= 5028\ J/°C$$

The heat flow to the system is thus:

$$q = 5028\ J/°C \times 3.826\ °C = 1.924 \times 10^4\ J = 1.924 \times 10^1\ kJ$$

and the heat of neutralization is the negative of this value, since the neutralization process is exothermic:

$$\Delta H = -1.924 \times 10^1\ kJ \div 0.33183\ mol = -57.98\ kJ/mol$$

5.43 This problem may be solved exactly as for problem 5.42:

total heat capacity:

$$(4.031\ J/g\ °C \times 610.28\ g) + (4.003\ J/g\ °C \times 619.69\ g) + 77.99\ J/°C$$

$$= 5019\ J/°C$$

common temperature change:

$$19.410\ °C - 15.533\ °C = 3.877\ °C$$

Heat flow to the system:

$$q = 5019\ J/°C \times 3.877\ °C = 1.946 \times 10^4\ J = 1.946 \times 10^1\ kJ$$

The enthalpy change is thus:

$$\Delta H = -1.946 \times 10^1\ kJ \div 0.3314\ mol = -58.72\ kJ/mol$$

Standard Heats of Reaction

5.44 Since the heats of reaction will in general depend on temperature and pressure, we need some standard set of values for temperature and pressure so that comparisons of various heats of reaction are made under identical conditions.

5.45 $\Delta H°$ applies only at STP.

5.46 A thermochemical equation contains the value for the associated ΔH.

5.47 A coefficient such as this always signifies the number of moles.

5.48 $4Al(s) + 2Fe_2O_3(s) \rightarrow 2Al_2O_3(s) + 4Fe(s)$, $\Delta H° = -1708$ kJ

5.49

$$1.500 \text{ mol } C_6H_6(\ell) \times \frac{-6542 \text{ kJ}}{2.000 \text{ mol } C_6H_6(\ell)} = -4907 \text{ kJ}$$

5.50 $CaO(s) + 1OH_2O(\ell) \rightarrow 1OCa(OH)_2(s)$, $\Delta H° = -653$ kJ

Hess's Law and Thermochemical Equations

5.51

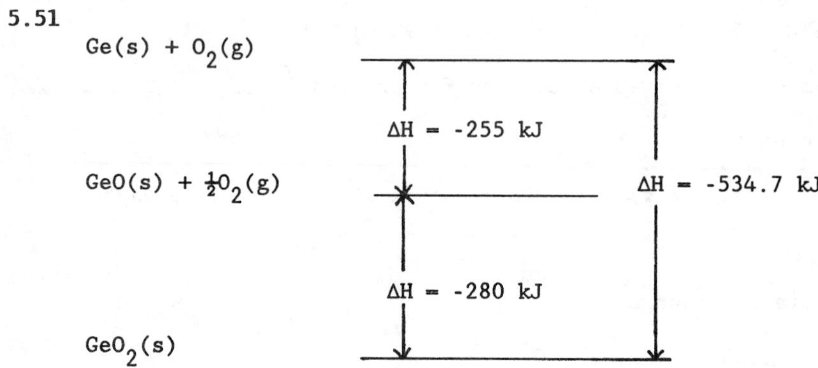

5.52

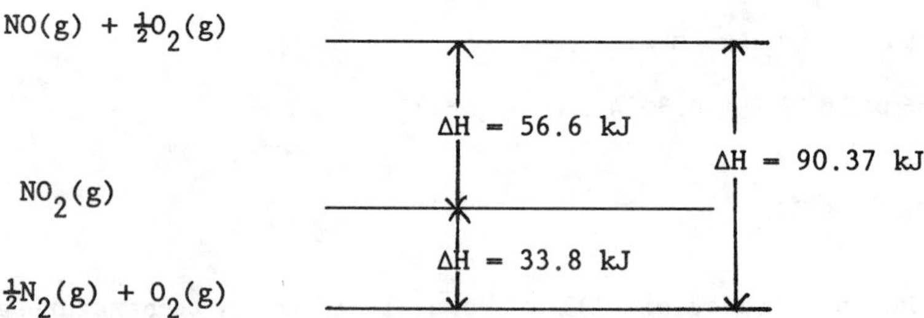

The enthalpy change for the desired reaction is the negative of that diagramed above, namely:

$NO(g) + \frac{1}{2}O_2(g) \rightarrow NO_2(g)$, $\Delta H = -56.6$ kJ

5.53 For any reaction that can be written in steps, the standard heat of reaction for the overall process is the sum of the standard heats of reaction of the individual steps.

5.54 Enthalpy is a state function; its value does not depend on the path or on the kinds of steps used to proceed from the initial state to the final state.

5.55 Since NO_2 does not appear in the desired overall reaction, the two steps are to be manipulated in such a manner so as to remove it by cancellation. Add the second equation to the inverse of the first, remembering to change the sign of the first equation, since it is to be reversed:

$$2NO_2(g) \rightarrow N_2O_4(g), \quad \Delta H° = -57.93 \text{ kJ}$$

$$2NO(g) + O_2(g) \rightarrow 2NO_2(g), \quad \Delta H° = -113.14 \text{ kJ}$$

Adding, we have:

$$2NO(g) + O_2(g) \rightarrow N_2O_4(g), \quad \Delta H° = -171.07 \text{ kJ}$$

5.56

$$NO(g) \rightarrow \tfrac{1}{2}O_2(g) + \tfrac{1}{2}N_2(g), \quad \Delta H° = -90.37 \text{ kJ}$$

$$\tfrac{1}{2}N_2(g) + O_2(g) \rightarrow NO_2(g), \quad \Delta H° = 33.8 \text{ kJ}$$

Adding, we get:

$$NO(g) + \tfrac{1}{2}O_2(g) \rightarrow NO_2(g), \quad \Delta H° = -56.6 \text{ kJ}$$

5.57 Reverse the first equation, multiply the result by two, and add it to the second equation:

$$2KCl(s) + 2H_2O(\ell) \rightarrow 2HCl(g) + 2KOH(s), \quad \Delta H° = 407.2 \text{ kJ}$$

$$H_2SO_4(\ell) + 2KOH(s) \rightarrow K_2SO_4(s) + 2H_2O(\ell), \quad \Delta H° = -342.4 \text{ kJ}$$

Adding gives us:

$$2KCl(s) + H_2SO_4(\ell) \rightarrow 2HCL(g) + K_2SO_4(s), \quad \Delta H° = 64.8 \text{ kJ}$$

5.58 If we label the four known thermochemical equations consecutively, 1, 2, 3, and 4, then the sum is made in the following way: Divide equation #3 by two, and reverse all of the other equations (#1, #2, and #4), while also dividing each by two:

$\frac{1}{2}Na_2O + HCl \rightarrow \frac{1}{2}H_2O + NaCl$, $\Delta H° = -253.66$ kJ

$NaNO_2 \rightarrow \frac{1}{2}NO + \frac{1}{2}NO_2 + \frac{1}{2}Na_2O$, $\Delta H° = 213.57$ kJ

$\frac{1}{2}NO + \frac{1}{2}NO_2 \rightarrow \frac{1}{2}N_2O + \frac{1}{2}O_2$, $\Delta H° = -21.34$ kJ

$\frac{1}{2}H_2O + \frac{1}{2}O_2 + \frac{1}{2}N_2O \rightarrow HNO_2$, $\Delta H° = -17.18$ kJ

Adding gives:

$HCl + NaNO_2 \rightarrow NaCl + HNO_2$, $\Delta H° = -78.61$ kJ

5.59 Add the reverse of the first equation to the second equation:

$H_2SO_4(\ell) \rightarrow SO_3(g) + H_2O(\ell)$, $\Delta H° = 78.2$ kJ

$BaO(s) + SO_3(g) \rightarrow BaSO_4(s)$, $\Delta H° = -213$ kJ

$BaO(s) + H_2SO_4(\ell) \rightarrow BaSO_4(s) + H_2O(\ell)$, $\Delta H° = -135$ kJ

5.60 The desired net equation is obtained by adding together the reverse of the two thermochemical equations:

$Zn(NO_3)_2(aq) + Cu(s) \rightarrow Cu(NO_3)_2(aq) + Zn(s)$, $\Delta H° = 61.7$ kcal

$Cu(NO_3)_2(aq) + 2Ag(s) \rightarrow 2AgNO_3(aq) + Cu(s)$, $\Delta H° = 25.3$ kcal

$2Ag(s) + Zn(NO_3)_2(aq) \rightarrow Zn(s) + 2AgNO_3(aq)$, $\Delta H° = 87.0$ kcal

Since this $\Delta H°$ has a positive value, it is the reverse reaction that occurs spontaneously.

5.61 Reverse the second equation, and then divide each by two before adding:

$CO(g) + \frac{1}{2}O_2(g) \rightarrow CO_2(g)$, $\Delta H° = -283.1$ kJ

$CuO(s) \rightarrow 2Cu(s) + \frac{1}{2}O_2(g)$, $\Delta H° = 155.3$ kJ

$CuO(s) + CO(g) \rightarrow Cu(s) + CO_2(g)$, $\Delta H° = -127.8$ kJ

5.62 Reverse the second and the third thermochemical equations and add them to the first:

$CaO(s) + 2HCl(aq) \rightarrow CaCl_2(aq) + H_2O(\ell)$, $\Delta H° = -186$ kJ

$Ca(OH)_2(s) \rightarrow CaO(s) + H_2O(\ell)$, $\Delta H° = 62.3$ kJ

$Ca(OH)_2(aq) \rightarrow Ca(OH)_2(s)$, $\Delta H° = 12.6$ kJ

$Ca(OH)_2(aq) + 2HCl(aq) \rightarrow CaCl_2(aq) + 2H_2O(\ell)$, $\Delta H° = -111$ kJ

5.63 The first thermochemical equation is reversed and multiplied by two, before adding it to the second equation:

$2LiOH(s) \rightarrow 2Li(s) + O_2(g) + H_2(g)$, $\Delta H° = 974.0$ kJ

$2Li(s) + Cl_2(g) \rightarrow 2LiCl(s)$, $\Delta H° = -815.0$ kJ

Next we add twice the standard heat of formation of liquid water:

$2H_2(g) + O_2(g) \rightarrow 2H_2O(\ell)$, $\Delta H° = -578.1$ kJ

and also we add the reverse of twice the enthalpy of formation of HCl(g):

$2HCl(g) \rightarrow H_2(g) + Cl_2(g)$, $\Delta H° = 184.6$ kJ

The last three equations that we need are obtained from those listed in the problem:

$2LiOH(aq) \rightarrow 2LiOH(s)$, $\Delta H° = 38.4$ kJ

$2HCl(aq) \rightarrow 2HCl(g)$, $\Delta H° = 154$ kJ

$2LiCl(s) \rightarrow 2LiCl(aq)$, $\Delta H° = -72.0$ kJ

where two of the last three have been reversed, and where all of the last three have been multiplied by two. The sum of all of the above equations is:

$2LiOH(aq) + 2HCl(aq) \rightarrow 2LiCl(aq) + 2H_2O(\ell)$, $\Delta H° = -53.8$ kJ

The desired thermochemical equation is half that given above:

$LiOH(aq) + HCl(aq) \rightarrow LiCl(aq) + H_2O(\ell)$, $\Delta H° = -26.9$ kJ

5.64 The first equation is reversed:

$Br_2(aq) + 2KCl(aq) \rightarrow Cl_2(g) + 2KBr(aq)$, $\Delta H° = 96.2$ kJ

The second equation is used "as is":

$H_2(g) + Cl_2(g) \rightarrow 2HCl(g)$, $\Delta H° = -184$ kJ

The third equation is multiplied by two:

$2HCl(aq) + 2KOH(aq) \rightarrow 2KCl(aq) + 2H_2O(\ell)$, $\Delta H° = -114.6$ kJ

Next we add twice the reverse of number four:

$2H_2O(\ell) + 2KBr(aq) \rightarrow 2KOH(aq) + 2HBr(aq)$, $\Delta H° = 114.6$ kJ

Number 5 is multiplied by two:

$2HCl(g) \rightarrow 2HCl(aq)$, $\Delta H° = -154.0$ kJ

Number six is used "as is":

$Br_2(g) \rightarrow Br_2(aq)$, $\Delta H° = -4.2$ kJ

Finally, we also use twice the reverse of number 7:

$2HBr(aq) \rightarrow 2HBr(g)$, $\Delta H° = 159.8$ kJ

The sum of all of the above gives:

$H_2(g) + Br_2(g) \rightarrow 2HBr(g)$, $\Delta H° = -86.2$ kJ

which is exactly twice the desired value:

$\frac{1}{2}H_2(g) + \frac{1}{2}Br_2(g) \rightarrow HBr(g)$, $\Delta H° = 43.1$ kJ

Hess's Law and Standard Heats of Formation

5.65 Only (b) should be labeled with $\Delta H_f°$.

5.66 (a) $2C(graphite) + 2H_2(g) + O_2(g) \rightarrow HC_2H_3O_2(\ell)$, $\Delta H_f° = -487.0$ kJ

(b) $Na(s) + \frac{1}{2}H_2(g) + C(graphite) + \frac{3}{2}O_2(g) \rightarrow NaHCO_3(s)$

$\Delta H_f° = -947.7$ kJ

(c) $Ca(s) + \frac{1}{8}S_8(s) + 3O_2(g) + 2H_2(g) \rightarrow CaSO_4 \cdot 2H_2O(s)$

$\Delta H_f° = -2021.1$ kJ

(d) $Ca(s) + \frac{1}{8}S_8(s) + \frac{5}{2}O_2(g) + \frac{1}{2}H_2(g) \rightarrow CaSO_4 \cdot \frac{1}{2}H_2O(s)$

$\Delta H^\circ_f = -1575.2$ kJ

(e) $C(graphite) + 2H_2(g) + \frac{1}{2}O_2(g) \rightarrow CH_3OH(\ell)$, $\Delta H^\circ_f = -238.6$ kJ

5.67 (a) $\Delta H^\circ = \Delta H^\circ_f[O_2(g)] + 2\Delta H^\circ_f[H_2O(\ell)] - 2\Delta H^\circ_f[H_2O_2(\ell)]$

$\Delta H^\circ = 0.0 + 2 \text{ mol} \times (-285.9 \text{ kJ/mol}) - 2 \times (-187.6 \text{ kJ/mol})$
$= -196.6$ kJ

(b) $\Delta H^\circ = \Delta H^\circ_f[H_2O(\ell)] + \Delta H^\circ_f[NaCl(s)] - \Delta H^\circ_f[HCl(g)] - \Delta H^\circ_f[NaOH(s)]$

$= 1 \text{ mol} \times (-285.9 \text{ kJ/mol}) + 1 \text{ mol} \times (-411.0 \text{ kJ/mol})$
$- 1 \text{ mol} \times (-92.30 \text{ kJ/mol}) - 1 \text{ mol} \times (-426.8 \text{ kJ/mol})$
$= -177.8$ kJ

(c) $\Delta H^\circ = \Delta H^\circ_f[HCl(g)] + \Delta H^\circ_f[CH_3Cl(g)] - \Delta H^\circ_f[CH_4(g)] - \Delta H^\circ_f[Cl_2(g)]$

$= 1 \text{ mol} \times (-92.30 \text{ kJ/mol}) + 1 \text{ mol} \times (-82.0 \text{ kJ/mol})$
$- 1 \text{ mol} \times (-74.85 \text{ kJ/mol}) - 1 \text{ mol} \times (0.0 \text{ kJ/mol})$
$= -99.5$ kJ

(d) $\Delta H^\circ = \Delta H^\circ_f[H_2O(\ell)] + \Delta H^\circ_f[CO(NH_2)_2(s)] - 2\Delta H^\circ_f[NH_3(g)]$

$- \Delta H^\circ_f[CO_2(g)]$

$= 1 \text{ mol} \times (-285.9 \text{ kJ/mol}) + 1 \text{ mol} \times (-333.19 \text{ kJ/mol})$
$- 2 \text{ mol} \times (-46.19 \text{ kJ/mol}) - 1 \text{ mol} \times (-393.5 \text{ kJ/mol})$
$= -133.2$ kJ

5.68 $\Delta H^\circ = \Delta H^\circ_f[H_2SO_4(\ell)] - \Delta H^\circ_f[SO_3(g)] - \Delta H^\circ_f[H_2O(\ell)]$

$= 1 \text{ mol} \times (-811.32 \text{ kJ/mol}) - 1 \text{ mol} \times (-395.2 \text{ kJ/mol})$
$- 1 \text{ mol} \times (-285.9 \text{ kJ/mol})$

$= -130.2$ kJ

5.69 $\Delta H^\circ = \Delta H^\circ_f[NO(g)] + 2\Delta H^\circ_f[HNO_3(\ell)] - 3\Delta H^\circ_f[NO_2(g)] - \Delta H^\circ_f[H_2O(\ell)]$

$= 1 \text{ mol} \times (90.37 \text{ kJ/mol}) + 2 \text{ mol} \times (-173.2 \text{ kJ/mol})$
$- 3 \text{ mol} \times (33.8 \text{ kJ/mol}) - 1 \text{ mol} \times (-285.9 \text{ kJ/mol})$
$= -71.5$ kJ

5.70 $\Delta H^\circ = 3\Delta H^\circ_f[CO(g)] + 2\Delta H^\circ_f[Fe(s)] - \Delta H^\circ_f[Fe_2O_3(s)] - 3\Delta H^\circ_f[C(s)]$

$= 3 \text{ mol} \times (-110.5 \text{ kJ/mol}) + 2 \text{ mol} \times (0.0 \text{ kJ/mol})$
$- 1 \text{ mol} \times (-822.2 \text{ kJ/mol}) - 3 \text{ mol} \times (0.0 \text{ kJ/mol})$
$= 490.7$ kJ

5.71 $\Delta H° = \Delta H_f°[NaHCO_3(s)] - \Delta H_f°[CO_2(g)] - \Delta H_f°[NaOH(s)]$

$= 1\ mol \times (-226.5\ kcal/mol) - 1\ mol \times (-94.05\ kcal/mol)$
$- 1\ mol \times (-102.0\ kcal/mol)$
$= -30.5\ kcal$

5.72 (a) $C_2H_2(g) + \frac{5}{2}O_2(g) \rightarrow 2CO_2(g) + H_2O(\ell)$ $\Delta H° = -1299.6\ kJ$

(b) $CH_3OH(\ell) + \frac{3}{2}O_2(g) \rightarrow CO_2(g) + 2H_2O(\ell)$ $\Delta H° = -726.51\ kJ$

(c) $C_4H_{10}O(\ell) + 6O_2(g) \rightarrow 4CO_2(g) + 5H_2O(\ell)$ $\Delta H° = -2751.1\ kJ$

(d) $C_7H_8(\ell) + 9O_2(g) \rightarrow 7CO_2(g) + 4H_2O(\ell)$ $\Delta H° = -3909\ kJ$

5.73 $\Delta H° = -2.82 \times 10^3\ kJ/mol$

$= 6\Delta H_f°[H_2O(\ell)] + 6\Delta H_f°[CO_2(g)] - \Delta H_f°[C_6H_{12}O_6(s)]$

$= 6\ mol \times (-285.9\ kJ/mol) + 6\ mol \times (-393.5\ kJ/mol)$

$- 1\ mol \times (\Delta H_f°[C_6H_{12}O_6(s)])$

Solving for the heat of formation of glucose, we get:

$\Delta H_f°[C_6H_{12}O_6(s)] = -1.26 \times 10^1\ kJ/mol$

5.74 (a) $C_{16}H_{32}O_2(s) + 23O_2(g) \rightarrow 16CO_2(g) + 16H_2O(\ell)$, $\Delta H° = -2380\ kcal$

(b) $\Delta H° = -2380\ kcal$

$= 16\Delta H_f°[H_2O(\ell)] + 16\Delta H_f°[CO_2(g)] - \Delta H_f°[C_{16}H_{32}O_2(s)]$

$= 16\ mol \times (-68.32\ kcal/mol) + 16\ mol \times (-94.05\ kcal/mol)$

$- 1\ mol \times (\Delta H_f°[C_{16}H_{32}O_2(s)])$

Solving for the heat of formation of palmitic acid:

$\Delta H_f°[C_{16}H_{32}O_2(s)] = -218\ kcal$

5.75 $\Delta H° = 2\Delta H_f°[H_2O(\ell)] + 2\Delta H_f°[CO_2(g)] - \Delta H_f°[C_2H_4(g)]$

$= 2\ mol \times (-285.9\ kJ/mol) + 2\ mol \times (-393.5\ kJ/mol)$
$- 1\ mol \times (52.28\ kJ/mol)$
$= -1411.1\ kJ$

5.76 These two thermochemical equations are added along with six times that for the formation of liquid water:

$P_4O_{10}(s) + 6H_2O(\ell) \rightarrow 4H_3PO_4(\ell)$, $\Delta H° = -257.2$ kJ

$4P(s) + 5O_2(g) \rightarrow P_4O_{10}(s)$, $\Delta H° = -3062$ kJ

$6H_2(g) + 3O_2(g) \rightarrow 6H_2O(\ell)$, $\Delta H° = -1715$ kJ

Adding gives:

$4P(s) + 6H_2(g) + 8O_2(g) \rightarrow 4H_3PO_4(\ell)$, $\Delta H° = -5034$ kJ

The above result should then be divided by four:

$\frac{1}{4}P_4(s) + \frac{3}{2}H_2(g) + 2O_2(g) \rightarrow H_3PO_4(\ell)$, $\Delta H° = -1258$ kJ

CHAPTER 6

PRACTICE EXERCISES

1. $\nu = c/\lambda$, $1\ Hz = 1\ s^{-1}$ and $1\ nm = 10^{-9}\ m$

$$\nu = \frac{3.00 \times 10^{8}\ m/s}{550 \times 10^{-9}\ m} = 5.45 \times 10^{14}\ s^{-1} = 5.45 \times 10^{14}\ Hz$$

2. $\lambda = c/\nu$ and $1\ Hz = 1\ s^{-1}$

$$\lambda = \frac{3.00 \times 10^{8}\ m/s}{93.5 \times 10^{6}\ s^{-1}} = 3.21\ m$$

3. Subshells for n = 3: 3s, 3p, 3d
Subshells for n = 4: 4s, 4p, 4d, 4f

4. There are nine orbitals, corresponding to the nine possible values for m_ℓ of $-\ell$ to $+\ell$ in integer steps, namely -4, -3, -2, -1, 0, 1, 2, 3, and 4.

5. (a) Mg $1s^2 2s^2 2p^6 3s^2$

With the same subshells grouped together:

$1s^2 2s^2 2p^6$

(b) Ge $1s^2 2s^2 2p^6 3s^2 3p^6 4s^2 3d^{10} 4p^2$

With the same subshells grouped together:

$1s^2 2s^2 2p^6 3s^2 3p^6 3d^{10} 4s^2 4p^2$

(c) Cd $1s^2 2s^2 2p^6 3s^2 3p^6 4s^2 3d^{10} 4p^6 5s^2 4d^{10}$

With the same subshells grouped together:

$1s^2 2s^2 2p^6 3s^2 3p^6 3d^{10} 4s^2 4p^6 4d^{10} 5s^2$

(d) Gd $1s^2 2s^2 2p^6 3s^2 3p^6 4s^2 3d^{10} 4p^6 5s^2 4d^{10} 5p^6 6s^2 5d^1 4f^7$

With the same subshells grouped together:

$1s^2 2s^2 2p^6 3s^2 3p^6 3d^{10} 4s^2 4p^6 4d^{10} 4f^7 5s^2 5p^6 5d^1 6s^2$

6. (a)

Na (↑↓) (↑↓) (↑↓)(↑↓)(↑↓) (↑)
1s 2s 2p 3s

(b)

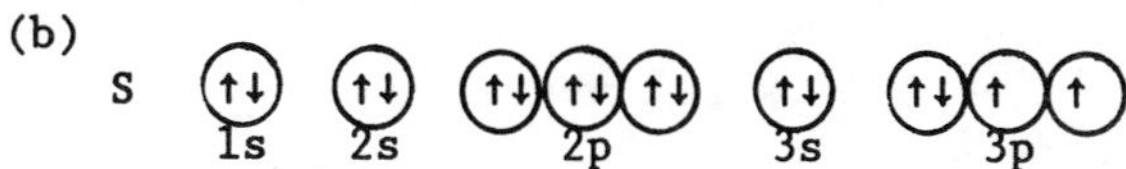

(c)

V ↑↓ ↑↓ ↑↓↑↓↑↓ ↑↓ ↑↓↑↓↑↓
1s 2s 2p 3s 3p

↑ ↑ ↑ ↑↓
3d 4s

7. (a) P $[Ne]3s^2 3p^3$

(b) Sn $[Kr]4d^{10} 5s^2 5p^2$

8. (a) N $2s^2 2p^3$ (b) Si $3s^2 3p^2$ (c) Sr $5s^2$

9. (a) Ni $[Ar]4s^2 3d^8$ or $[Ar]3d^8 4s^2$

(b) Ru $[Kr]5s^2 4d^6$ or $[Kr]4d^6 5s^2$

10. (a) Se $4s^2 4p^4$ (b) Sn $5s^2 5p^2$ (c) I $5s^2 5p^5$

11. (a) Sn (b) Ga (c) Fe (d) S^{2-}

12. (a) Be (b) C

REVIEW EXERCISES

Discovery of Subatomic Particles

6.1 (a) Thomson measured the charge-to-mass ratio of the electron, using the cathode ray tube shown in Figure 6.2.
(b) Millikan measured the charge of the electron using the oil drop experiment diagramed in Figure 6.3.
(c) Rutherford demonstrated that the atom has a dense, central nucleus, through the use of the gold foil experiment diagramed in Figure 6.5.
(d) Moseley discovered the existence of atomic numbers, as shown in Special Topic 6.3.
(e) Chadwick discovered the neutron.

6.2 Cathode rays are negatively charged particles that flow from the cathode to the anode in a gas discharge tube. They are much lighter than the positive particles that are also found in such gas discharge tubes.

6.3

$$e/m = \frac{1.60 \times 10^{-19}\ \text{coulombs}}{1.67 \times 10^{-24}\ \text{g}} = 9.58 \times 10^{4}\ \text{coulombs/g}$$

6.4 $e/m = 4.82 \times 10^{4}$ coulombs/g

and $m = e/4.82 \times 10^{4}$ coulombs/g

$$m = (+2)(1.60 \times 10^{-19}\ \text{coulombs}) \times \frac{1\ \text{g}}{4.82 \times 10^{4}\ \text{coulombs}}$$

$$m = 6.64 \times 10^{-24}\ \text{g}$$

Note that this is also equal to four times the mass of the proton.

6.5 Since the neutron has no charge, its charge to mass ratio would be zero, or undefined.

Electromagnetic Radiation

6.6 Light consists, as do all forms of electromagnetic radiation, of periodic fluctuations in the intensity of electrical and magnetic forces, or fields.

6.7 Frequency is given by the number of waves that pass a given point in space in a second. The symbol for frequency is the Greek letter ν ("nu"), and the SI unit of frequency is the Hertz, Hz. $1\ \text{Hz} = 1\text{s}^{-1}$

6.8 Wavelength is the distance between successive peaks of a wave, and it has the Greek symbol "lambda", λ.

6.9 See Figure 6.6.

6.10 Radio waves (short wave, TV, and microwaves), infrared, visible, ultraviolet, X rays, gamma rays

6.11 By the visible spectrum, we mean that narrow portion of the electromagnetic spectrum to which our eyes are sensitive.

6.12 $\nu\lambda = c = 3.00 \times 10^{8}$ m/s

6.13

$$\nu = \frac{c}{\lambda} = \frac{3.00 \times 10^{8}\ \text{m/s}}{425 \times 10^{-9}\ \text{m}} = 7.06 \times 10^{14}\ \text{s}^{-1} = 7.06 \times 10^{14}\ \text{Hz}$$

6.14 $6.50\ \mu\text{m} = 6.50 \times 10^{-5}$ m

8 ⊖6 .4645 × 10^14 14-1

$$\nu = \frac{c}{\lambda} = \frac{3.00 \times 10^{8} \text{ m/s}}{6.50 \times 10^{-6} \text{ m}} = 4.62 \times 10^{13} \text{ s}^{-1} = 4.62 \times 10^{13} \text{ Hz}$$

6.15 295 nm = 295×10^{-9} m

$$\nu = \frac{c}{\lambda} = \frac{3.00 \times 10^{8} \text{ m/s}}{295 \times 10^{-9} \text{ m}} = 1.02 \times 10^{15} \text{ s}^{-1} = 1.02 \times 10^{15} \text{ Hz}$$

6.16 3.19 cm = 3.19×10^{-2} m

$$\nu = \frac{c}{\lambda} = \frac{3.00 \times 10^{8} \text{ m/s}}{3.19 \times 10^{-2} \text{ m}} = 9.40 \times 10^{9} \text{ s}^{-1} = 9.40 \times 10^{9} \text{ Hz}$$

6.17 101.1 MHz = 101.1×10^{6} Hz = 101.1×10^{6} s^{-1}

$$\lambda = \frac{c}{\nu} = \frac{3.00 \times 10^{8} \text{ m/s}}{101.1 \times 10^{6} \text{ s}^{-1}} = 2.97 \text{ m}$$

6.18 5.09×10^{14} Hz = 5.09×10^{14} s^{-1}

$$\lambda = \frac{c}{\nu} = \frac{3.00 \times 10^{8} \text{ m/s}}{5.09 \times 10^{14} \text{ s}^{-1}} = 5.89 \times 10^{-7} \text{ m}$$

$$5.89 \times 10^{-7} \text{ m} \times \frac{1 \text{ nm}}{1 \times 10^{-9} \text{ m}} = 589 \text{ nm}$$

6.19 $E = h\nu$, where E is energy, h is Planck's constant, and ν is frequency.

6.20 First, $\nu = c/\lambda$

Also, $E = h\nu$, so that $\nu = E/h$

Setting both relationships for ν so that they are equal to one another:

$\nu = E/h = c/\lambda$ $\quad\quad$ $\therefore E = hc/\lambda$

6.21 The radiation that has the higher energy is that which has the shorter wavelength and the higher frequency:

(a) infrared (b) visible (c) X rays (d) ultraviolet

6.22 A photon is a particle of light having an energy given by the equation, $E = h\nu$, where ν is the frequency associated with the photon.

6.23 The quantum is the lowest possible packet of energy, that of a single photon.

6.24 $E = h\nu$, and 1 Hz = 1 s^{-1}

$$E = (6.63 \times 10^{-34}\ Js)(4.0 \times 10^{14}\ s^{-1}) = 2.7 \times 10^{-19}\ J$$

6.25 $E = h\nu = hc/\lambda$, and 550 nm = 550 $\times 10^{-9}$ m

$$E = \frac{(6.63 \times 10^{-34}\ Js)(3.00 \times 10^{8}\ m/s)}{550 \times 10^{-9}\ m} = 3.62 \times 10^{-19}\ J$$

6.26 We proceed by calculating the wavelength of a single photon:

$$E = \frac{hc}{\lambda} = \frac{(6.63 \times 10^{-34}\ Js)(3.00 \times 10^{8}\ m/s)}{3.00 \times 10^{-3}\ m} = 6.63 \times 10^{-23}\ J$$

Since the specific heat of water is 4.184 J/g °C, it will take 4.184 J of heat energy to raise the temperature of the water. The required number of photons is then:

$$4.184\ J \times \frac{1\ \text{photon}}{6.63 \times 10^{-23}\ J} = 6.31 \times 10^{22}\ \text{photons}$$

Atomic Spectra

6.27 An atomic spectrum consists of a series of discrete (selected, definite and reproducible) frequencies (and therefore of discrete energies) that are emitted by atoms that have been excited. The particular values for the emission frequencies are characteristic of the element at hand. In contrast, a continuous spectrum, such as that emitted by the sun or another hot, glowing object, contains all frequencies and, therefore, photons of all energies.

6.28 From the answer to question 6.27, it becomes obvious that an electron in an atom can have only selected or definite values for energy. Aside from these discrete energies, other energies are not allowed.

Bohr Atom

6.29 Bohr considered the atom to be composed of a central proton, with an electron traveling around it in circular orbits, one orbit differing from another by the size of its radius, and, hence, by the energy associated with the radius.

6.30 When an electron falls from an orbit of higher energy (larger radius) to an orbit of lower energy (smaller radius), the energy that is released appears as a photon with the appropriate frequency. The energy of the photon is the same as the difference in energy between the two orbits.

6.31 The lowest energy state of an atom is termed the ground state.

6.32 Bohr's model was a success because it accounted for the spectrum of the hydrogen atom, but it failed to account for spectra of more complex atoms.

Wave Nature of Matter

6.33 Very small particles "behave" more like waves than do large particles. More to the point, small particles are more aptly described also by their wave characteristics than are heavy particles. Large objects have very short wavelengths compared to subatomic particles, because wavelength is inversely proportional to the mass of the particle; the larger the mass, the shorter the wavelength. The very short wavelengths of large objects make their wave properties unnoticeable.

6.34 Defraction is a phenomenon caused by the constructive or the destructive interference of two or more waves. The fact that electrons and other subatomic particles exhibit diffraction supports the theory that matter is correctly considered to have wave nature.

6.35 In a traveling wave, the positions of the peaks and nodes change with time. In a standing wave, the peaks and nodes remain in the same positions.

Electron Waves in Atoms

6.36 This is wave or quantum mechanics.

6.37 This is the orbital of the electron.

6.38 First, we are interested in the energies of orbitals, because it is the energies of the various orbitals that determines which orbitals are occupied by the electrons of the atom. Secondly, we are interested in the shapes and orientations of the various orbitals, because this is important in determining how atoms form bonds in chemical compounds.

Quantum Numbers

6.39 $n = 1, 2, 3, 4, 5, \ldots\infty$

6.40 (a) $n = 1$ (b) $n = 3$

6.41 (a) p (b) f (c) h

6.42 (a) $n = 2$, $\ell = 0$ (b) $n = 3$, $\ell = 2$ (c) $n = 5$, $\ell = 3$

6.43 0, 1, 2, 3

6.44 Every shell contains the possibility that $\ell = 0$.

6.45 (a) $m_\ell = 1, 0,$ or -1 (b) $m_\ell = 3, 2, 1, 0, -1, -2,$ or -3

6.46 (a) 1 (b) 3 (c) 5 (d) 7

6.47 There are eleven values: -5, -4, -3, -2, -1, 0, 1, 2, 3, 4, and 5.

6.48

n	ℓ	m_ℓ	m_s
2	1	-1	$\frac{1}{2}$
2	1	-1	$-\frac{1}{2}$
2	1	0	$\frac{1}{2}$
2	1	0	$-\frac{1}{2}$
2	1	1	$\frac{1}{2}$
2	1	1	$-\frac{1}{2}$

6.49 The value corresponds to the row in which the element resides: (a) 5 (b) 4 (c) 4 (d) 6

Electron Spin

6.50 The electron behaves like a magnet, because the revolving charge (spin) of the electron creates a magnetic field.

6.51 Atoms with unpaired electrons are termed paramagnetic.

6.52 $m_s = +\frac{1}{2}$ or $-\frac{1}{2}$

Electron Configuration of Atoms

6.53 The electronic structure of the atom is the manner (order or pattern) in which the electrons are distributed among the various orbitals of the atom.

6.54 The orbitals within a given shell are arranged in the following order of increasing energy: $s < p < d < f$.

6.55 The energies of the subshells are quantized.

6.56 The orbitals of a given subshell have the same energy.

6.57 No two electrons in the same atom can have exactly the same set of values for all of the four quantum numbers. This limits the allowed number of electrons per orbital to two, since with other quantum numbers being necessarily the same, two electrons in the same orbital must at least have different values of m_s.

6.58

Li	$1s^2 2s^1$	N	$1s^2 2s^2 2p^3$
Be	$1s^2 2s^2$	O	$1s^2 2s^2 2p^4$
B	$1s^2 2s^2 2p^1$	F	$1s^2 2s^2 2p^5$
C	$1s^2 2s^2 2p^2$	Ne	$1s^2 2s^2 2p^6$

6.59 (a) S $1s^2 2s^2 2p^6 3s^2 3p^4$

(b) K $1s^2 2s^2 2p^6 3s^2 3p^6 4s^1$

(c) Ti $1s^2 2s^2 2p^6 3s^2 3p^6 4s^2 3d^2$

(d) Sn $1s^2 2s^2 2p^6 3s^2 3p^6 4s^2 3d^{10} 4p^6 5s^2 4d^{10} 5p^2$

6.60 (a) As $1s^2 2s^2 2p^6 3s^2 3p^6 4s^2 3d^{10} 4p^3$

(b) Cl $1s^2 2s^2 2p^6 3s^2 3p^5$

(c) Fe $1s^2 2s^2 2p^6 3s^2 3p^6 4s^2 3d^6$

(d) Si $1s^2 2s^2 2p^6 3s^2 3p^2$

6.61 (a) Cr [Ar]$4s^1 3d^5$ or [Ar]$3d^5 4s^1$

(b) Cu [Ar]$4s^1 3d^{10}$ or [Ar]$3d^{10} 4s^1$

6.62 (a) Mg (↑↓) 1s (↑↓) 2s (↑↓)(↑↓)(↑↓) 2p (↑↓) 3s

(b) Ti (↑↓) 1s (↑↓) 2s (↑↓)(↑↓)(↑↓) 2p (↑↓) 3s (↑↓)(↑↓)(↑↓) 3p (↑)(↑)( )( )( ) 3d (↑↓) 4s

6.63 (a) As (↑↓) 1s (↑↓) 2s (↑↓)(↑↓)(↑↓) 2p (↑↓) 3s (↑↓)(↑↓)(↑↓) 3p (↑↓)(↑↓)(↑↓)(↑↓)(↑↓) 3d (↑↓) 4s (↑)(↑)(↑) 4p

(b) Ni (↑↓) 1s (↑↓) 2s (↑↓)(↑↓)(↑↓) 2p (↑↓) 3s (↑↓)(↑↓)(↑↓) 3p (↑↓)(↑↓)(↑↓)(↑)(↑) 3d (↑↓) 4s

6.64 (a) Mg is $1s^2 2s^2 2p^6 3s^2$, $\therefore$ zero unpaired electrons

(b) P is $1s^2 2s^2 2p^6 3s^2 3p^3$, $\therefore$ three unpaired electrons

(c) V is $1s^2 2s^2 2p^6 3s^2 3p^6 4s^2 3d^3$, $\therefore$ three unpaired electrons

6.65 (a) Ni $[Ar]4s^2 3d^8$ or $[Ar]3d^8 4s^2$

(b) Cs $[Xe]6s^1$

(c) Ge $[Ar]4s^2 3d^{10} 4p^2$ or $[Ar]3d^{10} 4s^2 4p^2$

(d) Br $[Ar]4s^2 3d^{10} 4p^5$ or $[Ar]3d^{10} 4s^2 4p^5$

6.66 (a) Al $[Ne]3s^2 3p^1$

(b) Se $[Ar]4s^2 3d^{10} 4p^4$ or $[Ar]3d^{10} 4s^2 4p^4$

(c) Ba $[Xe]6s^2$

(d) Sb $[Kr]5s^2 4d^{10} 5p^3$ or $[Kr]4d^{10} 5s^2 5p^3$

6.67 Elements in a given group generally have the same electron configuration except that the value for n is different, and corresponds to the row in which the element is found.

6.68 The valence shell is the outermost shell for an atom, and the valence eletrons of a given atom are those in the outermost (valence) shell.

6.69 (a) Na $3s^1$ (b) Al $3s^2 3p^1$ (c) Ge $4s^2 4p^2$ (d) P $3s^2 3p^3$

6.70 (a) Mg $3s^2$ (b) Br $4s^2 4p^5$ (c) Ga $4s^2 4p^1$ (d) Pb $6s^2 6p^2$

6.71 (a) Mn is $[Ar]4s^2 3d^5$, $\therefore$ five unpaired electrons

(b) As is $[Ar]4s^2 3d^{10} 4p^3$, $\therefore$ three unpaired electrons

(c) S is $[Ne]3s^2 3p^4$, $\therefore$ two unpaired electrons

(d) Sr is $[Kr]5s^2$, $\therefore$ zero unpaired electrons

(e) Ar is $1s^2 2s^2 2p^6 3s^2 3p^6$, $\therefore$ zero unpaired electrons

6.72 (a) Ba is $[Xe]6s^2$, zero unpaired electrons: paramagnetic

(b) Se is $[Ar]4s^2 3d^{10} 4p^4$, two unpaired electrons: paramagnetic

(c) Zn is $[Ar]4s^2 3d^{10}$, zero unpaired electrons: diamagnetic

(d) Si is $[Ne]3s^2 3p^2$, two unpaired electrons: paramagnetic

Shapes of Electron Orbitals

6.73 The "locations" of electrons are described only in terms of the relative probability of their being found at various points.

6.74 See Figures 6.24 and 6.25.

6.75 As n becomes larger, the orbital becomes larger.

6.76 The p orbitals of a given subshell are oriented at right angles (90°) to one another.

6.77 A nodal plane is a plane in which there is zero probability of finding an electron.

Atomic and Ionic Size

6.78 The effective nuclear charge is the net nuclear charge that an electron actually experiences. It is different from the formal nuclear charge because of the varying imperfect ways in which one electron is shielded from the nuclear charge by the other electrons that are present. The effective nuclear charge increases from top to bottom in any one group of the periodic table, and it increases from left to right in any one row of the periodic table.

6.79 (a) Na (b) Sb

6.80 (a) Al (b) In

6.81 Sn

6.82 The larger atoms are found in the lower left corner of the periodic table; the smaller atoms are found in the upper right corner of the periodic table.

6.83 Since these atoms and ions all have the same number of electrons, the size should be inversely related to the positive charge:

$Mg^{2+} < Na^{+} < Ne < F^{-} < O^{2-} < N^{3-}$

6.84 The size changes within a transition series are more gradual because, whereas the "outer" electrons are in an s subshell, the electrons that are added from one element to another enter an inner (n - 1) d subshell.

6.85 Cations are generally smaller than the corresponding atom, and anions are generally larger than the corresponding atom:

(a) Na (b) Co^{2+} (c) Cl^-

Ionization Energy

6.86 Ionization energy is the energy that is needed in order to remove an electron from a gaseous atom or ion.

6.87 (a) C (b) O (c) Cl

6.88 Removing an electron from a completed or noble gas electron configuration is difficult, as is adding an electron to such an atom. The removal of an electron from a closed shell is difficult because the effective nuclear charge is characteristically very high. Adding an electron to a noble gas atom would require the use of the next quantum level, and this is unfavorable. Therefore, the noble gases do not form either cations or anions.

6.89 Ionization energy increases from left to right in a row of the periodic table because the effective nuclear charge increases from left to right. The latter trend occurs because of the consequences of the increasingly imperfect shielding of electrons by other electrons within the same level. The ionization energy decreases down a periodic table group because the electrons become farther from the nucleus with each successive quantum level that is occupied. The farther the electrons are from the nucleus, the less tightly they are held by the nucleus.

6.90 Removing a second electron involves pulling it away from a greater positive charge because of the positive charge created by the removal of the first electron. Hence, more energy must be spent to ionize the second electron than the first.

Electron Affinity

6.91 Electron affinity is the enthalpy change associated with the addition of an electron to a gaseous atom:

$$X(g) + e^- \rightarrow X^-(g)$$

6.92 (a) Cl (b) Br (c) Si

6.93 The second electron affinity is always unfavorable (endothermic) because it requires that a second electron be forced onto an ion that is already negative.

CHAPTER 7

PRACTICE EXERCISES

1. Two electrons are to be added to a sulfur atom to make the sulfide anion:

   $S([Ne]3s^2 3p^4) + 2e^- \rightarrow S^{2-}([Ne]3s^2 3p^6)$

   Two electrons are to be removed from a magnesium atom to make its cation:

   $Mg([Ne]3s^2) \rightarrow Mg^{2+}([Ne] + 2e^-)$

2. (a) $\dot{\ddot{Se}}$: (b) :Ï: (c) ·Ca·

3. Mg + ·Ö: → Mg^{2+} + $[\,:\ddot{O}:\,]^{2-}$

4. (a) H:P:H with H below P (b) :S:F: with :F: below S

5. O S O

   O N O O

   H O Cl O with O above Cl

   H O P O H with O above P, O below P, H below

6. Since S and O are both in Group VIA, each atom of either of these two elements contributes 6 valence electrons to the overall diagram: 3 × 6 = 18 electrons total are to be used in drawing SO_2.

   Since P is in Group VA, it will contribute 5 electrons to the diagram, whereas each oxygen atom contributes 6 electrons. The ion's charge of 3- must also be considered, and this requires the further use of three electrons: 5 + (4 × 6) + 3 = 32 electrons are to be used for the diagram of PO_4^{3-}.

   Since N is in Group VA, it will contribute 5 electrons, and O, being in Group VIA will contribibure 6 electrons. The ion's overall charge of 1+ requires the removal of one electron from the total: 5 + 6 - 1 = 10 electrons for NO^+.

7., 8.

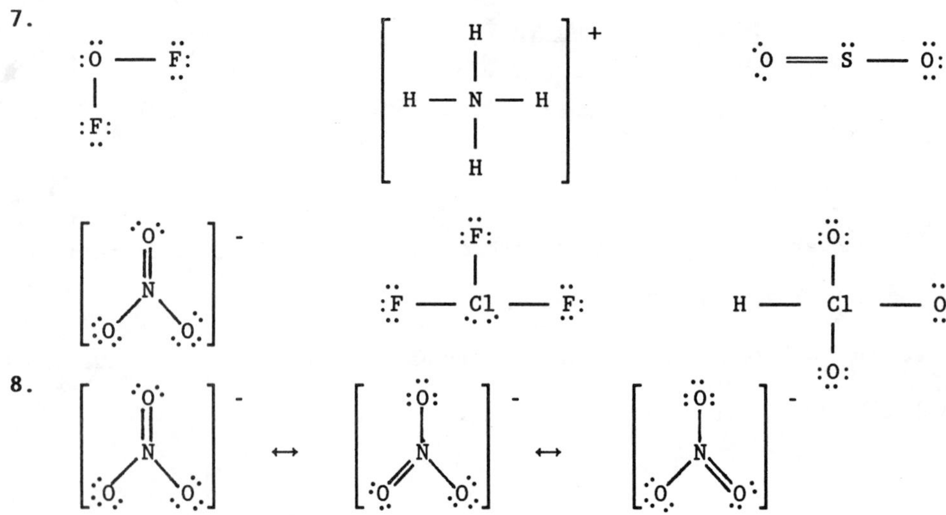

9. $SbCl_5$ should have a trigonal bipyramidal shape because, like PCl_5 shown in Figure 7.6 on page 247 of the text, it has five electron pairs around the central atom.

10. In ClO_3^-, there are three bond pairs and one lone pair of electrons at the chlorine atom, and as shown in Figure 7.7 on page 249 of the text, this ion has a pyramidal shape.
In XeO_4, there are four bond pairs of electrons around the Xe atom, and as shown in Figure 7.7 on page 249 of the text, this molecule is tetrahedral.
In OF_2, there are two bond pairs and two lone pairs of electrons around the oxygen atom, and as shown in Figure 7.7 on page 249 of the text, this molecule is bent.

11. Carbonate ion, CO_3^{2-}, is a planar, triangular ion, with a geometry analogous to that shown on page 253 of the text for SO_3.

12. In this problem, we choose the atom having the higher electronegativity, as listed in Figure 7.11 on page 255 of the text.

(a) Br (b) Cl (c) Cl

13. (a) SF_6 is octahedral, and it is not polar.

(b) SO_2 is bent, and it is polar.

(c) BrCl is polar because there is a difference in electronegativity between Br and Cl.

(d) AsH_3, like NH_3, is pyramidal, and it is polar.

(e) CF_2Cl_2 is polar, for much the same reason that $CHCl_3$, Figure 7.16, is polar.

REVIEW EXERCISES

Ionic Bonding

7.1 A stable compound forms from a collection of atoms when bonding results in a net lowering of the energy. The process of bonding the atoms together must release energy to make the bonded compound more stable than the original collection.

7.2 The ionic bond is the attraction between positive and negative ions in an ionic compound. It is largely an electrostatic attraction, and it gives rise to the lattice energy of the ionic compound. The lattice energy is the net gain in stability when the gaseous ions are brought together to form the crystalline ionic compound.

7.3 Ionic bonds tend to form upon combining an element having a high EA with an element having a low IE.

7.4 The lattice energy is the energy necessary to separate a mole of an ionic solid into its constituent ions in the gas phase. It is also the energy that is released on forming the ionic solid from the gaseous ions. As discussed in the answer to questions 7.1 and 7.2, it is the lattice energy that is primarily responsible for the stability of ionic compounds.

7.5 Magnesium loses two electrons:

$$Mg\ ([Ne]3s^2) \rightarrow Mg^{2+}\ ([Ne]) + 2e^-$$

Two bromine atoms each gain an electron:

$$2Br\ ([Ar]3d^{10}4s^24p^5) + 2e^- \rightarrow 2Br^-([Kr])$$

The net reaction is obtained by adding these two equations:

$$Mg + 2Br \rightarrow MgBr_2$$

7.6 Nitrogen gains three electrons and achieves the electron configuration of the next noble gas, Ne:

N^{3-} $1s^22s^22p^6$

7.7 Magnesium can achieve the electron configuration of the nearest noble gas (Ne) by losing only two electrons:

Mg^{2+} $1s^22s^22p^6$

Notice that this is the same electron configuration that was written in problem 7.6 for nitride ion.

7.8 Chlorine can achieve the electron configuration of the nearest noble gas (Ar) by gaining only one electron:

Cl^- $1s^22s^22p^63s^23p^6$

Notice that all of the other elements in Group VIIA also form mononegative anions.

7.9 Many of the transition metals have an ns^2 outer-shell electron configuration. Since these characteristically are the first electrons to be lost when the atom is ionized, it is common that a 2+ ion should be formed.

Lewis Symbols

7.10 (a) ·Si· (b) ·Sb· (c) ·Ba· (d) ·Al·

7.11 (a) Ca + 2 :Br: → Ca^{2+} + $2[:Br:]^-$

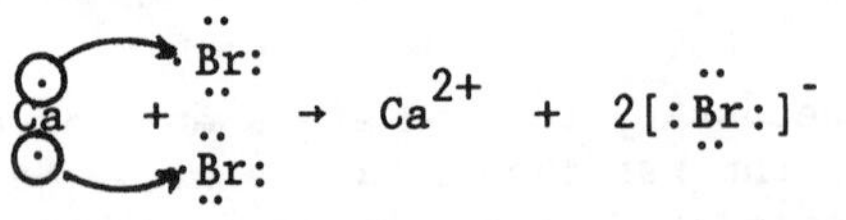

(b) 2 Al + 3 :O: → $2Al^{3+}$ + $3[:O:]^{2-}$

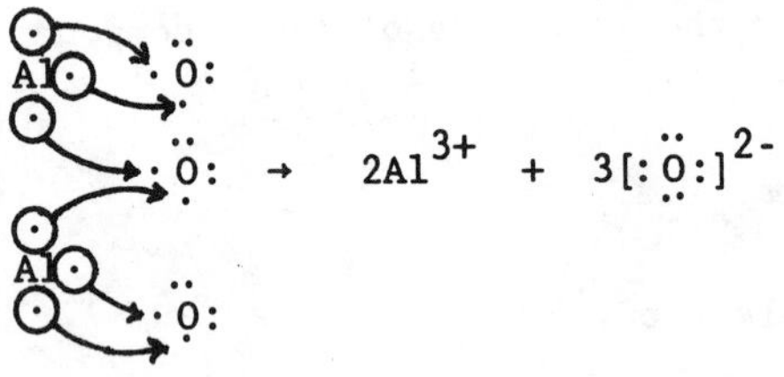

(c) 2 K + ·S: → $2K^+$ + $[:S:]^{2-}$

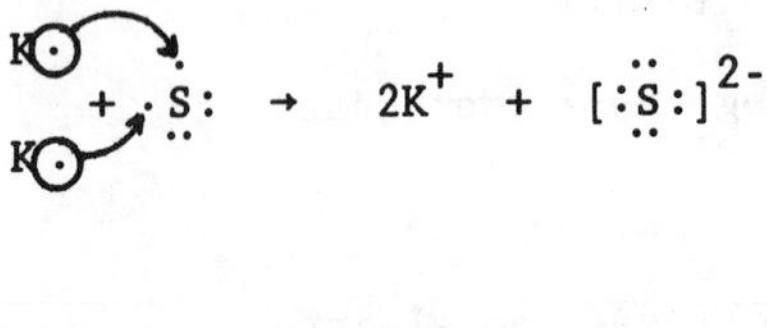

Chapter 7

Electron Sharing

7.12 Ionic bonding does not occur between two nonmetal elements because more energy must be provided in the form of IE and EA than can be recovered from the lattice energy.

7.13 As two hydrogen atoms approach each other in forming the H_2 molecule, the electron density of the two atoms shifts to the region between the two nuclei.

7.14 The force that holds the nuclei together in a covalently bonded pair of atoms is the attraction that the two nuclei have for negatively charged electron density that is found between the nuclei.

7.15 The energy drops to some optimum or minimum value when the nuclei have become separated by the distance called the bond distance.

7.16 The bond distance in a covalent bond is determined by a balance (compromise) between the separate attractions of the nuclei for the electron density that is found between them, and the repulsions between the like-charged nuclei and those bewteen the like-charged electrons. These attractions and repulsions oppose one another, and a bond distance is achieved that maximizes the attraction while minimizing the repulsions.

7.17 When a covalent bond is formed, the electrons that are shared become paired. Also, the normal, single covalent bond is accomplished by the sharing of two electrons.

7.18 Bond formation is always exothermic.

7.19

$$435\ \text{kJ/mol} \times \frac{1\ \text{mol}}{6.02 \times 10^{23}\ \text{molecules}} \times \frac{1000\ \text{J}}{1\ \text{kJ}} = 7.23 \times 10^{-19}\ \text{J/molecule}$$

7.20 Let q be the amount of energy released in the formation of 1 mol of H_2 molecules from H atoms: 435 kJ/mol, the single bond energy for hydrogen.

$$q = \text{specific heat} \times \text{mass} \times \Delta T$$

$$\therefore \text{mass} = q \div (\text{specific heat} \times \Delta T)$$

$$\text{mass} = \frac{435 \times 10^{3}\ \text{J}}{(4.184\ \text{J/g °C})(100\text{°C} - 25\text{°C})} = 1.4 \times 10^{3}\ \text{g}\ H_2O$$

Covalent Bonding and the Octet Rule

7.21 (a) $:\ddot{\underset{\cdot\cdot}{\mathrm{Br}}}\cdot \ + \ \cdot\ddot{\underset{\cdot\cdot}{\mathrm{Br}}}: \ \rightarrow \ :\ddot{\underset{\cdot\cdot}{\mathrm{Br}}}:\ddot{\underset{\cdot\cdot}{\mathrm{Br}}}:$

```
           ..           ..
(b) 2H· + ·O:   →    H:O:
           .            ..
                        H

          ..           ..
(c) 3H· + ·N· →    H:N:H
          .            ..
                       H
```

7.22 The octet rule is the expectation that atoms tend to lose or gain electrons until they achieve a noble gas-type electron configuration, namely eight electrons in the valence, or outer-most, shell. It is the stability of the closed-shell electron configuration of a noble gas that accounts for this.

7.23 (a) We predict the formula H_2Se because selenium, being in Group VIA, needs only two additional electrons (one each from two hydrogen atoms) in order to complete its octet.

(b) Arsenic, being in Group VA, needs three electrons from hydrogen atoms in order to complete its octet, and we predict the formula H_3As.

(c) Silicon is in Group IVB, and it needs four electrons (and hence four hydrogen atoms) to complete its octet: SiH_4.

7.24 (a) Each chlorine atom needs one further electron in order to achieve an octet, and the phosphorus atom requires three electrons from an appropriate number of chlorine atoms. We conclude that a phosphorus atom is bonded to three chlorine atoms, and that each chlorine atom is bonded only once to the phosphorus atom: PCl_3.

(b) Since carbon needs four additional electrons, the formula must be CF_4. In this arrangement, each fluorine acquires the one additional electron that is needed to reach its octet.

(c) Each halogen atom needs only one additional electron from the other: BrCl.

7.25 The valence shell of a period two element can hold only eight electrons, whereas the valence shell of elements in row three can hold as many as eighteen electrons.

7.26 (a) single bond: a covalent bond formed by the sharing of one pair of electrons.
(b) double bond: a covalent bond formed by the sharing of two pairs of electrons.
(c) triple bond: a covalent bond formed by the sharing of three pairs of electrons.

7.27 A structural formula shows which atoms are attached to one another in a molecule or polyatomic ion.

7.28

H — C ≡ N:

7.29 Since the outer shell or valence shell of hydrogen can hold only two electrons, hydrogen is not said to obey the octet rule. It does, however, still satisfy its requirement for a closed shell electron configuration through the formation of one covalent bond.

Failure of the Octet Rule

7.30 (a) 4 electrons or two bonding pairs
(b) six electrons, all in bonding pairs
(c) two electrons for each hydrogen, in each of two bonding pairs

7.31 ten electrons, in bonding pairs between As and each of five Cl atoms

Drawing Lewis Structures

7.32 (a)

```
        Cl
        |
  Cl — Si — Cl
        |
        Cl
```

(b)

```
F — P — F
    |
    F
```

(c)

```
H — P — H
    |
    H
```

(d) Cl — S — Cl

7.33 (a) 32 (b) 26 (c) 8 (d) 20

7.34 (a)

```
         ..
        :Cl:
         |
   ..          ..
  :Cl — Si — Cl:
   ..          ..
         |
        :Cl:
         ..
```

(b)

```
 ..   ..   ..
:F — P — F:
 ..   |    ..
     :F:
      ..
```

(c)

```
     ..
H — P — H
     |
     H
```

(d)

```
 ..   ..   ..
:Cl — S — Cl:
 ..   ..   ..
```

7.35 (a) S = C = S (each S with two lone pairs)

(b) [:C ≡ N:]$^-$

(c)

```
        :O:
         |
  ..             ..
 :O  —  Se  =  O
  ..             ..
```

(d) :O — Se = O (with lone pairs)

Chapter 7

7.36 (a)  H — O — N = O

(b) 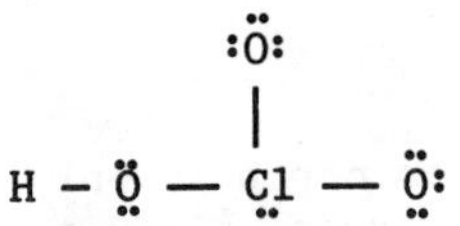 H — O — Cl(=O) — O

(c) 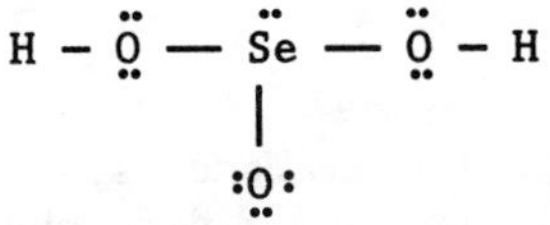 H — O — Se(—O) — O — H

7.37 (a) 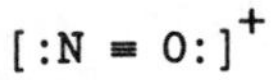 $[:N \equiv O:]^+$

(b) 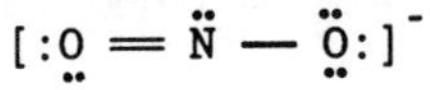 $[:O = N — O:]^-$

(c) 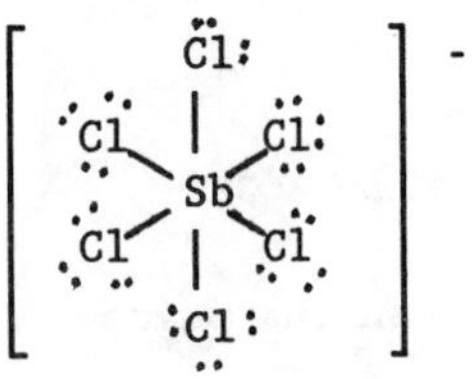 $[SbCl_6]^-$

(d) 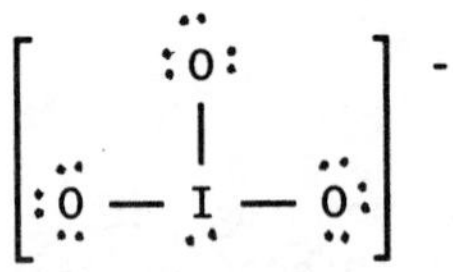 $[IO_3]^-$

7.38 (a) 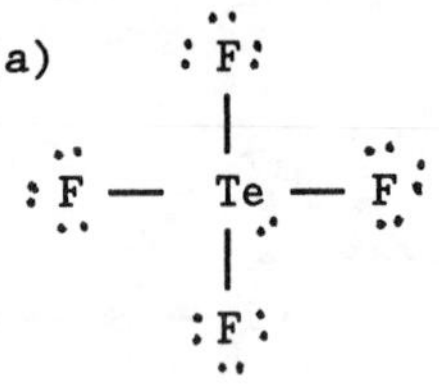 TeF_4

(b) 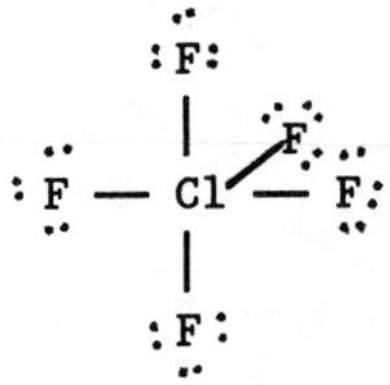 ClF_5

(c) 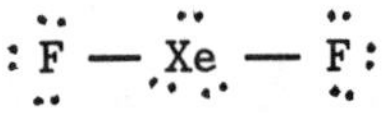 :F — Xe — F:

(d) 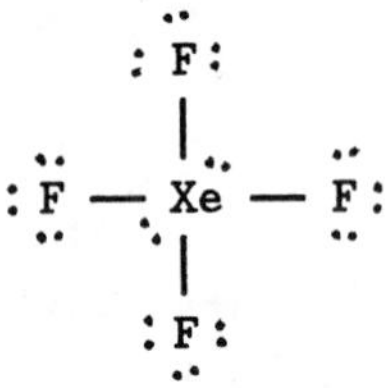 XeF_4

7.39 (a) 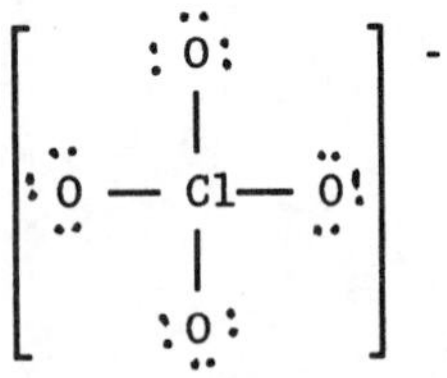 $[ClO_4]^-$

(b) 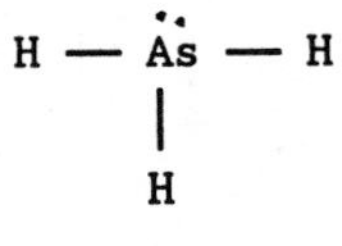 H — As — H, with H bonded to As

(c)

```
      ..
    : Cl :
      |
 ..          ..
:Cl — P — Cl :
 ..          ..
      |
    : Cl :
      ..
```
$^{+}$

(d)

```
        :Cl:
   :Cl\  |  /Cl:
         P
   :Cl/  |  \Cl:
        :Cl:
```
$^{-}$

7.40

```
H\
   C = O:
H/
```

7.41 (a)

```
       ..
     : Cl:
       |
 ..          ..
:Cl — Ge — Cl :
 ..          ..
       |
     : Cl:
       ..
```

(b)

```
      :O:
       ||
       C
      / \
   :O:   :O:
```
$^{2-}$

(c)

```
      ..
     :O:
      |
 ..         ..
:O — P — O:
 ..         ..
      |
     :O:
      ..
```
$^{3-}$

(d)

```
  ..     ..
[:O — O:]
  ..     ..
```
$^{2-}$

Bond Length and Bond Energy

7.42 Bond length - the distance between the nuclei of two atoms that are linked by a covalent bond.
Bond energy - the energy that is needed to break a chemical bond; conversely, the energy that is released when a chemical bond is formed.

7.43 The Lewis structure for NO_3^- is given in the answer to practice exercises 7 and 8, and that for NO_2^- is given in the answer to review exercise 7.37. Resonance causes the average number of bonds in each N—O linkage of NO_3^- to be 1.33. Resonance causes the average number of electron pair bonds in each linkage of NO_2^- to be 1.5. We conclude that the N—O bond in NO_2^-

should be shorter than that in NO_3^-. See also the answer to review exercise 7.45.

7.44 The H – Cl bond energy is defined to be the energy required to break the bond to give atoms of H and Cl, not the ions H^+ and Cl^-. In other words, when the bond is broken between the H and the Cl atoms, one of the two electrons of the bond must go to each of the atoms. When ions are obtained, however, both electrons of the bond have "gone with" the chlorine atom.

7.45 As the number of electron pairs in a covalent linkage increases, the bond energy increases and the bond length decreases.

Resonance

7.46

7.47 The Lewis structures of the three must be compared. The Lewis structure for CO_3^{2-} is given in the answer to review exercise 7.46 and the structure for CO_2 is given in the text, as well as in review exercise 7.49. Carbon monoxide has a triple bond, :O≡C:. The order of increasing C–O bond length is:

$$CO < CO_2 < CO_3^{2-}.$$

7.48

7.49 :O ≡ C – Ö: and :Ö – C ≡ O:

7.50 A resonance hybrid is the true structure of a molecule or polyatomic ion, whereas the various resonance structures that are used to depict the hybrid do not individually have any reality. The hybrid is a mix or average of the various resonance structures that compose it.

7.51 The resonance structures for NO_3^- are given in the answer to practice exercise 7.8 and the resonance structures for HNO_3 are given below:

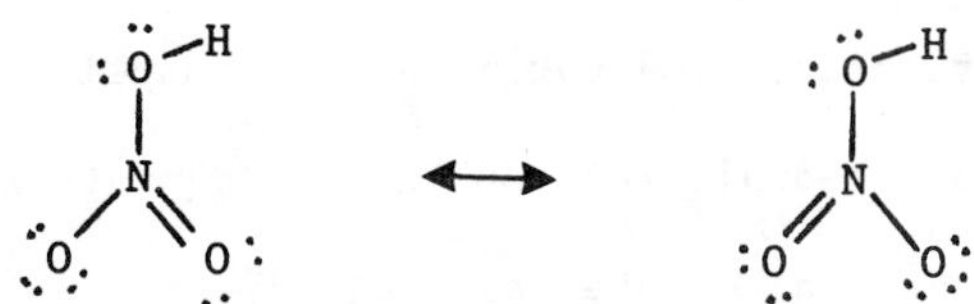

In the nitrate anion, there are three equivalent O atoms, and therefore we can write three resonance forms. In nitric acid, there are only two equivalent O atoms (the ones not having an H atom attached), and we may, therefore, write only two resonance forms.

Coordinate Covalent Bonds

7.52 A coordinate covalent bond is one in which both electrons of the bond are contributed by (or donated by) only one of the atoms that are linked by the bond.

7.53 Once formed, a coordinate covalent bond is no different than any other covalent bond.

7.54

```
        :Cl:          H                    :Cl:  H
         |            |                     |    |
  :Cl —  B ←———————— :O — H      →    :Cl — B ← O — H
         |                                  |
        :Cl:                               :Cl:
```

7.55

```
           H                  [      H      ] +
           |          →       [      |      ]
  H+ ←——— :O — H              [  H — O — H  ]
```

7.56 An addition compound is one that is formed by the use of a coordinate covalent bond. It results in the addition of one molecule to another, and it arises from the donation of an electron pair from one atom to another atom that can accept a pair of electrons.

VSEPR Theory

7.57 (a) planar triangular, otherwise known as trigonal planar

(b) octahedral (c) tetrahedral (d) trigonal bipyramidal

7.58 See Figure 7.6 on page 247 of the text.

7.59 (a) pyramidal (b) square planar (c) angular (nonlinear or bent)

7.60 (a) nonlinear (b) trigonal bipyramidal (c) pyramidal (d) nonlinear

7.61 (a) distorted tetrahedral (b) octahedral (c) nonlinear (d) tetrahedral

7.62 (a) tetrahedral (b) square planar (c) octahedral (d) linear

7.63 (a) linear (b) square planar (c) T-shaped (d) planar triangular

7.64 (a) pyramidal (b) tetrahedral (c) pyramidal (d) tetrahedral

7.65 180°

7.66 all angles 120°

7.67 planar triangular

Polar Bonds and Electronegativity

7.68 A polar covalent bond is one in which the electrons of the bond are not shared equally by the atoms of the bond, and this causes one end of the linkage to carry a partial negative charge while the other end carries a corresponding partial positive charge. In other words, there is a dipole in a polar bond. A dipole moment is the product of the amount of charge on one end of a polar bond (which behaves as a dipole) and the distance between the two partial charges that compose the dipole. It is also normally taken to be the product of the charge in the dipole of a polar bond and the internuclear distance in the polar bond.

7.69 Electronegativity is the attraction that an atom has for the electrons in chemical bonds to that atom.

7.70 Fluorine has the largest electronegativity, whereas oxygen has the second largest electronegativity.

7.71 The noble gases are assigned electronegativity values of zero.

7.72 Here we choose the atom with the smaller electronegativity:
(a) N (b) I (c) N

7.73 Here we choose the linkage that has the greatest difference in electronegativities between the atoms of the bond: N—S.

7.74 The most polar bond of the four is Si—F because it is the bond that has the greatest difference in electronegativities between the linked atoms.

The atom with the partial negative charge in the dipole of the bond is the one with the higher electronegativity of the linked atoms:

(a) I (b) Cl (c) F (d) N

7.75 A bond is more ionic than covalent (i.e. more than 50% ionic) when the atoms that are linked differ in electronegativity by more than 1.7 units. This is true of Si–F and Mg–N.

7.76 Elements having low electronegativities are metals.

7.77 (a) P (b) P (c) F

7.78 $\delta+$ H — F: $\delta-$

Predicting Molecular Polarity

7.79 Polar molecules attract one another, and that influences the physical and chemical properties of substances.

7.80 A bond's dipole moment is depicted with an arrow having a + sign on one end, where the "barb" of the arrow is taken to represent the location of the opposing negative charge of the dipole: ⟼.

7.81 A molecule having polar bonds will be nonpolar only if the bond dipoles are arranged so as to cancel one another's effect.

7.82

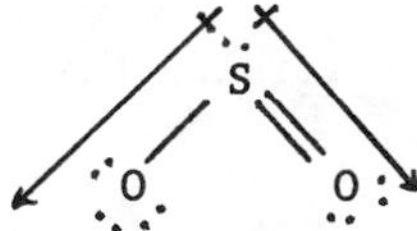

The individual bond dipoles do not cancel one another.

7.83 The ones that are polar are (a), (b), and (c). The last two have symmetrical structures, and although individual bonds in these substances are polar bonds, the geometry of the bonds serves to cause the individual dipole moments of the various bonds to cancel one another.

7.84 Two of these substances have planar triangular structures that are not polar because the individual bond dipole moments cancel one another: SO_3 and BCl_3. Two of these molecules have pyramidal structures, and are, therefore, polar: PBr_3 and $AsCl_3$. ClF_3 is T-shaped and, therefore, polar.

CHAPTER 8

PRACTICE EXERCISES

1. The H—Cl bond is formed by the overlap of the half-filled 1s atomic orbital of a H atom with the half-filled 3p valence orbital of a Cl atom:

   Cl (in HCl) 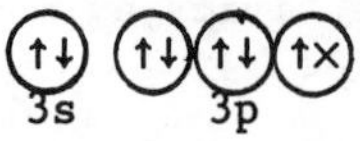 (↑↓) 3s (↑↓)(↑↓)(↑×) 3p (× = H electron)

2. The half-filled 1s atomic orbital of each H atom overlaps with a half-filled 3p atomic orbital of the P atom, to give three P—H bonds. This should give a bond angle of 90°.

   P (in PH_3) (↑↓) 3s (↑×)(↑×)(↑×) 3p (× = H electron)

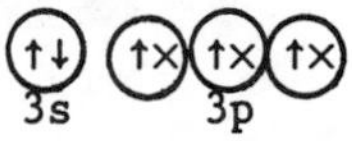

3. (a) sp^3 (b) sp^3d

4. (a) sp^3 (b) sp^3d

5. (a) sp^3d^2, since six atoms are bonded to the central atom

   (b) for P in PCl_6^-: (↑×)(↑×)(↑×)(↑×)(↑×)(↑↓) sp^3d^2 ( )( )( ) d (× = Cl electron)

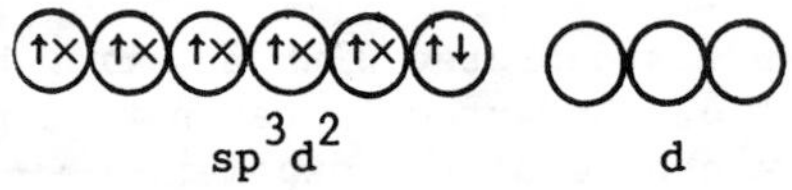

   (c) The ion is octahedral because six atoms and no lone pairs surround the central atom.

6. We first draw the Lewis structure and then determine the contribution from each bond in the molecule:

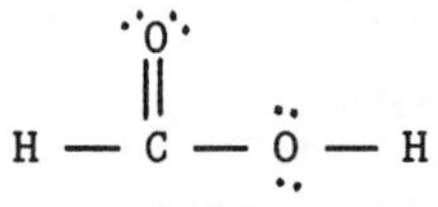

```
       ..
       O
       ||      ..
  H —  C  —  O  — H
               ..
```

   one C − H bond, 413 kJ/mol
   one C = O bond, 715 kJ/mol
   one C − O bond, 351 kJ/mol
   one O − H bond, 464 kJ/mol

   To atomize the molecule, each of these bonds must be broken. The atomization energy is thus the sum of all of the above bond energies:

   1.94×10^3 kJ/mol

7. $\Delta H_f^\circ[CH_3Cl(g)]$ refers to the enthalpy change under standard conditions for the following reaction:

$$C_{graphite} + \tfrac{3}{2}H_2(g) + \tfrac{1}{2}Cl_2(g) \rightarrow CH_3Cl(g)$$

We can arrive at this net reaction in an equivalent way, namely by vaporizing all of the necessary elements to give gaseous atoms, and then allowing the gaseous atoms to form all of the appropriate bonds. The overall enthalpy of formation by this route is numerically equal to that for the above reaction, and, conveniently, the enthalpy changes for each step are available in either Table 8.2 or Table 8.3:

$$\Delta H_f^\circ = \text{sum}(\Delta H_f^\circ[\text{gaseous atoms}]) - \text{sum}(\text{average bond energies in the molecule})$$

$$\Delta H_f^\circ[CH_3Cl(g)] = [715.0 + 3 \times 218.0 + 121.0] - [3 \times 413 + 328]$$

$$= -77 \text{ kJ/mol}$$

The actual value given in Table 5.2 (-82 kJ/mol) is different from that calculated here by only about 6%.

8. NO has 11 valence electrons, and the MO diagram is essentially like that in Table 8.4 for O_2, except that one fewer electron is employed at the highest energy level:

$\sigma^*_{2p_x}$	○
$\pi^*_{2p_y}, \pi^*_{2p_z}$	(↑)○
σ_{2p_x}	(↑↓)
π_{2p_y}, π_{2p_z}	(↑↓)(↑↓)
σ^*_{2s}	(↑↓)
σ_{2s}	(↑↓)○

The bond order is $\dfrac{8 - 3}{2} = 2.5$

Chapter 8

REVIEW EXERCISES

Modern Bonding Theories

8.1 Both VB and MO theory have wave mechanics as their theoretical basis.

8.2 Lewis structures do not explain how atoms share electrons, nor do they explain why molecules adopt particular shapes.

8.3 VB theory views atoms coming together with their orbitals already containing specific electrons. The bonds that are formed according to VB theory do so by the overlap of orbitals on neighboring atoms, and this is accompanied by the pairing (sharing) of the electrons that are contained in the orbitals.

MO theory, on the other hand, considers a molecule as a collection of positive nuclei surrounded by a set of molecular orbitals, which, by definition, belong to the molecule as a whole. The electrons of the molecule are distributed among the molecular orbitals according to the same rules that govern the filling of atomic orbitals.

Valence Bond Theory

8.4 Orbital overlap occurs when orbitals from different atoms share the same space.

8.5 According to VB theory, electrons are shared between atoms via the overlap of orbitals from the different atoms.

8.6 This is shown in Figure 8.1.

8.7 This is the same as Figure 8.2. A half-filled valence p orbital of Br overlaps with the half-filled 1s orbital of the hydrogen atom.

8.8 The 1s atomic orbitals of the hydrogen atoms overlap with the mutually perpendicular p atomic orbitals of the selenium atom.

Se(in H_2Se) (↑↓) (↑↓)(↑×)(↑×) (× = H electron)
4s 4p

8.9 This is shown in Figure 8.4. Another way to diagram it would be to show the orbitals of one of the fluorine atoms:

Chapter 8

(↑↓) (↑↓)(↑↓)(↑×)   (× = an electron from the other F atom)

Hybrid Orbitals

8.10 hybridization

8.11 Hybrid orbitals provide better overlap than do atomic orbitals, and this results in stronger bonds.

8.12 This is shown in Figure 8.9.

8.13 Elements in period two do not have a d subshell in the valence level.

8.14 Lewis structures are something like shorthand representations of VB structures.

8.15 atomic Be 

hybridized Be ↑× ↑×

sp p

where × = a Cl electron from a half-filled 3p orbital of a Cl atom.

8.16 (a) There are three bonds to the central Cl atom, plus one lone pair of electrons, so the Cl atom is to be sp^3 hybridized:

```
[ :Ö — C̈l — Ö: ] -
        |
       :Ö:
```

(b) There are three atoms bonded to the central sulfur atom, and no lone pairs on the central sulfur. The geometry of the molecule is that of a planar triangle, and the hybridization of the S atom is sp^2:

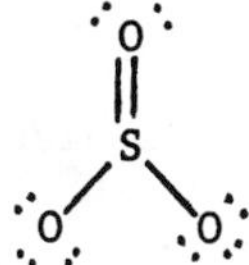

Two other resonance structures should also be drawn for SO_3.

(c) There are two bonds to the central O atom, as well as two lone pairs. The O atom is to be sp^3 hybridized, and the geometry of the molecule is nonlinear (bent).

```
 ..    ..
:O  —  F:
 |     ..
 |
:F:
 ..
```

(d) Six Cl atoms surround the central Sb atom in an octahedral geometry, and the hybridization of Sb is sp^3d^2.

```
[        :Cl:          ] -
   :Cl\   |   /Cl:
          Sb
   :Cl/   |   \Cl:
         :Cl:
```

(e) Three Cl atoms are bonded to the Br atom, plus the Br atom has two lone pairs of electrons. This requires the Br atom to be sp^3d hybridized, and the geometry is T-shaped.

```
 ..    . .    ..
:Cl  —  Br  —  Cl:
 ..     |     ..
       :Cl:
        ..
```

(f) The central Xe atom is bonded to four F atoms, plus it has two lone pairs of electrons. This requires sp^3d^2 hybridization of Xe.

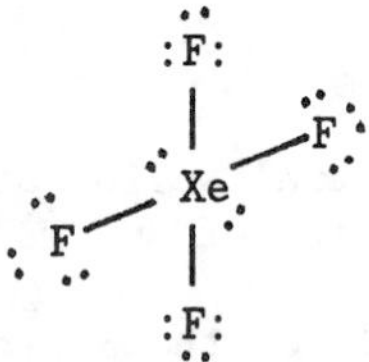

8.17 This is an octahedral ion with sp^3d^2 hybridized tin:

8.18 (a) Sn in $SnCl_4$

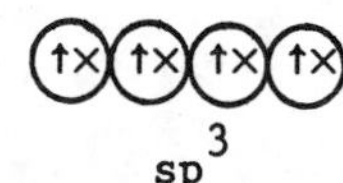

sp^3

(b) Sb in $SbCl_5$

sp^3d^2 5d (× = Cl electron)

8.19 (a) There are three bonds to As and one lone pair at As, requiring As to be sp^3 hybridized.

The Lewis diagram is:

```
[ :Cl: — As — :Cl: ]
          |
        :Cl:
```

The hybrid orbital diagram for As in $AsCl_3$ is:

(↑↓)(↑×)(↑×)(↑×) (× = Cl electron)

sp^3

(b) There are three atoms bonded to the central Cl atom, and it also has two lone pairs of electrons. The hybridization of Cl is thus sp^3d, and the Lewis diagram is:

```
:F: — Cl — :F:
       |
      :F:
```

The hybrid orbital diagram for Cl in ClF_3 is:

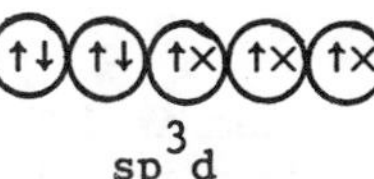

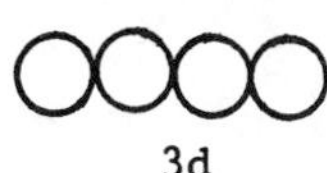

(× = F electron)

sp^3d 3d

(c) Five Cl atoms are bonded to the central Sb atom, which is therefore sp^3d^2 hybridized. The Lewis diagram is:

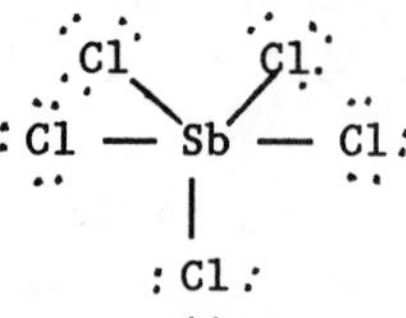

The hybrid orbital diagram for Sb in $SbCl_5$ is:

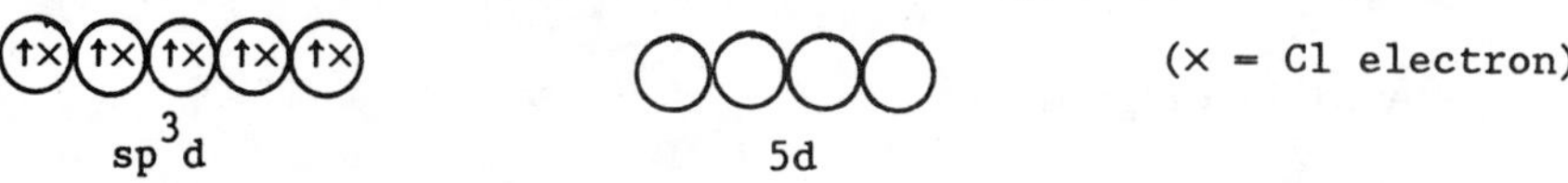

(d) Se has two bonds to Cl atoms and two lone pairs of electrons, requiring it to be sp^3 hybridized. The Lewis diagram is:

:C̈l — S̈e — C̈l:

and the hybrid orbital diagram is:

(× = Cl electron)

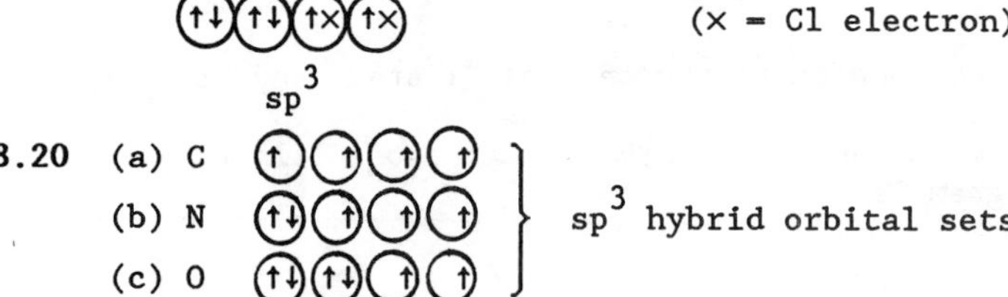

8.20 (a) C (b) N (c) O — sp^3 hybrid orbital sets

8.21 (a) See example 8.1 on page 276 of the text.
(b) See page 280 of the text.
(c) See page 280 of the text.

8.22 This angle would have to be 90°, the angle between one atomic p orbital and another.

8.23 The normal C—C—C angle for an sp^3 hybridized carbon atom is 109.5°. The 60° bond angle in cyclopropane is much less than this optimum bond angle. This means that the bonding within the ring cannot be accomplished through the desirable "head on" overlap of hybrid orbitals from each C atom. As a result, the overlap of the hybrid orbitals in cyclopropane is less effective than that in the more normal, noncyclic propane molecule, and this makes the C—C bonds in cyclopropane comparatively weaker than those in the noncyclic molecule.

8.24 Boron typically forms substances such as BCl_3, which have triangular geometries. Also, the valence shell of boron contains only three electrons. Since these three electrons of boron could half-fill only three hybrid orbitals, it is likely that only three hybrids should be formed. This leads to sp^2 hybridization.

8.25 For boron, see Figures 8.7 and 8.8. For carbon, see the discussion of page 284, as well as Figures 8.15 and 8.17.

8.26 (a) Phosphorus uses three half-filled sp^3 hybrid orbitals to overlap with a half-filled p atomic orbital from each of three F atoms. The fourth sp^3 hybrid orbital of phosphorus holds a lone pair of electrons.

(b) Unhybridized or atomic p orbitals of phosphorus would overlap with atomic p orbitals of each of the three fluorine atoms.

(c) Neither model works very well, because the actual bond angles are intermediate between the value that would be predicted using hybrid orbitals at phosphorus (109.5°) and the value that would result from the use of atomic phosphorus orbitals (90°).

8.27 90°

Coordinate Covalent Bonds and VB Theory

8.28 The ammonium ion has a tetrahedral geometry, with bond angles of 109.5°.

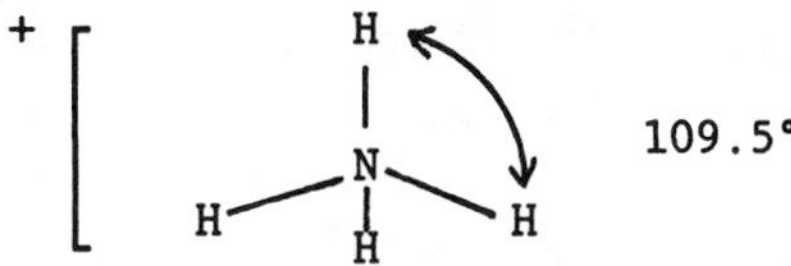

8.29 The oxygen atom in H_2O has two lone pairs of electrons, either one of which can form a coordinate covalent bond to a proton:

Chapter 8

Multiple Bonds and Hybrid Orbitals

8.30 σ bond - The electron density is concentrated along an imaginary straight line joining the nuclei of the bonded atoms.

π bond - The electron density lies above and below an imaginary straight line joining the bonded nuclei.

8.31 The characteristic side-to-side overlap of p atomic orbitals that characterizes π bonds is destroyed upon rotation about the bond axis. This is not the case for a σ bond, because regardless of rotation, a σ bond is still effective at overlap.

8.32 See Figure 8.15.

8.33 See Figure 8.18.

8.34 Whereas the arrangement around an sp^2 hybridized atom may have 120° bond angles (which well accommodates a six-membered ring), the geometrical arrangement around an atom that is sp hybridized must be linear, as in C – C ≡ C – C systems.

8.35 (a) (↑↓)(↑)(↑) (↑)

sp^2 2p

The arrangement of the above orbitals is:

(b)

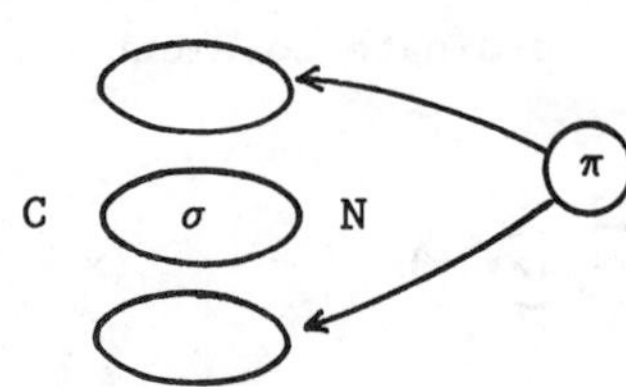

(c)

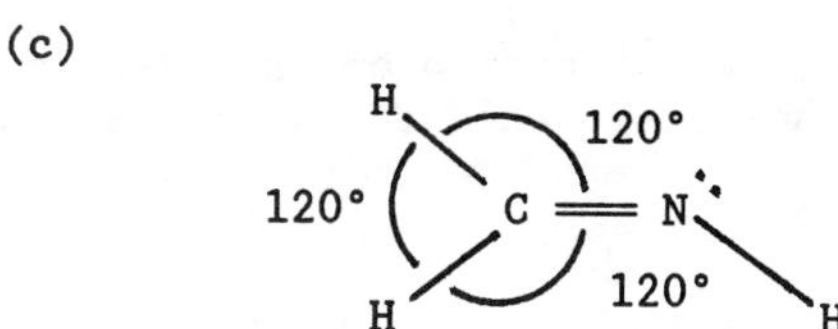

8.36 (a) N (↑↓)(↑) (↑)(↑)

sp 2p

(b)

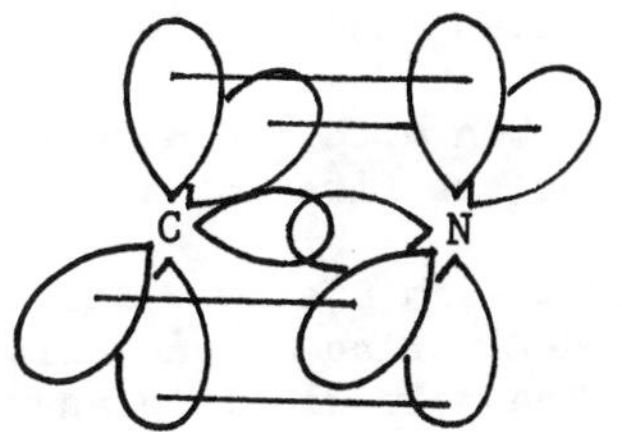

(c) The σ bonds:

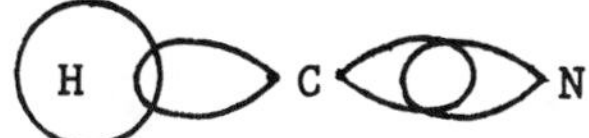

The π bonds:

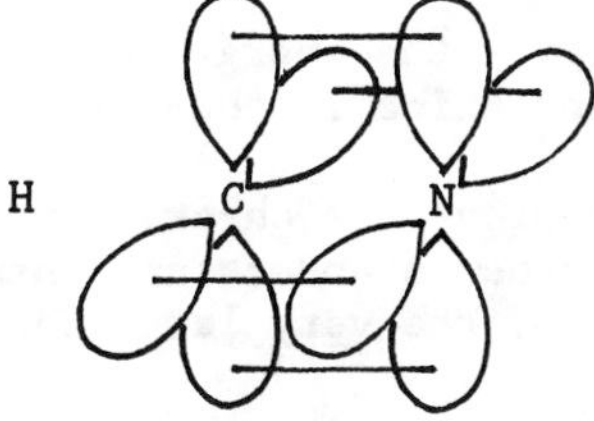

(d) The HCN bond angle should be 180°.

8.37 Two resonance hybrids should be drawn, as discussed on page 292 of the text. The π bond system in one of these resonance forms is shown below:

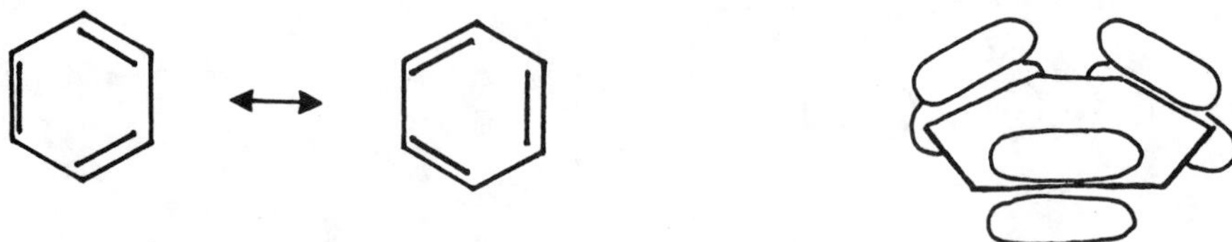

8.38 Each carbon atom is sp^2 hybridized, and each C—Cl bond is formed by the overlap of an sp^2 hybrid of carbon with a p atomic orbital of a chlorine atom. The C=C double bond consists first of a C—C σ bond formed by "head on" overlap of sp^2 hybrids from each C atom. Secondly, the C=C double bond consists of a side-to-side overlap of unhybridized p orbitals of each C atom, to give one π bond. The molecule is planar, and the expected bond angles are all 120°.

8.39 The bonding in phosgene is the same as that diagramed for H_2CO in Figure 8.17, except that H atom 1s orbitals are replaced by Cl atom p atomic orbitals.

Bond Energies

8.40 Bond energy - the amount of energy that is released when two atoms become joined by a bond. It is also numerically equal to the amount of energy needed to break the bond.

Atomization energy - the sum of all of the bond energies in a molecule. It is the amount of energy that is required to break all of the bonds in a molecule, giving neutral, gaseous atoms.

Bond dissociation energy - the energy required to break a bond, giving gaseous atoms. It is numerically equal to the bond energy.

8.41 Measured bond energies provide a check of predictions made by bonding theories, and they help us to understand chemical properties of substances. For example, the very large bond energy of N_2 helps to explain the low reactivity of this molecule.

8.42 The value of the standard heat of formation of H(g) is equal to half the energy needed to break a mole of H—H bonds:

$$\tfrac{1}{2}H_2(g) \rightarrow H(g), \qquad \Delta H_f^\circ[H(g)] = 218.0 \text{ kJ/mol}$$

The bond energy of H_2 is thus 2 × 218.0 = 436.0 kJ/mol.

8.43 In each case, as in review exercise 8.42, the bond energy is equal to twice the heat of formation of the gaseous atom:

$$\frac{472.7 \text{ kJ}}{\text{mol N(g)}} \times \frac{2 \text{ mol N(g)}}{1 \text{ mol } N_2(g)} = 945.4 \text{ kJ/mol, the } N_2 \text{ bond energy.}$$

$$\frac{249.2 \text{ kJ}}{\text{mol O(g)}} \times \frac{2 \text{ mol O(g)}}{1 \text{ mol } O_2} = 498.4 \text{ kJ/mol, the } O_2 \text{ bond energy.}$$

8.44 This requires the breaking of three N—H single bonds:

$$NH_3 \rightarrow N + 3H$$

The enthalpy of atomization of NH_3 is thus three times the average N—H single bond energy:

$3 \times 391 \text{ kJ/mol} = 1.17 \times 10^3 \text{ kJ/mol}$

8.45 The energy released during the formation of 1 mol of acetone is equal to the sum of all of the bond energies in the molecule:

for the six C—H bonds: 413 kJ/mol × 6 mol C—H bonds
for the two C—C bonds: 348 kJ/mol × 2 mol C—C bonds
for the one C=O bond: 715 kJ/mol × 1 mol C=O bonds

Adding the above contributions we get 3.89×10^3 kJ released per mole of acetone formed.

8.46 $\Delta H_f^\circ[C_2H_4(g)]$ refers to the enthalpy change under standard conditions for the following reaction:

$$2C_{graphite} + 2H_2(g) \rightarrow C_2H_4(g), \qquad \Delta H_f^\circ[C_2H_4(g)] = 52.284 \text{ kJ/mol}$$

We can arrive at this net reaction in an equivalent way, namely, by vaporizing all of the necessary elements to give gaseous atoms, and then allowing the gaseous atoms to form all of the appropriate bonds. The overall enthalpy of formation by this route is numerically equal to that for the above reaction, and, conveniently, the enthalpy changes for each step are available in either Table 8.2 or Table 8.3:

ΔH_f° = sum(ΔH_f°[gaseous atoms])
- sum(average bond energies in the molecule)

$\Delta H_f^\circ[C_2H_4(g)]$ = 52.284 kJ/mol = [2 × 715.0 + 4 × 218.0] - [4 × 413 + C=C]

from which we can calculate the C=C bond energy: 598 kJ/mol.

8.47 We proceed as in the answer to review exercise 8.46.
The various bonds that compose the molecule are:
5 C—H bonds
1 C—C bond
1 C—O bond
1 O—H bond

ΔH_f° = sum(ΔH_f°[gaseous atoms])
- sum(average bond energies in the molecule)

$\Delta H_f^\circ[C_2H_5OH(g)]$ = -235.3 kJ/mol

= [2 × 715.0 + 6 × 218.0 + 249.2] - [5 × 413 + 348 + 464 + C—O]

from which we can calculate that the C—O bond energy is 345 kJ/mol.

8.48 This problem is solved in the fashion of review exercises 8.46 and 8.47. There are two C=S double bonds to be considered:

ΔH_f° = sum(ΔH_f°[gaseous atoms])
- sum(average bond energies in the molecule)

$\Delta H_f^\circ[CS_2(g)]$ = 115.3 kJ/mol = [715.0 + 2 × 274.7] - [2 × C=S]

The C=S double bond energy is therefore given by the equation:

C=S = -(115.3 - 715.0 - 2 × 274.7) ÷ 2 = 574.6 kJ/mol

8.49 See the method of review exercises 8.46 through 8.48. There are six S—F bonds in the molecule:
ΔH_f° = sum(ΔH_f°[gaseous atoms])
- sum(average bond energies in the molecule)

$\Delta H_f^\circ[SF_6(g)]$ = -1096 kJ/mol = [274.7 + 6 × 78.91] - [6 × S—F]

S—F = (1096 + 274.7 + 6 × 78.91) ÷ 6 = 307.4 kJ/mol

8.50 See exercises 8.46 through 8.48.
ΔH_f° = sum(ΔH_f°[gaseous atoms])
- sum(average bond energies in the molecule)

$\Delta H_f^\circ[H_2S(g)]$ = -20.15 kJ/mol = [274.7 + 2 × 218.0] - [2 × H—S]

H—S = (20.15 + 274.7 + 2 × 218.0) ÷ 2 = 365.4 kJ/mol

8.51 We can again use the general method of review exercises 8.46 through 8.50 to approach this question.

ΔH_f° = sum(ΔH_f°[gaseous atoms])
- sum(average bond energies in the molecule)

$\Delta H_f^\circ[SF_4(g)]$ = [274.7 + 4 × 78.91] - [4 × 307.4]

= -639.3 kJ/mol = -152.7 kcal/mol

The % difference is [(172 - 153) ÷ 172] × 100 = 11 %

8.52 Using the approach of review exercises 8.46 through 8.50, we must consider two S=O bonds and two S—F bonds:

ΔH_f° = sum(ΔH_f°[gaseous atoms])
- sum(average bond energies in the molecule)

$\Delta H_f^\circ[SO_2F_2(g)]$ = -858 kJ/mol = [274.7 + 2 × 249.2 + 2 × 78.91]
- [2 × 307.4 + 2 × S=O]

S=O = 587 kJ/mol

8.53 See review exercises 8.46 through 8.52.

ΔH_f° = sum(ΔH_f°[gaseous atoms])
- sum(average bond energies in the molecule)

$\Delta H_f^\circ[C_2H_2(g)]$ = [2 × 715.0 + 2 × 218.0] - [2 × 413 + 812]

= 228 kJ/mol

8.54 Proceed as in the answers to review exercises 8.46 through 8.53.

ΔH_f° = sum(ΔH_f°[gaseous atoms])
- sum(average bond energies in the molecule)

$\Delta H_f^\circ[CCl_4(g)]$ = [715.0 + 4 × 121.0] - [4 × 328] = -113 kJ/mol

8.55 The heat of formation of CF_4 should be more exothermic than that of CCl_4 because more energy is released on formation of a C—F bond than on formation of a C—Cl bond. Also, less energy is needed to form gaseous F atoms than to form gaseous Cl atoms.

8.56 No. The bonding in SO_3 is described in terms of resonance; the true structure is that of a hybrid of three resonance forms. Since no single resonance form completely describes the real molecule, the additional resonance energy is not taken into account when any one resonance form is used as the basis for a calculation like that done in review exercises 8.46 and following. The molecule is actually more stable than would be expected based on only one resonance form.

8.57 The extra stabilization energy is termed resonance energy, or delocalization energy.

Molecular Orbital Theory

8.58 If an electron is forced to occupy this MO, the molecule loses stability and the bond is made weaker than if an electron, or a pair of electrons, is made to occupy the lower energy (bonding) MO.

8.59 See Figure 8.22.

8.60 In the hypothetical molecule He_2, both the bonding and the antibonding MO are double occupied, and the net bond order is zero, as shown in Figure 8.24. See Figure 8.23 for the MO diagram for H_2.

8.61 As shown in Table 8.4, the highest energy electrons in dioxygen occupy the double degenerate π-antibonding level:

$$\underset{\pi^*_{2p_x}}{\underline{\uparrow}} \qquad \underset{\pi^*_{2p_y}}{\underline{\uparrow}}$$

Since their spins are unpaired, the molecule is paramagnetic.

8.62 As shown in Table 8.4, the bond order of Li_2 is 1.0. The bond order of Be_2 would be zero.

8.63 Here we pick the one with the higher bond order.

(a) O_2^+ (b) O_2 (c) N_2

8.64 (a) 2.5 (b) 1.5 (c) 1.5

8.65 As bond order increases, bond energy (and strength) increases.

8.66 See Figure 8.25.

8.67 A delocalized MO is one that extends over more than two nuclei.

8.68 

8.69 (a) The C—C single bonds are formed from head-to-head overlap of C atom sp^2 hybrids. This leaves one unhybridized atomic p orbital on each carbon atom, and each such atomic orbital is oriented perpendicular to the plane of the molecule.

(b) Sideways or π type overlap is expected between the first and the second carbon atoms, as well as between the third and the fourth carbon atoms. However, since all of these atomic p orbitals are properly aligned, there can be continuous π type overlap between all four carbon atoms.

(c) The situation described in part (b) is delocalized. We expect completely delocalized π type bonding among the carbon atoms.

(d) The bond is shorter because of extra stability that arises from delocalization energy.

8.70 MO theory avoids the need in VB theory for the cumbersome and numerous resonance forms.

8.71 This is shown at the bottom of page 299 of the text.

8.72 Delocalization increases stability.

8.73 Delocalization energy is a term used in MO theory to mean essentially the same thing as the term resonance energy, which derives from VB theory. They both represent the additional stability associated with a spreading out of electron density.

CHAPTER NINE

PRACTICE EXERCISES

1. The largest surface has an area of (5.00 in.)(4.72 in.) = 23.6 in.^{2}. Hence it exerts the following pressure:

$$\text{pressure} = \text{force/unit area} = \frac{1\ \text{lb}}{23.6\ \text{in.}^2} = 0.0424\ \text{lb/in.}^2$$

The surface of the long and narrow side is (5.00 in.)(1.22 in.) = 6.10 in.^{2}. The pressure is thus:

$$\text{pressure} = 1\ \text{lb}/6.10\ \text{in.}^2 = 0.164\ \text{lb/in.}^2$$

2.

$$740\ \text{torr} \times \frac{1\ \text{atm}}{760\ \text{torr}} = 0.974\ \text{atm}$$

$$740\ \text{torr} \times \frac{133.3224\ \text{Pa}}{1\ \text{torr}} \times \frac{1\ \text{kPa}}{1000\ \text{Pa}} = 98.7\ \text{kPa}$$

$$740\ \text{torr} \times \frac{1\ \text{mm Hg}}{1\ \text{torr}} = 740\ \text{mm Hg}$$

3. $P_{bulb} = P_{atm} + 66\ \text{torr} = (752 + 66)\ \text{torr} = 818\ \text{torr}$

4.

$$719\ \text{torr} \times \frac{1\ \text{atm}}{760\ \text{torr}} = 0.946\ \text{torr}$$

$$818\ \text{torr} \times \frac{1\ \text{atm}}{760\ \text{torr}} = 1.08\ \text{atm}$$

5. In general, $P_1V_1 = P_2V_2$, and if we know three of these, we can calculate the fourth:

$$P_2 = P_1 \times V_1/V_2 = 740\ \text{torr} \times \frac{880\ \text{mL}}{440\ \text{mL}} = 1.48 \times 10^3\ \text{torr}$$

Note that this is sensible because if the volume is smaller ($V_2 = 440$ mL), the pressure must increase. Thus we would have concluded that the original pressure must be multiplied by a volume ratio greater than one.

6. Since the pressure increases, the volume must decrease, and we conclude that we must multiply the original volume by a pressure ratio less than one in order to obtain the new volume. Also, in general, $P_1V_1 = P_2V_2$, and if we know three of these, we can calculate the fourth:

$$V_2 = V_1 \times P_1/P_2 = 200\ \text{mL} \times \frac{760\ \text{torr}}{800\ \text{torr}} = 190\ \text{mL}$$

7. $P_{total} = P_{O_2} + P_{N_2} = 760$ torr

$\therefore P_{O_2} = (760 - 601)$ torr $= 159$ torr

8. $P_{total} = 745$ torr $= P_{N_2} + P_{water}$

The vapor pressure of water at 15 °C is available in Table 9.1: 12.79 torr. Hence:

$P_{N_2} = (745 - 13)$ torr $= 732$ torr

Also, since $P_1V_1 = P_2V_2$, then the new volume is given by the formula:

$$V_2 = V_1 \times P_1/P_2 = 310 \text{ mL} \times \frac{732 \text{ torr}}{760 \text{ torr}} = 299 \text{ mL}$$

Notice that it is sensible that the second volume be smaller than the first, because the second pressure (760) is larger than the first (732).

9. Here we wish to decrease the volume from the value $V_1 = 300$ mL to the value $V_2 = 250$ mL. Recall that 20 °C = (20 + 273) K. By applying Charles' Law we have $V_1/T_1 = V_2/T_2$, or:

$$T_2 = 293 \text{ K} \times \frac{250 \text{ mL}}{300 \text{ mL}} = 244 \text{ K, or } (244 - 273) = -29 \text{ °C}$$

It is sensible that we should multiply the initial temperature by a volume ratio less than one, because if the volume must decrease, the temperature must also decrease, according to Charles' Law.

10. Under these circumstances, the pressure is constant, and $P_1 = P_2$. Hence, both sides of the combined gas law can be divided by the value P_1, giving: $\frac{P_1V_1}{P_1T_1} = \frac{P_2V_2}{P_1T_2}$, and one obtains $\frac{V_1}{T_1} = \frac{V_2}{T_2}$, which is Equation 9.15.

11. Since this condition requires that $V_1 = V_2$, we can divide both sides of Equation 9.20 by V_1 or by V_2 to obtain:

$P_1/T_1 = P_2/T_2$

Hence, for both sets of pressure and temperature, it is true that P_1/T_1 and P_2/T_2 are both simultaneously equal to a constant, C''. Thus, we have the relationship in general that:

$P = C'' \times T$

For a gas sample at constant mass, and at constant volume, the pressure is directly proportional to the Kelvin temperature of the gas, i.e. the pressure increases as the Kelvin temperature increases. This can be also written as follows:

$P \propto T$ (at constant m, V) where the constant of proportionality is C''.

12. If the temperature has increased, so too must the pressure increase. Thus we must multiply the original pressure by a temperature ratio that is greater than one. Also it is generally true that $P_1/T_1 = P_2/T_2$, and solving for P_2 gives:

$$P_2 = P_1 \times T_2/T_1 = 760 \text{ torr} \times \frac{(800 + 273)\text{ K}}{(25 + 273)\text{ K}} = 2.74 \times 10^3 \text{ torr}$$

$$2.74 \times 10^3 \text{ torr} \times \frac{1 \text{ atm}}{760 \text{ torr}} = 3.61 \text{ atm}$$

13. (a)
$$P_2 = P_1 \times \frac{V_1}{V_2} \times \frac{T_2}{T_1}$$

(b) $$V_2 = V_1 \times \frac{P_1}{P_2} \times \frac{T_2}{T_1}$$

(c)
$$T_2 = T_1 \times \frac{V_2}{V_1} \times \frac{P_2}{P_1}$$

14.
$$P_2 = P_1 \times \frac{V_1}{V_2} \times \frac{T_2}{T_1}$$

$$P_2 = 745 \text{ torr} \times \frac{950 \text{ m}^3}{1150 \text{ m}^3} \times \frac{(60 + 273)\text{ K}}{(25 + 273)\text{ K}} = 688 \text{ torr}$$

15. We begin by solving the ideal gas law for volume:

$V = nRT/P$

and by calculating the number of moles of gas that are involved:

$$n = 10.2 \text{ g } CH_4 \times \frac{1 \text{ mol}}{16.0 \text{ g}} = 0.638 \text{ mol}$$

We next use the following values in the above equation for volume:

$R = 0.0821$ L atm/K mol
$T = (25 + 273) = 298$ K
$P = 755$ torr ÷ 760 torr/atm = 0.993 atm

$$V = \frac{(0.638 \text{ mol})(0.0821 \text{ L atm/K mol})(298 \text{ K})}{0.993 \text{ atm}} = 15.7 \text{ L}$$

16. The ideal gas law is first used to determine the number of moles of the gas:

$n = PV/RT$

where the values to be used are the following:

$P = 745 \text{ torr} \div 760 \text{ torr/atm} = 0.980 \text{ atm}$
$V = 0.817 \text{ L}$
$R = 0.0821 \text{ L atm/K mol}$
$T = (25 + 273) \text{ K} = 298 \text{ K}$

$$n = \frac{(0.980 \text{ atm})(0.817 \text{ L})}{(0.0821 \text{ L atm/K mol})((298 \text{ K})} = 0.0327 \text{ mol}$$

The molecular weight is:

$$\frac{0.0682 \text{ g}}{0.0327 \text{ mol}} = 2.09 \text{ g/mol, which is most likely } H_2.$$

17. $n = 0.015 \text{ mol}$ $R = 0.0821 \text{ L atm/K mol}$

$T = (28.0 + 273) \text{ K} = 301 \text{ K}$ $P = 744 \text{ torr} \div 760 \text{ torr/atm}$

$$V = \frac{(0.015 \text{ mol})(0.0821 \text{ L atm/K mol})(301 \text{ K})}{0.979 \text{ atm}} = 0.38 \text{ L} = 3.8 \times 10^2 \text{ mL}$$

18. The balanced equation is: $N_2(g) + 3H_2(g) \rightarrow 2NH_3(g)$

The number of moles of NH_3 that form are:

$$28.9 \text{ mol } N_2 \times \frac{2 \text{ mol } NH_3}{1 \text{ mol } N_2} = 57.8 \text{ mol } NH_3$$

The number of grams are:

$$57.8 \text{ mol } NH_3 \times \frac{17.0 \text{ g } NH_3}{1 \text{ mol } NH_3} = 983 \text{ g } NH_3$$

Finally, the number of liters is:

$$V = \frac{(57.8 \text{ mol})(0.0821 \text{ L atm/K mol})(273 \text{ K})}{1 \text{ atm}} = 1.30 \times 10^3 \text{ L}$$

19. We make use of equation 9.25:

$$\sqrt{\frac{0.654\ g/L}{0.0818\ g/L}} = \frac{(\text{effusion rate})_{\text{hydrogen}}}{(\text{effusion rate})_{\text{methane}}} = 2.83$$

We conclude that hydrogen will effuse 2.83 times faster than methane. This suggests that any leak in the flow system for hydrogen gas will be more severe than for methane, and the valves and fittings might be more susceptible to leaks of hydrogen than they are to leaks of methane.

20. We make use of equation 9.27:

$$\frac{(\text{effusion rate})_{\text{hydrogen}}}{(\text{effusion rate})_{\text{oxygen}}} = \sqrt{\frac{32.0\ g/mol}{2.01\ g/mol}} = 3.98$$

We conclude that hydrogen effuses 3.98 times faster than does oxygen.

REVIEW EXERCISES

Concept of Pressure

9.1 Pressure is the ratio of force per unit area, and it is calculated by dividing the force that is acting on an area by the area.

9.2 The pressure of the atmosphere arises from the gravitational force of the earth, which acts on the gases of the atmosphere to give weight to the gases. The gases in turn "press down" on the earth, exerting a certain force over a given area, i.e. pressure.

9.3 If the liquid in the Torricelli barometer had a lower density, the barometer tube would have to be longer (in order to accommodate a higher tower of liquid) for purposes of measuring the same pressure. A liquid half as dense would have to stand in the barometer at twice the height in order to press down with the same force.

9.4 The low vapor pressure of mercury is advantageous because only a small amount of mercury vapor is present in the closed chamber of the barometer, and, therefore, very little pressure from mercury vapor arises in the barometer. The result is that the mercury vapor has a negligible effect on the height of the column that is supported by the atmosphere.

9.5 Since the density of water is approximately 13 times smaller than that of mercury, a barometer constructed with water as the moveable liquid would have to be some 13 times longer than one constructed using mercury. Also, the vapor pressure of water is large enough that the closed end of the barometer may fill with sufficient water vapor so as to affect atmospheric pressure readings. In fact, the measurement of atmospheric pressure would be about 18 torr too low, due to the presence of water vapor in the closed end of the barometer.

9.6 1 torr = 1 mm Hg

9.7 760 mm Hg = 1 atm

9.8

$$\frac{200 \text{ lb}}{1 \text{ ft}^2} \times \left(\frac{1 \text{ ft}}{12 \text{ in.}}\right)^2 = 1.39 \text{ lb/in.}^2$$

9.9 pressure = force/unit area

For the 100 lb force, P = 100 lb/25 in.^{2} = 4.0 lb/in.^{2}

For the 25 lb force, P = 25 lb/5 in.^{2} = 5 lb/in.^{2}

9.10 744 mm Hg × 1 torr/1 mm Hg = 744 torr

9.11 755 torr × 1 mm Hg/1 torr = 755 mm Hg

9.12 750 torr × 1 atm/760 torr = 0.987 atm

9.13 0.850 atm × 760 torr/1 atm = 646 torr

9.14 0.445 atm × 760 torr/atm = 338 torr

9.15

$$755 \text{ torr} \times \frac{133.3224 \text{ Pa}}{1 \text{ torr}} \times \frac{1 \text{ kPa}}{1000 \text{ Pa}} = 101 \text{ kPa}$$

9.16 For nitrogen we have:

$$594.70 \text{ torr} \times \frac{133.3224 \text{ Pa}}{1 \text{ torr}} \times \frac{1 \text{ kPa}}{1000 \text{ Pa}} = 79.287 \text{ kPa}$$

For oxygen:

$$160.00 \text{ torr} \times \frac{133.3224 \text{ Pa}}{1 \text{ torr}} \times \frac{1 \text{ kPa}}{1000 \text{ Pa}} = 21.332 \text{ kPa}$$

9.17 For nitrogen:

$$569 \text{ torr} \times \frac{133.3224 \text{ Pa}}{1 \text{ torr}} \times \frac{1 \text{ kPa}}{1000 \text{ Pa}} = 75.9 \text{ kPa}$$

$$116 \text{ torr} \times \frac{133.3224 \text{ Pa}}{1 \text{ torr}} \times \frac{1 \text{ kPa}}{1000 \text{ Pa}} = 15.5 \text{ kPa}$$

9.18 The most perfect vacuum pump can only at most cause a column of mercury to rise to a height of that day's atmospheric pressure, since that is the maximum height of a column of mercury that the atmosphere can support. That is to say that a vacuum pump cannot induce a mercury column to rise any farther than what is forced up by atmospheric pressure. Correspondingly, the highest a pump could cause a column of water to rise will be only that amount of water that will be supported by atmospheric pressure, that is the weight of water that is equivalent to the weight of 760 mm of mercury. Since the density of mercury is 13.6 g/cm^3 whereas that of water is only 1.00 g/cm^3, the heights of the two columns are also related by the proportion 13.6 - to - 1.00. Thus the height of an equivalent column of water would be:

$$760 \text{ mm Hg} \times 13.6 = 1.03 \times 10^4 \text{ mm}$$

This is next converted to a value in feet:

$$1.03 \times 10^4 \text{ mm} \times \frac{1 \text{ in.}}{25.4 \text{ mm}} \times \frac{1 \text{ ft}}{12 \text{ in.}} = 33.9 \text{ ft}$$

This means that the best conceivable vacuum pump (one capable of producing a perfect vacuum on the column of water being pulled up) could cause the water in the pipe attached to the pump to rise only 33.9 ft above the surface of the water in the pit. The water at the bottom of a 35 ft pit could not be removed by use of a vacuum pump.

9.19 In review exercise 9.18 we found that 1 atm = 33.9 ft of water. This is equivalent to 33.9 ft × 12 in./ft = 407 in. of water, which in this problem is equal to the height of a water column that is uniformly 1.00 in.^{2} in diameter. Next, we convert the given density of water from the units g/mL to the units lb/in.^{3}:

$$\frac{1.00 \text{ g}}{1.00 \text{ mL}} \times \frac{1 \text{ mL}}{1 \text{ cm}^3} \times \left(\frac{2.54 \text{ cm}}{1 \text{ in.}}\right)^3 \times \frac{1 \text{ lb}}{454 \text{ g}} = 0.0361 \text{ lb/in.}^3$$

The area of the total column of water is now calculated:

$$1.00 \text{ in.}^2 \times 407 \text{ in.} = 407 \text{ in.}^3$$

along with the mass of the total column of water:

$$407 \text{ in.}^3 \times 0.0361 \text{ lb/in.}^3 = 14.7 \text{ lb}$$

Finally, we can determine the pressure (force/unit area) that corresponds to one atm:

$$1 \text{ atm} = 14.7 \text{ lb} \div 1.00 \text{ in.}^2 = 14.7 \text{ lb/in.}^2$$

Chapter 9

Specific Gas Laws

9.20 Pressure-Volume law: The volume of a given mass of a gas is inversely proportional to its pressure, provided the temperature is held constant.

$V \propto 1/P$ or $P_1V_1 = P_2V_2$

9.21 Temperature-Volume law: The volume of a given mass of a gas is directly proportional to the Kelvin temperature, provided the pressure is held constant.

$V \propto T$ or $V_1/T_1 = V_2/T_2$

9.22 The Law of Partial Pressures: The total pressure of a mixture of gases is equal to the sum of the partial pressures of the gases that compose the mixture.

$P_{total} = P_a + P_b + P_c + \ldots$

9.23 The fact that a simple sum of partial pressures gives the total pressure of a collection of different gases indicates that the individual gas molecules do not physically interact with one another, either to attract or to repel each other. This is one of the postulates of the Kinetic Theory of Gases.

9.24 (a) mass and temperature
(b) mass and pressure
(c) No stipulation other than the presence of various gases is necessary.
(d) mass and volume
(e) pressure and temperature
(f) pressure and temperature

9.25 Since volume is to decrease, pressure must increase, and we multiply the starting pressure by a volume ratio that is larger than one. Also, since $P_1V_1 = P_2V_2$, we can solve for P_2:

$$P_2 = P_1 \times V_1/V_2 = 760 \text{ torr} \times \frac{750 \text{ mL}}{500 \text{ mL}} = 1.14 \times 10^3 \text{ torr}$$

9.26 Since volume is to increase, we want to multiply the original volume by a pressure ratio that is greater than one:

$$V_2 = V_1 \times P_1/P_2 = V_1 \times \frac{950 \text{ torr}}{760 \text{ torr}} = 1.25 \times V_1$$

9.27 (a) Since $P_1V_1 = P_2V_2$, we can solve for P_2, keeping in mind that the new pressure must be lower than the original pressure, since the volume is larger after the expansion of the gas:

$$P_2 = P_1 \times V_1/V_2 = 500 \text{ torr} \times \frac{100 \text{ mL}}{250 \text{ mL}} = 200 \text{ torr}$$

(b) The Combined Gas Law equation is needed here:

$$P_2 = P_1 \times V_1/V_2 \times T_2/T_1$$

$$P_2 = 500 \text{ torr} \times \frac{100 \text{ mL}}{250 \text{ mL}} \times \frac{(15 + 273) \text{ K}}{(20 + 273) \text{ K}} = 197 \text{ torr}$$

9.28 We start with the Combined Gas Law and derive a relationship for the new temperature:

$$T_2 = T_1 \times P_2/P_1 \times V_2/V_1$$

$$T_2 = (25 + 273) \text{ K} \times \frac{900 \text{ mL}}{1000 \text{ mL}} \times \frac{814 \text{ torr}}{740 \text{ torr}} = 295 \text{ K}$$

In other words, the temperature of the sample will have to decrease from 298 K to 295 K.

9.29 For each new pressure, the volume should decrease. At each stage the volume may be calculated from the last using the relationship:

$$P_1V_1 = P_2V_2, \quad \text{or} \quad V_2 = V_1 \times P_1/P_2$$

$$V_{780 \text{ torr}} = 100 \text{ mL} \times \frac{760 \text{ torr}}{780 \text{ torr}} = 97.4 \text{ mL}$$

Pressure (torr)	Volume (mL)
780	97.4
800	95.0
820	92.7
840	90.5
860	88.4
880	86.4
900	84.5
920	82.7
940	80.9
960	79.2
980	77.6
1000	76.0

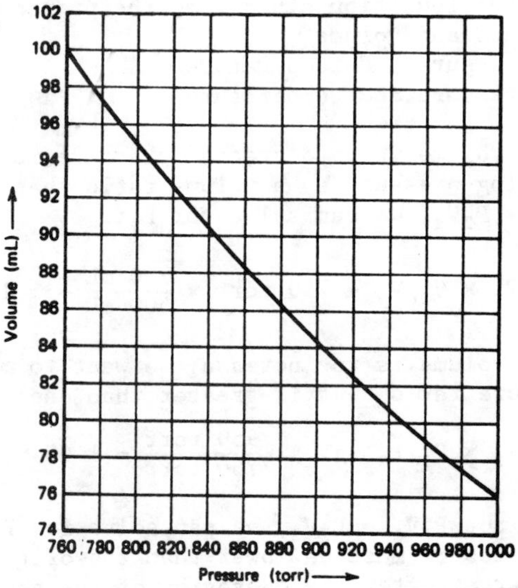

9.30 At each successively higher temperature, the pressure should increase. Thus, for each new temperature, the last pressure (P_1) may be used in the calculation of the new pressure (P_2) using the following relationship:

$P_1/T_1 = P_2/T_2$, or $P_2 = P_1 \times T_2/T_1$

$$P_{313\ K} = 760 \text{ torr} \times \frac{313\ K}{293\ K} = 812 \text{ torr}$$

Temperature (K)	Pressure (torr)
313	812
333	864
353	916
373	968
393	102×10^1

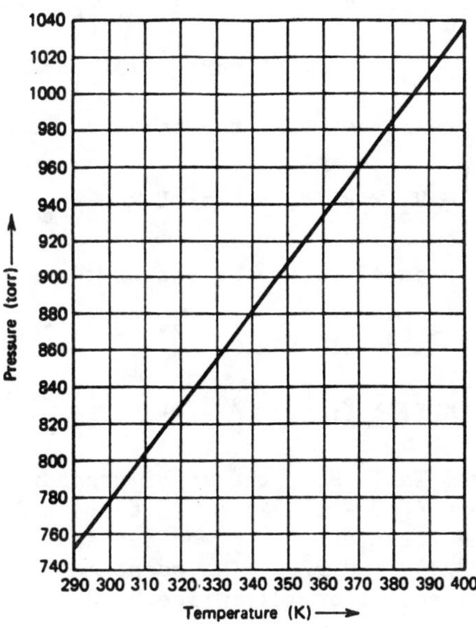

9.31 $P_{total} = 738 \text{ torr} = (P_{N_2} + P_{H_2O})$

From table 9.1 we find: $P_{H_2O} = 31.82$ torr at 30 °C.

$P_{N_2} = 738 - 31.82 = 706$ torr

9.32 $P_{total} = 751 \text{ torr} = (P_{O_2} + P_{H_2O})$

$P_{H_2O} = 9.209$ torr at 10 °C, from Table 9.1.

$P_{O_2} = 751 \text{ torr} - 9.209 = 742$ torr

9.33 $P_{total} = (P_{CO} + P_{H_2O})$

$P_{H_2O} = 17.54$ torr at 20 °C, from Table 9.1

$P_{CO} = 749 - 17.54 \text{ torr} = 731$ torr

If we assume that the temperature stays constant, then $P_1V_1 = P_2V_2$, and

$$V_2 = V_1 \times P_1/P_2 = 275 \text{ mL} \times \frac{731 \text{ torr}}{760 \text{ torr}} = 265 \text{ mL of dry CO}$$

9.34 We proceed as in the answer to review question 9.33:

$$P_{H_2} = P_{total} - P_{H_2O} = 736 - 23.76 = 712 \text{ torr}$$

$$V_2 = V_1 \times P_1/P_2 = 295 \text{ mL} \times \frac{712 \text{ torr}}{760 \text{ torr}} = 276 \text{ mL dry } H_2$$

9.35 From Table 9.1, the vapor pressure of water at 20 °C is 17.54 torr. Thus only (740 - 17.54) = 722 torr is due to "dry" methane. In other words, the fraction of the wet methane sample that is pure or dry methane is 722/740 = 0.976. The question can now be phrased: What volume of wet methane, when multiplied by 0.976, equals 240 mL?

$$\text{Volume}_{\text{"wet" methane}} \times 0.976 = 240 \text{ mL}$$

$$\text{Volume}_{\text{"wet" methane}} = 240 \text{ mL}/0.976 = 246 \text{ mL}$$

In other words, one must collect 246 total mL of "wet methane" gas in order to have collected 240 mL of methane.

9.36 First convert the needed amount of oxygen at 760 torr to the volume that would correspond to the laboratory conditions, 746 torr:

$$P_1V_1 = P_2V_2 \quad \text{or} \quad V_2 = P_1V_1/P_2$$

$$V_2 = 260 \text{ mL} \times 760 \text{ torr}/746 \text{ torr} = 265 \text{ mL of dry oxygen gas}$$

The wet sample of oxygen gas will also be collected at atmospheric pressure, 746 torr. The vapor pressure of water at 15 °C is equal to 12.8 torr (from Table 9.1), and the wet sample will have the following partial pressure of oxygen, once it is collected:

$P_{O_2} = P_{total} - P_{H_2O} = 746 - 12.8 = 733$ torr of oxygen in the wet sample.

Thus the wet sample of oxygen is composed of the following % oxygen:

% oxygen in the wet sample = 733/746 × 100 = 98.3 %

The question now becomes what amount of a wet sample of oxygen will contain the equivalent of 265 mL of pure oxygen, if the wet sample is only 98.3 % oxygen (and 1.7 % water)?

$0.983 \times V_{wet} = 265$ mL, hence $V_{wet} = 270$ mL

This means that 270 mL of a wet sample of oxygen must be collected in order to obtain as much oxygen as would be present in 265 mL of a pure sample of oxygen.

9.37 Effusion rates for gases are normally inversely proportional to the square root of the gas density, and the gas with the lower density ought to effuse more rapidly. Nitrogen in this problem has the higher effusion rate because it has the lower density:

$$\frac{\text{rate}(N_2)}{\text{rate}(CO_2)} = \sqrt{\frac{1.96 \text{ g/L}}{1.25 \text{ g/L}}} = 1.25$$

9.38

$$\text{rate(U-235)} = \text{rate(U-238)} \times \sqrt{\frac{\text{Density(U-238)}}{\text{Density(U-235)}}}$$

$$\text{rate(U-235)} = \text{rate(U-238)} \times \sqrt{\frac{2.3119 \text{ g/mL}}{2.2920 \text{ g/mL}}} = \text{rate(U-283)} \times 1.0043$$

Meaning that the rate of effusion of the less dense UF_6 derivative having the U-235 isotope effuses only 1.0043 times faster than the more dense derivative having the U-238 isotope.

Combined and Ideal Gas Laws

9.39 Since in general PV = nRT, then R = PV/nT

Let P = 760 torr, T = 273 K, and n = 1, and express the volume of the standard mole using the units mL instead of L: 22.4 L = 22,400 mL.

$$R = \frac{(760 \text{ torr})(22{,}400 \text{ mL})}{(1 \text{ mol})(273 \text{ K})} = 6.24 \times 10^4 \text{ mL torr/K mol}$$

9.40 *Molecule*: the smallest unit of a gas, except those gases such as argon that are monatomic.
Mole: Avogadro's number of gas molecules (or of gaseous atoms if the gas is monatomic).
Molar volume: the volume occupied by one mole of a gas at STP.

9.41 Since this is the volume of one mole of any gas at STP, the sample contains one mole of H_2: 6.02×10^{23} molecules of H_2.

9.42 (a) If in general PV = nRT, and if n and T are held constant, then the product P_1V_1 is equal to the constant nRT. Also, another product P_2V_2 is equal to the same constant nRT, meaning that the two pressure-volume products must be equal to one another: $P_1V_1 = P_2V_2$. Alternatively, we can divide one equation P_1V_1 = nRT by the other equation P_2V_2 = nRT, arriving at the same result by realizing that the product nRT in one equation is equal to the product nRT in the other equation.
(b) For constant n and P, $PV_1 = nRT_1$ and $PV_2 = nRT_2$. Dividing one equation by the other, we get: $V_1/T_1 = V_2/T_2$
(c) For constant n and V, $P_1V = nRT_1$ and $P_2V = nRT_2$. Dividing one equation by the other we get: $P_1/T_1 = P_2/T_2$
(d) For constant P and T, $PV_1 = n_1RT$ and $PV_2 = n_2RT$. Dividing one equation by the other gives: $V_1/V_2 = n_1/n_2$. Also, if $V_1 = V_2$, then we realize that n_1 must be equal to n_2.

9.43 In general the combined gas law equation is:

$$\frac{P_1V_1}{T_1} = \frac{P_2V_2}{T_2}$$

and in particular, for this problem, we have:

$$P_2 = P_1 \times \frac{T_2}{T_1} \times \frac{V_1}{V_2} = 745 \text{ torr} \times \frac{(75 + 273)\text{ K}}{(25 + 273)\text{ K}} \times \frac{2.55\text{ L}}{2.75\text{ L}} = 807 \text{ torr}$$

9.44 In general the combined gas law equation is:

$$\frac{P_1V_1}{T_1} = \frac{P_2V_2}{T_2}$$

and in particular, for this problem, we have:

$$P_2 = P_1 \times \frac{T_2}{T_1} \times \frac{V_1}{V_2} = 1.00 \text{ atm} \times \frac{(65 + 273)\text{ K}}{(15 + 273)\text{ K}} \times \frac{665\text{ mL}}{695\text{ mL}} = 1.12 \text{ atm}$$

9.45 In general the combined gas law equation is:

$$\frac{P_1V_1}{T_1} = \frac{P_2V_2}{T_2}$$

and in particular, for this problem, we have:

$$V_2 = V_1 \times \frac{T_2}{T_1} \times \frac{P_1}{P_2} = 2.75\ \text{L} \times \frac{(273 + 37)\ \text{K}}{(273 + 25)\ \text{K}} \times \frac{742\ \text{torr}}{760\ \text{torr}} = 2.79\ \text{L}$$

9.46 In general the combined gas law equation is:

$$\frac{P_1V_1}{T_1} = \frac{P_2V_2}{T_2}$$

and in particular, for this problem, we have:

$$V_2 = V_1 \times \frac{T_2}{T_1} \times \frac{P_1}{P_2} = 275\ \text{mL} \times \frac{(273 + 30)\ \text{K}}{(273 + 15)\ \text{K}} \times \frac{742\ \text{torr}}{760\ \text{torr}} = 282\ \text{mL}$$

9.47 In general the combined gas law equation is:

$$\frac{P_1V_1}{T_1} = \frac{P_2V_2}{T_2}$$

and in particular, for this problem, we have:

$$T_2 = T_1 \times \frac{P_2}{P_1} \times \frac{V_2}{V_1} = (20 + 273)\ \text{K} \times \frac{375\ \text{torr}}{765\ \text{torr}} \times \frac{9.55\ \text{L}}{6.25\ \text{L}} = 219\ \text{K}$$

$(219\ \text{K} - 273\ \text{K}) = -54\ °\text{C}$

9.48 In general the combined gas law equation is:

$$\frac{P_1V_1}{T_1} = \frac{P_2V_2}{T_2}$$

and in particular, for this problem, we have:

$$T_2 = T_1 \times \frac{P_2}{P_1} \times \frac{V_2}{V_1} = (25 + 273)\ \text{K} \times \frac{2.00\ \text{atm}}{1.50\ \text{atm}} \times \frac{225\ \text{mL}}{445\ \text{mL}} = 201\ \text{K}$$

$201\ \text{K} - 273\ \text{K} = -72\ °\text{C}$

9.49 In general the combined gas law equation is:

$$\frac{P_1V_1}{T_1} = \frac{P_2V_2}{T_2}$$

and in particular, for this problem, we have:

$$P_1 = P_2 \times \frac{V_2}{V_1} \times \frac{T_1}{T_2} = 795 \text{ torr} \times \frac{585 \text{ mL}}{525 \text{ mL}} \times \frac{(273 + 21) \text{ K}}{(273 + 85) \text{ K}} = 727 \text{ torr}$$

9.50 In general the combined gas law equation is:

$$\frac{P_1V_1}{T_1} = \frac{P_2V_2}{T_2}$$

and in particular, for this problem, we have:

$$T_1 = T_2 \times \frac{V_1}{V_2} \times \frac{P_1}{P_2} = 300 \text{ K} \times \frac{675 \text{ mL}}{340 \text{ mL}} \times \frac{1.00 \text{ atm}}{2.00 \text{ atm}} = 298 \text{ K}$$

9.51 First calculate the initial volume (V_1) and the final volume (V_2) of the cylinder using the given geometrical data, noting that the radius is half the diameter (10.8/2 = 5.4 cm):

$$V_1 = 3.14 \times (5.4 \text{ cm})^2 \times 13.3 \text{ cm} = 1.22 \times 10^3 \text{ cm}^3$$

$$V_2 = 3.14 \times (5.4 \text{ cm})^2 \times (13.3 \text{ cm} - 12.7 \text{ cm}) = 54.9 \text{ cm}^3$$

In general the combined gas law equation is:

$$\frac{P_1V_1}{T_1} = \frac{P_2V_2}{T_2}$$

and in particular, for this problem, we have:

$$T_2 = T_1 \times \frac{P_2}{P_1} \times \frac{V_2}{V_1} = 363 \text{ K} \times \frac{34.0 \text{ atm}}{1.00 \text{ atm}} \times \frac{54.9 \text{ cm}^3}{1.22 \times 10^3 \text{ cm}^3} = 555 \text{ K}$$

555 K - 273 K = 282 °C

9.52 First convert the temperature data to the Kelvin scale:

273 + 5/9(61 - 32) = 289 K and 273 + 5/9(105 - 32) = 314 K

Next, calculate the final pressure at the gauge, taking into account the temperature change only:

$$P_2 = P_1 \times \frac{T_2}{T_1} = 64.7 \text{ lb/in.}^2 \times \frac{314 \text{ K}}{289 \text{ K}} = 70.3 \text{ lb/in.}^2$$

This represents the actual pressure inside the tire. The pressure gauge measures only the difference between the pressure inside the tire and the pressure outside the tire (atmospheric pressure). Hence the gauge reading is equal to the internal pressure of the tire less atmospheric pressure:

$(70.3 - 14.7) \text{ lb/in.}^2 = 55.6 \text{ lb/in.}^2$

9.53 Since PV = nRT, then V = nRT/P

$$V = \frac{(1.00 \text{ mol})(0.0821 \text{ L atm/K mol})(298 \text{ K})}{760 \text{ torr} \times \frac{1 \text{ atm}}{760 \text{ torr}}} = 24.5 \text{ L}$$

9.54 Since PV = nRT, then V = nRT/P

$$V = \frac{(1.00 \text{ mol})(0.0821 \text{ L atm/K mol})(323 \text{ K})}{740 \text{ torr} \times \frac{1 \text{ atm}}{760 \text{ torr}}} = 27.2 \text{ L}$$

9.55 Since PV = nRT, then P = nRT/V

$$P = \frac{(4.25 \text{ mol})(0.0821 \text{ L atm/K mol})(293 \text{ K})}{25.0 \text{ L}} = 4.09 \text{ atm}$$

9.56 Since PV = nRT, then P = nRT/V

$$P = \frac{(10.0 \text{ mol})(0.0821 \text{ L atm/K mol})(300 \text{ K})}{1.50 \text{ L}} = 164 \text{ atm}$$

9.57 Since PV = nRT, then P = nRT/V

$$P = \frac{(83.4 \text{ mol})(0.0821 \text{ L atm/K mol})(298 \text{ K})}{10.0 \text{ L}} = 204 \text{ atm}$$

9.58 Since PV = nRT, then P = nRT/V

$$P = \frac{(40.0 \text{ mol})(0.0821 \text{ L atm/K mol})(298 \text{ K})}{4.50 \text{ L}} = 217 \text{ atm}$$

The cylinder has a 500 atm safety limit, and is in no danger of exploding.

9.59 Since PV = nRT, then n = PV/RT

$$n = \frac{(150 \text{ atm})(25.0 \text{ L})}{(0.0821 \text{ L atm/K mol})(24 + 273 \text{ K})} = 154 \text{ mol } N_2$$

9.60 Since PV = nRT, then n = PV/RT

$$n = \frac{(1.00\ \text{atm})(18.4\ \text{L})}{(0.0821\ \text{L atm/K mol})(298\ \text{K})} = 0.752\ \text{mol He remain}$$

9.61 Since PV = nRT, then n = PV/RT

$$n = \frac{(208\ -\ 201\ \text{atm})(14.5\ \text{L})}{(0.0821\ \text{L atm/K mol})(298\ \text{K})} = 4.15\ \text{mol } O_2 \text{ consumed}$$

$$4.15\ \text{mol } O_2 \times \frac{32.0\ \text{g } O_2}{1\ \text{mol } O_2} = 133\ \text{g } O_2 \text{ consumed}$$

9.62 Since PV = nRT, then n = PV/RT

$$n = \frac{(115 \div 2\ \text{atm})(0.855\ \text{L})}{(0.0821\ \text{L atm/K mol})(299\ \text{K})} = 2.00\ \text{mol } H_2 \text{ consumed}$$

$$2.00\ \text{mol } H_2 \times \frac{2.02\ \text{g } H_2}{1\ \text{mol } H_2} = 4.04\ \text{g } H_2 \text{ consumed}$$

9.63 (a) $$\frac{16.0\ \text{g } CH_4}{1\ \text{mol } CH_4} \times \frac{1\ \text{mol}}{22.4\ \text{L at STP}} = 0.714\ \text{g } CH_4 \text{ per L}$$

(b) $$\frac{32.0\ \text{g } O_2}{1\ \text{mol } O_2} \times \frac{1\ \text{mol}}{22.4\ \text{L at STP}} = 1.43\ \text{g } O_2 \text{ per L}$$

(c) $$\frac{2.02\ \text{g } H_2}{1\ \text{mol } H_2} \times \frac{1\ \text{mol}}{22.4\ \text{L at STP}} = 0.0902\ \text{g } H_2 \text{ per L}$$

9.64 (a) $$\frac{4.00\ \text{g He}}{1\ \text{mol He}} \times \frac{1\ \text{mol}}{22.4\ \text{L at STP}} = 0.179\ \text{g He/L} = 0.179\ \text{mg/mL}$$

(b) $$\frac{28.0\ \text{g } N_2}{1\ \text{mol } N_2} \times \frac{1\ \text{mol}}{22.4\ \text{L at STP}} = 1.25\ \text{g } N_2 \text{ per L} = 1.25\ \text{mg/mL}$$

(c) $$\frac{30.1\ \text{g } C_2H_6}{1\ \text{mol } C_2H_6} \times \frac{1\ \text{mol}}{22.4\ \text{L at STP}} = 1.34\ \text{g } C_2H_6 \text{ per L} = 1.34\ \text{mg/mL}$$

9.65 In general PV = nRT, where n = mass ÷ f.wt.

Thus PV = mass/(f.wt.) × RT

and we arive at the formula for the density of a gas:

D = (f.wt.)(P)/RT

$$D = \frac{(32.0\ g/mol)(745/760\ atm)}{(0.0821\ L\ atm/K\ mol)(298\ K)} = 1.28\ g/L \text{ for } O_2$$

9.66 In general PV = nRT, where n = mass ÷ f.wt.

Thus PV = mass/(f.wt.) × RT

and we arive at the formula for the density of a gas:

D = (f.wt.)(P)/RT

$$D = \frac{(39.9\ g/mol)(752/760\ atm)}{(0.0821\ L\ atm/K\ mol)(293\ K)} = 1.64\ g/L \text{ for Ar}$$

9.67 If PV = nRT, then PV = mass/(f.wt.) × RT

Hence f.wt. = (mass)RT/PV = mass/V × RT/P = D × RT/P

and, if we are given the density, the f.wt. can be calculated:

$$f.wt. = \frac{(1.14\ g/L)(0.0821\ L\ atm/K\ mol)(295\ K)}{(755/760\ atm)} = 27.8\ g/mol$$

9.68 If PV = nRT, then PV = mass/(f.wt.) × RT

Hence f.wt. = (mass)RT/PV = mass/V × RT/P = D × RT/P

and, if we are given the density, the f.wt. can be calculated:

$$f.wt. = \frac{(1.76\ g/L)(0.0821\ L\ atm/K\ mol)(297\ K)}{(741/760\ atm)} = 44.0\ g/mol$$

9.69 If PV = nRT, then PV = mass/(f.wt.) × RT

Hence f.wt. = (mass)RT/PV = mass/V × RT/P = D × RT/P

and, if we are given the density, the f.wt. can be calculated:

$$f.wt. = \frac{(12.1\ g)(0.0821\ L\ atm/K\ mol)(298\ K)}{(255\ L)(10/760\ atm)} = 88.2\ g/mol$$

9.70 (a) If PV = nRT, then PV = mass/(f.wt.) × RT

Hence f.wt. = (mass)RT/PV = mass/V × RT/P = D × RT/P

and, if we are given the density, the f.wt. can be calculated:

$$f.wt. = \frac{(6.3\ mg/385\ mL)(0.0821\ L\ atm/K\ mol)(298\ K)}{(11/760\ atm)} = 28\ g/mol$$

(b) The formula weights of the boron hydrides are:

BH_3, 13.8

B_2H_6, 27.6

B_4H_{10}, 53.3

and we conclude that the sample must have been B_2H_6.

9.71 When gases are held at the same temperature and pressure, and dispensed in this fashion during chemical reactions, then they react in a ratio of volumes that is equal to the ratio of the coefficients (moles) in the balanced chemical equation for the given reaction. We can, therefore, directly use the stoichiometry of the balanced chemical equation to determine the combining ratio of the gas volumes:

$$5.00 \text{ L } H_2 \times \frac{1 \text{ volume } Cl_2}{1 \text{ volume } H_2} = 5.00 \text{ L } Cl_2$$

9.72 When gases are held at the same temperature and pressure, and dispensed in this fashion during chemical reactions, then they react in a ratio of volumes that is equal to the ratio of the coefficients (moles) in the balanced chemical equation for the given reaction. We can, therefore, directly use the stoichiometry of the balanced chemical equation to determine the combining ratio of the gas volumes:

$$45.0 \text{ L } H_2 \times \frac{1 \text{ volume } N_2}{3 \text{ volumes } H_2} = 15.0 \text{ L } N_2$$

9.73 When gases are held at the same temperature and pressure, and dispensed in this fashion during chemical reactions, then they react in a ratio of volumes that is equal to the ratio of the coefficients (moles) in the balanced chemical equation for the given reaction. We can, therefore, directly use the stoichiometry of the balanced chemical equation to determine the combining ratio of the gas volumes:

$$48 \text{ L } NH_3 \times \frac{5 \text{ volumes } O_2}{4 \text{ volumes } NH_3} = 60 \text{ L } O_2$$

9.74

$$14.0 \text{ g } C_2H_4 \times \frac{1 \text{ mol } C_2H_4}{28.1 \text{ g } C_2H_4} = 0.498 \text{ mol } C_2H_4$$

$$0.498 \text{ mol } C_2H_4 \times \frac{1 \text{ mol } H_2}{1 \text{ mol } C_2H_4} \times \frac{22.4 \text{ L at STP}}{1 \text{ mol}} = 11.2 \text{ L } H_2 \text{ at STP}$$

9.75 (a) $Zn(s) + 2HCl(aq) \rightarrow H_2(g) + ZnCl_2(aq)$

(b) Calculate the number of moles of hydrogen:

$$n = PV/RT = \frac{(1.00 \text{ atm})(10.0 \text{ L})}{(0.0821 \text{ L atm/K mol})(298 \text{ K})} = 0.409 \text{ mol } H_2$$

and the number of moles of zinc:

$$0.409 \text{ mol H} \times \frac{1 \text{ mol Zn}}{1 \text{ mol } H_2} = 0.409 \text{ mol Zn}$$

The number of grams of zinc that are needed are, therefore:

$$0.409 \text{ mol Zn} \times \frac{65.4 \text{ g Zn}}{1 \text{ mol Zn}} = 26.7 \text{ g Zn}$$

(c)

$$0.409 \text{ mol } H_2 \times \frac{2 \text{ mol HCl}}{1 \text{ mol } H_2} = 0.818 \text{ mol HCl are needed}$$

(d)

$$0.818 \text{ mol HCl} \times \frac{1.00 \text{ L HCl}}{6.00 \text{ mol HCl}} = 0.136 \text{ L} = 136 \text{ mL of } 6.00 \text{ M HCl}$$

9.76 $P_{total} = 745 \text{ torr} = P_{H_2} + P_{water}$

The vapor pressure of water at 20 °C is available in Table 9.1: 17.54 torr.
Hence:

$$P_{H_2} = (745 - 18) \text{ torr} = 727 \text{ torr}$$

Next, we calculate the number of moles of hydrogen gas that this represents:

$$n = \frac{(727/760 \text{ atm})(0.325 \text{ L})}{(0.0821 \text{ L atm/K mol})(293 \text{ K})} = 0.0129 \text{ mol } H_2$$

The balanced chemical equation is:

$$Zn(s) + 2HCl(aq) \rightarrow H_2(g) + ZnCl_2(aq)$$

and the quantities of the reagents that are needed are:

$$0.0129 \text{ mol } H_2 \times \frac{1 \text{ mol Zn}}{1 \text{ mol } H_2} \times \frac{65.4 \text{ g Zn}}{1 \text{ mol Zn}} = 0.844 \text{ g Zn}$$

$$0.0129 \text{ mol } H_2 \times \frac{2 \text{ mol HCl}}{1 \text{ mol } H_2} \times \frac{1.00 \text{ L HCl}}{8.00 \text{ mol HCl}} = 0.00323 \text{ L} = 3.23 \text{ mL of HCl}$$

9.77 (a)

$$225 \text{ kg } (NH_4)_2SO_4 \times \frac{1000 \text{ g}}{1 \text{ kg}} \times \frac{1 \text{ mol } (NH_4)_2SO_4}{132 \text{ g } (NH_4)_2SO_4} \times \frac{2 \text{ mol } NH_3}{1 \text{ mol } (NH_4)_2SO_4}$$

$$\times \frac{22.4 \text{ L at STP}}{1 \text{ mol}} = 7.64 \times 10^4 \text{ L } NH_3 \text{ at STP}$$

(b)

$$7.64 \times 10^4 \text{ L } NH_3 \times \frac{1 \text{ mol}}{22.4 \text{ L STP}} \times \frac{1 \text{ mol } H_2SO_4}{2 \text{ mol } NH_3} = 1.71 \times 10^3 \text{ mol } H_2SO_4$$

(c)

$$1.71 \times 10^3 \text{ mol } H_2SO_4 \times \frac{1 \text{ L } H_2SO_4}{6 \text{ mol } H_2SO_4} = 285 \text{ L of 6 M } H_2SO_4$$

9.78 $P_{total} = 755 \text{ torr} = P_{CO_2}s + P_{water}$

The vapor pressure of water at 20 °C is available in Table 9.1: 17.54 torr.
Hence:

$$P_{CO_2} = (755 - 18) \text{ torr} = 737 \text{ torr}$$

Next, we calculate the number of moles of carbon dioxide gas that this represents:

$$n = \frac{(737/760 \text{ atm})(0.475 \text{ L})}{(0.0821 \text{ L atm/K mol})(293 \text{ K})} = 0.0191 \text{ mol } CO_2$$

Next we determine the mass of the reagents needed:

$$0.0191 \text{ mol } CO_2 \times \frac{1 \text{ mol } CaCO_3}{1 \text{ mol } CO_2} \times \frac{100 \text{ g } CaCO_3}{1 \text{ mol } CaCO_3} = 1.91 \text{ g } CaCO_3$$

$$0.0191 \text{ mol } CO_2 \times \frac{2 \text{ mol HCl}}{1 \text{ mol } CO_2} \times \frac{1 \text{ L HCl}}{6.00 \text{ mol HCl}} = 0.00637 \text{ L HCl} = 6.37 \text{ mL HCl}$$

9.79 (a)

$$28.3 \times 10^3 \text{ L } CH_4 \times \frac{16.0 \text{ mol } NH_3}{7.00 \text{ mol } CH_4} = 6.47 \times 10^4 \text{ L } NH_3$$

(b)

$$6.47 \times 10^4 \text{ L } NH_3 \times \frac{1 \text{ mol}}{22.4 \text{ L}} = 2.89 \times 10^3 \text{ mol } NH_3$$

(c)

$$2.89 \times 10^3 \text{ mol } NH_3 \times \frac{17.0 \text{ g } NH_3}{1 \text{ mol } NH_3} \times \frac{1 \text{ kg } NH_3}{1000 \text{ g } NH_3} = 49.1 \text{ kg } NH_3$$

9.80 (a)

$$100 \text{ g } CaC_2 \times \frac{1 \text{ mol } CaC_2}{64.1 \text{ g } CaC_2} \times \frac{1 \text{ mol } C_2H_2}{1 \text{ mol } CaC_2} = 1.56 \text{ mol } C_2H_2$$

$$1.56 \text{ mol } C_2H_2 \times \frac{22.4 \text{ L}}{1 \text{ mol}} = 34.9 \text{ L } C_2H_2$$

(b)

$$1.00 \times 10^6 \text{ L } C_2H_2 \times \frac{1 \text{ mol}}{22.4 \text{ L}} \times \frac{1 \text{ mol } CaC_2}{1 \text{ mol } C_2H_2} = 4.46 \times 10^4 \text{ mol } CaC_2$$

$$4.46 \times 10^4 \text{ mol } CaC_2 \times \frac{64.1 \text{ g } CaC_2}{1 \text{ mol } CaC_2} \times \frac{1 \text{ kg}}{1000 \text{ g}} = 2.86 \times 10^3 \text{ kg } CaC_2$$

9.81 (a) First determine the % by weight S and O in the sample:

% S = 1.448 g/3.620 g × 100 = 40.00

% O = 2.172 g/3.620 g × 100 = 60.00

(b) Next, determine the number of moles of S and O in a sample of the material weighing 100 g exactly, in order to make the conversion from % by weight to grams straightforward:

In 100 g of the material, there are 40.00 g S and 60.00 g O:

40.00 g S ÷ 32.01 g/mol = 1.250 mol S

60.00 g O ÷ 16.00 g/mol = 3.750 mol O

Dividing each of these mole amounts by the smaller of the two gives the relative mole amounts of S and O in the material:

for S, 1.250 ÷ 1.250 = 1.000 relative moles

for O, 3.750 ÷ 1.250 = 3.000 relative moles

and the empirical formula is, therefore, SO_3.

(c) We determine the formula weight of the material by use of the ideal gas law:

$$n = PV/RT = \frac{(0.987 \text{ atm})(1.12 \text{ L})}{(0.0821 \text{ L atm/K mol})(298 \text{ K})} = 0.0452 \text{ mol}$$

The formula weight is given by the mass in grams (given in the problem) divided by the moles determined here:

f.wt. = 3.620 g ÷ 0.0452 mol = 80.1

Since this is equal to the formula weight of the empirical unit determined in step (b) above, namely SO_3, then the molecular formula is also SO_3.

9.82 We proceed as in review exercise 9.81:

(a) First determine the % by weight C and H in the sample:

% C = 1.389 g/1.620 g × 100 = 85.74

% H = 0.2314 g/1.620 g × 100 = 14.28

(b) Next, determine the number of moles of C and H in a sample of the material weighing 100 g exactly, in order to make the conversion from % by weight to grams straightforward:

In 100 g of the material, there are 85.74 g C and 14.28 g H:

85.74 g C ÷ 12.01 g/mol = 7.139 mol C

14.28 g H ÷ 1.008 g/mol = 14.17 mol H

Dividing each of these mole amounts by the smaller of the two gives the relative mole amounts of C and H in the material:

for C, 7.139 ÷ 7.139 = 1.000 relative moles

for H, 14.17 ÷ 7.139 = 1.985 relative moles

and the empirical formula is, therefore, CH_2.

(c) We determine the molecular weight of the material by use of the ideal gas law:

$$n = PV/RT = \frac{(0.984 \text{ atm})(0.941 \text{ L})}{(0.0821 \text{ L atm/K mol})(293 \text{ K})} = 0.0385 \text{ mol}$$

The molecular weight is given by the mass in grams (given in the problem) divided by the moles determined here:

mol.wt. = 1.620 g ÷ 0.0385 mol = 42.1

Since this is equal to some whole number multiple of the formula weight of the empirical unit determined in step (b) above, namely CH_2 (f.wt. = 14.0), then it follows that:

$42.1 = n \times 14.0, \qquad \therefore n = 3$

and the molecular formula is three times the empirical formula, namely C_3H_6.

9.83 (a) $P_{total} = 746.0 \text{ torr} = P_{H_2O} + P_{N_2}$

$P_{N_2} = 746.0 \text{ torr} - 22.1 \text{ torr} = 723.9 \text{ torr}$

Now use the ideal gas equation to determine the moles of N_2 that have been collected:

$$n = \frac{(723.9/760 \text{ atm})(0.01891 \text{ L})}{(0.0821 \text{ L atm/K mol})(297 \text{ K})} = 7.39 \times 10^{-4} \text{ mol } N_2$$

Then the mass of nitrogen that has been collected is determined:

$7.39 \times 10^{-4} \text{ mol } N_2 \times 28.0 \text{ g/mol} = 2.07 \times 10^{-2} \text{ g } N_2$

Next the % by weight nitrogen in the material is calculated:

$\% \text{ N} = (0.0207 \text{ g})/(0.2394 \text{ g}) \times 100 = 8.65 \% \text{ N}$

(b) mass of C in the sample:

$$17.57 \times 10^{-3} \text{g } CO_2 \times \frac{12.01 \text{ g C}}{44.01 \text{ g } CO_2} = 4.795 \times 10^{-3} \text{ g C}$$

mass of H in the sample:

$$4.319 \times 10^{-3} \text{ g } H_2O \times \frac{2.016 \text{ g H}}{18.02 \text{ g } H_2O} = 4.832 \times 10^{-4} \text{ g H}$$

mass of N in the sample:

$8.65 \% \text{ of } 6.478 \times 10^{-3} \text{ g total} = 5.60 \times 10^{-4} \text{ g N}$

mass of O in the sample = total mass - (mass C + H + N)

mass of O = 6.40×10^{-4} g O

Next we convert each of these mass amounts into the corresponding mole values:

for C, $4.795 \times 10^{-3} \text{ g} \div 12.01 \text{ g/mol} = 3.993 \times 10^{-4} \text{ mol C}$

for H, $4.832 \times 10^{-4} \text{ g} \div 1.008 \text{ g/mol} = 4.794 \times 10^{-4} \text{ mol H}$

for N, $5.60 \times 10^{-4} \text{ g} \div 14.01 \text{ g/mol} = 4.00 \times 10^{-5} \text{ mol N}$

for O, 6.40×10^{-4} g ÷ 16.00 g/mol = 4.00×10^{-5} mol O

Last, we convert these mole amounts into relative mole amounts by dividing each by the smallest of the four:

for C, $3.993 \times 10^{-4} / 4.00 \times 10^{-5} = 9.98$

for H, $4.794 \times 10^{-4} / 4.00 \times 10^{-5} = 12.0$

for N, $4.00 \times 10^{-5} / 4.00 \times 10^{-5} = 1.00$

for O, $4.00 \times 10^{-5} / 4.00 \times 10^{-5} = 1.00$

The empirical formula is therefore $C_{10}H_{12}N_1O_1$

(c) The formula weight of the empirical unit is 162. Since this is half the value of the known molecular weight, the molecular formula must be twice the empirical formula, $C_{20}H_{24}N_2O_2$.

9.84 (a) The equation can be rearranged to give:

$$0.04489 \times \frac{V(P - P_{H_2O})}{(273 + °C)} = \%N \times W$$

This means that the left side of the above equation should be obtainable simply from the ideal gas law, applied to the nitrogen case. If PV = nRT, then for nitrogen:

PV = (mass nitrogen)/(28.01 g/mol) × RT

and the mass of nitrogen that is collected is given by:

(mass nitrogen) = PV(28.01)/RT,

where R = 82.1 mL atm/K mol × 760 torr/atm = 6.24×10^4 mL torr/K mol

Using this value for R in the above equation, we have the following result for the mass of nitrogen, remembering that the pressure of nitrogen is less than the total pressure, by an amount equal to the vapor pressure of water:

$$\text{(mass nitrogen)} = \frac{(28.01) \times V \times (P_{total} - P_{H_2O})}{(6.24 \times 10^4 \text{ mL torr/K mol})(273 + °C)}$$

Finally, it is only necessary to realize that the value

$$\frac{(28.01)}{6.24 \times 10^4} \times 100 = 0.04489$$

is exactly the value given in the problem.

(b)
$$\%N = \frac{(0.04489)(18.90)(723.9)}{(0.2394)(296.95)} = 8.64\ \%N$$

Kinetic Theory of Gases

9.85 A gas consists of hard, round particles in random motion, and the particles neither attract nor repel one another.

9.86 (a) T ∝ average kinetic energy of the particles

(b) T ∝ PV

9.87 The net effect of any migration must cause the distribution of the gas particles to become increasingly random. This requires that the entirely random motions and collisions of the particles lead to a random distribution of the particles, i.e. the particles will spread out into the largest volume possible.

9.88 the randomness of the motion of the particles

9.89 The temperature will decrease.

9.90 The increase in temperature requires an increase in kinetic energy. This can happen only if the gas velocities increase. Higher velocities cause the gas particles to strike the walls of the container with more force, and this in turn causes the container to expand.

9.91 As in the answer to review exercise 9.90, the increase in temperature causes an increase in the force with which the gas particles strike the container walls. If the container cannot expand, an increase in pressure must result.

9.92 It is not true that the gas particles occupy no volume themselves, apart from the volume between the gas particles. Also, it is not true that the gas particles exert no force on one another.

9.93 the ideal gas law

9.94 A large value for the constant $\underline{b}$ suggests that the gas molecules are large.

9.95 A small value for the constant $\underline{a}$ suggests that the gas molecules have weak forces of attraction among themselves.

CHAPTER TEN

PRACTICE EXERCISES

1. The number of molecules in the vapor will increase, and the number of molecules in the liquid will decrease, but the sum of the molecules in the vapor and the liquid remains the same.

2. Adding heat will shift the equilibrium to the right, producing more vapor. This increase in the amount of vapor causes a corresponding increase in the pressure, such that the vapor pressure generally increases with increasing temperature.

3. We use the curve for water, and find that at 330 torr, the boiling point is approximately 75 °C.

4. Because this is a high melting, hard material, it must be a covalent or network solid. Covalent bonds link the various atoms of the crystal.

5. Since the melt does not conduct electricity, it is not an ionic substance. The softness and the low melting point suggest that this is a molecular solid, and indeed the formula is most properly written S_8.

6. Refer to the phase diagram for water, Figure 10.30 on page 378 of the text. We "move" along a horizontal line marked for a pressure of 2.15 torr. At -20 °C, the sample is a solid. If we bring the temperature from -20 °C to 50 °C, keeping the pressure constant at 2.15 torr, the sample becomes a gas. The process is thus solid → gas, i.e. sublimation.

7. As diagramed in Figure 10-30, this is a liquid.

REVIEW EXERCISES

Comparing the States of Matter

10.1 student answer

10.2 This would be the condition that minimizes intermolecular interactions, high temperature and low pressure.

10.3 Gases are able to behave ideally, that is behave in ways that are unrelated to chemical composition, because the molecules of a gas sample are far apart (ideally), and intermolecular interactions are negligible.

10.4 Intermolecular forces in liquids and solids are more important than in gases because the molecules and atoms of liquid and solid samples are so much closer together than is true in a gas.

Intermolecular Attractions

10.5 Dipole-dipole interactions arise from the attraction of the permanent dipole moment of one molecule with that of an adjacent molecule, the positive end of one dipole being drawn to the negative end of the other dipole. This is diagramed in Figure 10.2 of the text.

10.6 London forces are diagramed on page 352. These weak forces of attraction are caused by instantaneous dipoles that attract induced dipoles in neighboring molecules. London forces increase in strength with increasing molecular size, as illustrated in Figure 10.4 and as shown by the data of Table 10.2 and Table 10.3.

10.7 Hydrogen bonds are special types of dipole-dipole attractions, and they have greater strength than other types of intermolecular forces. A hydrogen bond arises when a hydrogen atom is covalently bonded to fluorine, oxygen, or nitrogen. Compounds in which hydrogen bonding is important have boiling points that are higher than might be otherwise expected.

10.8 fluorine, oxygen and nitrogen

10.9 Since these are both nonpolar molecular substances, the only type of intermolecular force that we need to consider is London forces. The larger molecule has the greater London force of attraction and hence the higher boiling point: C_8H_{18}.

10.10 Whereas the ether has no O-H linkage, ethanol does. Therefore ethanol can have hydrogen bonding between molecules and ether cannot. Ethanol thus has stronger intermolecular forces, and its boiling point is consequently higher.

10.11 Covalent bonds are normally about 100 times stronger than normal dipole-dipole attractions; hydrogen bonds are about 5-10 times stronger than dipole-dipole attractions.

10.12 The instantaneous dipole moment in an otherwise nonpolar substance arises from a momentary imbalance in the electron distribution within the molecule. This creates an induced dipole in a neighboring molecule.

10.13 London forces are possible in them all. Where another intermolecular force can operate, it is generally stronger than London forces, and this other type of interaction overshadows the importance of the London force.

The substances in the list that can have dipole-dipole attractions are those with permanent dipole moments: (a), (c), and (e). Both CS_2 and SF_6 are nonpolar molecular substances.

General Properties of Liquids and Solids

10.14 Physical properties that depend on tightness of packing: compressibility and diffusion.

Physical properties that depend on the strengths of intermolecular interactions: retention of volume and shape, surface tension, and evaporation.

10.15 The particles of a gas are free to move randomly, and thus to diffuse readily. Diffusion in liquids is comparatively slower because of the more numerous collisions that a molecule in a liquid sample must undergo in traveling from place to place. The particles of a solid are not free to move from place to place in a solid sample.

10.16 The transfer of a gas from one container to another may be accompanied by either a change in shape or volume, or both. Only the shape of a liquid may change when its container is altered; the volume of a liquid does not change when the liquid is transferred to a new container. A solid changes neither its shape nor its volume when it is transferred into a new container.

10.17 Liquids and solids are difficult to compress because there is not much extra space between the particles of the solid or liquid. The compressibility of a gas, however, arises because there may be a great deal of room between the particles, and compression reduces the amount of this empty space.

10.18 The rate of diffusion should increase because the molecules move faster at the higher temperature.

10.19 Surface tension is related to the energy needed to increase the surface area of a liquid. Molecules at the surface of a liquid have no other molecules above them, and they consequently are attracted only to those molecules that are next to them - namely, those in the interior of the liquid. This is illustrated in Figure 10.8.

10.20 Surface tension is a property that causes a liquid to wet another surface and to form rounded droplets.

10.21 The greater the intermolecular attractions, the greater the surface tension.

10.22 Water should have the greater surface tension because it has the stronger intermolecular force, i.e. hydrogen bonding.

10.23 Wetting - spreading a liquid across a surface.
Surfactant - a substance that lowers surface tension in a liquid and thereby promotes wetting.

10.24 There is no intermolecular force common to both polyethylene and water that can allow for wetting. The surface tension of water, which is high, is not disrupted by any effective interaction between water and polyethylene.

10.25 Glycerin ought to wet the surface of glass quite nicely, because the oxygen atoms at the surface of glass can form effective hydrogen bonds to the O-H groups of glycerin.

10.26 Since it is the high energy molecules in a sample that are the first to evaporate, the remaining molecules have a lower average kinetic energy. A reduction in kinetic energy corresponds to a decrease in temperature.

10.27 Raising the temperature of the sample increases the fraction of molecules in the sample that have enough kinetic energy to escape by evaporation.

10.28 An increase in surface area causes an increase in the rate of evaporation because there are more molecules in position at the surface, and that are hence capable of evaporation. The stronger the intermolecular forces, the less readily a substance can evaporate.

10.29 The snow evaporates by sublimation.

10.30 Freeze drying is accomplished by sublimation. It offers the advantage that the sample does not have to be subjected to high temperatures.

10.31 The rate of evaporation of a liquid increases with increasing temperature.

Changes of State and Equilibrium

10.32 A change of state is a change from one physical form to another, e.g. from a solid to a liquid.

10.33 This happens because of the loss in kinetic energy that the colliding molecule experiences when it hits the surface molecules. The colliding molecule has less kinetic energy after striking the surface, and its ability to escape subsequently from the liquid is momentarily diminished.

10.34 A dynamic equilibrium is established if the liquid evaporates into a sealed container. It is termed a dynamic equilibrium because opposing processes (evaporation and condensation) continue to take place, once the condition of equilibrium has been achieved. At equilibrium, the rate of condensation is equal to the rate of evaporation, and there is consequently no net change in the number of molecules in the vapor or in the liquid.

10.35 A dynamic equilibrium is achieved when a solid is held at its melting temperature.

10.36 This is the melting point.

Energy Changes that Accompany Changes of State

10.37 (a) molar heat of vaporization
(b) molar heat of sublimation
(c) molar heat of fusion

10.38

$$1.00 \text{ kg} \times \frac{1000 \text{ g}}{1 \text{ kg}} \times \frac{1 \text{ mol}}{18.0 \text{ g}} \times \frac{43.9 \text{ kJ}}{1 \text{ mol}} = 2.44 \times 10^3 \text{ kJ}$$

10.39

$$1.00 \text{ g} \times \frac{1 \text{ mol}}{58.1 \text{ g}} \times \frac{30.3 \text{ kJ}}{1 \text{ mol}} = 0.522 \text{ kJ}$$

10.40 The heat of vaporization of a molecular substance is generally larger than the heat of fusion, because in vaporization, the molecules undergo much larger changes in their distance of separation (and require the disruption of much stronger intermolecular forces) than is true of melting. The heat of sublimation is typically larger than the heat of vaporization of a liquid because sublimation involves a greater change in intermolecular separation, a larger disruption of intermolecular forces of attraction, and hence a larger change in potential energy.

10.41 The heat of condensation is exothermic, and it is equal in magnitude to the heat of vaporization (which is endothermic).

10.42 The condensation process is exothermic. The energy that is released warms the air, which then rises. This warming and rising of the air creates wind currents.

10.43 The substance with the larger heat of vaporization has the stronger intermolecular forces. This is ethanol, which has hydrogen bonding, whereas ethyl acetate does not.

10.44 In general, the enthalpy of sublimation is equal to the sum of the enthalpy change for fusion and vaporization. This can be seen by adding the equations for fusion and vaporization:

$HC_2H_3O_2(s) \rightarrow HC_2H_3O_2(\ell), \quad \Delta H°_{fus} = 10.8$ kJ/mol

$HC_2H_3O_2(\ell) \rightarrow HC_2H_3O_2(g), \quad \Delta H°_{vap} = 24.3$ kJ/mol

---

$HC_2H_3O_2(s) \rightarrow HC_2H_3O_2(g), \quad \Delta H°_{sub} = 35.1$ kJ/mol

10.45 We can approach this problem by first asking either of two equivalent questions about the system: how much heat energy (q) is needed in order to melt the entire sample of solid water (100 g) or, how much energy is lost when the liquid water (50.0 g) is cooled to the freezing point? Regardless, there is only one final temperature for the combined (150.0 g) sample, and we need to know if this temperature is at the melting point (0 °C, at which temperature some solid water remains in equilibrium with a certain amount of liquid water) or above the melting point (at which temperature all of the solid water will have melted).

Heat flow on melting all of the solid water:

$$q = 6.01 \text{ kJ/mole} \times 100 \text{ g} \times 1 \text{ mol}/18.0 \text{ g} = 33.4 \text{ kJ}$$

Heat flow on cooling the liquid water to the freezing point:

$$q = 50.0 \text{ g} \times 4.18 \text{ J/g °C} \times 80 \text{ °C} = 1.67 \times 10^4 \text{ J} = 16.7 \text{ kJ}$$

The lesser of these two values is the correct one, and we conclude that 16.7 kJ of heat energy will be transferred from the liquid to the solid, and that the final temperature of the mixture will be 0 °C. The system will be an equilibrium mixture weighing 150 g and having some solid and some liquid in equilibrium with one another. The amount of solid that must melt in order to decrease the temperature of 50.0 g of water from 80 °C to 0 °C is:

$$16.7 \text{ kJ} \div 6.01 \text{ kJ/mol} = 2.78 \text{ mol of solid water}$$

$$2.78 \text{ mol} \times 18.0 \text{ g/mol} = 50.0 \text{ g of water must melt.}$$

10.46 Steam can have a temperature greater than 100 °C.

10.47 $CF_4 < CH_4 < HCl < HF$

Vapor Pressure

10.48 Equilibrium vapor pressure is the pressure exerted by a vapor that is in equilibrium with its liquid. It is a dynamic equilibrium because events have not ceased. Liquid continues to evaporate once the state of equilibrium has been reached, but the rate of evaporation is equal to the rate of condensation. These two opposing processes occur at equal rates, such that there is no further change in the amount of either the liquid or the gas.

10.49 Changing the volume only upsets the equilibrium for a moment. After sufficient time has elapsed, the rates of evaporation and condensation again become equal to one another, and the same condition of equilibrium is achieved. The vapor pressure (or the ease of evaporation) only depends on the strength of intermolecular forces in the liquid sample.

10.50 Raising the temperature increases the vapor pressure by imparting enough kinetic energy (for evaporation) to more of the liquid molecules.

10.51 ether < acetone < benzene < water < acetic acid

10.52 Air with 100% humidity is saturated with water vapor, meaning that the partial pressure of water in the air has become equal to the vapor pressure of water at that temperature. Since vapor pressure increases with increasing temperature, so too should the total amount of water vapor in humid (saturated) air increase with increasing temperature.

10.53 As the air rises, it becomes cooler, and eventually the amount of moisture in the air becomes greater than is required for equilibrium with the liquid. The air is less able to hold a given amount of vapor at the lower temperature, and condensation occurs.

10.54 At the temperature of the cool glass, the equilibrium vapor pressure of the water is lower than the partial pressure of water in the air. The air in contact with the cool glass is induced to relinquish some of its water, and condensation occurs.

10.55 In humid air, the rate of condensation on the skin is more nearly equal to the rate of evaporation from the skin, and the net rate of evaporation of perspiration from the skin is low. The cooling effect of the evaporation of perspiration is low, and our bodies are cooled only slowly under such conditions. In dry air, however, perspiration evaporates more rapidly, and the cooling effect is high.

10.56 Although the cold air outside the building may be nearly saturated at the low temperature, the same air inside at a higher temperature is not now saturated. In fact the water content of the air at the higher temperature may be only a fraction of the maximum % humidity for that temperature. The % humidity of the air at the indoor temperature is, therefore, comparatively low.

10.57 The equilibrium vapor pressure is governed only by the strength of the attractive forces within the liquid and by the temperature.

10.58 As the cooling takes place, the average kinetic energy of the gas molecules decreases, and the attractive forces that can operate among the gas molecules become able to bind the various molecules together in a process that leads to condensation. The air thus loses much of its moisture on the ascending side of the mountain. On descending the other side of the mountain, the air is compressed, and the temperature rises according to Charles' Law. The relative humidity of the air is now very low for two reasons: much of the moisture was released on the other side of the mountain range, and now that the temperature is higher, there is far less than the maximum allowable water content.

Le Chatelier's Principle

10.59 See page 367.

10.60 By "position of equilibrium" we mean the relative amounts of the various reactants and products that exist in the equilibrium mixture.

10.61 This is an endothermic system, and adding heat to the system will shift the position of the equilibrium to the right, producing a new equilibrium mixture having more liquid and less solid. Some of the solid melts when heat is added to the system.

Boiling Points of Liquids

10.62 Boiling point - the temperature at which the vapor pressure of a liquid has become equal to the prevailing atmospheric pressure.

Normal boiling point - the temperature at which the vapor pressure of a liquid is equal to 1 atm, or the boiling point when the atmospheric pressure is 1 atm.

10.63 This happens because boiling is a process that must occur in opposition to pressure. Since the vapor pressure varies with temperature, the boiling point must also change as the pressure changes.

10.64 at about 73 °C

10.65 Boiling point is an easily measured property that varies considerably from one substance to another. It is thus characteristic of a substance, and useful as an identifying property of various substances.

10.66 diethyl ether < ethanol < water < ethylene glycol

10.67 In a sealed container, the water reaches the boiling point only at a temperature that is higher than the normal boiling point. This is because the vaporization of the liquid into a closed container causes an increase in pressure and a corresponding increase in the boiling point. This means that the food is cooked at a higher temperature.

10.68 Even at higher temperatures, the contents of the radiator do not boil, because the pressure in the system increases, since the system is closed. The boiling point of the liquid is higher since the pressure is higher.

10.69 Inside the lighter, the liquid butane is in equilibrium with its vapor, which exerts a pressure somewhat above normal atmospheric pressure. If the liquid butane were spilled on a desk top, it would vaporize (boil) because it would be at a pressure of about 1 atm only, which is somewhat less than the vapor pressure of butane. The intermolecular forces in butane are weak.

10.70 Water should have the higher heat of vaporization; its boiling point is higher and, hence, its intermolecular forces are stronger.

10.71 Since H_2Se is larger than H_2S, its London forces are stronger than in H_2S.

Since water is capable of hydrogen bonding, whereas H_2S is not, its boiling point is higher than that of H_2S.

10.72 The hydrogen bond network in HF is less extensive than in water, because it is a monohydride not a dihydride.

Crystalline Solids and X-Ray Diffraction

10.73 Crystals possess flat surfaces that intersect at characteristic angles.

10.74 Crystal lattice - an infinite and uniform array of particles in crystalline solid.

Unit cell - the simplest portion of a crystal lattice, having all of the geometrical properties of the substance's crystal lattice, such that the entire lattice of the solid can be generated by repeated use of the unit cell only.

10.75 Simple cubic - possesses lattice points only at the eight corners of a cube.

Face-centered cubic - possesses the eight lattice points of the simple cubic cell, plus one in the center of each of the six faces of the cube.

Body-centered cubic - possesses the eight lattice points of the simple cubic cell, plus one at the center of the cube.

10.76 The following diagram of one layer in NaCl depicts the face-centered arrangement of Na^+ ions, which are circled. The Cl^- ions are also arranged in a face-centered fashion.

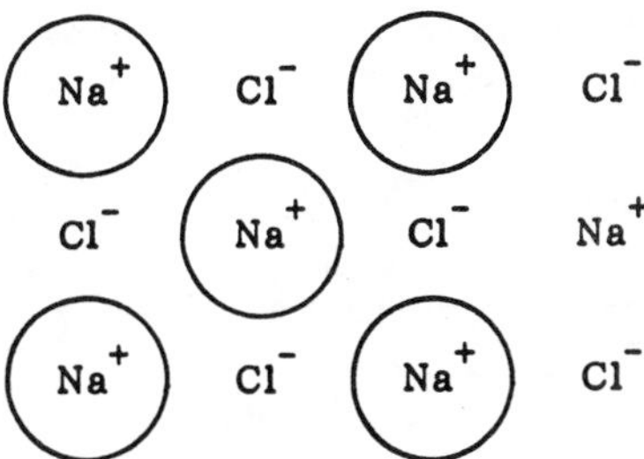

10.77 These structures are both of the face-centered cubic variety. They differ only in the length of an edge for a unit cell, that is the length of the cube edge is different in the two metals. Silver might be expected to have a face-centered unit cell, also.

10.78 Although there are only fourteen different kinds of lattice geometries that can fill space, there are essentially an infinite variety of cell dimensions that can be adopted by substances.

10.79 $n\lambda = 2d \sin\theta$

n = an integer (1, 2, 3, . . .)
λ = wavelength of the X rays
d = the interplane spacing in the crystal
θ = the angle of incidence and the angle of reflectance of X rays to the various crystal planes.

10.80 By measuring the various angles θ, one can compute the d-spacings between planes of atoms in the crystal lattice. This, plus the intensities of the reflected X rays, is used to deduce the locations of the atoms in the unit cell. Some chemical intuition is then needed in order to decide which atoms are bonded together.

10.81 Each atom of a cubic unit cell is shared by eight unit cells. This means that only one eighth of each corner atom can be assigned to a given unit cell. Eight corner atoms times 1/8 each assigned to a given unit cell yields one atom per unit cell.

10.82 A cube has six faces and eight corners. Each of the six face atoms is shared by two adjacent unit cells: $6 \times 1/2 = 3$ atoms. The eight corner atoms are each shared by eight unit cells: $8 \times 1/8 = 1$ atom. The total number atoms to be assigned to any one cell is thus $3 + 1 = 4$.

10.83 In the following diagram, the dashed line is a face diagonal of the unit cell. The unit cell edge is 362 pm. By the Pythagorean theorem, we have:

$\text{diagonal}^2 = \text{edge}^2 + \text{edge}^2$, where an edge is equal to 362 pm.

Also, as shown in the diagram below, the face diagonal is equal to $4 \times r$, where r equals the atom's radius.

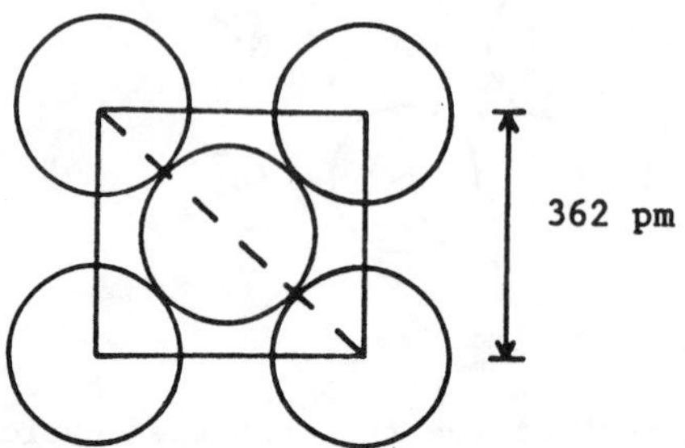

This means that $[4r]^2 = (362)^2 + (362)^2$. Taking the square root of both sides of this equation gives:

$4r = 1.414 \times 362$ pm, and we get $r = 128$ pm.

10.84 The following diagram is appropriate. As in problem 10.83 above, the face diagonal is 4 times the radius of the atom. The Pythagorean theorem is:

$\text{diagonal}^2 = \text{edge}^2 + \text{edge}^2$

Hence we have:

$[4(144 \text{ pm})]^2 = 2 \times \text{edge}^2$

Solving for the edge length we get 407 pm.

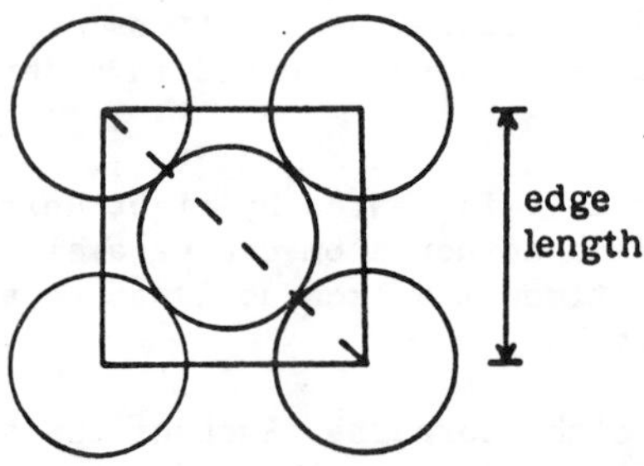

Chapter 10

Crystal Types

10.85 The lattice positions are occupied by metal cations, which are then surrounded by the electrons of the metal.

10.86 (a) dipole-dipole, London forces or hydrogen bonds
(b) electrostatic forces
(c) covalent bonds

10.87 Covalent crystals are also termed network solids because they are constructed of atoms that are covalently bonded to one another, giving a giant interlocking network.

10.88 This must be a molecular solid, because, if it were ionic, it would be high melting and the melt would conduct.

10.89 We expect ionic substances in general to be hard, brittle, high melting, and nonconducting in the solid but conducting when melted.

10.90 This is a covalent solid.

10.91 This is a metallic solid.

10.92 This is a covalent solid.

10.93 This is metallic.

10.94 This is a molecular, or covalent solid.

Amorphous Solids

10.95 Amorphous means, literally, without form. It is taken here to represent a solid that does not have the regular, repeating geometrical form normally associated with a crystal lattice.

10.96 An amorphous solid is a noncrystalline solid. As discussed in the answer to review question 10.95, it is a solid that lacks the long-range order that characterizes a crystalline substance. When cooled, a liquid that will form an amorphous solid gradually becomes viscous and slowly hardens to give a glass, or a supercooled liquid.

10.97 Supercooling is the cooling of a liquid to a temperature below its normal freezing point, without obtaining a crystalline solid. If a solid results, it is a supercooled liquid, i.e. an amorphous solid.

10.98 When heated in a flame, glass (like any other amorphous solid) gradually softens over a wide temperature range. A crystalline solid, however, remains solid up to the melting temperature, and then on further heating, is found to melt entirely within a narrow temperature range.

Phase Diagrams

10.99

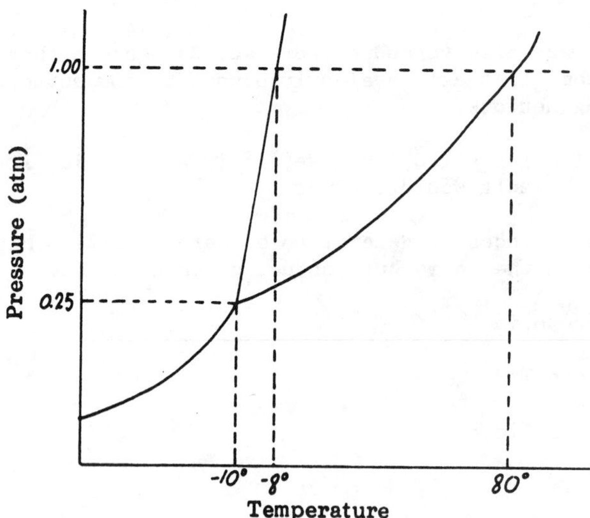

10.100 Sublimation is possible only below a pressure of 0.25 atm, as marked on the phase diagram.

10.101 The density of the solid is higher than that of the liquid. Notice that the line separating the solid from the liquid slopes to the right, in contrast to the diagram for water, Figure 10.30 of the text.

10.102 An increase in pressure should favor the system with the lower volume, i.e. the solid. Therefore, if the substance is at its melting point at a pressure of one atm, and then if the pressure were to be increased, more solid would form at the expense of liquid - that is, more of the substance would freeze. If melting were to be accomplished at the higher pressure, it would require a temperature that is higher than the normal melting temperature.

10.103 Critical temperature - the temperature above which the substance can not exist as a liquid, regardless of the applied pressure. It is, therefore, the temperature above which a gas cannot be made to liquify, regardless of the amount of pressure that is applied.

Critical pressure - the vapor pressure of a liquid at the liquid's critical temperature.

A critical temperature and critical pressure together constitute a substance's critical point.

10.104 solid, liquid and gas

10.105 Carbon dioxide does not have a normal boiling point because its triple point lies above one atmosphere. Thus, the liquid-vapor equilibrium that is taken to represent the boiling point does not exist at the pressure (1 atm) conventionally used to designate the "normal" boiling point.

10.106 (a) solid (b) gas (c) liquid (d) solid, liquid, and gas

10.107 The solid's temperature increases until the solid-liquid line is reached, at which point the solid sublimes. After it has completely vaporized, the vapor's temperature increases to 0 °C.

10.108 The vapor is compressed until the liquid-vapor line is reached, at which point the vapor condenses to a liquid. As the pressure is increased further, the solid-liquid line is reached, and the liquid freezes. At -58°C, the gas is compressed until the solid-vapor line is reached, at which point the vapor condenses directly to a solid.

10.109 The solid-liquid line slants toward the right.

10.110 The triple point must be below room temperature and atmospheric pressure.

10.111 The critical temperature of hydrogen is below room temperature because, at room temperature, it cannot be liquified by the application of pressure. The critical temperature of butane is above room temperature, because butane can be liquified by the application of pressure. See also the answer to review exercise 10.69.

CHAPTER ELEVEN

PRACTICE EXERCISES

1. (a) $MgCl_2(s) \rightarrow Mg^{2+}(aq) + 2Cl^-(aq)$

   (b) $Al(NO_3)_3(s) \rightarrow Al^{3+}(aq) + 3NO_3^-(aq)$

   (c) $Na_2CO_3(s) \rightarrow 2Na^+(aq) + CO_3^{2-}(aq)$

   (d) $(NH_4)_2SO_4(s) \rightarrow 2NH_4^+(aq) + SO_4^{2-}(aq)$

2. $HCHO_2(aq) + H_2O(\ell) \rightleftharpoons H_3O^+(aq) + CHO_2^-(aq)$

3. $NH_3(aq) + H_2O(\ell) \rightleftharpoons NH_4^+(aq) + OH^-(aq)$

4. molecular:

   $2NaOH(aq) + MgCl_2(aq) \rightarrow Mg(OH)_2(s) + 2NaCl(aq)$

   ionic:

   $2Na^+(aq) + 2OH^-(aq) + Mg^{2+}(aq) + 2Cl^-(aq) \rightarrow Mg(OH)_2(s) + 2Na^+(aq) + 2Cl^-(aq)$

   net ionic:

   $Mg^{2+}(aq) + 2OH^-(aq) \rightarrow Mg(OH)_2(s)$

5. (a) $Ag^+(aq) + Cl^-(aq) \rightarrow AgCl(s)$

   (b) $S^{2-}(aq) + Pb^{2+}(aq) \rightarrow PbS(s)$

   (c) no reaction

6. (a) molecular:

   $Ni(OH)_2(s) + 2HClO_4(aq) \rightarrow Ni(ClO_4)_2(aq) + 2H_2O(\ell)$

   ionic:

   $Ni(OH)_2(s) + 2H_3O^+(aq) + 2ClO_4^-(aq) \rightarrow Ni^{2+}(aq) + 2ClO_4^-(aq) + 4H_2O(\ell)$

   net ionic:

   $Ni(OH)_2(s) + 2H_3O^+(aq) \rightarrow Ni^{2+}(aq) + 4H_2O(\ell)$

   (b) molecular:

   $Al_2O_3(s) + 3H_2SO_4(aq) \rightarrow Al_2(SO_4)_3(aq) + 3H_2O(\ell)$

ionic:

$Al_2O_3(s) + 6H_3O^+(aq) + 3SO_4^{2-}(aq) \rightarrow 2Al^{3+}(aq) + 3SO_4^{2-}(aq) + 9H_2O(\ell)$

net ionic:

$Al_2O_3(s) + 6H_3O^+(aq) \rightarrow 2Al^{3+}(aq) + 9H_2O(\ell)$

(c) molecular:

$H_2CO_3(aq) + 2KOH(aq) \rightarrow K_2CO_3(aq) + 2H_2O(\ell)$

ionic:

$2H_3O^+(aq) + CO_3^{2-}(aq) + 2K^+(aq) + 2OH^-(aq) \rightarrow$

$2K^+(aq) + CO_3^{2-}(aq) + 4H_2O(\ell)$

net ionic:

$H_3O^+(aq) + OH^-(aq) \rightarrow 2H_2O(\ell)$

7. It is possible to form a weak electrolyte, so we write:

molecular:

$HCl(aq) + NaF(aq) \rightarrow NaCl(aq) + HF(aq)$

ionic:

$H^+(aq) + Cl^-(aq) + Na^+(aq) + F^-(aq) \rightarrow Na^+(aq) + Cl^-(aq) + HF(aq)$

net ionic:

$H^+(aq) + F^-(aq) \rightarrow HF(aq)$

8. molecular:

$Ba(OH)_2(aq) + H_2SO_4(aq) \rightarrow BaSO_4(s) + 2H_2O(\ell)$

ionic:

$Ba^{2+}(aq) + 2OH^-(aq) + 2H_3O^+(aq) + SO_4^{2-}(aq) \rightarrow BaSO_4(s) + 4H_2O(\ell)$

net ionic:

$Ba^{2+}(aq) + 2OH^-(aq) + 2H_3O^+(aq) + SO_4^{2-}(aq) \rightarrow BaSO_4(s) + 4H_2O(\ell)$

9. (a) molecular:

$(NH_4)_2SO_4(aq) + 2NaOH(aq) \rightarrow Na_2SO_4(aq) + 2NH_3(g) + 2H_2O(\ell)$

ionic:

$2NH_4^+(aq) + SO_4^{2-}(aq) + 2Na^+(aq) + 2OH^-(aq) \rightarrow$

$2Na^+(aq) + SO_4^{2-}(aq) + 2NH_3(g) + 2H_2O(\ell)$

net ionic:

$NH_4^+(aq) + OH^-(aq) \rightarrow NH_3(g) + H_2O(\ell)$

(b) molecular:

$(NH_4)_2SO_4(aq) + Ba(OH)_2(aq) \rightarrow BaSO_4(s) + 2NH_3(g) + 2H_2O(\ell)$

ionic:

$2NH_4^+(aq) + SO_4^{2-}(aq) + Ba^{2+}(aq) + 2OH^-(aq) \rightarrow$

$BaSO_4(s) + 2NH_3(g) + 2H_2O(\ell)$

net ionic:

$2NH_4^+(aq) + SO_4^{2-}(aq) + Ba^{2+}(aq) + 2OH^-(aq) \rightarrow$

$BaSO_4(s) + 2NH_3(g) + 2H_2O(\ell)$

10. (a)

$$0.0500 \text{ L} \times \frac{0.600 \text{ mol HCl}}{1 \text{ L HCl soln}} = 3.00 \times 10^{-2} \text{ mole HCl}$$

(b)

$$4.00 \times 10^{-2} \text{ mol HCl} \times \frac{1 \text{ L HCl soln}}{0.600 \text{ mol HCl}} = 0.0667 \text{ L} = 66.7 \text{ mL of HCl soln}$$

11. $FeCl_2(aq) + 2KOH(aq) \rightarrow Fe(OH)_2(s) + 2KCl(aq)$

$$60.0 \text{ mL} \times \frac{0.250 \text{ mol FeCl}_2}{1000 \text{ mL FeCl}_2} \times \frac{2 \text{ mol KOH}}{1 \text{ mol FeCl}_2} \times \frac{1000 \text{ mL KOH}}{0.500 \text{ mol KOH}} = 60.0 \text{ mL KOH soln}$$

12.

$$75.0 \text{ mL } Fe_2(SO_4)_3 \times \frac{0.230 \text{ mol } Fe_2(SO_4)_3}{1000 \text{ mL } Fe_2(SO_4)_3} \times \frac{2 \text{ mol } Fe^{3+}}{1 \text{ mol } Fe_2(SO_4)_3}$$

$$= 3.45 \times 10^{-2} \text{ mol } Fe^{3+}$$

$$75.0 \text{ mL } Fe_2(SO_4)_3 \times \frac{0.230 \text{ mol } Fe_2(SO_4)_3}{1000 \text{ mL } Fe_2(SO_4)_3} \times \frac{3 \text{ mol } SO_4^{2-}}{1 \text{ mol } Fe_2(SO_4)_3}$$

$$= 5.18 \times 10^{-2} \text{ mol } SO_4^{2-}$$

13. First calculate the number of moles of each reactant that are available:

mol $BaCl_2$ = mol Ba^{2+} = 0.600 M × 0.0200 L = 0.0120

mol $MgSO_4$ = mol SO_4^{2-} = 0.500 M × 0.0300 L = 0.0150

The two reagents react in a 1:1 mole ratio:

$BaCl_2(aq) + MgSO_4(aq) \rightarrow BaSO_4(s) + MgCl_2(aq)$

Since there are fewer moles of $BaCl_2$ than of $MgSO_4$, we conclude that $BaCl_2$ is the limiting reagent (i.e. $MgSO_4$ is present in excess) and that the amount of $BaSO_4$ that will form is to be based on the amount of $BaCl_2$ that is available: mole $BaSO_4$ = 0.0120 = mol Ba^{2+}. Since $BaSO_4$ is insoluble, its concentration will be zero (or nearly so) in the final solution. The final volume of the solution (0.050 L = 50.0 mL) is to be used in calculating the concentration of the remaining substances. We calculate first the moles of the various substances that remain after reaction:

for SO_4^{2-}, (0.0150 - 0.0120) = 0.0030 mol remain unreacted

for Mg^{2+}, 0.0150 mol remain unreacted

for Cl^-, 2 × 0.0120 = 0.0240 mol remain unreacted

Next we calculate the concentrations of these materials in the final volume of the product solution:

for SO_4^{2-}, 0.0030 mol ÷ 0.050 L = 0.060 M

for Mg^{2+}, 0.0150 mol ÷ 0.050 L = 0.300 M

for Cl^-, 0.0240 mol ÷ 0.050 L = 0.480 M

14. (a)

$$0.736 \text{ g } CaSO_4 \times \frac{1 \text{ mol } CaSO_4}{136 \text{ g } CaSO_4} \times \frac{1 \text{ mol } Ca^{2+}}{1 \text{ mol } CaSO_4} = 5.41 \times 10^{-3} \text{ mol } Ca^{2+}$$

(b) Since all calcium had to come from the original 2.000 g sample, this is the same as the number of moles calculated in part (a): 0.00541 mol.

(c) as in the answer to part (b), 0.00541 mol

(d)

$$0.00541 \text{ mol } CaCl_2 \times \frac{111 \text{ g}}{1 \text{ mol}} = 0.601 \text{ g } CaCl_2$$

(e)

$$\% \ CaCl_2 = \frac{\text{mass } CaCl_2}{\text{total mass of the sample}} \times 100$$

$$\% \ CaCl_2 = \frac{0.601}{2.000} \times 100 = 30.1 \ \%$$

15. First calculate the number of moles of NaOH and OH^- that are present in the given amount of sodium hydroxide solution:

$NaOH(aq) \rightarrow Na^+(aq) + OH^-(aq)$

$36.42 \times 10^{-3} \text{ L} \times 0.147 \text{ M} = 5.35 \times 10^{-3} \text{ mol NaOH} = 5.35 \times 10^{-3} \text{ mol } OH^-$

Next, we calculate the number of moles of H_2SO_4 and of H_3O^+ that are present in the original sulfuric acid solution by realizing that at neutralization, the number of moles of H_3O^+ is equal to the number of moles of OH^-:

$H_3O^+ + OH^- \rightarrow 2H_2O$

$H_2SO_4(aq) + 2H_2O(\ell) \rightarrow 2H_3O^+ + SO_4^{2-}$

$$5.35 \times 10^{-3} \text{ mol } H_3O^+ \times \frac{1 \text{ mol } H_2SO_4}{2 \text{ mol } H_3O^+} \times \frac{1}{0.01500 \text{ L } H_2SO_4} = 0.178 \text{ mol/L } H_2SO_4$$

16. We proceed as in practice exercise 15.

$KOH(aq) \rightarrow K^+(aq) + OH^-(aq)$

$0.0100 \text{ M KOH} \times 0.01100 \text{ L} = 1.10 \times 10^{-4} \text{ mol KOH} = 1.10 \times 10^{-4} \text{ mol } OH^-$

This will be neutralized by the same number of moles of H_3O^+. The acid is:

$HCl(aq) + H_2O(\ell) \rightarrow H_3O^+ + Cl^-$, whose molarity is, therefore:

$$\frac{1.10 \times 10^{-4} \text{ mol } H_3O^+}{5.00 \times 10^{-3} \text{ L HCl}} \times \frac{1 \text{ mol HCl}}{1 \text{ mol } H_3O^+} = \frac{0.0220 \text{ mol HCl}}{1 \text{ L HCl}} = 0.0220 \text{ M HCl}$$

Alternatively, we can write:

1.10×10^{-4} mol HCl ÷ 5.00×10^{-3} L = 0.0220 M HCl

Next we can calculate the mass of HCl that is present in 1000 ml of solution:

mass HCl = 0.0220 mol/L × 36.5 g/mol = 0.803 g HCl in one L of solution

Also, if the density of the solution is 1 g/mL, then this is numerically equal to 1000 g/L. One L of the solution has a total mass of 1000 g.

The % HCl is determined as follows:

$$\% \text{ HCl} = \frac{\text{mass of HCl}}{\text{total mass of the solution}} \times 100$$

$$\% \text{ HCl} = \frac{0.803 \text{ g}}{1000 \text{ g}} \times 100 = 0.0803 \% \text{ HCl}$$

17. The equation for complete neutralization is:

$H_2C_2O_4 + 2NaOH \rightarrow Na_2C_2O_4 + 2H_2O$

Since two hydrogen ions are neutralized per molecule of $H_2C_2O_4$, then we conclude that 1 mol of oxalic acid contains 2 equivalents.

The equation for partial neutralization is:

$H_2C_2O_4 + NaOH \rightarrow NaHC_2O_4 + H_2O$

in which case we conclude that 1 mol of oxalic acid contains 1 equivalent.

18. Since there are three mol of OH^- ions per mol of $Al(OH)_3$, there are three equivalents of base in aluminum hydroxide.

19. (a) Complete neutralization of one molecule requires the reaction of 3 eq of hydrogen. There are thus 3 eq per mol, and the equivalent weight is 1/3 of the molecular weight:

$$1 \text{ eq } H_3PO_4 \times \frac{1 \text{ mol } H_3PO_4}{3 \text{ eq } H_3PO_4} \times \frac{98.0 \text{ g } H_3PO_4}{1 \text{ mol } H_3PO_4} = 32.7 \text{ g } H_3PO_4$$

(b) Partial neutralization of H_3PO_4 to form HPO_4^{2-} requires the neutralization of 2 hydrogen ions per molecule. There are thus two equivalents per mol, and the equivalent weight in this reaction is 1/2 of the molecular weight:

$$1 \text{ eq } H_3PO_4 \times \frac{1 \text{ mol } H_3PO_4}{2 \text{ eq } H_3PO_4} \times \frac{98.0 \text{ g } H_3PO_4}{1 \text{ mol } H_3PO_4} = 49.0 \text{ g } H_3PO_4$$

20. The equation for complete neutralization is:

$$2KOH + H_2SO_4 \rightarrow K_2SO_4 + 2H_2O$$

$$14.0 \text{ g KOH} \times \frac{1 \text{ eq KOH}}{56.1 \text{ g KOH}} \times \frac{1 \text{ eq } H_2SO_4}{1 \text{ eq KOH}} \times \frac{49.0 \text{ g } H_2SO_4}{1 \text{ eq } H_2SO_4} = 12.2 \text{ g } H_2SO_4$$

21. (a) for $Ba(OH)_2$ we have:

$$0.10 \text{ mol/L} \times 2 \text{ eq/mol} = 0.20 \text{ eq } Ba(OH)_2 = 0.20 \text{ N } Ba(OH)_2$$

(b) for H_2SO_4 we have:

$$0.60 \text{ eq/L} \times 1 \text{ mol/2 eq} = 0.30 \text{ mol } H_2SO_4\text{/L} = 0.30 \text{ M } H_2SO_4$$

22. When neutralization is accomplished, the number of equivalents of acid that have reacted will be equal to the number of equivalents of base that have reacted. Thus we have the following general equation for neutralization:

$V_A \times N_A = V_B \times N_B$, which in this case becomes:

$$V_{NaOH} = V_{H_2SO_4} \times N_{H_2SO_4} \div N_{NaOH} = 45.0 \text{ mL} \times 0.150 \text{ N} \div 0.200 \text{ N} = 33.8 \text{ mL}$$

23. 0.035 L × 0.600 eq/L = 0.0210 eq

24. (a) $0.200 \text{ eq/L} \times 0.0316 \text{ L} = 6.32 \times 10^{-3}$ eq NaOH

(b) The number of equivalents of acid is equal to the number of equivalents of base, 6.32×10^{-3} eq.

(c) The equivalent weight is simply the mass divided by the number of equivalents:

$$\text{Eq. wt.} = 1.000 \text{ g} / 6.32 \times 10^{-3} \text{ eq} = 158 \text{ g/eq}$$

25. In each case the conjugate base is obtained by removal of a proton from the acid:

(a) OH^- (b) I^- (c) NO_2^- (d) $H_2PO_4^-$ (e) HPO_4^{2-} (f) PO_4^{3-} (g) H^- (h) NH_3

26. In each case the conjugate acid is obtained by addition of a proton to the conjugate base:

(a) H_2O_2 (b) HSO_4^- (c) HPO_4^{2-} (d) $HC_2H_3O_2$ (e) NH_3 (f) NH_4^+ (g) H_3PO_4

(h) $H_2PO_4^-$

27. The Brønsted acids are $H_2PO_4^-(aq)$ and $H_2CO_3(aq)$.

The Brønsted bases are HCO_3^- and $HPO_4^{2-}(aq)$.

conjugate pair

$$HCO_3^- + H_2PO_4^- \rightarrow H_2CO_3 + HPO_4^{2-}$$

conjugate pair

28. base: PO_4^{3-}, conjugate acid: HPO_4^{2-}

base: $C_2H_3O_2^-$, conjugate acid: $HC_2H_3O_2$

29. The anion HPO_4^{2-} can react either as an acid:

$HPO_4^{2-} + H_2O \rightarrow PO_4^{3-} + H_3O^+$

or as a base:

$HPO_4^{2-} + H_2O \rightarrow H_2PO_4^- + OH^-$

Alternatively, the above two equations could be written:

$HPO_4^{2-} + OH^- \rightarrow PO_4^{3-} + H_2O$

$HPO_4^{2-} + H_3O^+ \rightarrow H_2PO_4^- + H_2O$

30. (a) $HBrO_4$, since Cl is above Br in the periodic table

(b) H_2SeO_4, since Se is to the right of As in the periodic table

31. The acidity increases in the series:

$HClO < HClO_2 < HClO_3 < HClO_4$

32. (a) HBr - since Br lies to the right of Se in the periodic table

(b) H_2Te - since Te lies below Se in the periodic table

(c) CH_3SH - since S lies beneath O in the periodic table

33. CO_2 is the Lewis acid, and OH^- is the Lewis base.

34. $[Ag(S_2O_3)_2]^{3-}$ and $(NH_4)_3[Ag(S_2O_3)_2]$

35. $AlCl_3 \cdot 6H_2O$ and $[Al(H_2O)_6]^{3+}$

REVIEW EXERCISES

Electrolytes

11.1 Electrolyte - a substance that dissolves in water to give an electrically conducting solution.

Nonelectrolyte - a substance that dissolves in water to give a solution that does not conduct electricity.

11.2 dissociation - the dissolving of an ionic compound in water such that the individual ions that compose the ionic compound become separated from one another (via hydration), and move about freely in solution, acting more or less independently of one another.

hydration - the process that aids in dissociation of an ionic solid in water, i.e. the surrounding of individual ions by water molecules.

11.3 (a) $LiCl(s) \rightarrow Li^+(aq) + Cl^-(aq)$

(b) $BaCl_2(s) \rightarrow Ba^{2+}(aq) + 2Cl^-(aq)$

(c) $Al(C_2H_3O_2)_3(s) \rightarrow Al^{3+}(aq) + 3C_2H_3O_2^-(aq)$

(d) $(NH_4)_2CO_3(s) \rightarrow 2NH_4^+(aq) + CO_3^{2-}(aq)$

(e) $FeCl_3(s) \rightarrow Fe^{3+}(aq) + 3Cl^-(aq)$

11.4 (a) $CuSO_4(s) \rightarrow Cu^{2+}(aq) + SO_4^{2-}(aq)$

(b) $Al_2(SO_4)_3(s) \rightarrow 2Al^{3+}(aq) + 3SO_4^{2-}(aq)$

(c) $CrCl_3(s) \rightarrow Cr^{3+}(aq) + 3Cl^-(aq)$

(d) $(NH_4)_2HPO_4(s) \rightarrow 2NH_4^+(aq) + HPO_4^{2-}(aq)$

(e) $KMnO_4(s) \rightarrow K^+(aq) + MnO_4^-(aq)$

Acids and Bases as Electrolytes

11.5 According to the definition of Arrhenius, an acid gives H_3O^+ ions in water, and a base gives OH^- ions in water.

11.6 Ionic substances already contain the ions that dissociate when the substance dissolves in water. On the other hand, molecular compounds such as HCl, form ions by reaction with water (ionization), although such substances do not contain the resulting ions to start with. Both dissociation of an ionic compound and ionization of a molecular compound in water solutions lead to the formation of solutions that are electrically conducting.

11.7 This is an ionization reaction:

$$HClO_4(\ell) + H_2O(\ell) \rightarrow H_3O^+(aq) + ClO_4^-(aq)$$

11.8 A strong electrolyte becomes completely dissociated or ionized in water solution. A weak electrolyte becomes only partially ionized in water solution.

11.9 For a weak electrolyte, there is a strong tendency for the ions to recombine to form their parent molecular compound. For a strong electrolyte, the converse is true; a strong electrolyte has a great tendency to be fully ionized in water solution.

11.10 $HNO_2(aq) + H_2O(\ell) \rightleftharpoons H_3O^+(aq) + NO_2^-(aq)$

11.11 $N_2H_4(\ell) + H_2O(\ell) \rightleftharpoons N_2H_5^+(aq) + OH^-(aq)$

11.12 These are not reversible reactions, that is the reverse reaction has practically no tendency to occur.

11.13 $HClO_3(aq) + H_2O(\ell) \rightarrow H_3O^+(aq) + ClO_3^-(aq)$

11.14 $H_2CO_3(aq) + H_2O(\ell) \rightleftharpoons H_3O^+(aq) + HCO_3^-(aq)$

$$HCO_3^-(aq) + H_2O(\ell) \rightleftharpoons H_3O^+(aq) + CO_3^{2-}(aq)$$

11.15 $H_3PO_4(\ell) + H_2O(\ell) \rightleftharpoons H_3O^+(aq) + H_2PO_4^-(aq)$

$$H_2PO_4^-(aq) + H_2O(\ell) \rightleftharpoons H_3O^+(aq) + HPO_4^{2-}(aq)$$

$$HPO_4^{2-}(aq) + H_2O(\ell) \rightleftharpoons H_3O^+(aq) + PO_4^{3-}(aq)$$

Ionic Reactions

11.16 Ionic reaction - a reaction involving ions.
Metathesis reaction - a double replacement reaction.
Precipitate - a solid that forms in solution as a result of a chemical reaction.
Spectator ions - ions that are not involved in a chemical reaction in solution, but that are present in solution at the time of the reaction.

11.17 A molecular equation is a chemical equation in which one has written complete molecular formulas for all reactants and products of a reaction. An ionic equation is a chemical equation in which all of the soluble strong electrolytes are listed in their dissociated form. A net ionic equation is written by listing only those ions and molecules that are involved in the reaction at hand. A net ionic equation differs from an ionic equation in that all of the spectator ions are omitted from the former.

11.18 In a balanced ionic equation, both the mass and the electrical charge must be balanced.

11.19 (a) ionic:

$$2NH_4^+(aq) + CO_3^{2-}(aq) + Mg^{2+}(aq) + 2Cl^-(aq) \rightarrow 2NH_4^+(aq) + 2Cl^-(aq) + MgCO_3(s)$$

net ionic:

$$Mg^{2+}(aq) + CO_3^{2-}(aq) \rightarrow MgCO_3(s)$$

(b) ionic:

$$Cu^{2+}(aq) + 2Cl^-(aq) + 2Na^+(aq) + 2OH^-(aq) \rightarrow Cu(OH)_2(s) + 2Na^+(aq) + 2Cl^-(aq)$$

net ionic:

$$Cu^{2+}(aq) + 2OH^-(aq) \rightarrow Cu(OH)_2(s)$$

(c) ionic:

$$3Fe^{2+}(aq) + 3SO_4^{2-}(aq) + 6Na^+(aq) + 2PO_4^{3-}(aq) \rightarrow Fe_3(PO_4)_2(s) + 6Na^+(aq) + 3SO_4^{2-}(aq)$$

net ionic:

$3Fe^{2+}(aq) + 2PO_4^{3-}(aq) \rightarrow Fe_3(PO_4)_2(s)$

(d) ionic:

$2Ag^+(aq) + 2C_2H_3O_2^-(aq) + Ni^{2+}(aq) + 2Cl^-(aq) \rightarrow$

$2AgCl(s) + 2C_2H_3O_2^-(aq) + Ni^{2+}(aq)$

net ionic:

$Ag^+(aq) + Cl^-(aq) \rightarrow AgCl(s)$

11.20 (a) ionic:

$Cu^{2+}(aq) + SO_4^{2-}(aq) + Ba^{2+}(aq) + 2Cl^-(aq) \rightarrow$

$BaSO_4(s) + Cu^{2+}(aq) + 2Cl^-(aq)$

net ionic:

$Ba^{2+}(aq) + SO_4^{2-}(aq) \rightarrow BaSO_4(s)$

(b) $Fe^{3+}(aq) + 3NO_3^-(aq) + 3Li^+(aq) + 3OH^-(aq) \rightarrow$

$Fe(OH)_3(s) + 3Li^+(aq) + 3NO_3^-(aq)$

net ionic:

$Fe^{3+}(aq) + 3OH^-(aq) \rightarrow Fe(OH)_3(s)$

(c) $6Na^+(aq) + 2PO_4^{3-}(aq) + 3Ca^{2+}(aq) + 6Cl^-(aq) \rightarrow$

$Ca_3(PO_4)_2(s) + 6Na^+(aq) + 6Cl^-(aq)$

net ionic:

$3Ca^{2+}(aq) + 2PO_4^{3-}(aq) \rightarrow Ca_3(PO_4)_2(s)$

(d) $2Na^+(aq) + S^{2-}(aq) + 2Ag^+(aq) + 2C_2H_3O_2^-(aq) \rightarrow$

$2Na^+(aq) + 2C_2H_3O_2^-(aq) + Ag_2S(s)$

net ionic:

$2Ag^+(aq) + S^{2-}(aq) \rightarrow Ag_2S(s)$

11.21 The charge is not balanced.

11.22 $Cu^{2+}(aq) + S^{2-}(aq) \rightarrow CuS(s)$

11.23 Since AgBr is insoluble, the concentrations of $Ag^+(aq)$ and $Br^-(aq)$ in a saturated solution of AgBr are very small, practically zero. When solutions of $AgNO_3$ and NaBr are mixed, the concentrations of Ag^+ and Br^-

are momentarily larger than those in a saturated AgBr solution. Since this solution is immediately supersaturated in the moment of mixing, a precipitate of AgBr forms spontaneously.

11.24 molecular:

$AgNO_3(aq) + NaBr(aq) \rightarrow AgBr(s) + NaNO_3(aq)$

ionic:

$Ag^+(aq) + NO_3^-(aq) + Na^+(aq) + Br^-(aq) \rightarrow AgBr(s) + Na^+(aq) + NO_3^-(aq)$

net ionic:

$Ag^+(aq) + Br^-(aq) \rightarrow AgBr(s)$

11.25 molecular:

$2Na_3PO_4(aq) + 3CaCl_2(aq) \rightarrow Ca_3(PO_4)_2(s) + 6NaCl(aq)$

ionic:

$6Na^+(aq) + 2PO_4^{3-}(aq) + 3Ca^{2+}(aq) + 6Cl^-(aq) \rightarrow$

$Ca_3(PO_4)_2(s) + 6Na^+(aq) + 6Cl^-(aq)$

net ionic:

$3Ca^{2+}(aq) + 2PO_4^{3-}(aq) \rightarrow Ca_3(PO_4)_2(s)$

Predicting Double Replacement Reactions

11.26 The soluble ones are (a), (b), and (d).

11.27 The insoluble ones are (a), (d), and (f).

11.28 (a) $Na_2SO_3(aq) + Ba(NO_3)_2(aq) \rightarrow BaSO_3(s) + 2NaNO_3(aq)$

ionic:

$2Na^+(aq) + SO_3^{2-}(aq) + Ba^{2+}(aq) + 2NO_3^-(aq) \rightarrow$

$BaSO_3(s) + 2Na^+(aq) + 2NO_3^-(aq)$

net ionic:

$Ba^{2+}(aq) + SO_3^{2-}(aq) \rightarrow BaSO_3(s)$

(b) $K_2S(aq) + ZnCl_2(aq) \rightarrow ZnS(s) + 2KCl(aq)$

ionic:

$2K^+(aq) + S^{2-}(aq) + Zn^{2+}(aq) + 2Cl^-(aq) \rightarrow ZnS(s) + 2K^+(aq) + 2Cl^-(aq)$

net ionic:

$Zn^{2+}(aq) + S^{2-}(aq) \rightarrow ZnS(s)$

(c) $2NH_4Br(aq) + Pb(C_2H_3O_2)_2(aq) \rightarrow 2NH_4C_2H_3O_2(aq) + PbBr_2(s)$

ionic:

$2NH_4^+(aq) + 2Br^-(aq) + Pb^{2+}(aq) + 2C_2H_3O_2^-(aq) \rightarrow$

$2NH_4^+(aq) + 2C_2H_3O_2^-(aq) + PbBr_2(s)$

net ionic:

$Pb^{2+}(aq) + 2Br^-(aq) \rightarrow PbBr_2(s)$

(d) $2NH_4ClO_4(aq) + Cu(NO_3)_2(aq) \rightarrow Cu(ClO_4)_2(aq) + 2NH_4NO_3(aq)$

ionic:

$2NH_4^+(aq) + 2ClO_4^-(aq) + Cu^{2+}(aq) + 2NO_3^-(aq) \rightarrow$

$Cu^{2+}(aq) + 2ClO_4^-(aq) + 2NO_3^-(aq) + 2NH_4^+(aq)$

net ionic: N.R.

11.29 (a) $(NH_4)_2S + 2NaCl \rightarrow 2NH_4Cl + Na_2S$

ionic:

$2NH_4^+(aq) + S^{2-}(aq) + 2Na^+(aq) + 2Cl^-(aq) \rightarrow$

$2NH_4^+(aq) + 2Cl^-(aq) + 2Na^+(aq) + S^{2-}(aq)$

net ionic: N.R.

(b) $Cr_2(SO_4)_3 + 3K_2CO_3 \rightarrow Cr_2(CO_3)_3 + 3K_2SO_4$

ionic:

$2Cr^{3+}(aq) + 3SO_4^{2-}(aq) + 6K^+(aq) + 3CO_3^{2-}(aq) \rightarrow$

$Cr_2(CO_3)_3(s) + 6K^+(aq) + 3SO_4^{2-}(aq)$

net ionic:

$2Cr^{3+}(aq) + 3CO_3^{2-}(aq) \rightarrow Cr_2(CO_3)_3(s)$

(c) $Sr(OH)_2(aq) + MgCl_2(aq) \rightarrow SrCl_2(aq) + Mg(OH)_2(s)$

ionic:

$Sr^{2+}(aq) + 2OH^-(aq) + Mg^{2+}(aq) + 2Cl^-(aq) \rightarrow$

$Mg(OH)_2(s) + Sr^{2+}(aq) + 2Cl^-(aq)$

net ionic:

$Mg^{2+}(aq) + 2OH^-(aq) \rightarrow Mg(OH)_2(s)$

(d) $Ba(NO_3)_2(aq) + Na_2SO_3(aq) \rightarrow BaSO_3(s) + 2NaNO_3(aq)$

ionic:

$Ba^{2+}(aq) + 2NO_3^-(aq) + 2Na^+(aq) + SO_3^{2-}(aq) \rightarrow$

$BaSO_3(s) + 2NO_3^-(aq) + 2Na^+(aq)$

net ionic:

$Ba^{2+}(aq) + SO_3^{2-}(aq) \rightarrow BaSO_3(s)$

11.30 (a) $3HNO_3(aq) + Cr(OH)_3(s) \rightarrow Cr(NO_3)_3(aq) + 3H_2O(\ell)$

ionic:

$3H_3O^+(aq) + 3NO_3^-(aq) + Cr(OH)_3(s) \rightarrow Cr^{3+}(aq) + 3NO_3^-(aq) + 6H_2O(\ell)$

net ionic:

$3H_3O^+(aq) + Cr(OH)_3(s) \rightarrow Cr^{3+}(aq) + 6H_2O(\ell)$

(b) $HClO_4(aq) + NaOH(aq) \rightarrow NaClO_4(aq) + H_2O(\ell)$

ionic:

$H_3O^+(aq) + ClO_4^-(aq) + Na^+(aq) + OH^-(aq) \rightarrow$

$Na^+(aq) + ClO_4^-(aq) + 2H_2O(\ell)$

net ionic:

$H_3O^+(aq) + OH^-(aq) \rightarrow 2H_2O(\ell)$

(c) $Cu(OH)_2(s) + 2HC_2H_3O_2(aq) \rightarrow Cu(C_2H_3O_2)_2(aq) + 2H_2O(\ell)$

ionic:

$Cu(OH)_2(s) + 2HC_2H_3O_2(aq) \rightarrow Cu^{2+}(aq) + 2C_2H_3O_2^-(aq) + 2H_2O(\ell)$

net ionic:

$Cu(OH)_2(s) + 2HC_2H_3O_2(aq) \rightarrow Cu^{2+}(aq) + 2C_2H_3O_2^-(aq) + 2H_2O(\ell)$

(d) $ZnO(s) + 2HBr(aq) \rightarrow ZnBr_2(aq) + H_2O(\ell)$

ionic:

$ZnO(s) + 2H_3O^+(aq) + 2Br^-(aq) \rightarrow Zn^{2+}(aq) + 2Br^-(aq) + 3H_2O(\ell)$

net ionic:

$ZnO(s) + 2H_3O^+(aq) \rightarrow Zn^{2+}(aq) + 3H_2O(\ell)$

11.31 (a) $NaHSO_3(aq) + HBr(aq) \rightarrow SO_2(g) + NaBr(aq) + H_2O(\ell)$

ionic:

$Na^+(aq) + HSO_3^-(aq) + H_3O^+(aq) + Br^-(aq) \rightarrow$

$SO_2(g) + Na^+(aq) + Br^-(aq) + H_2O(\ell)$

net ionic:

$HSO_3^-(aq) + H_3O^+(aq) \rightarrow SO_2(g) + 2H_2O(\ell)$

(b) $(NH_4)_2SO_4(aq) + 2NaOH(aq) \rightarrow 2NH_3(g) + Na_2SO_4(aq) + 2H_2O(\ell)$

ionic:

$2NH_4^+(aq) + SO_4^{2-}(aq) + 2Na^+(aq) + 2OH^-(aq) \rightarrow$

$2NH_3(g) + SO_4^{2-}(aq) + 2Na^+(aq) + 2H_2O(\ell)$

net ionic:

$NH_4^+(aq) + OH^-(aq) \rightarrow NH_3(g) + H_2O(\ell)$

(c) $(NH_4)_2SO_4(aq) + Ba(OH)_2(aq) \rightarrow BaSO_4(s) + 2BH_3(g) + 2H_2O(\ell)$

ionic:

$2NH_4^+(aq) + SO_4^{2-}(aq) + Ba^{2+}(aq) + 2OH^-(aq) \rightarrow$

$BaSO_4(s) + 2NH_3(g) + 2H_2O(\ell)$

net ionic:

$2NH_4^+(aq) + SO_4^{2-}(aq) + Ba^{2+}(aq) + 2OH^-(aq) \rightarrow$

$BaSO_4(s) + 2NH_3(g) + 2H_2O(\ell)$

(d) $FeS(s) + 2HCl(aq) \rightarrow FeCl_2(aq) + H_2S(g)$

ionic:

$FeS(s) + 2H_3O^+(aq) + 2Cl^-(aq) \rightarrow$

$Fe^{2+}(aq) + 2Cl^-(aq) + H_2S(g) + 2H_2O(\ell)$

net ionic:

$FeS(s) + 2H_3O^+(aq) \rightarrow Fe^{2+}(aq) + H_2S(g) + 2H_2O(\ell)$

11.32 The electrical conductivity would decrease regularly, until one solution had neutralized the other, forming a nonelectrolyte:

$Ba^{2+}(aq) + 2OH^-(aq) + 2H_3O^+(aq) + SO_4^{2-}(aq) \rightarrow BaSO_4(s) + 4H_2O(\ell)$

Once the point of neutralization had been reached, the addition of excess sulfuric acid would cause the conductivity to increase, because sulfuric acid is a strong electrolyte itself.

11.33 The substance is an electrolyte that dissolves readily in water:

$Na_2CO_3 \cdot 10H_2O(s) \rightarrow 2Na^+(aq) + CO_3^{2-}(aq) + 10H_2O(\ell)$

The carbonate anion then serves to cause the precipitation of calcium cations:

$Ca^{2+}(aq) + CO_3^{2-}(aq) \rightarrow CaCO_3(s)$

11.34 $3Ca^{2+}(aq) + 2PO_4^{3-}(aq) \rightarrow Ca_3(PO_4)_2(s)$

$3Mg^{2+}(aq) + 2PO_4^{3-}(aq) \rightarrow Mg_3(PO_4)_2(s)$

11.35 There are numerous possible answers. One of many possible sets of answers would be:

(a) $NaHCO_3 + HCl \rightarrow H_2O + CO_2 + NaCl$

(b) $FeCl_2 + 2NaOH \rightarrow Fe(OH)_2 + 2NaCl$

(c) $Ba(NO_3)_2 + K_2SO_3 \rightarrow BaSO_3 + 2KNO_3$

(d) $2AgNO_3 + Na_2S \rightarrow Ag_2S + 2NaNO_3$

(e) $ZnO + 2HCl \rightarrow ZnCl_2 + H_2O$

11.36 We need to choose a set of reactants that are both soluble and that react to yield only one solid product. Choose (b) and (d).

11.37 These reactions have the following "driving forces":

(a) formation of insoluble $Cr(OH)_3$

(b) formation of water, a weak electrolyte
(c) formation of a gas, CO_2

(d) formation of a weak electrolyte, $H_2C_2O_4$

Solution Stoichiometry

11.38 (a) $KOH \rightarrow K^+ + OH^-$

$1.30\ mol/L \times 0.0250\ L = 0.0325\ mol\ KOH$

$0.0325\ mol\ KOH \times 1\ mol\ OH^-/mol\ KOH = 0.0325\ mol\ OH^-$

$0.0325\ mol\ KOH \times 1\ mol\ K^+/mol\ KOH = 0.0325\ mol\ K^+$

(b) $CaCl_2 \rightarrow Ca^{2+} + 2Cl^-$

$0.50\ mol/L \times 0.0375\ L = 0.019\ mol\ CaCl_2$

$0.019\ mol\ CaCl_2 \times 1\ mol\ Ca^{2+}/mol\ CaCl_2 = 0.019\ mol\ Ca^{2+}$

$0.019\ mol\ CaCl_2 \times 2\ mol\ Cl^-/mol\ CaCl_2 = 0.038\ mol\ Cl^-$

(c) $(NH_4)_2CO_3 \rightarrow 2NH_4^+ + CO_3^{2-}$

$0.50\ mol/L \times 0.0200\ L = 0.010\ mol\ (NH_4)_2CO_3$

$0.010\ mol\ (NH_4)_2CO_3 \times 2\ mol\ NH_4^+/mol\ (NH_4)_2CO_3 = 0.020\ mol\ NH_4^+$

$0.010\ mol\ (NH_4)_2CO_3 \times 1\ mol\ CO_3^{2-}/mol\ (NH_4)_2CO_3 = 0.010\ mol\ CO_3^{2-}$

(d) $Al_2(SO_4)_3 \rightarrow 2Al^{3+} + 3SO_4^{2-}$

$0.40\ mol/L \times 0.0350\ L = 0.014\ mol\ Al_2(SO_4)_3$

$0.014\ mol\ Al_2(SO_4)_3 \times 2\ mol\ Al^{3+}/mol\ Al_2(SO_4)_3 = 0.028\ mol\ Al^{3+}$

$$0.014 \text{ mol } Al_2(SO_4)_3 \times 3 \text{ mol } SO_4^{2-}/\text{mol } Al_2(SO_4)_3 = 0.042 \text{ mol } SO_4^{2-}$$

11.39 First determine the number of moles of Na_2CO_3 that are to react:

$$\frac{0.10 \text{ mol}}{1000 \text{ mL}} \times 25.0 \text{ mL} = 2.5 \times 10^{-3} \text{ mol } Na_2CO_3$$

Next calculate the number of moles of $NiCl_2$ that are required based on the coefficients in the balanced equation:

$$NiCl_2 + Na_2CO_3 \rightarrow NiCO_3 + 2NaCl$$

$$2.5 \times 10^{-3} \text{ mol } Na_2CO_3 \times \frac{1 \text{ mol } NiCl_2}{1 \text{ mol } Na_2CO_3} = 2.5 \times 10^{-3} \text{ mol } NiCl_2$$

Next calculate the volume of 0.30 M $NiCl_2$ solution that will contain this number of moles:

$$2.5 \times 10^{-3} \text{ mol } NiCl_2 \times \frac{1000 \text{ mL}}{0.30 \text{ mol}} = 8.3 \text{ mL}$$

Alternatively we can write:

$$2.5 \times 10^{-3} \text{ mol} \div 0.30 \text{ mol/L} = 0.0083 \text{ L}$$

The number of grams of product to expect is calculated as follows:

$$2.5 \times 10^{-3} \text{ mol } Na_2CO_3 \times \frac{1 \text{ mol } NiCO_3}{1 \text{ mol } Na_2CO_3} \times \frac{119 \text{ g } NiCO_3}{1 \text{ mol } NiCO_3} = 0.30 \text{ g } NiCO_3$$

11.40 First determine the number of moles of $BaCl_2$ that are to react:

$$0.1000 \text{ L} \times 0.100 \text{ mol/L} = 0.0100 \text{ mol } BaCl_2$$

Next determine the number of moles of Epsom salts that are required, based on the balanced equation:

$$MgSO_4 \cdot 7H_2O + BaCl_2 \rightarrow BaSO_4 + MgCl_2 + 7H_2O$$

$$0.0100 \text{ mol } BaCl_2 \times 1 \text{ mol } MgSO_4 \cdot 7H_2O/\text{mol } BaCl_2 = 0.0100 \text{ mol } MgSO_4 \cdot 7H_2O$$

Finally, calculate the required mass:

$$0.0100 \text{ mol } MgSO_4 \cdot 7H_2O \times 246 \text{ g/mol} = 2.46 \text{ g } MgSO_4 \cdot 7H_2O$$

11.41 First determine the number of moles of H_3PO_4 that are to react:

$0.0150 \text{ L} \times 0.300 \text{ mol/L} = 4.50 \times 10^{-3} \text{ mol } H_3PO_4$

Next determine the number of moles of NaOH that are required by the balanced chemical equation:

$H_3PO_4 + 3NaOH \rightarrow Na_3PO_4 + 3H_2O$

$4.50 \times 10^{-3} \text{ mol } H_3PO_4 \times 2 \text{ mol NaOH/mol } H_3PO_4 = 0.0135 \text{ mol NaOH}$

Last calculate the volume of NaOH solution that will deliver this required number of moles of NaOH:

$0.0135 \text{ mol NaOH} \times 1000 \text{ mL}/0.200 \text{ mol} = 67.5 \text{ mL}$

Alternatively, we can write:

$0.0135 \text{ mol} \div 0.200 \text{ mol/L} = 0.0675 \text{ L}$

11.42 $NaHCO_3 + HCL \rightarrow NaCl + H_2O + CO_2$

$$150 \text{ mL HCl} \times \frac{0.050 \text{ mol HCl}}{1000 \text{ mL HCl}} \times \frac{1 \text{ mol } NaHCO_3}{1 \text{ mol HCl}} \times \frac{84.0 \text{ g } NaHCO_3}{1 \text{ mol } NaHCO_3} = 0.63 \text{ g } NaHCO_3$$

11.43 First calculate the number of moles of CO_2 that are formed:

$$150 \text{ mL HCl} \times \frac{0.050 \text{ mol HCl}}{1000 \text{ mL}} \times \frac{1 \text{ mol } CO_2}{1 \text{ mol HCl}} = 7.5 \times 10^{-3} \text{ mol } CO_2$$

Next determine the volume that this much CO_2 would occupy:

$V = nRT/P$

$$V = \frac{(0.0075 \text{ mol})(0.0821 \text{ L atm/K mol})(310 \text{ K})}{1.00 \text{ atm}} = 0.19 \text{ L}$$

11.44

$$35.0 \text{ mL } AgNO_3 \times \frac{0.600 \text{ mol } AgNO_3}{1000 \text{ mL } AgNO_3} \times \frac{1 \text{ mol } CaCl_2}{2 \text{ mol } AgNO_3} \times \frac{1000 \text{ mL } CaCl_2}{0.400 \text{ mol } CaCl_2} = 26.3 \text{ mL}$$

Alternatively, we can write:

$$0.0350 \text{ L} \times (0.600 \text{ mol/L}) \times \frac{1 \text{ mol } CaCl_2}{2 \text{ mol } AgNO_3} \times 1 \text{ L}/0.400 \text{ mol} = 0.0263 \text{ L}$$

11.45 First calculate the number of moles of NaOH that are involved:

$0.0600\ L \times 1.00\ mol/L = 0.0600\ mol\ NaOH$

Next determine the number of moles of NH_4^+ ion that are required, and the number of moles of $(NH_4)_2SO_4$ that are required:

$0.0600\ mol\ NaOH \times 1\ mol\ NH_4^+/mol\ NaOH = 0.0600\ mol\ NH_4^+$

$0.0600\ mol\ NH_4^+ \times 1\ mol\ (NH_4)_2SO_4/2\ mol\ NH_4^+ = 0.0300\ mol\ (NH_4)_2SO_4$

$0.0300\ mol\ (NH_4)_2SO_4 \div 0.300\ mol/L = 0.100\ L$

11.46 $Fe_2O_3 + 6HCl \rightarrow 2FeCl_3 + 3H_2O$

$0.0200\ L\ HCl \times 0.500\ mol/L = 1.00 \times 10^{-2}\ mol\ HCl$

$$1.00 \times 10^{-2}\ mol\ HCl \times \frac{1\ mol\ Fe_2O_3}{6\ mol\ HCl} \times \frac{2\ mol\ Fe^{3+}}{1\ mol\ Fe_2O_3} = 3.33 \times 10^{-3}\ mol\ Fe^{3+}$$

$$\frac{3.33 \times 10^{-3}\ mol\ Fe^{3+}}{0.0200\ L\ soln} = 0.167\ M\ Fe^{3+}$$

$$3.33 \times 10^{-3}\ mol\ Fe^{3+} \times \frac{1\ mol\ Fe_2O_3}{2\ mol\ Fe^{3+}} \times \frac{160\ g\ Fe_2O_3}{1\ mol\ Fe_2O_3} = 0.267\ g\ Fe_2O_3$$

Therefore, the mass of Fe_2O_3 that remains unreacted is:

$(5.00\ g - 0.267\ g) = 4.73\ g$

11.47 The equation for the reaction indicates that the two materials react in equimolar amounts, i.e. the stoichiometry is 1 to 1:

$AgNO_3(aq) + NaCl(aq) \rightarrow AgCl(s) + NaNO_3(aq)$

Also we are given a mole excess of sodium chloride, so $AgNO_3$ is the limiting reagent.

(a)

$$0.0300\ L\ AgNO_3 \times \frac{0.300\ mol\ AgNO_3}{1\ L\ AgNO_3} \times \frac{1\ mol\ AgCl}{1\ mol\ AgNO_3} = 9.00 \times 10^{-3}\ mol\ AgCl$$

(b) Assuming that AgCl is essentially insoluble, the concentration of silver ion can be said to be zero. The number of moles of chloride ion would be reduced by the precipitation of 9.00×10^{-3} mol AgCl, such that the final number of moles of chloride ion would be:

$0.0300 \text{ L} \times 0.400 \text{ mol/L} - 9.00 \times 10^{-3} \text{ mol} = 3.0 \times 10^{-3} \text{ mol } Cl^-$ remain

The final concentration of Cl^- is, therefore:

$3.0 \times 10^{-3} \text{ mol} \div 0.0600 \text{ L} = 0.050 \text{ M } Cl^-$

All of the original number of moles of NO_3^- and of Na^+ would still be present in solution, and their concentrations would be:

For NO_3^-:

$$0.0300 \text{ L } AgNO_3 \times \frac{0.300 \text{ mol } AgNO_3}{1 \text{ L } AgNO_3} \times \frac{1 \text{ mol } NO_3^-}{1 \text{ mol } AgNO_3} = 9.00 \times 10^{-3} \text{ mol } NO_3^-$$

$9.00 \times 10^{-3} \text{ mol}/0.0600 \text{ L} = 0.150 \text{ M } NO_3^-$

For Na^+:

$$0.0300 \text{ L NaCl} \times \frac{0.400 \text{ mol NaCl}}{1 \text{ L NaCl}} \times \frac{1 \text{ mol } Na^+}{1 \text{ mol NaCl}} = 1.20 \times 10^{-2} \text{ mol } Na^+$$

$1.20 \times 10^{-2} \text{ mol}/0.0600 \text{ L} = 0.200 \text{ M } Na^+$

11.48 (a) $3Ca(NO_3)_2 + 2Na_3PO_4 \rightarrow Ca_3(PO_4)_2 + 6NaNO_3$

First determine the initial number of moles of Ca^{2+} ion that are present:

$$0.0300 \text{ L } Ca(NO_3)_2 \times \frac{0.150 \text{ mol } Ca(NO_3)_2}{1 \text{ L } Ca(NO_3)_2 \text{ soln}} \times \frac{1 \text{ mol } Ca^{2+}}{1 \text{ mol } Ca(NO_3)_2}$$

$$= 4.50 \times 10^{-3} \text{ mol } Ca^{2+}$$

Next determine the initial number of moles of phosphate ion that are present:

$$0.0200 \text{ L } Na_3PO_4 \times \frac{0.200 \text{ mol } Na_3PO_4}{1 \text{ L } Na_3PO_4 \text{ soln}} \times \frac{1 \text{ mol } PO_4^{3-}}{1 \text{ mol } Na_3PO_4}$$

$$= 4.00 \times 10^{-3} \text{ mol } PO_4^{3-}$$

Now determine the number of moles of calcium ion that are required to react with this much phosphate ion, and compare the result to the amount of calcium ion that is available:

$$4.00 \times 10^{-3} \text{ mol } PO_4^{3-} \times \frac{3 \text{ mol } Ca^{2+}}{2 \text{ mol } PO_4^{3-}} = 6.00 \times 10^{-3} \text{ mol } Ca^{2+}$$

Since there is not this much Ca^{2+} available according to the above calculation, then we can conclude that Ca^{2+} must be the limiting reagent, and that subsequent calculations should be based on the number of moles of it that are present:

$$4.50 \times 10^{-3} \text{ mol } Ca^{2+} \times \frac{1 \text{ mol } Ca_3(PO_4)_2}{3 \text{ mol } Ca^{2+}} \times \frac{310 \text{ g } Ca_3(PO_4)_2}{1 \text{ mol } Ca_3(PO_4)_2}$$

$$= 0.465 \text{ g } Ca_3(PO_4)_2 \text{ are formed.}$$

(b) If we assume that the $Ca_3(PO_4)_2$ is completely insoluble, then its concentration may be said to be essentially zero. The concentrations of the other ions are determined as follows:

For nitrate:

$$0.0300 \text{ L } Ca(NO_3)_2 \times \frac{0.150 \text{ mol } Ca(NO_3)_2}{1 \text{ L } Ca(NO_3)_2} \times \frac{2 \text{ mol } NO_3^-}{1 \text{ mol } Ca(NO_3)_2}$$

$$= 9.00 \times 10^{-3} \text{ mol } NO_3^-$$

$$\frac{9.00 \times 10^{-3} \text{ mol } NO_3^-}{0.0500 \text{ L soln}} = 0.180 \text{ mol/L, } NO_3^-$$

For Na^+:

$$0.0200 \text{ L } Na_3PO_4 \times \frac{0.200 \text{ mol } Na_3PO_4}{1 \text{ L } Na_3PO_4} \times \frac{3 \text{ mol } Na^+}{1 \text{ mol } Na_3PO_4} =$$

$$= 1.20 \times 10^{-2} \text{ mol } Na^+$$

$$\frac{1.20 \times 10^{-2} \text{ mol } Na^+}{0.0500 \text{ L soln}} = 0.240 \text{ mol/L, } Na^+$$

For phosphate, we determine the number of moles that react with calcium:

$$4.50 \times 10^{-3} \text{ mol } Ca^{2+} \times \frac{2 \text{ mol } PO_4^{3-}}{3 \text{ mol } Ca^{2+}} = 3.00 \times 10^{-3} \text{ mol } PO_4^{3-}$$

and subtract from the original number of moles that were present:

$(4.00 - 3.00) \times 10^{-3}$ mol = 1.00×10^{-3} mol PO_4^{3-} that are unreacted.

This allows a calculation of the final phosphate concentration:

1.00×10^{-3} mol/0.0500 L = 0.0200 mol/L, PO_4^{3-}

11.49 $CuCO_3 + H_2SO_4 \rightarrow CuSO_4 + H_2O + CO_2(g)$

$$0.500 \text{ g } CuCO_3 \times \frac{1 \text{ mol } CuCO_3}{124 \text{ g } CuCO_3} \times \frac{1 \text{ mol } H_2SO_4}{1 \text{ mol } CuCO_3} \times \frac{1 \text{ L } H_2SO_4}{0.500 \text{ mol } H_2SO_4}$$

$$0.00806 \text{ L } H_2SO_4 = 8.06 \text{ mL}$$

11.50 $Mg(OH)_2 + 2HCl \rightarrow MgCl_2 + 2H_2O$

The number of moles of HCl that are added are:

0.0100 L HCl × 0.100 mol/L = 1.00×10^{-3} mol HCl

The number of moles of $Mg(OH)_2$ that react (dissolve) are:

1.00×10^{-3} mol HCl × 1 mol HCl/2 mol $Mg(OH)_2$ = 0.500×10^{-3} mol $Mg(OH)_2$

The amount of unreacted $Mg(OH)_2$ is thus:

2.00×10^{-3} mol - 0.500×10^{-3} mol = 1.50×10^{-3} mol $Mg(OH)_2$ unreacted.

The concentration of $Mg(OH)_2$ is essentially zero, since this material is said to be insoluble.

The concentration of chloride is :

1.00×10^{-3} mol/0.0700 L = 0.0143 mol/L Cl^-

The concentration of Mg^{2+} is:

0.500×10^{-3} mol/0.0700 L = 7.14×10^{-3} mol/L Mg^{2+}

Chapter 11

Titrations and Chemical Analysis

11.51 (a) Buret - a long glass tube fitted with a stopcock, graduated in mL, and used for the controlled, measured addition of a volume of a solution to a receiving flask.

(b) Titrant - the solution delivered from a buret during a titration.

(c) Indicator - a substance whose color change is taken to mark the end point in a titration.

(d) End point - that point during a titration when the indicator changes color, the titration is stopped, and the total added volume of the titrant is recorded.

(e) Stopcock - a valve at the end of a buret that is used to control the delivery of the titrant.

(f) Standard solution - a solution of accurately known concentration that can be used in titrations to determine the concentration of other solutions by reaction of those solutions with the standard until an end point is reached.

11.52 First calculate the number of moles HCl based on the titration according to the following equation:

$$NaOH(aq) + HCl(aq) \rightarrow NaCl(aq) + H_2O(\ell)$$

$$0.02400 \text{ L NaOH} \times \frac{0.1000 \text{ mol NaOH}}{1.00 \text{ L}} \times \frac{1 \text{ mol HCl}}{1 \text{ mol NaOH}} = 2.40 \times 10^{-3} \text{ mol HCl}$$

Next determine the concentration of the HCl solution:

$$2.40 \times 10^{-3} \text{ mol} \div 0.02000 \text{ L} = 0.120 \text{ M HCl}$$

11.53 (a) The balanced equation for the titration is:

$$NaOH(aq) + HC_2H_3O_2(aq) \rightarrow NaC_2H_3O_2(aq) + H_2O(\ell)$$

$$0.01340 \text{ L NaOH} \times \frac{0.500 \text{ mol NaOH}}{1 \text{ L NaOH}} \times \frac{1 \text{ mol } HC_2H_3O_2}{1 \text{ mol NaOH}} = 6.70 \times 10^{-3} \text{ mol } HC_2H_3O_2$$

$$6.70 \times 10^{-3} \text{ mol}/0.0100 \text{ L} = 0.670 \text{ M } HC_2H_2O_2$$

(b) First convert the density of vinegar to a value appropriate for one L of solution:

$$1.0 \text{ g/mL} \times 1000 \text{ mL/L} = 1000 \text{ g/L}$$

Next determine the mass of acetic acid that is present in this vinegar:

$$0.670 \text{ mol } HC_2H_3O_2 \times \frac{60.1 \text{ g } HC_2H_3O_2}{1 \text{ mol } HC_2H_3O_2} = 40.3 \text{ g } HC_2H_3O_2 \text{ 1 L vinegar}$$

The % by weight of acetic acid in vinegar solution is then given by the following:

$40.3 \text{ g } HC_2H_3O_2/L \div 1000 \text{ g/L} \times 100 = 4.03$ % acetic acid

This is the mass of acetic acid in one L of solution divided by the total mass of one L of solution, multiplied by 100.

11.54 Since lactic acid is monoprotic, it reacts with sodium hydroxide on a one to one mole basis:

$$0.0185 \text{ L NaOH} \times \frac{0.160 \text{ mol NaOH}}{1 \text{ L NaOH}} = 2.94 \times 10^{-3} \text{ mol NaOH}$$

$$= 2.94 \times 10^{-3} \text{ mol lactic acid}$$

11.55 (a)

$$\frac{1.174 \text{ g } BaSO_4}{233.4 \text{ g/mol}} = 5.030 \times 10^{-3} \text{ mol } BaSO_4$$

(b) as many moles as of $BaSO_4$, namely 5.030×10^{-3} mol $MgSO_4$

(c) First determine the mass of $MgSO_4$ that was present:

5.030×10^{-3} mol $MgSO_4 \times 120.4$ g/mol $= 0.6056$ g $MgSO_4$

and subtract this from the total mass of the sample to find the mass of water in the original sample:

$1.240 \text{ g} - 0.6056 \text{ g} = 0.63 \text{ g } H_2O$

(d) We need to know the number of moles of water that are involved:

$0.63 \text{ g} \div 18.0 \text{ g/mol} = 3.5 \times 10^{-2}$ mol H_2O

and the relative mole amounts of water and $MgSO_4$:

For $MgSO_4$, $5.030 \times 10^{-3}/5.030 \times 10^{-3} = 1.000$

For water, $3.5 \times 10^{-2}/5.030 \times 10^{-3} = 7.0$

Hence the formula is $MgSO_4 \cdot 7H_2O$

11.56 (a)

$$0.678 \text{ g AgCl} \times \frac{35.5 \text{ g Cl}}{143 \text{ g AgCl}} = 0.168 \text{ g Cl}$$

(b) Determine moles and then relative moles of Fe and Cl:

$0.168 \text{ g Cl} \div 35.5 \text{ g/mol} = 4.73 \times 10^{-3} \text{ mol Cl}$

$(0.300 - 0.168) \text{ g Fe} \div 55.8 \text{ g/mol Fe} = 2.37 \times 10^{-3} \text{ mol Fe}$

For Cl, the relative number of moles is:

$4.73 \times 10^{-3} \text{ mol}/2.37 \times 10^{-3} \text{ mol} = 2.00$

For Fe, the relative number of moles is:

$2.37 \times 10^{-3}/2.37 \times 10^{-3} = 1.00$

and the formula is seen to be $FeCl_2$.

11.57 (a) $NaOH(aq) + HCl(aq) \rightarrow NaCl(aq) + H_2O(\ell)$

$$0.03460 \text{ L} \times \frac{0.0500 \text{ mol}}{1 \text{ L}} = 1.73 \times 10^{-3} \text{ mol NaOH}$$

(b) The number of moles of HCl remaining after reaction with $CaCO_3$ is the same as the number neutralized by the NaOH titration:

1.73×10^{-3} moles.

(c) The number of moles of HCl that have reacted with $CaCO_3$ is equal to the difference between the initial number of moles of HCl and the number of moles that are neutralized by the subsequent titration with NaOH.

The initial number of moles of HCl:

$0.0500 \text{ L} \times 0.200 \text{ mol HCl/L solution} = 0.0100 \text{ mol HCl}$

The number of moles of HCl that have reacted with $CaCO_3$:

$0.0100 \text{ mol} - 1.73 \times 10^{-3} \text{ mol} = 8.3 \times 10^{-3} \text{ mol}$

The number of moles of $CaCO_3$ are then determined using the coefficients in the balanced equation:

$CaCO_3 + 2HCl \rightarrow CaCl_2 + H_2O + CO_2$

8.3×10^{-3} mol HCl × 1 mol $CaCO_3$/2 mol HCl = 4.2×10^{-3} mol $CaCO_3$

(d) First determine the mass of $CaCO_3$ that is represented:

4.2×10^{-3} mol $CaCO_3$ × 100 g/mol = 0.42 g $CaCO_3$

Next determine the % by weight of $CaCO_3$ in the rock:

0.42 g $CaCO_3$ ÷ 1.500 g total × 100 = 28 % $CaCO_3$

11.58

0.02940 L KOH × 0.0300 mol/L = 8.82×10^{-4} mol KOH

8.82×10^{-4} mol KOH × 1 mol aspirin/mol KOH = 8.82×10^{-4} mol aspirin

8.82×10^{-4} mol aspirin × 180 g/mol = 0.159 g aspirin

0.159 g aspirin/0.250 g total × 100 = 63.6 % aspirin

Equivalents and Equivalent Weights

11.59 Equivalents always react in a one-to-one mole ratio because one equivalent is defined to be the amount of an acid or a base that supplies one mole of H^+ (for an acid) and one mole of OH^- (for a base).

11.60 For complete neutralization, the number of eq/mol is determined by the formula of the acid or base. The total number of eq/mol for an acid is equal to the number of ionizable hydrogen atoms in the formula. Also see the answer to review exercise 11.61.

(a) 1 (b) 2 (c) 3 (d) 1

11.61 The number of eq/mol for an acid or a base depends on the number of H^+ or OH^- ions that are reacted. Although H_2SO_4 has two ionizable protons, only one of them is neutralized in (a) below.

(a) 1 (b) 3 (c) 2 (d) 2

11.62 Unlike mole amounts, which must be determined by reference to the balanced equation, and which depend on the coefficients in the balanced equation, equivalent amounts are defined to be the amounts of reagents that react with one another on a 1:1 equivalent basis.

(a) 0.250 eq $Ba(OH)_2$

(b) 0.150 eq $Ba(OH)_2$

(c) 0.440 eq $Ba(OH)_2$

(d) 0.350 eq $Ba(OH)_2$

11.63 (a) For complete neutralization, all 3 hydrogen ions are reacted:

$$1 \text{ eq } H_3PO_4 \times \frac{1 \text{ mol } H_3PO_4}{3 \text{ eq } H_3PO_4} \times \frac{98.0 \text{ g } H_3PO_4}{1 \text{ mol } H_3PO_4} = 32.7 \text{ g } H_3PO_4$$

(b) For neutralization of only one of the three ionizable protons:

$$1 \text{ eq } H_3PO_4 \times \frac{1 \text{ mol } H_3PO_4}{1 \text{ eq } H_3PO_4} \times \frac{98.0 \text{ g } H_3PO_4}{1 \text{ mol } H_3PO_4} = 98.0 \text{ g } H_3PO_4$$

11.64 (a)

$$15.0 \text{ g NaOH} \times \frac{1 \text{ mol NaOH}}{40.0 \text{ g NaOH}} \times \frac{1 \text{ eq NaOH}}{1 \text{ mol NaOH}} = 0.375 \text{ eq NaOH}$$

(b)

$$15.0 \text{ g } H_3AsO_4 \times \frac{1 \text{ mol } H_3AsO_4}{142 \text{ g } H_3AsO_4} \times \frac{3 \text{ eq } H_3AsO_4}{1 \text{ mol } H_3AsO_4} = 0.317 \text{ eq } H_3AsO_4$$

(c)

$$15.0 \text{ g } H_2S \times \frac{1 \text{ mol } H_2S}{34.1 \text{ g } H_2S} \times \frac{2 \text{ eq } H_2S}{1 \text{ mol } H_2S} = 0.880 \text{ eq } H_2S$$

(d)

$$15.0 \text{ g } Ca(OH)_2 \times \frac{1 \text{ mol } Ca(OH)_2}{74.1 \text{ g } Ca(OH)_2} \times \frac{2 \text{ eq } Ca(OH)_2}{1 \text{ mol } Ca(OH)_2} = 0.405 \text{ eq } Ca(OH)_2$$

11.65

$$45.0 \text{ g } Ba(OH)_2 \times \frac{2 \text{ eq } Ba(OH)_2}{171 \text{ g } Ba(OH)_2} \times \frac{1 \text{ eq } H_3PO_4}{1 \text{ eq } Ba(OH)_2} \times \frac{98.0 \text{ g } H_3PO_4}{3 \text{ eq } H_3PO_4}$$

$$= 17.2 \text{ g } H_3PO_4$$

<u>Normality</u>

11.66 The normality of a solution is the number of equivalents per liter of solution. Normality, like molarity, is a concentration, giving the amount of solute in a liter of solution. The difference between normality and molarity, however, is that normality is based on the number of equivalents per liter, and normality is generally some whole number multiple of molarity. This is because a given solute may react or ionize to deliver more than one equivalent per mole.

11.67

$$\frac{0.500 \text{ mol } H_2SO_4}{L} \times \frac{2 \text{ eq } H_2SO_4}{1 \text{ mol } H_2SO_4} \times \frac{1.00 \text{ eq } H_2SO_4}{L} = 1.00 \text{ N } H_2SO_4$$

11.68 (a) 0.200 mol/L H_3PO_4 × 3 eq/mol = 0.600 eq/L = 0.600 N H_3PO_4

(b) 0.200 mol/L H_3PO_4 × 1 eq/mol = 0.200 eq/L = 0.200 N H_3PO_4

(c) 0.200 mol/L H_3PO_4 × 2 eq/mol = 0.400 eq/L = 0.400 N H_3PO_4

11.69

$$0.250 \text{ L} \times \frac{0.0100 \text{ eq Ba(OH)}_2}{1 \text{ L Ba(OH)}_2} \times \frac{1 \text{ mol Ba(OH)}_2}{2 \text{ eq Ba(OH)}_2} \times \frac{171 \text{ g Ba(OH)}_2}{1 \text{ mol Ba(OH)}_2}$$

$$= 0.214 \text{ g Ba(OH)}_2$$

11.70 (0.500 N)(0.0350 L) = 0.0175 eq

11.71 When neutralization is accomplished, the number of equivalents of acid that have reacted will be equal to the number of equivalents of base that have reacted. Thus we have the following general equation for neutralization:

$$V_A \times N_A = V_B \times N_B$$

(a)

$$N_{H_2SO_4} = \frac{(0.03540 \text{ L})(0.140 \text{ N})}{0.04020 \text{ L}} = 0.123 \text{ N}$$

(b) 0.123 eq/L × 1 mol/2 eq = 0.0615 mol/L = 0.0615 M H_2SO_4

11.72 The number of equivalents of acid is equal to the number of equivalents of base, i.e. $V_b \times N_b$.

(a) 8.20×10^{-3} L × 0.100 eq/L = 8.20×10^{-4} eq of acid

eq. wt. = 0.100 g/8.20×10^{-4} eq = 122 g/eq of benzoic acid

(b) In this case the equivalent weight and the molecular weight are the same, and the acid is monoprotic. If the acid had been diprotic, the equivalent weight would have been equal to one half of the molecular weight.

11.73 (a) The number of equivalents of acid is determined as follows:

$(0.03125\ L)(0.100\ eq/L) = 3.13 \times 10^{-3}$ eq of acid

The equivalent weight is given by:

$0.200\ g/3.13 \times 10^{-3}\ eq = 63.9\ g/eq$

(b) An acid that is polyprotic has an equivalent weight that is some fraction of the molecular weight, whereas a monoprotic acid has an equivalent weight that is equal to the molecular weight. We found the latter case in review exercise 11.72. In review exercise 11.73, however, the equivalent weight is less than the molecular weight, and we must determine the whole number that, when multiplied by the equivalent weight, gives the molecular weight.

$192\ g/mol \div 63.9\ g/eq = 3.00\ eq/mol$

The acid is found to be triprotic, and citric acid can furnish three hydrogen ions per molecule.

Brønsted Acids and Bases

11.74 A Brønsted acid is a proton donor, whereas a Brønsted base is a proton acceptor.

11.75 (a) NH_4^+ (b) $N_2H_5^+$ (c) $C_5H_5NH^+$ (d) HCO_3^- (e) NH_3

11.76 (a) NH_2^- (b) CO_3^{2-} (c) CN^- (d) $H_4IO_6^-$ (e) NH_3

11.77 (a) HNO_3 (acid) $\Leftrightarrow$ NO_3^- (base) ; N_2H_4 (base) $\Leftrightarrow$ $N_2H_5^+$ (acid)

(b) NH_3 (base) $\Leftrightarrow$ NH_4^+ (acid) ; $N_2H_5^+$ (acid) $\Leftrightarrow$ N_2H_4 (base)

(c) $H_2PO_4^-$ (acid) $\Leftrightarrow$ HPO_4^{2-} (base) ; CO_3^{2-} (base) $\Leftrightarrow$ HCO_3^- (acid)

(d) HIO_3 (acid) $\Leftrightarrow$ IO_3^- (base) ; $HC_2O_4^-$ (base) $\Leftrightarrow$ $H_2C_2O_4$ (acid)

11.78 (a) HSO_4^- (acid) ⇔ SO_4^{2-} (base) ; SO_3^{2-} (base) ⇔ HSO_3^- (acid)

(b) S^{2-} (base) ⇔ HS^- (acid) ; H_2O (acid) ⇔ OH^- (base)

(c) CN^- (base) ⇔ HCN (acid) ; H_3O^+ (acid) ⇔ H_2O (base)

(d) H_2Se (acid) ⇔ HSe^- (base) ; H_2O (base) ⇔ H_3O^+ (acid)

11.79 An amphoteric substance can act either as an acid or as a base.

$HCl + H_2O \rightarrow H_3O^+ + Cl^-$, in which water serves as a base

$NH_3 + H_2O \rightarrow OH^- + NH_4^+$, in which water serves as an acid

11.80 HOCl (acid) ⇔ OCl^- (base); and OCl^- is a stronger base than water

H_2O (base) ⇔ H_3O^+ (acid), and H_3O^+ is a stronger acid than HOCl

11.81 The equilibrium would lie to the left, more so than to the right, so as to lie in the direction which favors formation of the weaker acid and base.

11.82 $C_2H_3O_2^-$ is a stronger base than NO_2^-.

11.83 This equilibrium should lie to the left, because if HNO_3 is a strong acid, then, by definition, NO_3^- must be a weak base.

Trends in Acid-Base Strength

11.84 (a) H_2Se (b) HI (c) HIO_4 (d) $HClO_4$

11.85 (a) HBr (b) HF (c) H_3AsO_4 (d) HBr

11.86 The element would be classified as a metal if it had a characteristic metallic luster, was a good conductor of heat and electricity, was ductile and malleable, formed ionic compounds with nonmetals and formed an oxide that was a basic anhydride.

11.87 (a) The relative strength of the binary acids increases from top to bottom in a group of the periodic table, so we expect HAt to be a stronger acid than HI.

(b) $HAtO_4$

11.88 In nitric acid, there are more oxygen atoms bound to the nitrogen atom than in nitrous acid.

11.89 The relative strengths of the binary acids increases from top to bottom in a group of the periodic table, so we expect H_2S to be a stronger acid than H_2O.

11.90 There are more oxygen atoms not attached to protons in $HClO_4$ than in H_2SeO_4.

Lewis Acids and Bases

11.91 Acid - in the formation of a coordinate covalent bond, an electron pair acceptor.

Base - in the formation of a coordinate covalent bond, an electron pair donor.

11.92 The H^+ ion accepts a pair of electrons from the oxygen atom of the water molecule. This makes the H^+ ion the Lewis acid, whereas the H_2O molecule is the Lewis base.

11.93

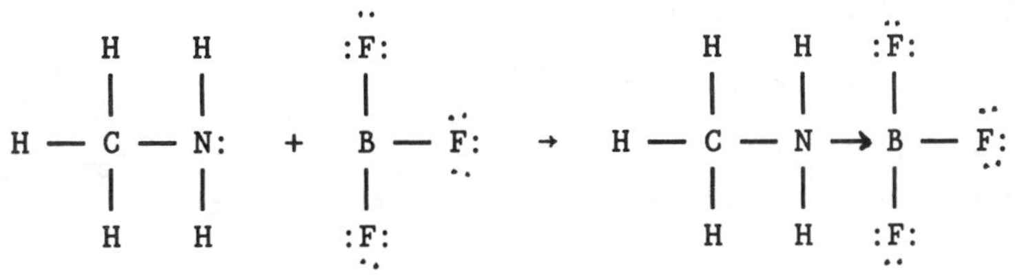

```
        H   H        :F:                 H    H   :F:
        |   |         |                  |    |    |
    H — C — N:   +    B — F:    →    H — C —  N →  B — F:
        |   |         |                  |    |    |
        H   H        :F:                 H    H   :F:
```

11.94

```
 Cl          Cl          Cl
    \      ↙(:)  \      /
      Al           Al
    /     \(:)↗      \
 Cl          Cl          Cl
```

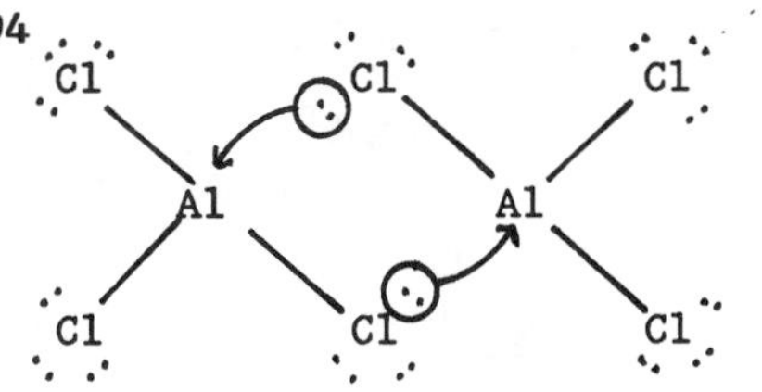

11.95 The acid is SO_2, and the base is H_2O:

$$H-\ddot{O}:(-H) \rightarrow S(=O)(-O) \rightarrow \left[H-\ddot{O}(-H)-S(-O)(-O)\right] \rightarrow H-\ddot{O}-S(-O)-O-H$$

11.96 The O^{2-} ion has a complete valence shell. It can donate electrons and serve, thereby, as a Lewis base. It cannot, however, serve as a Lewis acid by accepting more electrons.

Complex Ions

11.97 The metal ion is a Lewis acid, accepting a pair of electrons from the ligand, which serves as a Lewis base.

(a) The Lewis acid is Cu^{2+}, and the Lewis base is H_2O.

(b) The ligand is H_2O.

(c) Water provides the donor atom, oxygen.

(d) Oxygen is the donor atom, because it is attached to the copper ion.

(e) The copper ion is the acceptor.

11.98 A ligand serves as a Lewis base because it donates the two electrons that are required for the formation of the coordinate covalent bond with the metal ion.

11.99 They are formed by the use of coordinate covalent bonds.

11.100 water, ammonia

11.101 F^-, Cl^-, Br^-, I^-

11.102

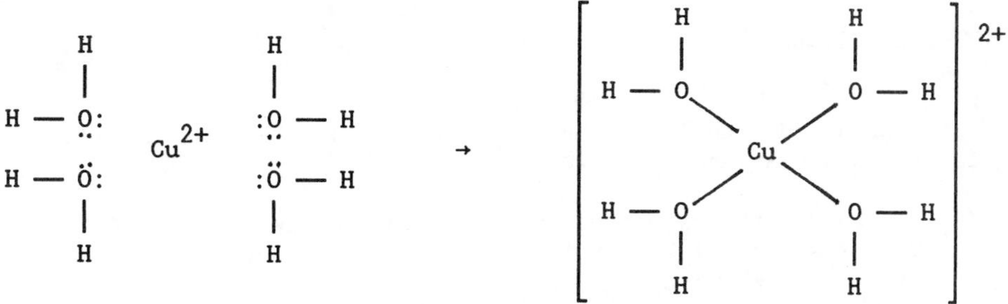

11.103 A bidentate ligand must have two atoms that can form a coordinate covalent bond with a metal to give a ring including the metal.

11.104 EDTA has six donor atoms.

11.105 EDTA forms complexes with the sorts of metal ions that would otherwise promote spoilage.

11.106 EDTA forms complexes with the metal ions that are otherwise responsible for hardness.

11.107 The net charge is 1-, and the formula is $[Co(EDTA)]^-$.

11.108 The net charge is 3-, and the formula is $[Fe(CN)_6]^{3-}$.

11.109 This is the ion $[Ag(NH_3)_2]^+$, which can exist as the chloride salt: $[Ag(NH_3)_2]Cl$.

11.110 The bonds to NH_3 are known to be stronger, because the test for copper(II) ion requires the ready displacement of water ligands by ammonia ligands to give the ion $[Cu(NH_3)_4]^{2+}$, which has a recognizable deep blue color.

CHAPTER TWELVE

PRACTICE EXERCISES

1. Ca is oxidized, and it is the reducing agent. Chlorine is reduced, and it is the oxidizing agent.

2. (a) The sum of the oxidation numbers should be zero:

   Cl $2 \times (-1) = -2$
   Ni $1 \times (+2) = +2$

   $\therefore$ Ni^{2+} and Cl^{-}

   (b) The sum of the oxidation numbers should be zero:

   O $4 \times (-2) = -8$
   Mg $2 \times (+2) = +4$
   Ti $1 \times (+4) = +4$

   $\therefore$ Mg^{2+}, O^{2-}, and Ti^{4+}

   (c) The sum of the oxidation numbers should be zero:

   K $2 \times (+1) = +2$
   O $7 \times (-2) = -14$
   Cr $2 \times (+6) = +12$

   $\therefore$ K^{+}, O^{2-}, and Cr^{6+}

   (d) The sum of the oxidation numbers should be -2:

   O $4 \times (-2) = -8$
   S $1 \times (+6) = +6$

   $\therefore$ S^{6+} and O^{2-}

3. The sum of the oxidation numbers should be zero:

   O $1 \times (-2) = -2$
   H $8 \times (+1) = +8$
   C $3 \times (n) = -6$

   $\therefore$ n = -2, and we have O^{2-}, H^{+}, and on the average, C^{2-}

4. (a) List the oxidation number above each element in the various compounds:

$$\overset{1+}{K}\,\overset{1-}{Cl} + \overset{4+}{Mn}\,\overset{2-}{O_2} + \overset{1+}{H_2}\,\overset{6+}{S}\,\overset{2-}{O_4} \rightarrow \overset{1+}{K_2}\,\overset{6+}{S}\,\overset{2-}{O_4} + \overset{2+}{Mn}\,\overset{6+}{S}\,\overset{2-}{O_4} + \overset{0}{Cl_2} + \overset{1+}{H_2}\,\overset{2-}{O}$$

   We learn that manganese is reduced from (4+) to (2+) and that chlorine is oxidized from (1-) to (0). We therefore need two Cl on the left, in order to make oxidation and reduction changes equal to one another (i.e. 2 e^{-}). Once oxidation/reduction is balanced, the rest can be balanced by inspection:

$2KCl + MnO_2 + 2H_2SO_4 \rightarrow K_2SO_4 + MnSO_4 + Cl_2 + H_2O$

(b)

$$\overset{1+}{K}\ \overset{7+}{Mn}\ \overset{2-}{O_4} + \overset{2+}{Fe}\ \overset{6+}{S}\ \overset{2-}{O_4} + \overset{1+}{H_2}\ \overset{6+}{S}\ \overset{2-}{O_4} \rightarrow \overset{1+}{K_2}\ \overset{6+}{S}\ \overset{2-}{O_4} + \overset{2+}{Mn}\ \overset{6+}{S}\ \overset{2-}{O_4} + \overset{3+}{Fe_2}(\overset{6+}{S}\ \overset{2-}{O_4})_3 + \overset{1+}{H_2}\ \overset{2-}{O}$$

Manganese is reduced from (7+) to (2+), a five-electron change; iron is oxidized from (2+) to (3+), a one-electron change. Now the oxidation/reduction could be balanced by multiplying the oxidation part alone by (5). However, we note by inspection that the right hand side of the equation contains an even number of Fe atoms, and we realize that so, too, will the left hand side, once balanced. Hence, we balance oxidation/reduction by multiplying the reduction part by (2) and the oxidation part by (10):

$2KMnO_4 + 10FeSO_4 + H_2SO_4 \rightarrow K_2SO_4 + 2MnSO_4 + 5Fe_2(SO_4)_3 + H_2O$

Now that the oxidation/reduction parts are balanced, we balance the remainder by inspection:

$2KMnO_4 + 10FeSO_4 + 8H_2SO_4 \rightarrow K_2SO_4 + 2MnSO_4 + 5Fe_2(SO_4)_3 + 8H_2O$

(c)

$$\overset{0}{Zn} + \overset{1+}{H}\ \overset{5+}{N}\ \overset{2-}{O_3} + \overset{1+}{H}\ \overset{5+}{N}\ \overset{2-}{O_3} \rightarrow \overset{2+}{Zn}(\overset{5+}{N}\ \overset{2-}{O_3})_2 + \overset{3-}{N}\ \overset{1+}{H_4}\ \overset{5+}{N}\ \overset{2-}{O_3} + \overset{1+}{H_2}\ \overset{2-}{O}$$

We conclude that Zn is oxidized from (0) to (2+), a 2-electron change. The nitrogen from one of the HNO_3 molecules is reduced from (5+) to

(3-), an 8-electron change. We balance the oxidation/reduction part of the equation, therefore, by multiplying the oxidation part by (4), so that both the oxidation and the reduction parts involve changes in oxidation numbers of eight:

$4Zn + HNO_3 + HNO_3 \rightarrow 4Zn(NO_3)_2 + NH_4NO_3 + H_2O$

The rest of the equation is balanced by inspection:

$4Zn + HNO_3 + 9HNO_3 \rightarrow 4Zn(NO_3)_2 + NH_4NO_3 + 3H_2O$

or simply:

$4Zn + 10HNO_3 \rightarrow 4Zn(NO_3)_2 + NH_4NO_3 + 3H_2O$

5. $3 \times (Cu \rightarrow Cu^{2+} + 2e^-)$

$2 \times (4H^+ + 3e^- + NO_3^- \rightarrow NO + 2H_2O)$

$3Cu + 8H^+ + 2NO_3^- \rightarrow 3Cu^{2+} + 2NO + 4H_2O$

6. $2 \times (4H^+ + 3e^- + MnO_4^- \rightarrow MnO_2 + 2H_2O)$

$3 \times (2H_2O + C_2O_4^{2-} \rightarrow 2CO_3^{2-} + 4H^+ + 2e^-)$

$8H^+ + 2MnO_4^- + 6H_2O + 3C_2O_4^{2-} \rightarrow 2MnO_4 + 4H_2O + 6CO_3^{2-} + 12H^+$

which simplifies to:

$2MnO_4^- + 2H_2O + 3C_2O_4^{2-} \rightarrow 2MnO_2 + 6CO_3^{2-} + 4H^+$

Next, we convert to basic conditions by adding (titrating) OH^- to both sides of the equation:

$4OH^- + 2MnO_4^- + 2H_2O + 3C_2O_4^{2-} \rightarrow 2MnO_2 + 6CO_3^{2-} + 4H_2O$

which simplifies to:

$4OH^- + 2MnO_4^- + 3C_2O_4^{2-} \rightarrow 2MnO_2 + 2H_2O + 6CO_3^{2-}$

7. (a) molecular:

$Mg(s) + 2HCl(aq) \rightarrow MgCl_2(aq) + H_2(g)$

ionic:

$Mg(s) + 2H^+(aq) + 2Cl^-(aq) \rightarrow Mg^{2+}(aq) + 2Cl^-(aq) + H_2(g)$

net ionic:

$Mg(s) + 2H^+(aq) \rightarrow Mg^{2+}(aq) + H_2(g)$

(b) molecular:

$2Al(s) + 6HCl(aq) + 2AlCl_3(aq) + 3H_2(g)$

ionic:

$2Al(s) + 6H^+(aq) + 6Cl^-(aq) \rightarrow 2Al^{3+}(aq) + 6Cl^-(aq) + 3H_2(g)$

net ionic:

$2Al(s) + 6H^+(aq) \rightarrow 2Al^{3+}(aq) + 3H_2(g)$

8. (a) $Mg(s) + Cu^{2+}(aq) \rightarrow Cu(s) + Mg^{2+}(aq)$

(b) no reaction

9. $2C_4H_{10} + 13O_2 \rightarrow 8CO_2 + 10H_2O$

10. $C_2H_5OH + 3O_2 \rightarrow 2CO_2 + 3H_2O$

11. $4Fe + 3O_2 \rightarrow 2Fe_2O_3$

12. $P_4 + 5O_2 \rightarrow P_4O_{10}$

13. (a) $5 \times (Fe^{2+}(aq) \rightarrow Fe^{3+}(aq) + e^-)$

$$\underline{(MnO_4^-(aq) + 8H^+(aq) + 5e^- \rightarrow Mn^{2+}(aq) + 4H_2O)}$$

$$5Fe^{2+}(aq) + MnO_4^-(aq) + 8H^+(aq) \rightarrow 5Fe^{3+}(aq) + Mn^{2+}(aq) + 4H_2O$$

(b) $6 \times (Fe^{2+}(aq) \rightarrow Fe^{3+}(aq) + e^-)$

$$\underline{Cr_2O_7^{2-}(aq) + 14H^+(aq) + 6e^- \rightarrow 2Cr^{3+}(aq) + 7H_2O}$$

$$6Fe^{2+}(aq) + Cr_2O_7^{2-}(aq) + 14H^+(aq) \rightarrow 6Fe^{3+}(aq) + 2Cr^{3+}(aq) + 7H_2O$$

14. (a) $H_2O + HSO_3^-(aq) \rightarrow SO_4^{2-}(aq) + 3H^+(aq) + 2e^-$

(b) $H_2O + H_2SO_3(aq) \rightarrow SO_4^{2-}(aq) + 4H^+(aq) + 2e^-$

15. $3 \times (S_2O_8^{2-}(aq) + 2e^- \rightarrow 2SO_4^{2-}(aq))$

$$\underline{7H_2O + 2Cr^{3+}(aq) \rightarrow Cr_2O_7^{2-}(aq) + 14H^+(aq) + 6e^-}$$

$$3S_2O_8^{2-}(aq) + 7H_2O + 2Cr^{3+}(aq) \rightarrow Cr_2O_7^{2-}(aq) + 6SO_4^{2-}(aq) + 14H^+(aq)$$

$$34.0\text{ g } Cr(NO_3)_3 \times \frac{1\text{ mol } Cr(NO_3)_3}{238\text{ g } Cr(NO_3)_3} \; \frac{3\text{ mol } K_2S_2O_8}{2\text{ mol } Cr(NO_3)_3} \times \frac{270\text{ g } K_2S_2O_8}{1\text{ mol } K_2S_2O_8} = 57.9\text{ g } K_2S_2O_8$$

16. $6 \times (Fe^{2+}(aq) \rightarrow Fe^{3+}(aq) + e^-)$

$$\frac{Cr_2O_7^{2-}(aq) + 14H^+(aq) + 6e^- \rightarrow 2Cr^{3+}(aq) + 7H_2O}{6Fe^{2+}(aq) + Cr_2O_7^{2-}(aq) + 14H^+(aq) \rightarrow 6Fe^{3+}(aq) + 2Cr^{3+}(aq) + 7H_2O}$$

$$0.0500\ L\ Cr_2O_7^{2-} \times \frac{0.400\ mol\ Cr_2O_7^{2-}}{1\ L\ Cr_2O_7^{2-}} \times \frac{6\ mol\ Fe^{2+}}{1\ mol\ Cr_2O_7^{2-}} \times \frac{1\ L\ Fe^{2+}}{0.200\ mol\ Fe^{2+}}$$

$$= 0.600\ L\ FeSO_4$$

17. (a) $5 \times (Sn^{2+}(aq) \rightarrow Sn^{4+}(aq) + 2e^-)$

$$\frac{2 \times (MnO_4^-(aq) + 8H^+(aq) + 5e^- \rightarrow Mn^{2+}(aq) + 4H_2O)}{5Sn^{2+}(aq) + 2MnO_4^-(aq) + 16H^+(aq) \rightarrow 5Sn^{4+}(aq) + 2Mn^{2+}(aq) + 8H_2O(aq)}$$

(b)

$$8.08\ mL\ KMnO_4 \times \frac{0.0500\ mol\ KMnO_4}{1000\ mL\ KMnO_4} \times \frac{5\ mol\ Sn}{2\ mol\ KMnO_4} \times \frac{119\ g\ Sn}{1\ mol\ Sn} = 0.120\ g\ Sn$$

(c) $(0.120/0.300) \times 100 = 40.0$ % Sn in the sample

(d)

$$0.120\ g\ Sn \times \frac{1\ mol\ Sn}{119\ g\ Sn} \times \frac{1\ mol\ SnO_2}{1\ mol\ Sn} \times \frac{151\ g\ SnO_2}{1\ mol\ SnO_2} = 0.152\ g\ SnO_2$$

Hence the percent SnO_2 in the original sample would be:

$(0.152\ g\ SnO_2/0.300\ g\ total) \times 100 = 50.7$ % SnO_2

18. (a) $I_3^-(aq) + 2S_2O_3^{2-}(aq) \rightarrow 3I^-(aq) + S_4O_6^{2-}(aq)$

$$.02820\ L\ S_2O_3^{2-} \times \frac{0.0500\ mol\ S_2O_3^{2-}}{1\ L\ S_2O_3^{2-}} \times \frac{1\ mol\ I_3^-}{2\ mol\ S_2O_3^{2-}}$$

$$= 7.05 \times 10^{-4}\ mol\ I_3^-$$

(b) Since the given equation has a stoichiometry of 1:1, we know that the number of moles of I_3^- that are formed are equal to the number of moles of OCl^- that have reacted: 7.05×10^{-4} mol OCl^-.

(c) 7.05×10^{-4} mol/1.00×10^{-3} L = 0.705 mol/L OCl^-

19. The oxidation number of each carbon atom in $H_2C_2O_4$ is 3+, whereas the oxidation state of each carbon atom in CO_2 is 4+. The reaction thus represents a change in oxidation state by one unit. Since there are 2 carbon atoms in $H_2C_2O_4$, there are then 1 × 2 = 2 eq/mol when $H_2C_2O_4$ is oxidized to CO_2.

1 mol $H_2C_2O_4$ = 2 eq $H_2C_2O_4$

20. The Cr atom of Na_2CrO_4 is in the +6 oxidation state, whereas the product ion has the 3+ oxidation state. There is hence a 3 e^- reduction of the Cr atom on going from Na_2CrO_4 to Cr^{3+}. ∴ 1 mol Na_2CrO_4 = 3 eq Na_2CrO_4

$$1 \text{ eq } Na_2CrO_4 \times \frac{1 \text{ mol } Na_2CrO_4}{3 \text{ eq } Na_2CrO_4} \times \frac{162 \text{ g } Na_2CrO_4}{1 \text{ mol } Na_2CrO_4} = 54.0 \text{ g } Na_2CrO_4$$

21. As determined in Practice exercise 20, the equivalent weight of Na_2CrO_4, when it is reduced by three electrons, is 54.0 g/eq. Also in this reaction we have the oxidation of sulfur in SO_3^{2-} (oxidation state 4+) to sulfur in SO_4^{2-} (oxidation state 6+). Thus, the equivalent weight of Na_2SO_3 is equal to one-half the molecular weight, i.e. 126 ÷ 2 = 63 g/eq. Remember also that, whereas redox reagents may not react on a 1:1 mole basis, by definition they always react on a 1:1 eq equivalent basis:

$$12.0 \text{ g } Na_2CrO_4 \times \frac{1 \text{ eq } Na_2CrO_4}{54.0 \text{ g } Na_2CrO_4} \times \frac{1 \text{ eq } Na_2SO_3}{1 \text{ eq } Na_2CrO_4} \times \frac{1 \text{ mol } Na_2SO_3}{2 \text{ eq } Na_2SO_3}$$

$$\times \frac{126 \text{ g } Na_2SO_3}{1 \text{ mol } Na_2SO_3} = 14.0 \text{ g } Na_2SO_3$$

Alternatively, we can write:

$$12.0 \text{ g } Na_2CrO_4 \times \frac{1 \text{ eq } Na_2CrO_4}{54.0 \text{ g } Na_2CrO_4} \times \frac{1 \text{ eq } Na_2SO_3}{1 \text{ eq } Na_2CrO_4} \times \frac{63 \text{ g } Na_2SO_3}{1 \text{ eq } Na_2SO_3}$$

$= 14.0 \text{ g } Na_2SO_3$

22. (a) Mn goes from oxidation state (+7) to oxidation state (+2), and since this is a change of 5 units, there are thus 5 eq/mol in $KMnO_4$.

(b) 0.200 mol/L × 5 eq/mol = 1.00 eq/L = 1.00 N $KMnO_4$

(c) No, because this would represent a different case, wherein the oxidation state change of manganese is only 7 - 4 = 3 units, meaning only 3 eq/mol.

23. $N_AV_A = N_BV_B$, where it is alsways true that reagents react on a 1:1 equivalent basis.

$Na_{Na_2SO_3} = (N_{KMnO_4})(V_{KMnO_4})/(V_{Na_2SO_3})$

$N_{Na_2SO_3} = (0.231 \text{ N})(31.05 \text{ mL})/25.00 \text{ mL} = 0.287 \text{ N}$

24. The tin reagent reacts as below:

$Sn^{2+} \rightarrow Sn^{4+} + 2e^-$

and there are thus seen to be 2 eq/mol for this reagent.

$$0.2168 \text{ L } KMnO_4 \times \frac{0.231 \text{ eq } KMnO_4}{1 \text{ L } KMnO_4} \times \frac{1 \text{ eq Sn}}{1 \text{ eq } KMnO_4} \times \frac{1 \text{ mol Sn}}{2 \text{ eq Sn}} = 0.00250 \text{ mol Sn}$$

25. (a) 0.02730 L × 0.216 eq/L = 5.90×10^{-3} eq $KMnO_4$

(b) Since reagents always combine in equal amounts of equivalents by definition, then the number of equivalents of $KMnO_4$ is equal to the number of equivalents of Na_2SO_3, i.e. 5.90×10^{-3} eq of Na_2SO_3.

(c) The number of equivalents in the original 50 mL of Na_2SO_3 solution is:

$0.0500 \text{ L } Na_2SO_3 \times 0.400 \text{ eq/L} = 2.00 \times 10^{-2} \text{ eq } Na_2SO_3$.

(d) This is equal to the number of equivalents of Na_2SO_3 that have reacted originally with OCl^-, namely:

$2.00 \times 10^{-2} \text{ eq} - 5.90 \times 10^{-3} \text{ eq} = 1.41 \times 10^{-2} \text{ eq } OCl^-$ have reacted.

(e) The reaction of the OCl^- with Na_2SO_3 requires a 2-electron change in oxidation state for OCl^-, because the oxidation state of Cl in OCl^- is

(+1), whereas that in Cl^- is (-1). There are thus 2 eq/mol of OCl^- in this process. The number of moles of OCl^- that have reacted are thus:

1.41×10^{-2} eq OCl^- $\times$ 1 mol/2 eq = 7.05×10^{-3} mol OCl^-

(f) First determine the mass of NaOCl that is involved:

7.05×10^{-3} mol $\times$ 74.4 g/mol = 0.525 g NaOCl

and next the % by weight:

0.525 g NaOCl/10.0 g total $\times$ 100 = 5.25 %

REVIEW EXERCISES

Oxidation-Reduction

12.1 Oxidation - loss of electrons, which corresponds to an increase in oxidation number.

Reduction - gain of electrons, which corresponds to a decrease in oxidation number.

12.2 The hydrogen atom loses electron density to the more electronegative chlorine atom as the polar covalent bond is formed.

12.3 The Mg atom loses two electrons and each chlorine atom gains one of these electrons as the Mg^{2+} and Cl^- ions are formed.

12.4 In both of these cases, one element loses electron density (is oxidized) and the other gains electron density (is reduced).

12.5 Hydrogen is oxidized and is the reducing agent; chlorine is reduced and is the oxidizing agent.

12.6 The number of electrons involved in both the reduction and the oxidation must be the same; only those electrons that come from the reductant and go to the oxidant are involved. No electrons from external, or uninvolved sources are allowed to enter the process, and there cannot be any electrons left unreacted at the end.

Chapter 12

Oxidation Numbers

12.7 An oxidation number is the "charge" an atom would have if all of the bond electrons were assigned to atoms so as to give both electrons of any given bond to the more electronegative atom.

12.8 The sum of the oxidation numbers should be zero:

O 6 × (2-) = -12
As 4 × (n) = +12

∴ n = 3+ and we have O^{2-} and As^{3+}

12.9 (a) 2-

(b) The sum of the oxidation numbers should be zero:

O 2 × (2-) = 4-
S 1 × (4+) = 4+ ∴ S is 4+

(c) The oxidation state of an element is always zero, by definition.

(d) The sum of the oxidation numbers should be zero:

H 3 × (+1) = 3+
P 1 × (-3) = 3- ∴ P is 3-

12.10 (a) The sum of the oxidation numbers should be -1:

O 4 × (2-) = 8-
Cl 1 × (n) = 7+ ∴ n = 7+ and Cl = 7+

(b) The sum of the oxidation numbers should be zero:

Cl 3 × (1-) = 3-
Cr 1 × (n) = 3+ ∴ n = 3+ and Cr = 3+

(c) The sum of the oxidation numbers should be zero:

S 2 × (2-) = 4-
Sn 1 × (n) = 4+ ∴ n = 4+ and Sn = 4+

(d) The sum of the oxidation numbers should be zero:

NO_3^- 3 × (1-) = 3-

Au 1 × (n) = 3+ ∴ n = 3+ and Au = 3+

12.11 The sum of the oxidation numbers should be zero:

(a) O $4 \times (2-) = 8-$
Na $2 \times (1+) = 2+$
H $1 \times (1+) = 1+$
P $1 \times (5+) = 5+$

(b) Ba $1 \times (2+) = 2+$
O $4 \times (2-) = 8-$
Mn $1 \times (6+) = 6+$

(c) Na $2 \times (1+) = 2+$
O $6 \times (2-) = 12-$
S $4 \times (n) = 10+$ $\therefore$ $n = 2.5+$ and S $= 2.5+$

(d) F $3 \times (1-) = 3-$
Cl $1 \times (3+) = 3+$

12.12 The sum of the oxidation numbers should be equal to the total, or overall charge on the ion:

(a) O $3 \times (2-) = 6-$
N $1 \times (5+) = 5+$

(b) O $3 \times (2-) = 6-$
S $1 \times (4+) = 4+$

(c) O $1 \times (2-) = 2-$
N $1 \times (3+) = 3+$

(d) O $7 \times (2-) = 14-$
Cr $2 \times (6+) = 12+$

12.13 The sum of the oxidation numbers should be zero:

(a) O $1 \times (2-) = 2-$
N $1 \times (2+) = 2+$

(b) O $2 \times (2-) = 4-$
N $1 \times (4+) = 4+$

(c) O $3 \times (2-) = 6-$
N $2 \times (3+) = 6+$

(d) O $5 \times (2-) = 10-$
N $2 \times (5+) = 10+$

(e) H $4 \times (1+) = 4+$
N $2 \times (2-) = 4-$

(f) zero, since this is the element

(g) O $1 \times (2-) = 2-$
H $3 \times (1+) = 3+$
N $1 \times (1-) = 1-$

(h) H $3 \times (1+) = 3+$
N $1 \times (3-) = 3-$

(i) Na $1 \times (1+) = 1+$
N $3 \times (n) = 1- \quad \therefore n = \frac{1}{3}-$

12.14 This is not a redox reaction since there is no change in oxidation number.

12.15 The sum of the oxidation numbers should be zero:

(a) O $1 \times (2-) = 2-$
Na $1 \times (1+) = 1+$
Cl $1 \times (1+) = 1+$

(b) O $2 \times (2-) = 4-$
Na $1 \times (1+) = 1+$
Cl $1 \times (3+) = 3+$

(c) O $3 \times (2-) = 6-$
Na $1 \times (1+) = 1+$
Cl $1 \times (5+) = 5+$

(d) O $4 \times (2-) = 8-$
Na $1 \times (1+) = 1+$
Cl $1 \times (7+) = 7+$

12.16 (a) O $6 \times (2-) = 12-$
Ca $1 \times (2+) = 2+$
V $2 \times (n) = 10+ \quad \therefore n = 5$ and $V = 5+$

(b) Cl $4 \times (1-) = 4-$
Sn $1 \times (4+) = 4+$

(c) The sum of the oxidation numbers should be 2-:

O $4 \times (2-) = 8-$
Mn $1 \times (6+) = 6+$

(d) O $2 \times (2-) = 4-$
Mn $1 \times (4+) = 4+$

12.17 The sum of the oxidation numbers should be zero:

(a) S $1 \times (2-) = 2-$
Pb $1 \times (2+) = 2+$

(b) Cl $4 \times (1-) = 4-$
Ti $1 \times (4+) = 4+$

(c) O $6 \times (2-) = 12-$
Sr $1 \times (2+) = 2+$
I $2 \times (n) = 10+$ $\therefore$ n = 5 and I = 5+

(d) S $3 \times (2-) = 6-$
Cr $2 \times (n) = 6+$ $\therefore$ n = 3 and Cr = 3+

12.18 (a) O $1 \times (2-) = 2-$
H $6 \times (1+) = 6+$
C $2 \times (n) = 4-$ $\therefore$ n = 2- and C = 2-

(b) O $11 \times (2-) = 22-$
H $22 \times (1+) = 22+$
C $12 \times (n) = 0$ $\therefore$ n = 0 and C = 0

(c) O $3 \times (2-) = 6-$
Ca $1 \times (2+) = 2+$
C $1 \times (4+) = 4+$

(d) O $3 \times (2-) = 6-$
Na $1 \times (1+) = 1+$
H $1 \times (1+) = 1+$
C $1 \times (4+) = 4+$

12.19 (a) Assign the two electrons of the bond to the more electronegative atom, Cl. This gives the artificial charges Cl^- and Br^+.

(b) Assign the electrons of the two S—Cl linkages to the more electronegative atom, Cl. This gives two Cl^- and one S^{2+}.

12.20 This change in oxidation number represents a reduction of nitrogen by 5 units, and it requires that nitrogen gain 5 electrons.

Chapter 12

<u>Balancing Redox Equations Using Oxidation Numbers</u>

12.21 (a) First, diagram the changes in oxidation state that take place for N and for As:

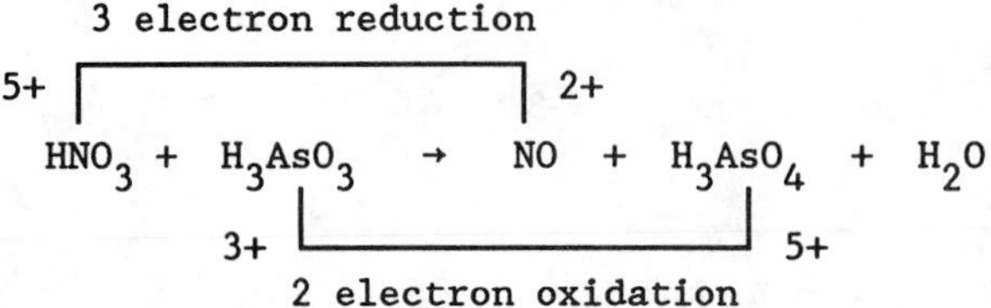

Multiply the reduction part by 2 and the oxidation part by 3 and add the results:

$$2HNO_3 + 3H_3AsO_3 \rightarrow 2NO + 3H_3AsO_4 + H_2O$$

(b) First, diagram the changes in oxidation state for iodine and chlorine:

6 electron oxidation

1- 5+

$$NaI + HOCl \rightarrow NaIO_3 + HCl$$

1+ 1-

2 electron reduction

Multiply the reduction part by 3 and add the result to the oxidation part:

$$NaI + 3HOCl \rightarrow NaIO_3 + 3HCl$$

(c) First, diagram the oxidation state changes for manganese and carbon:

5 electron reduction

7+ 2+

$$KMnO_4 + H_2C_2O_4 + H_2SO_4 \rightarrow 2CO_2 + K_2SO_4 + MnSO_4 + H_2O$$

3+ 4+

1 electron oxidation per carbon atom = 2 electrons overall

Multiply the reduction part by 2 and the oxidation part by 5, add the result, and balance the rest by inspection:

$$2KMnO_4 + 5H_2C_2O_4 + 3H_2SO_4 \rightarrow 10CO_2 + K_2SO_4 + 2MnSO_4 + 8H_2O$$

(d) First, diagram the oxidation state changes for S and Al:

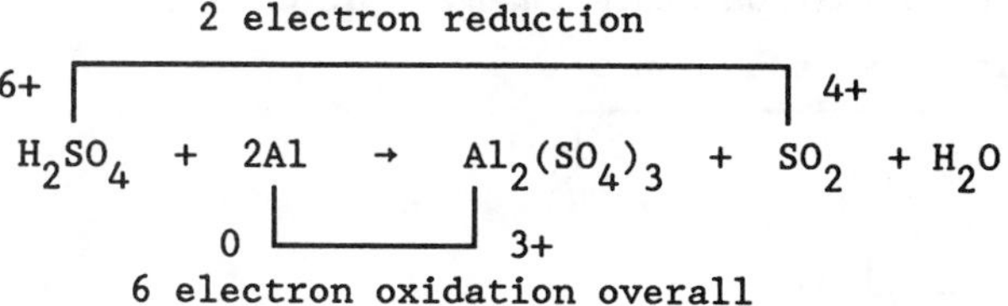

Multiply the reduction part by 3 and add to the oxidation part, balancing the remainder by inspection:

$6H_2SO_4 + 2Al \rightarrow Al_2(SO_4)_3 + 3SO_2 + 6H_2O$

12.22 (a) First, diagram the oxidation state changes for Cr and Cl:

2 electron oxidation overall (1 electron per Cl)

1- ... 0

$K_2Cr_2O_7 + 2HCl \rightarrow KCl + CrCl_3 + Cl_2 + H_2O$

6+ ... 3+

6 electron reduction overall (3 electrons per Al)

Multiply the oxidation part by 3, add to the reduction part, and balance the remainder by inspection:

$K_2Cr_2O_7 + 14HCl \rightarrow 2KCl + 2CrCl_3 + 3Cl_2 + 7H_2O$

(b) First, diagram the oxidation state changes for iodine:

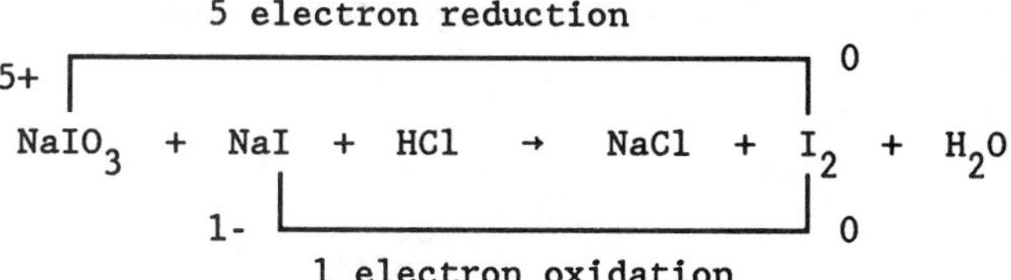

Multiply the oxidation part by 5, add to the reduction part, and balance the remainder by inspection:

$NaIO_3 + 5NaI + 6HCl \rightarrow 6NaCl + 3I_2 + 3H_2O$

(c) First, diagram the oxidation state changes for Cu and N:

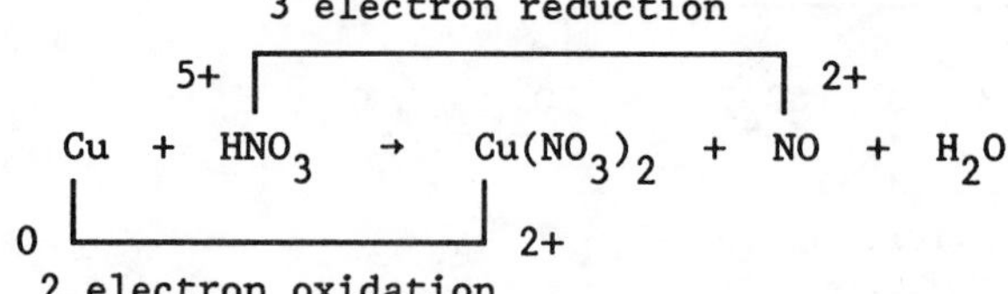

Multiply the reduction part by 2 and the oxidation part by 3, add these together and balance the remainder by inspection:

$3Cu + 8HNO_3 \rightarrow 3Cu(NO_3)_2 + 2NO + 4H_2O$

(d) First, diagram the oxidation state changes for Cu and N:

2 electron oxidation

0 2+

$Cu + HNO_3 \rightarrow Cu(NO_3)_2 + NO_2 + H_2O$

5+ 4+

1 electron reduction

Multiply the reduction part by 2, add to the oxidation part, and balance the remainder by inspection:

$Cu + 4HNO_3 \rightarrow Cu(NO_3)_2 + 2NO_2 + 2H_2O$

12.23 (a) First, diagram the oxidation state changes for Cu and S:

2 electron oxidation

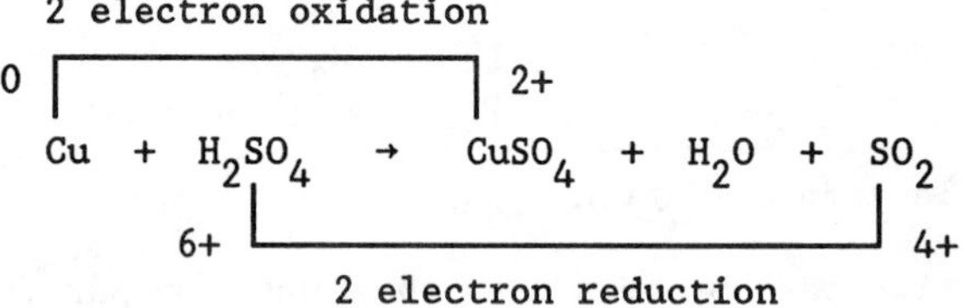

The redox part is already balanced, and the rest is balanced by inspection:

$Cu + 2H_2SO_4 \rightarrow CuSO_4 + SO_2 + 2H_2O$

(b) First, diagram the oxidation state changes for S and N:

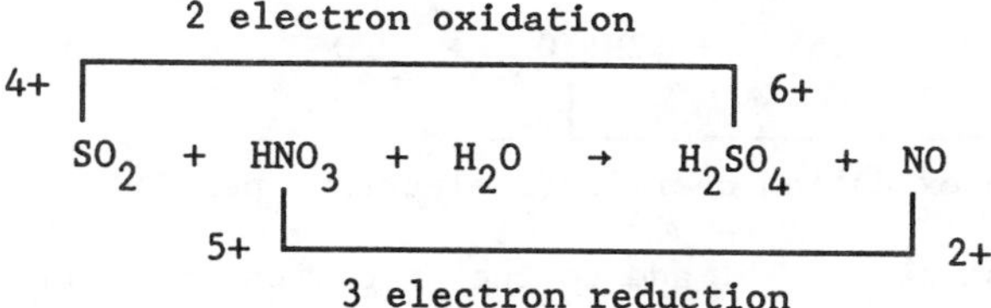

Multiply the oxidation part by 3 and the reduction part by 2, and add the results together. Balance the remainder by inspection:

$3SO_2 + 2HNO_3 + 2H_2O \rightarrow 3H_2SO_4 + 2NO$

(c) First, diagram the oxidation state changes for Zn and S:

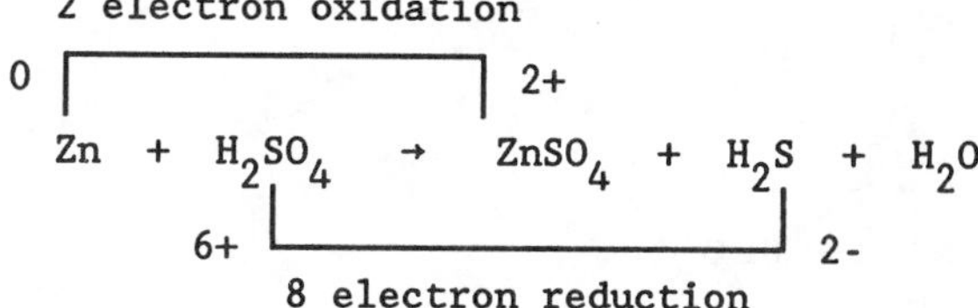

Multiply the oxidation part by 4, add to the reduction part, and balance the remainder by inspection:

$5H_2SO_4 + 4Zn \rightarrow 4ZnSO_4 + H_2S + 4H_2O$

(d) First, diagram the oxidation state changes for I and N:

10 electron oxidation overall (5 electrons per I)

0 — 5+

$I_2 + HNO_3 \rightarrow 2HIO_3 + NO_2 + H_2O$

5+ — 4+

1 electron reduction

Multiply the reduction part by 10, add to the oxidation part, and balance the remainder by inspection:

$I_2 + 10HNO_3 \rightarrow 2HIO_3 + 10NO_2 + 4H_2O$

12.24 Because I_2 is both oxidized and reduced, it is convenient to diagram it twice as follows:

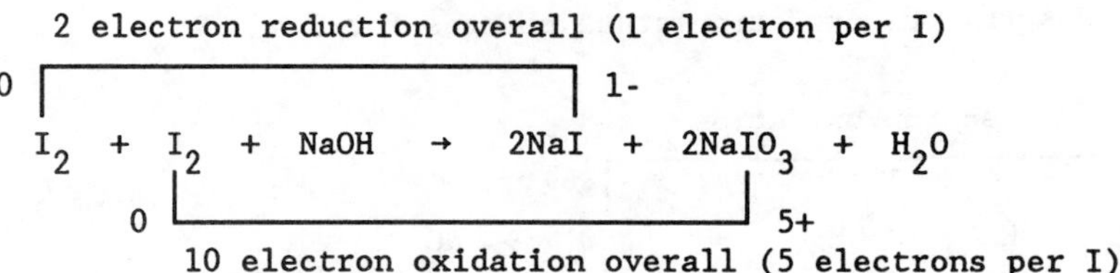

Multiply the reduction part by 5, add to the oxidation part, and balance the rest by inspection:

$3I_2 + 6NaOH \rightarrow 5NaI + NaIO_3 + 3H_2O$

Ion Electron Method

12.25 (a) $6H^+ + BiO_3^- + 2e^- \rightarrow Bi^{3+} + 3H_2O$

This is reduction of BiO_3^{2-}.

(b) $10H^+ + NO_3^- + 8e^- \rightarrow NH_4^+ + 3H_2O$

This constitutes reduction of NO_3^-.

(c) $2H_2O + Pb^{2+} \rightarrow PbO_2 + 4H^+ + 2e^-$

This is oxidation of Pb^{2+}.

(d) $6H_2O + Cl_2 \rightarrow 2ClO_3^- + 12H^+ + 10e^-$

This constitutes oxidation of Cl_2.

12.26 (a) $Fe + 2OH^- \rightarrow Fe(OH)_2 + 2e^-$

This is oxidation of Fe.

(b) $2e^- + 2OH^- + SO_2Cl_2 \rightarrow SO_3^{2-} + 2Cl^- + H_2O$

This constitutes reduction of SO_2Cl_2.

(c) $6OH^- + Mn(OH)_2 \rightarrow MnO_4^{2-} + 4H_2O + 4e^-$

This is oxidation of $Mn(OH)_2$.

(d) $14e^- + 4H_2O + 2H_4IO_6^- \rightarrow I_2 + 16OH^-$

This constitutes reduction of $H_4IO_6^-$.

12.27 (a) $2S_2O_3^{2-} \rightarrow S_4O_6^{2-} + 2e^-$

$\underline{2H^+ + OCl^- + 2e^- \rightarrow Cl^- + H_2O}$

$2H^+ + OCl^- + 2S_2O_3^{2-} \rightarrow S_4O_6^{2-} + Cl^- + H_2O$

(b) $2 \times (e^- + 2H^+ + NO_3^- \rightarrow NO_2 + H_2O)$

$\underline{Cu \rightarrow Cu^{2+} + 2e^-}$

$4H^+ + 2NO_3^- + Cu \rightarrow 2NO_2 + Cu^{2+} + 2H_2O$

(c) $6H^+ + 6e^- + IO_3^- \rightarrow I^- + 3H_2O$

$\underline{3 \times (H_2O + AsO_3^{3-} \rightarrow AsO_4^{3-} + 2H^+ + 2e^-)}$

$6H^+ + 3H_2O + 3AsO_3^{3-} + IO_3^- \rightarrow I^- + 3H_2O + 3AsO_4^{3-} + 6H^+$

which simplifies to give:

$3AsO_3^{3-} + IO_3^- \rightarrow I^- + 3AsO_4^{3-}$

(d) $4H^+ + 2e^- + SO_4^{2-} \rightarrow SO_2 + 2H_2O$

$\underline{Zn \rightarrow Zn^{2+} + 2e^-}$

$4H^+ + Zn + SO_4^{2-} \rightarrow Zn^{2+} + SO_2 + 2H_2O$

(e) $10H^+ + 8e^- + NO_3^- \rightarrow NH_4^+ + 3H_2O$

$\underline{4 \times (Zn \rightarrow Zn^{2+} + 2e^-)}$

$10H^+ + NO_3^- + 4Zn \rightarrow 4Zn^{2+} + NH_4^+ + 3H_2O$

(f) $7H_2O + 2Cr^{3+} \rightarrow Cr_2O_7^{2-} + 14H^+ + 6e^-$

$\underline{3 \times (2e^- + 6H^+ + BiO_3^- \rightarrow Bi^{3+} + 3H_2O)}$

$7H_2O + 2Cr^{3+} + 18H^+ + 3BiO_3^- \rightarrow Cr_2O_7^{2-} + 14H^+ + 3Bi^{3+} + 9H_2O$

which simplifies to give:

$2Cr^{3+} + 4H^+ + 3BiO_3^- \rightarrow Cr_2O_7^{2-} + 3Bi^{3+} + 2H_2O$

(g) $I_2 + 6H_2O \rightarrow 2IO_3^- + 12H^+ + 10e^-$

$5 \times (2e^- + 2H^+ + OCl^- \rightarrow Cl^- + H_2O)$

$I_2 + H_2O + 5OCl^- \rightarrow 2IO_3^- + 2H^+ + 5Cl^-$

(h) $2 \times (Mn^{2+} + 4H_2O \rightarrow MnO_4^- + 8H^+ + 5e^-)$

$5 \times (2e^- + BiO_3^- + 6H^+ \rightarrow Bi^{3+} + 3H_2O)$

$2Mn^{2+} + 5BiO_3^- + 30\ H^+ + 8H_2O \rightarrow 2MnO_4^- + 16H^+ + 5Bi^{3+} + 15H_2O$

which simplifies to:

$2Mn^{2+} + 5BiO_3^- + 14H^+ \rightarrow 2MnO_4^- + 5Bi^{3+} + 7H_2O$

(i) $3 \times (H_2O + H_3AsO_3 \rightarrow H_3AsO_4 + 2H^+ + 2e^-)$

$14H^+ + 6e^- + Cr_2O_7^{2-} \rightarrow 2Cr^{3+} + 7H_2O$

$3H_2O + H_3AsO_3 + 14H^+ + Cr_2O_7^{2-} \rightarrow 3H_3AsO_4 + 6H^+ + 2Cr^{3+} + 7H_2O$

which simplifies to give:

$3H_3AsO_3 + 8H^+ + Cr_2O_7^{2-} \rightarrow 3H_3AsO_4 + 2Cr^{3+} + 4H_2O$

(j) $2I^- \rightarrow I_2 + 2e^-$

$2e^- + 3H^+ + HSO_4^- \rightarrow SO_2 + 2H_2O$

$2I^- + 3H^+ + HSO_4^- \rightarrow SO_2 + 2H_2O + I_2$

12.28 (a) $3 \times (2H_2O + Sn \rightarrow SnO_2 + 4H^+ + 4e^-)$

$4 \times (3e^- + 4H^+ + NO_3^- \rightarrow NO + 2H_2O)$

$6H_2O + 3Sn + 16H^+ + 4NO_3^- \rightarrow 3SnO_2 + 12H^+ + 4NO + 8H_2O$

which simplifies to:

$3Sn + 4H^+ + 4NO_3^- \rightarrow 3SnO_2 + 4NO + 2H_2O$

(b) $2e^- + 4H^+ + 2Cl^- + PbO_2 \rightarrow PbCl_2 + 2H_2O$

$2Cl^- \rightarrow Cl_2 + 2e^-$

$4H^+ + 4Cl^- + PbO_2 \rightarrow PbCl_2 + Cl_2 + 2H_2O$

(c) $Ag \rightarrow Ag^+ + e^-$

$e^- + 2H^+ + NO_3^- \rightarrow NO_2 + H_2O$

$Ag + 2H^+ + NO_3^- \rightarrow NO_2 + H_2O + Ag^+$

(d) $4 \times (Fe^{3+} + e^- \rightarrow Fe^{2+})$

$2NH_3OH^+ \rightarrow H_2O + N_2O + 4e^- + 6H^+$

$4Fe^{3+} + 2HN_3OH^+ \rightarrow 4Fe^{2+} + H_2O + N_2O + 6H^+$

(e) $2I^- \rightarrow I_2 + 2e^-$

$2 \times (e^- + H^+ + HNO_2 \rightarrow NO + H_2O)$

$2I^- + 2H^+ + 2HNO_2 \rightarrow I_2 + 2NO + 2H_2O$

(f) $C_2O_4^{2-} \rightarrow 2CO_2 + 2e^-$

$2 \times (e^- + H^+ + HNO_2 \rightarrow NO + H_2O)$

$C_2O_4^{2-} + 2H^+ + 2HNO_2 \rightarrow 2CO_2 + 2NO + 2H_2O$

(g) $5 \times (H_2O + HNO_2 \rightarrow 3H^+ + NO_3^- + 2e^-)$

$2 \times (8H^+ + 5e^- + MnO_4^- \rightarrow Mn^{2+} + 4H_2O)$

$5H_2O + 5HNO_2 + 16H^+ + 2MnO_4^- \rightarrow 15H^+ + 5NO_3^- + 2Mn^{2+} + 8H_2O$

which simplifies to give:

$5HNO_2 + H^+ + 2MnO_4^- \rightarrow 5NO_3^- + 2Mn^{2+} + 3H_2O$

(h) $3 \times (2H_2O + H_3PO_2 \rightarrow H_3PO_4 + 4H^+ + 4e^-)$

$2 \times (14H^+ + Cr_2O_7^{2-} + 6e^- \rightarrow 2Cr^{3+} + 7H_2O)$

$6H_2O + 3H_3PO_2 + 28H^+ + 2Cr_2O_7^{2-} \rightarrow 3H_3PO_4 + 12H^+ + 4Cr^{3+} + 14\ H_2O$

which simplifies to:

$3H_3PO_2 + 16H^+ + 2Cr_2O_7^{2-} \rightarrow 3H_3PO_4 + 4Cr^{3+} + 8H_2O$

(i) $2 \times (e^- + 2H^+ + VO_2^+ \rightarrow VO^{2+} + H_2O)$

$Sn^{2+} \rightarrow Sn^{4+} + 2e^-$

$4H^+ + 2VO_2^+ + Sn^{2+} \rightarrow 2VO^{2+} + 2H_2O + Sn^{4+}$

(j) $XeF_2 + 2e^- \rightarrow Xe + 2F^-$

$2Cl^- \rightarrow Cl_2 + 2e^-$

$XeF_2 + 2Cl^- \rightarrow Xe + Cl_2 + 2F^-$

12.29 For redox reactions in basic solution, we proceed to balance the half reactions as if they were in acid solution, and then add enough OH^- to each side of the resulting equation in order to neutralize (titrate) all of the H^+. This gives a corresponding amount of water ($H^+ + OH^- \rightarrow H_2O$) on one side of the equation, and an excess of OH^- on the other side of the equation, as befits a reaction in basic solution.

(a) $2 \times (4H^+ + 3e^- + CrO_4^{2-} \rightarrow CrO_2^- + 2H_2O)$

$3 \times (S^{2-} \rightarrow S + 2e^-)$

$8H^+ + 3S^{2-} + 2CrO_4^{2-} \rightarrow 2CrO_2^- + 2S + 4H_2O$

Adding $8OH^-$ to both sides of the above equation we obtain:

$8H_2O + 3S^{2-} + 2CrO_4^{2-} \rightarrow 2CrO_2^- + 8OH^- + 3S + 4H_2O$

which simplifies to:

$4H_2O + 3S^{2-} + 2CrO_4^{2-} \rightarrow 2CrO_2^- + 3S + 8OH^-$

(b) $3 \times (C_2O_4^{2-} \rightarrow 2CO_2 + 2e^-)$

$2 \times (4H^+ + 3e^- + MnO_4^- \rightarrow MnO_2 + 2H_2O)$

$3C_2O_4^{2-} + 8H^+ + 2MnO_4^- \rightarrow 6CO_2 + 2MnO_2 + 4H_2O$

Adding $8OH^-$ to both sides of the above equation we get:

$3C_2O_4^{2-} + 8H_2O + 2MnO_4^- \rightarrow 6CO_2 + 2MnO_2 + 4H_2O + 8OH^-$

which simplifies to give:

$3C_2O_4^{2-} + 4H_2O + 2MnO_4^- \rightarrow 6CO_2 + 2MnO_2 + 8OH^-$

(c) $4 \times (6e^- + 6H^+ + ClO_3^- \rightarrow Cl^- + 3H_2O)$

$3 \times (2H_2O + N_2H_4 \rightarrow 2NO + 8H^+ + 8e^-)$

$24H^+ + 4ClO_3^- + 6H_2O + 3N_2H_4 \rightarrow 4Cl^- + 12H_2O + 6NO + 24H^+$

which needs no OH^-, because it simplifies directly to:

$4ClO_3^- + 3N_2H_4 \rightarrow 4Cl^- + 6NO + 6H_2O$

(d) $2e^- + NiO_2 + 2H^+ \rightarrow Ni(OH)_2$

$2Mn(OH)_2 \rightarrow Mn_2O_3 + H_2O + 2H^+ + 2e^-$

$NiO_2 + 2Mn(OH)_2 \rightarrow Ni(OH)_2 + Mn_2O_3 + H_2O$

(e) $3 \times (SO_3^{2-} + H_2O \rightarrow SO_4^{2-} + 2H^+ + 2e^-)$

$2 \times (3e^- + 4H^+ + MnO_4^- \rightarrow MnO_2 + 2H_2O)$

$3SO_3^{2-} + 3H_2O + 8H^+ + 2MnO_4^- \rightarrow 3SO_4^{2-} + 6H^+ + 2MnO_2 + 4H_2O$

Adding $8OH^-$ to both sides of the equation we obtain:

$3SO_3^{2-} + 11H_2O + 2MnO_4^- \rightarrow 3SO_4^{2-} + 10H_2O + 2MnO_2 + 2OH^-$

which simplifies to:

$3SO_3^{2-} + H_2O + 2MnO_4^- \rightarrow 3SO_4^{2-} + 2MnO_2 + 2OH^-$

(f) $2 \times (2H_2O + CrO_2^- \rightarrow CrO_4^{2-} + 4H^+ + 3e^-)$

$3 \times (2e^- + S_2O_8^{2-} \rightarrow 2SO_4^{2-})$

$4H_2O + 3S_2O_8^{2-} + 2CrO_2^- \rightarrow 2CrO_4^{2-} + 6SO_4^{2-} + 8H^+$

Adding $8OH^-$ to both sides of this equation:

$4H_2O + 8OH^- + 3S_2O_8^{2-} + 2CrO_2^- \rightarrow 2CrO_4^{2-} + 6SO_4^{2-} + 8H_2O$

which simplifies to give:

$8OH^- + 3S_2O_8^{2-} + 2CrO_2^- \rightarrow 2CrO_4^{2-} + 6SO_4^{2-} + 4H_2O$

(g) $3 \times (H_2O + SO_3^{2-} \rightarrow SO_4^{2-} + 2H^+ + 2e^-)$

$2 \times (3e^- + CrO_4^{2-} + 4H^+ \rightarrow CrO_2^- + 2H_2O)$

$3H_2O + 3SO_3^{2-} + 2CrO_4^{2-} + 8H^+ \rightarrow 3SO_4^{2-} + 6H^+ + 2CrO_2^- + 4H_2O$

Adding $8OH^-$ to both sides of the equation we get:

$11H_2O + 3SO_3^{2-} + 2CrO_4^{2-} \rightarrow 3SO_4^{2-} + 10H_2O + 2CrO_2^- + 2OH^-$

which simplifies to:

$H_2O + 3SO_3^{2-} + 2CrO_4^{2-} \rightarrow 3SO_4^{2-} + 2CrO_2^- + 2OH^-$

(h) $2 \times (O_2 + 2H^+ + 2e^- \rightarrow H_2O_2)$

$N_2H_4 \rightarrow N_2 + 4H^+ + 4e^-$

$2O_2 + 4H^+ + N_2H_4 \rightarrow 2H_2O_2 + N_2 + 4H^+$

which simplifies to:

$2O_2 + N_2H_4 \rightarrow 2H_2O_2 + N_2$

(i) $4 \times (Fe(OH)_2 + OH^- \rightarrow Fe(OH)_3 + e^-)$

$4e^- + 2H_2O + O_2 \rightarrow 4OH^-$

$4Fe(OH)_2 + 4OH^- + 2H_2O + O_2 \rightarrow 4Fe(OH)_3 + 4OH^-$

which simplifies to:

$4Fe(OH)_2 + 2H_2O + O_2 \rightarrow 4Fe(OH)_3$

(j) $4 \times (Au + 4CN^- \rightarrow Au(CN)_4^- + 3e^-)$

$3 \times (4e^- + 2H_2O + O_2 \rightarrow 4OH^-)$

$4Au + 16CN^- + 6H_2O + 3O_2 \rightarrow 4Au(CN)_4^- + 12OH^-$

12.30 (a) $6e^- + 2NBr_3 \rightarrow N_2 + 6Br^-$

$6H_2O + 2NBr_3 \rightarrow N_2 + 6HOBr + 6H^+ + 6e^-$

$6H_2O + 4NBr_3 \rightarrow 2N_2 + 6HOBr + 6Br^- + 6H^+$

Add $6OH^-$ to both sides of the above equation:

$6H_2O + 6OH^- + 4NBr_3 \rightarrow 2N_2 + 6HOBr + 6Br^- + 6H_2O$

which simplifies to give:

$6OH^- + 4NBr_3 \rightarrow 2N_2 + 6HOBr + 6Br^-$

(b) $H_2O + ClNO_2 \rightarrow NO_3^- + Cl^- + 2H^+$

Adding $2OH^-$ to both sides we get:

$H_2O + 2OH^- + ClNO_2 \rightarrow NO_3^- + Cl^- + 2H_2O$

which simplifies to:

$2OH^- + ClNO_2 \rightarrow NO_3^- + Cl^- + H_2O$

(c) $5 \times (2e^- + Cl_2 \rightarrow 2Cl^-)$

$6H_2O + Cl_2 \rightarrow 2ClO_3^- + 12H^+ + 10e^-$

$6H_2O + 6Cl_2 \rightarrow 2ClO_3^- + 10Cl^- + 12H^+$

Adding $12OH^-$ to both sides gives:

$6H_2O + 6Cl_2 + 12OH^- \rightarrow 2ClO_3^- + 10Cl^- + 12H_2O$

which simplifies to:

$3Cl_2 + 6OH^- \rightarrow ClO_3^- + 5Cl^- + 3H_2O$

(d) $4H^+ + 4e^- + H_2SeO_3 \rightarrow Se + 3H_2O$

$2 \times (H_2S \rightarrow S + 2H^+ + 2e^-)$

$4H^+ + 2H_2S + H_2SeO_3 \rightarrow 2S + 4H^+ + Se + 3H_2O$

which simplifies to give:

$2H_2S + H_2SeO_3 \rightarrow 2S + Se + 3H_2O$

(e) $2e^- + 4H^+ + MnO_2 \rightarrow Mn^{2+} + 2H_2O$

$2SO_3^{2-} \rightarrow S_2O_6^{2-} + 2e^-$

$4H^+ + 2SO_3^{2-} + MnO_2 \rightarrow Mn^{2+} + S_2O_6^{2-} + 2H_2O$

Add $4OH^-$ to both sides and simplify:

$2H_2O + 2SO_3^{2-} + MnO_2 \rightarrow Mn^{2+} + S_2O_6^{2-} + 4OH^-$

(f) $H_2O + BrO_3F \rightarrow BrO_4^- + F^- + 2H^+$

Add OH^- to both sides:

$H_2O + 2OH^- + BrO_3F \rightarrow BrO_4^- + F^- + 2H_2O$

which simplifies to:

$2OH^- + BrO_3F \rightarrow BrO_4^- + F^- + H_2O$

(g) $6H^+ + 6e^- + XeO_3 \rightarrow Xe + 3H_2O$

$3 \times (2I^- \rightarrow I_2 + 2e^-)$

$6H^+ + 6I^- + XeO_3 \rightarrow 3I_2 + Xe + 3H_2O$

Upon adding $6OH^-$ to both sides we obtain:

$6H_2O + 6I^- + XeO_3 \rightarrow 3I_2 + Xe + 3H_2O + 6OH^-$

which simplifies to:

$3H_2O + 6I^- + XeO_3 \rightarrow 3I_2 + Xe + 6OH^-$

(h) $2HXeO_4^- \rightarrow XeO_6^{4-} + Xe + O_2 + 2H^+$

Adding $2OH^-$ to both sides gives:

$2HXeO_4^- + 2OH^- \rightarrow XeO_6^{4-} + Xe + O_2 + 2H_2O$

(i) $(CN)_2 + 2e^- \rightarrow 2CN^-$

$\underline{(CN)_2 + 2H_2O \rightarrow 2OCN^- + 4H^+ + 2e^-}$

$2(CN)_2 + 2H_2O \rightarrow 2CN^- + 2OCN^- + 4H^+$

Adding $4OH^-$ to both sides gives:

$2(CN)_2 + 4OH^- + 2H_2O \rightarrow 2CN^- + 2OCN^- + 4H_2O$

which simplifies to:

$(CN)_2 + 2OH^- \rightarrow CN^- + OCN^- + H_2O$

Reactions of Metals with Acids

12.31 (a) ionic:

$Mn(s) + 2H^+(aq) + 2Cl^-(aq) \rightarrow Mn^{2+}(aq) + H_2(g) + 2Cl^-(aq)$

net ionic:

$Mn(s) + 2H^+(aq) \rightarrow Mn^{2+}(aq) + H_2(g)$

(b) ionic:

$Cd(s) + 2H^+(aq) + 2Cl^-(aq) \rightarrow Cd^{2+} + H_2(g) + 2Cl^-(aq)$

net ionic:

$Cd(s) + 2H^+(aq) \rightarrow Cd^{2+} + H_2(g)$

(c) ionic:

$Sn(s) + 2H^+(aq) + 2Cl^-(aq) \rightarrow Sn^{2+}(aq) + H_2(g) + 2Cl^-(aq)$

net ionic:

$Sn(s) + 2H^+(aq) \rightarrow Sn^{2+}(aq) + H_2(g)$

(d) ionic:

$Ni(s) + 2H^+(aq) + 2Cl^-(aq) \rightarrow Ni^{2+}(aq) + H_2(g) + 2Cl^-(aq)$

net ionic:

$Ni(s) + 2H^+(aq) \rightarrow Ni^{2+}(aq) + H_2(g)$

(e) ionic:

$2Cr(s) + 6H^+(aq) + 6Cl^-(aq) \rightarrow 2Cr^{3+}(aq) + 3H_2(g) + 6Cl^-(aq)$

net ionic:

$2Cr(s) + 6H^+(aq) \rightarrow 2Cr^{3+}(aq) + 3H_2(g)$

12.32 (a) ionic:

$Mn(s) + 2H^+(aq) + SO_4^{2-}(aq) \rightarrow Mn^{2+}(aq) + H_2(g) + SO_4^{2-}(aq)$

net ionic:

$Mn(s) + 2H^+(aq) \rightarrow Mn^{2+}(aq) + H_2(g)$

(b) ionic:

$Cd(s) + 2H^+(aq) + SO_4^{2-}(aq) \rightarrow Cd^{2+}(aq) + H_2(g) + SO_4^{2-}(aq)$

net ionic:

$Cd(s) + 2H^+(aq) \rightarrow Cd^{2+}(aq) + H_2(g)$

(c) ionic:

$Sn(s) + 2H^+(aq) + SO_4^{2-}(aq) \rightarrow Sn^{2+}(aq) + H_2(g) + SO_4^{2-}(aq)$

net ionic:

$Sn(s) + 2H^+(aq) \rightarrow Sn^{2+}(aq) + H_2(g)$

(d) ionic:

$Ni(s) + 2H^+(aq) + SO_4^{2-}(aq) \rightarrow Ni^{2+}(aq) + H_2(g) + SO_4^{2-}(aq)$

net ionic:

$Ni(s) + 2H^+(aq) \rightarrow Ni^{2+}(aq) + H_2(g)$

(e) ionic:

$$2Cr(s) + 6H^+(aq) + 3SO_4^{2-}(aq) \rightarrow 2Cr^{3+}(aq) + 3H_2(g) + 3SO_4^{2-}(aq)$$

net ionic:

$$2Cr(s) + 6H^+(aq) \rightarrow 2Cr^{3+}(aq) + 3H_2(g)$$

12.33 A "nonoxidizing acid" is one in which the H^+ ion is the strongest oxidizing agent. That is, the anion of the acid is not itself a better oxidizing agent than H^+. Examples are HCl and H_2SO_4.

12.34 (a) $3Ag + 4HNO_3 \rightarrow 3AgNO_3 + 2H_2O + NO$

(b) $Ag + 2HNO_3 \rightarrow AgNO_3 + H_2O + NO_2$

12.35 $NO_3^-(aq)$

12.36 $Cu(s) \rightarrow Cu^{2+}(aq) + 2e^-$

$2e^- + 2H^+ + H_2SO_4 \rightarrow SO_2(g) + 2H_2O(\ell)$

$Cu(s) + 2H^+ + H_2SO_4 \rightarrow SO_2(g) + 2H_2O(\ell) + Cu^{2+}(aq)$

Displacement Reactions and the Activity Series

12.37 Reaction (a) demonstrates that Al is more readily oxidized than Cu.
Reaction (b) demonstrates that Al is more readily oxidized than Fe.
Reaction (c) demonstrates that Fe is more readily oxidized than Pb.
Reaction (d) demonstrates that Fe is more readily oxidized than Cu.
Reaction (e) demonstrates that Al is more readily oxidized than Pb.
Reaction (f) demonstrates that Pb is more readily oxidized than Cu.

Altogether, the above facts constitute the following trend of increasing ease of oxidation:

$Cu < Pb < Fe < Al$

12.38 No. Reactions (a), (d), and (e) were not necessary.

12.39 Any metal that is lower than hydrogen in the activity series shown in Table 12.1 of the text will react with H^+: (c) zinc and (d) magnesium.

12.40 We choose the metal that is lower (more reactive) in the activity series shown in Table 12.1:

(a) aluminum (b) zinc (c) magnesium

12.41 In each case, the reaction should proceed to give the less reactive of the two metals, together with the ion of the more reactive of the two metals. The reactivity is taken from the reactivity series, Table 12.1.

(a) N.R.

(b) $2Cr(s) + 3Pb^{2+}(aq) \rightarrow 2Cr^{3+}(aq) + 3Pb(s)$

(c) $2Ag^{+}(aq) + Fe(s) \rightarrow 2Ag(s) + Fe^{2+}(aq)$

(d) $3Ag(s) + Au^{3+}(aq) \rightarrow Au(s) + 3Ag^{+}(aq)$

12.42 In each case, the reaction should proceed to give the less reactive of the two metals, together with the ion of the more reactive of the two metals. The reactivity is taken from the reactivity series, Table 12.1.

(a) $Mg(s) + Fe^{2+}(aq) \rightarrow Mg^{2+}(aq) + Fe(s)$

(b) N.R.

(c) N.R.

(d) $Mn(s) + Co^{2+}(aq) \rightarrow Mn^{2+}(aq) + Co(s)$

(e) $2Cr(s) + 3Sn^{2+}(aq) \rightarrow 2Cr^{3+}(aq) + 3Sn(s)$

12.43 In each case, the reaction should proceed to give the less reactive of the two metals, together with the ion of the more reactive of the two metals. The reactivity is taken from the reactivity series, Table 12.1.

(a) $Zn(s) + Sn^{2+}(aq) \rightarrow Zn^{2+}(aq) + Sn(s)$

(b) $2Cr(s) + 6H^{+}(aq) \rightarrow 2Cr^{3+}(aq) + 3H_2(g)$

(c) N.R.

(d) $Mn(s) + Pb^{2+}(aq) \rightarrow Mn^{2+}(aq) + Pb(s)$

(e) $Zn(s) + Co^{2+}(aq) \rightarrow Zn^{2+}(aq) + Co(s)$

12.44 This would be any metal higher (less reactive) than hydrogen, i.e. gold, mercury, silver, and copper.

12.45 We have metals of two types, those that give monopositive cations (designated M^1) and those that give dipositive cations (designated M^2). The M^1 metals are the most reactive alkali metals from Group I, and the M^2 metals are the most reactive alkaline earth metals of Group II.

For the M^1 metals: $2M^1(s) + 2H_2O(\ell) \rightarrow 2M^1(OH)(aq) + H_2(g)$

where M^1 = Cs, Rb, K, or Na

For the M^2 metals: $M^2(s) + 2H_2O(\ell) \rightarrow M^2(OH)_2(aq) + H_2(g)$

where M^2 = Ca, Sr, Ba, or Mg (the latter only slowly)

Trends in Reactivity of Metals

12.46 This has to do with the ease of oxidation.

12.47 The most reactive metals are in the left-most groups of the periodic table, namely the metals of Groups IA and IIA. The least reactive metals are found in the second and third rows of the transition elements.

12.48 The lower the ionization energy, the more reactive is the metal.

12.49 calcium > iron > silver > iridium

Oxidation by O_2

12.50 Combustion is the rapid reaction of a substance with oxygen, which is accompanied by the evolution of light and heat.

12.51 Historically, the reaction of a substance with oxygen was termed oxidation. Now we realize that reaction with oxygen most typically means that oxygen acquires electrons from the substance with which it reacts. The oxidation of a substance is, therefore, taken to represent the loss of electrons by a substance, whether the substance has reacted with oxygen or with another oxidizing agent.

12.52 The nonmetals form covalent oxides because there is not much difference in the electronegativity of oxygen and the typical nonmetal.

12.53 (a) $2C_6H_6(\ell) + 15O_2(g) \rightarrow 12CO_2(g) + 6H_2O(g)$

(b) $C_3H_8(g) + 5O_2(g) \rightarrow 3CO_2(g) + 4H_2O(g)$

(c) $C_{21}H_{44}(s) + 32O_2(g) \rightarrow 21CO_2(g) + 22H_2O(g)$

(d) $2C_{12}H_{26}(\ell) + 37O_2(g) \rightarrow 24CO_2(g) + 26H_2O(g)$

(e) $2C_{18}H_{38}(\ell) + 55O_2(g) \rightarrow 36CO_2(g) + 38H_2O(g)$

12.54 $2CH_3OH(\ell) + 3O_2(g) \rightarrow 2CO_2(g) + 4H_2O(g)$

12.55 $C_6H_{12}O_6 + 6O_2 \rightarrow 6CO_2 + 6H_2O$

12.56 $C_{12}H_{22}O_{11} + 12O_2 \rightarrow 12CO_2 + 11H_2O$

12.57 We use the equation from Review Exercise 12.56:

$$\Delta H° = \{11 \times \Delta H°_f[H_2O(\ell)] + 12 \times \Delta H°_f[CO_2(g)]\} - \{12 \times \Delta H°_f[O_2(g)] + 1 \times \Delta H°_f[C_{12}H_{22}O_{11}(s)]\}$$

$$= \{11 \text{ mol} \times (-285.9 \text{ kJ/mol}) + 12 \text{ mol} \times (-393.5 \text{ kJ/mol})\} - \{12 \text{ mol} \times (0.00 \text{ kJ/mol}) + 1 \text{ mol} \times (-2230 \text{ kJ/mol})\}$$

$$= -5637 \text{ kJ/mol}$$

Next, determine the amount of heat energy involved (released) in the metabolism of 28.4 g (1 oz):

$$28.4 \text{ g } C_{12}H_{22}O_{11} \times \frac{1 \text{ mol } C_{12}H_{22}O_{11}}{342 \text{ g } C_{12}H_{22}O_{11}} \times \frac{5637 \text{ kJ released}}{1 \text{ mol } C_{12}H_{22}O_{11}} = 468 \text{ kJ}$$

12.58 $C_2H_5OH(\ell) + 3O_2(g) \rightarrow 2CO_2(g) + 3H_2O(\ell)$

$$\Delta H° = \{3 \times \Delta H°_f[H_2O(\ell)] + 2 \times \Delta H°_f[CO_2(g)]\} - \{1 \times \Delta H°_f[C_2H_5OH(\ell)] + 3 \times \Delta H°_f[O_2(g)]\}$$

$$= \{3 \text{ mol} \times (-285.9 \text{ kJ/mol}) + 2 \text{ mol} \times (-393.5 \text{ kJ/mol})\} - \{1 \text{ mol} \times (-277.63 \text{ kJ/mol}) + 3 \text{ mol} \times (0.00 \text{ kJ/mol})\}$$

$= -1367.1$ kJ/mol, for the combustion of C_2H_5OH.

Next determine the amount of heat released by the combustion of 6.56 kg ($= 6.5 \times 10^3$ g) of C_2H_5OH:

$$6.56 \times 10^3 \text{ g } C_2H_5OH \times \frac{1 \text{ mol } C_2H_5OH}{46.07 \text{ g } C_2H_5OH} \times \frac{1367.1 \text{ kJ released}}{1 \text{ mol } C_2H_5OH}$$

$$= 1.95 \times 10^5 \text{ kJ released}$$

12.59 (a) CO_2 and H_2O (b) CO and H_2O (c) C and H_2O

12.60 The products will be CO_2, H_2O, and SO_2:

$2C_2H_6S + 9O_2 \rightarrow 4CO_2 + 2SO_2 + 6H_2O$

12.61 The other product is water:

$4NH_3 + 3O_2 \rightarrow 2N_2 + 6H_2O$

12.62 (a) $2Zn + O_2 \rightarrow 2ZnO$

(b) $4Al + 3O_2 \rightarrow 2Al_2O_3$

(c) $2Mg + O_2 \rightarrow 2MgO$

(d) $2Fe + O_2 \rightarrow 2FeO$

alternatively we have:

$4Fe + 3O_2 \rightarrow 2Fe_2O_3$

(e) $2Ca + O_2 \rightarrow 2CaO$

Common Chemicals for Redox Reactions

12.63 Answers could include the following:

(a) potassium permanganate: $KMnO_4$

sodium or potassium chromate: Na_2CrO_4, K_2CrO_4

sodium or potassium dichromate: $Na_2Cr_2O_7$, $K_2Cr_2O_7$

(b) sodium or potassium sulfite: Na_2SO_3, K_2SO_3

sodium or potassium bisulfite: $NaHSO_3$, $KHSO_3$

oxalic acid: $H_2C_2O_4$

sodium oxalate: $Na_2C_2O_4$

sodium thiosulfate: $Na_2S_2O_3$

tin(II): Sn^{2+}

12.64 (a) $H_2O(\ell) + Cr_2O_7^{2-}(aq) \rightarrow 2CrO_4^{2-}(aq) + 2H^+(aq)$

(b) $2CrO_4^{2-}(aq) + 2H^+(aq) \rightarrow Cr_2O_7^{2-}(aq) + H_2O(\ell)$

12.65 (a) $Mn^{2+}(aq)$ (b) MnO_2

12.66 (a) $Cr^{3+}(aq)$ (b) $Cr(OH)_3$

12.67 $SO_4^{2-}(aq)$

12.68 We can consult the answers to Review Exercises 12.65 and 12.67 for part of the information that we need to know:

$$2MnO_4^-(aq) + H^+(aq) + 5HSO_3^-(aq) \rightarrow 5SO_4^{2-}(aq) + 2Mn^{2+}(aq) + 3H_2O(\ell)$$

12.69 $2H^+(aq) + OCl^-(aq) + 2S_2O_3^{2-}(aq) \rightarrow S_4O_6^{2-}(aq) + Cl^-(aq) + H_2O(\ell)$

12.70 $3H_2C_2O_4(aq) + 8H^+(aq) + Cr_2O_7^{2-}(aq) \rightarrow 6CO_2(g) + 2Cr^{3+}(aq) + 7H_2O(\ell)$

12.71 $I_2 + 2S_2O_3^{2-}(aq) \rightarrow 2I^-(aq) + S_4O_6^{2-}(aq)$

12.72 $2MnO_4^-(aq) + 16H^+(aq) + 5Sn^{2+}(aq) \rightarrow 5Sn^{4+}(aq) + 2Mn^{2+}(aq) + 8H_2O(\ell)$

Stoichiometry of Redox Reactions

12.73

$$15.0 \text{ g } PbO_2 \times \frac{1 \text{ mol } PbO_2}{239 \text{ g } PbO_2} \times \frac{1 \text{ mol } Cl_2}{1 \text{ mol } PbO_2} \times \frac{70.9 \text{ g } Cl_2}{1 \text{ mol } Cl_2} = 4.45 \text{ g } Cl_2$$

12.74 $Mg(s) + 2AgNO_3(aq) \rightarrow 2Ag(s) + Mg(NO_3)_2(aq)$

$$25.0 \text{ g } AgNO_3 \times \frac{1 \text{ mol } AgNO_3}{170 \text{ g } AgNO_3} \times \frac{1 \text{ mol Mg}}{2 \text{ mol } AgNO_3} \times \frac{24.3 \text{ g Mg}}{1 \text{ mol Mg}} = 1.79 \text{ g Mg}$$

12.75 (a) $2 \times (Mn^{2+}(aq) + 4H_2O(\ell) \rightarrow MnO_4^-(aq) + 8H^+(aq) + 5e^-)$

$5 \times (2e^- + 6H^+(aq) + BiO_3^-(aq) \rightarrow Bi^{3+}(aq) + 3H_2O(\ell))$

$$2Mn^{2+}(aq) + 8H_2O(\ell) + 30H^+(aq) + 5BiO_3^-(aq) \rightarrow$$
$$5\ Bi^{3+}(aq) + 15H_2O(\ell) + 2MnO_4^-(aq) + 16H^+(aq)$$

which simplifies to give:

$$2Mn^{2+}(aq) + 14H^+(aq) + 5BiO_3^-(aq) \rightarrow 5Bi^{3+}(aq) + 7H_2O(\ell) + 2MnO_4^-(aq)$$

(b)

$$15.0\ \text{g}\ Mn(NO_3)_2 \times \frac{1\ \text{mol}\ Mn(NO_3)_2}{179\ \text{g}\ Mn(NO_3)_2} \times \frac{5\ \text{mol}\ NaBiO_3}{2\ \text{mol}\ Mn(NO_3)_2} \times \frac{280\ \text{g}\ NaBiO_3}{1\ \text{mol}\ NaBiO_3} = 58.7\ \text{g}\ NaBiO_3$$

12.76 (a) The oxidizing agent is reduced by gaining electrons, and this is accompanied by a decrease (reduction) in the oxidation state. The reducing agent is oxidized by the loss of electrons, and this is accompanied by an increase (oxidation) in oxidation number. In this equation, the iodine in $NaIO_3$ has the 5+ oxidation state, whereas the iodine in the product I^- has the -1 oxidation state. Thus $NaIO_3$ is reduced and it serves as the oxidizing agent.

(b)

$$5.00\ \text{g}\ Na_2SO_3 \times \frac{1\ \text{mol}\ Na_2SO_3}{126\ \text{g}\ Na_2SO_3} \times \frac{1\ \text{mol}\ NaIO_3}{3\ \text{mol}\ Na_2SO_3} \times \frac{198\ \text{g}\ NaIO_3}{1\ \text{mol}\ NaIO_3} = 2.62\ \text{g}\ NaIO_3$$

12.77 (a) $2 \times (3e^- + CrO_4^{2-}(aq) + 5H^+(aq) \rightarrow Cr(OH)_3 + H_2O(\ell))$

$3 \times (H_2O(\ell) + SO_3^{2-}(aq) \rightarrow SO_4^{2-}(aq) + 2H^+(aq) + 2e^-)$

$$3H_2O(\ell) + 3SO_3^{2-}(aq) + 2CrO_4^{2-}(aq) + 10H^+(aq) \rightarrow 3SO_4^{2-}(aq) + 6H^+(aq) + 2Cr(OH)_3 + 2H_2O(\ell)$$

which simplifies to:

$$H_2O(\ell) + 3SO_3^{2-}(aq) + 2CrO_4^{2-}(aq) + 4H^+(aq) \rightarrow 3SO_4^{2-}(aq) + 2Cr(OH)_3$$

On adding base to both sides of the above equation, we obtain:

$$5H_2O(\ell) + 3SO_3^{2-}(aq) + 2CrO_4^{2-}(aq) \rightarrow 3SO_4^{2-}(aq) + 2Cr(OH)_3 + 4OH^-(aq)$$

(b)

$$3.09\ \text{g}\ Na_2SO_3 \times \frac{1\ \text{mol}\ Na_2SO_3}{126\ \text{g}\ Na_2SO_3} \times \frac{2\ \text{mol}\ CrO_4^{2-}}{3\ \text{mol}\ Na_2SO_3} = 0.0163\ \text{mol}\ CrO_4^{2-}$$

(c) Since there is one mole of Cr in each mole of CrO_4^{2-}, then the above number of moles of CrO_4^{2-} is also equal to the number of moles of Cr that were present:

0.0163 mol Cr × 52.0 g/mol = 0.848 g Cr in the original alloy

(d) 0.848/3.000 × 100 = 28.3 % Cr

12.78 (a) $Cr_2O_7^{2-}(aq) + 14H^+(aq) + 6e^- \rightarrow 2Cr^{3+}(aq) + 7H_2O(\ell)$

$3 \times (Sn^{2+}(aq) \rightarrow Sn^{4+}(aq) + 2e^-)$

$Cr_2O_7^{2-}(aq) + 14H^+(aq) + 3Sn^{2+}(aq) \rightarrow 3Sn^{4+}(aq) + 2Cr^{3+}(aq) + 7H_2O(\ell)$

(b)

$$0.368 \text{ g } Na_2Cr_2O_7 \times \frac{1 \text{ mol } Na_2Cr_2O_7}{262 \text{ g } Na_2Cr_2O_7} \times \frac{3 \text{ mol Sn}}{1 \text{ mol } Na_2Cr_2O_7} \times \frac{119 \text{ g Sn}}{1 \text{ mol Sn}}$$

= 0.501 g Sn in the alloy

(c) 0.501/1.50 × 100 = 33.4 % Sn

<u>Redox Reactions in Solution</u>

12.79

$$35.0 \text{ mL } SnCl_2 \times \frac{0.150 \text{ mol } SnCl_2}{1000 \text{ mL } SnCl_2} \times \frac{2 \text{ mol } KMnO_4}{5 \text{ mol } SnCl_2} \times \frac{1000 \text{ mL } KMnO_4}{0.150 \text{ mol } KMnO_4}$$

= 14.0 mL $KMnO_4$

12.80 (a) $3 \times (H_2O(\ell) + HSO_3^-(aq) \rightarrow SO_4^{2-}(aq) + 3H^+(aq) + 2e^-)$

$6e^- + 6H^+(aq) + ClO_3^-(aq) \rightarrow Cl^-(aq) + 3H_2O(\ell)$

$3H_2O(\ell) + 3HSO_3^-(aq) + 6H^+(aq) + ClO_3^-(aq) \rightarrow$
$3SO_4^{2-}(aq) + 9H^+(aq) + Cl^-(aq) + 3H_2O(\ell)$

which simplifies to give:

$3HSO_3^-(aq) + ClO_3^-(aq) \rightarrow 3SO_4^{2-}(aq) + 3H^+(aq) + Cl^-(aq)$

(b)

$$25.0\text{ mL NaHSO}_3 \times \frac{0.400\text{ mol NaHSO}_3}{1000\text{ mL NaHSO}_3} \times \frac{1\text{ mol NaClO}_3}{3\text{ mol NaHSO}_3} \times \frac{1000\text{ mL NaClO}_3}{0.100\text{ mol NaClO}_3}$$

$$= 33.3\text{ mL NaClO}_3$$

12.81 (a)

$$0.01760\text{ L KMnO}_4 \times \frac{0.200\text{ mol KMnO}_4}{1\text{ L KMnO}_4} \times \frac{5\text{ mol H}_2\text{O}_2}{2\text{ mol KMnO}_4} \times \frac{34.0\text{ g H}_2\text{O}_2}{1\text{ mol H}_2\text{O}_2}$$

$$= 0.2994\text{ g H}_2\text{O}_2\text{ reacted}$$

(b) $0.2994\text{ g}/1.00\text{ g} \times 100 = 29.94\ \%\ H_2O_2$

12.82

$$0.0116\text{ L KMnO}_4 \times \frac{0.0100\text{ mol KMnO}_4}{1\text{ L KMnO}_4} \times \frac{5\text{ mol NaNO}_2}{2\text{ mol KMnO}_4} \times \frac{69.0\text{ g NaNO}_2}{1\text{ mol NaNO}_2}$$

$$= 0.0200\text{ g NaNO}_2$$

$$\%\text{ NaNO}_2 = \frac{\text{mass NaNO}_2}{\text{total mass}} \times 100 = \frac{0.0200}{1.000} \times 100 = 2.00\ \%\text{ NaNO}_2$$

12.83 (a)

$$0.02388\text{ L KMnO}_4 \times \frac{0.1000\text{ mol KMnO}_4}{1\text{ L KMnO}_4} \times \frac{5\text{ mol C}_2\text{O}_4^{2-}}{2\text{ mol KMnO}_4}$$

$$= 5.970 \times 10^{-3}\text{ mol C}_2\text{O}_4^{2-}$$

(b) The stoichiometry for calcium is as follows:

mol $C_2O_4^{2-}$ = mol Ca^{2+} = mol $CaCl_2$

Thus the number of grams of $CaCl_2$ is given simply by:

5.970×10^{-3} mol $CaCl_2$ $\times$ 111.0 g/mol = 0.6627 g $CaCl_2$

(c) 0.6627/2.651 $\times$ 100 = 25.00 % $CaCl_2$

12.84 (a)

$$0.0290\text{ L Na}_2\text{S}_2\text{O}_3 \times \frac{0.300\text{ mol Na}_2\text{S}_2\text{O}_3}{1\text{ L Na}_2\text{S}_2\text{O}_3} \times \frac{1\text{ mol I}_3^-}{2\text{ mol Na}_2\text{S}_2\text{O}_3}$$

$$= 4.35 \times 10^{-3} \text{ mol } I_3^-$$

(b) $4.35 \times 10^{-3} \text{ mol } I_3^- \times 2 \text{ mol } NO_2^-/\text{mol } I_3^- = 8.70 \times 10^{-3} \text{ mol } NO_2^-$

(c)

$$8.70 \times 10^{-3} \text{ mol } NO_2^- \times \frac{1 \text{ mol } NaNO_2}{1 \text{ mol } NO_2^-} \times \frac{69.0 \text{ g } NaNO_2}{1 \text{ mol } NaNO_2} = 0.600 \text{ g } NaNO_2$$

$$\% \ NaNO_2 = \frac{\text{mass } NaNO_2}{\text{total mass}} \times 100 = \frac{0.600}{0.947} \times 100 = 63.4 \ \% \ NaNO_2$$

Equivalents and Equivalent Weights

12.85 In redox reactions, the number of equivalents per mole for any reagent is equal to the change in oxidation state for the reagent during the reaction at hand. This number may change from one reaction to another, depending on stoichiometry. Thus, we must determine the number of equivalents per mole for a reagent by examining the particular reaction of interest.

(a) The oxidation state of chlorine in $KClO_3$ is 5+, whereas that in Cl_2 is zero. Since this is a change of 5 units, there are 5 eq/mol for $KClO_3$.

(b) The oxidation state of chlorine in $KClO_3$ is 5+, whereas that in Cl^- is -1. Since this is a change of 6 units, there are 6 eq/mol for $KClO_3$.

(c) The oxidation state of nickel in NiO_2 is +4, whereas that in Ni^{2+} is +2. Since this is a change in oxidation state of 2 units, there are 2 eq/mol for NiO_2.

(d) The oxidation state of each chromium atom in Cr_2O_3 is +3, whereas that in CrO_4^{2-} is +6. Since two chromium atoms are involved, and since each undergoes a change of 3 units, there are 6 eq/mol in Cr_2O_3.

12.86 In redox reactions, the number of equivalents per mole for any reagent is equal to the change in oxidation state for the reagent during the reaction at hand. This number may change from one reaction to another, depending on stoichiometry. Thus, we must determine the number of equivalents per mole for a reagent by examining the particular reaction of interest.

(a) The oxidation state of chlorine in NaOCl is +1, whereas that in ClO_3^- is +5. This represents a change of 4 units, and there are 4eq/mol for NaOCl.

(b) The oxidation state of arsenic in H_3AsO_3 is +3, and the oxidation state of arsenic in H_3AsO_4 is +5. Since this is a change of 2 units, the number of eq/mol for H_3AsO_3 is 2.

(c) The oxidation state in I_2 is zero, and the oxidation state of iodine in IO_3^- is +5. The change is 5 units per iodine atom, but since there are two iodine atoms in I_2, there are 10 eq/mol in I_2.

(d) The oxidation state of chlorine in $NaClO_4$ is +7, and the oxidation state of Cl^- is -1. The change gives the number of equivalents per mole: 8 eq/mol.

12.87 We have a reaction that transforms chromium in CrO_4^{2-} (oxidation state +6) to CrO_2^- (oxidation state +3), i.e. a change of 3 units, meaning 3 eq/mol for CrO_4^{2-}.

$$1 \text{ eq } Na_2CrO_4 \times \frac{1 \text{ mol } Na_2CrO_4}{3 \text{ eq } Na_2CrO_4} \times \frac{162 \text{ g } Na_2CrO_4}{1 \text{ mol } Na_2CrO_4} = 54.0 \text{ g } Na_2CrO_4$$

12.88 The oxidation state of each manganese atom in Mn_2O_3 is +3, and the oxidation state of manganese in MnO_4^- is +7. Each manganese, therefore, changes by 4 units, and since there are two manganese atoms involved, the reagent Mn_2O_3 has 8 eq/mol.

$$1 \text{ eq } Mn_2O_3 \times \frac{1 \text{ mol } Mn_2O_3}{8 \text{ eq } Mn_2O_3} \times \frac{158 \text{ g } Mn_2O_3}{1 \text{ mol } Mn_2O_3} = 19.8 \text{ g } Mn_2O_3$$

12.89 The sulfur changes from oxidation state +4 (SO_2) to +6 (SO_4^{2-}), and we conclude that there are 2 eq/mol.

$$0.250 \text{ eq } SO_2 \times \frac{1 \text{ mol } SO_2}{2 \text{ eq } SO_2} \times \frac{64.1 \text{ g } SO_2}{1 \text{ mol } SO_2} = 8.01 \text{ g } SO_2$$

12.90 The nitrogen changes from +5 (HNO_3) to +2 (NO), and there are, therefore, 3 eq/mol in this reaction.

$$1.45 \text{ eq } HNO_3 \times \frac{1 \text{ mol } HNO_3}{3 \text{ eq } HNO_3} \times \frac{63.0 \text{ g } HNO_3}{1 \text{ mol } HNO_3} = 30.5 \text{ g } HNO_3$$

12.91 The change in oxidation state for manganese is from +7 (in $KMnO_4$) to +2 (in Mn^{2+}), meaning that we have 5 eq/mol for this reagent. Similarly, for sulfur we have a change from +4 (in SO_2) to +6 (in SO_4^{2-}), giving 2 eq/mol for this reagent. The equivalent weights of the two reagents are determined as follows:

For $KMnO_4$: 158 g/mol ÷ 5 eq/mol = 31.6 g/eq

For SO_2: 64.1 g/mol ÷ 2 eq/mol = 32.1 g/eq

$$35.0 \text{ g } KMnO_4 \times \frac{1 \text{ eq } KMnO_4}{31.6 \text{ g } KMnO_4} \times \frac{1 \text{ eq } SO_2}{1 \text{ eq } KMnO_4} \times \frac{32.1 \text{ g } SO_2}{1 \text{ eq } SO_2} = 35.5 \text{ g } SO_2$$

12.92 The oxidation state change for iodine is from +5 in $NaIO_3$ to zero in I_2, or a change of 5 units, meaning that this reagent has 5 eq/mol. For nitrogen in $NaNO_2$ we have a change from oxidation state +3 in NO_2^- to +5 in NO_3^-, and we conclude that this reagent has 2 eq/mol. The equivalent weights of the reagents are determined by dividing the molecular weight by the number of equivalents per mole:

For $NaIO_3$: 198 g/mol ÷ 5 eq/mol = 39.6 g/eq

For $NaNO_2$: 69.0 g/mol ÷ 2 eq/mol = 34.5 g/eq

$$0.250 \text{ g NaNO}_2 \times \frac{1 \text{ eq NaNO}_2}{34.5 \text{ g NaNO}_2} \times \frac{1 \text{ eq NaIO}_3}{1 \text{ eq NaNO}_2} \times \frac{39.6 \text{ g NaIO}_3}{1 \text{ eq NaIO}_3} = 0.287 \text{ g NaIO}_3$$

12.93 Each sulfur in $S_2O_3^{2-}$ has oxidation state +2, whereas sulfur in SO_4^{2-} has oxidation state +6. This represents a change of 4 units per sulfur atom, meaning 8 eq/mol since two sulfur atoms are involved. For chlorine we have a change in oxidation state from +1 (in OCl^-) to -1 (in Cl^-), and this reagent, therefore, has 2 eq/mol. The equivalent weight of $Na_2S_2O_3$ is calculated as follows:

158 g/mol ÷ 8 eq/mol = 19.8 g/eq for $Na_2S_2O_3$.

The number of moles of OCl^- that react are:

0.100 L × 0.70 mol/L = 0.070 mol

and the mass of $Na_2S_2O_3$ that is required is:

$$0.070 \text{ mol NaOCl} \times \frac{2 \text{ eq NaOCl}}{1 \text{ mol NaOCl}} \times \frac{1 \text{ eq Na}_2\text{S}_2\text{O}_3}{1 \text{ eq NaOCl}} \times \frac{19.8 \text{ g Na}_2\text{S}_2\text{O}_3}{1 \text{ eq Na}_2\text{S}_2\text{O}_3}$$

$$= 2.8 \text{ g Na}_2\text{S}_2\text{O}_3$$

Normality

12.94 The change in oxidation state for iodine is 5 units altogether, i.e. from +5 in IO_3^- to zero in I_2. There are, thus, 5 eq/mol for this reagent in this particular reaction.

0.200 mol/L × 5 eq/mol = 1.00 eq/L = 1.00 N $NaIO_3$

12.95 For the original reaction we have 5 eq/mol, as determined in the answer to Review Exercise 12.94 above. For the new reaction, the number of eq/mol changes because the product changes; we have oxidation state +5 for iodine in IO_3^- and -1 for iodine in I^-, giving overall, 6 eq/mol.

We can convert to the new normality (eq/L) by first determing the old molarity:

0.500 eq/L $NaIO_3 \div 5$ eq/mol = 0.100 mol/L $NaIO_3$

The new normality is then:

0.100 mol $NaIO_3 \times 6$ eq/mol = 0.600 eq/L = 0.600 N $NaIO_3$

12.96 The number of equivalents per mole for an OH-containing base is equal to the number of OH^- ions in the formula, here 2 eq/mol.

0.100 L $\times$ 0.0100 eq/L $\times$ 1 mol/2 eq $\times$ 171 g/mol = 0.0855 g $Ba(OH)_2$

12.97 The oxidation state of each chromium atom in $Na_2Cr_2O_7$ is +6. The

oxidation state of each product chromium is +3. The change is 3 units per chromium atom, and since two chromium atoms are involved, there are 6 eq/mol for this reagent in this reaction.

0.250 L $\times$ 0.100 eq/L $\times$ 1 mol/6 eq $\times$ 294 g/mol = 1.23 g $K_2Cr_2O_7$

12.98 0.025 L $\times$ 0.100 eq/L = 2.50×10^{-3} eq

or

25.0 mL $\times$ 0.100 eq/1000 mL = 2.50×10^{-3} eq

12.99 0.0350 L $\times$ 0.500 eq/L = 1.75×10^{-2} eq

or

35.0 mL $\times$ 0.500 eq/1000 mL = 1.75×10^{-2} eq

12.100 Since reagents react by definition on a 1:1 equivalent basis, the number of equivalents of $H_2C_2O_4$ that have reacted is equal to the number of

equivalents of $KMnO_4$ that are present:

25.00 mL $H_2C_2O_4 \times 0.1000$ eq/1000 mL = 2.500×10^{-3} eq $H_2C_2O_4$

= 2.500×10^{-3} eq $KMnO_4$

The normality of the $KMnO_4$ solution is thus:

2.500×10^{-3} eq/22.45×10^{-3} L = 0.1114 N

12.101 (a) Since reagents always react on a 1:1 equivalent basis, the number of eq of HNO_2 is equal to the number of equivalents of $K_2Cr_2O_7$:

$$0.02174 \text{ L } K_2Cr_2O_7 \times \frac{0.200 \text{ eq } K_2Cr_2O_7}{1 \text{ L } K_2Cr_2O_7} \times \frac{1 \text{ eq } HNO_2}{1 \text{ eq } K_2Cr_2O_7}$$

$$= 4.35 \times 10^{-3} \text{ eq } HNO_2 \text{ have reacted}$$

(b) The change in oxidation state for nitrogen is from +3 in HNO_2 to +5 in NO_3^-, meaning that there are to be 2 eq/mol for HNO_2 in this reaction.

4.35×10^{-3} eq HNO_2 × 1 mol/2 eq = 2.18×10^{-3} mol HNO_2

This is also equal to the number of moles of $NaNO_2$ in the original sample.

(c) The mass of $NaNO_2$ is:

2.18×10^{-3} mol × 69.0 g/mol = 0.150 g $NaNO_2$

and the % by weight is:

0.150 g/1.000 g × 100 = 15.0 % $NaNO_2$

12.102 (a) Since reagents react on a 1:1 equivalent basis, the number of equivalents of MnO_4^- that have been titrated is equal to the number of equivalents of $H_2C_2O_4$ that have been used in the titration:

0.0288 L × 0.200 eq/L = 5.76×10^{-3} eq $H_2C_2O_4$

= 5.76×10^{-3} eq $KMnO_4$

(b) The oxidation state change for manganese is from +7 (in MnO_4^-) to +2 (in Mn^{2+}), and we conclude that MnO_4^- has 5 eq/mol. The number of moles of MnO_4^- that have been found by the titration is thus:

5.76×10^{-3} eq × 1 mol/5 eq = 1.15×10^{-3} mol $KMnO_4$

(c)

$$1.15 \times 10^{-3} \text{ mol } KMnO_4 \times \frac{1 \text{ mol Mn}}{1 \text{ mol } KMnO_4} \times \frac{54.9 \text{ g Mn}}{1 \text{ mol Mn}} = 6.32 \times 10^{-2} \text{ g Mn}$$

(d) 6.32×10^{-2} g Mn/0.500 g total × 100 = 12.6 % Mn

12.103 (a)

$$18.9 \text{ mL } Na_2S_2O_3 \times \frac{0.100 \text{ eq } Na_2S_2O_3}{1000 \text{ mL } Na_2S_2O_3} \times \frac{1 \text{ eq } I_3^-}{1 \text{ eq } Na_2S_2O_3}$$

$= 1.89 \times 10^{-3}$ eq I_3^- have been titrated

(b) The average oxidation state of an iodine atom in I_3^- is $-{}^1/_3$. In I^-, the oxidation number is -1, meaning that each iodine atom changes by ${}^2/_3$. Since there are three iodine atoms involved in each I_3^- ion, there is an overall change of $3 \times {}^2/_3 = 2$ units, and we conclude that there are to be 2 eq/mol for this reaction.

1.89×10^{-3} eq I_3^- × 1 mol/2 eq = 9.45×10^{-4} mol I_3^-

(c)

$$9.45 \times 10^{-4} \text{ mol } I_3^- \times \frac{2 \text{ mol } Cu^{2+}}{1 \text{ mol } I_3^-} \times \frac{63.5 \text{ g Cu}}{1 \text{ mol } Cu^{2+}} = 0.120 \text{ g Cu}$$

(d) % Cu = mass Cu/total mass × 100 = 0.120/0.847 × 100 = 14.2 % Cu

CHAPTER THIRTEEN

PRACTICE EXERCISES

1. We have here the independent operation of Henry's Law for both gases:

$C_{O_2} = k_{O_2} \times P_{O_2}$ and $C_{N_2} = k_{N_2} \times P_{N_2}$

The solubilities that are given in the problem represent the maximum or saturated concentrations that are established for pure oxygen or nitrogen at a total pressure from the pure gas of 760 torr. These values can be used to determine the Henry's Law constants, k_{O_2} and k_{N_2} as follows:

For O_2: 4.30×10^{-3} g O_2/100 g H_2O = $k_{O_2} \times 760$ torr

Thus $k_{O_2} = 5.66 \times 10^{-6}$ g O_2/100 g H_2O torr

Similarly, for N_2: 1.90×10^{-3} g N_2/100 g H_2O = $k_{N_2} \times 760$ torr

Thus $k_{N_2} = 2.50 \times 10^{-6}$ g O_2/100 g H_2O torr

Notice that the units of k in this problem are:

(g gas)(100 g $H_2O)^{-1}$(torr)$^{-1}$.

In a sample of air, the partial pressure of each gas must be used to determine the concentration of a particular gas in the water. Henry's Law still applies, but each partial pressure is generally less than the total pressure.

For O_2 from air, the concentration of O_2 (dissolved in 100 g of H_2O) is then given by the following:

C_{O_2} = (5.66×10^{-6} g O_2/100 g H_2O torr) $\times$ 156 torr =

8.83×10^{-4} g O_2/100 g H_2O

For N_2 from air, the concentration of N_2 (dissolved in 100 g of H_2O) is then given by the following:

C_{N_2} = (2.50×10^{-6} g N_2/100 g H_2O torr) $\times$ 586 torr =

1.47×10^{-3} g N_2/100 g H_2O

2. The number of moles of each component is first determined:

For CH_3OH, 500 g ÷ 32.0 g/mol = 15.6 mol CH_3OH

For H_2O, 500 g ÷ 18.0 g/mol = 27.8 mol H_2O

The mole fractions are given by the following:

X_{CH_3OH} = mol CH_3OH/total moles and X_{H_2O} = mol H_2O/total moles

$X_{CH_3OH} = 15.6/(15.6 + 27.8) = 0.359$

$X_{H_2O} = 27.8/(15.6 + 27.8) = 0.641$

Notice that the sum of all mole fractions in a mixture is 1. Also, since the mole percent is given by the mole fraction × 100, the sum of all the mole percents is 100:

for CH_3OH, mole % = 0.359 × 100 = 35.9 %

for H_2O, mole % = 0.641 × 100 = 64.1 %

3. The number of moles of each component is first determined:

For NaCl, 43.88 g ÷ 58.44 g/mol = 0.7509 mol NaCl

For H_2O, 1000 g ÷ 18.02 g/mol = 55.49 mol H_2O

The mole fractions are given by the following:

X_{NaCl} = mol NaCl/total moles and X_{H_2O} = mol H_2O/total moles

$X_{NaCl} = 0.7509/(0.7509 + 55.49) = 0.01335$

$X_{H_2O} = 55.49/(0.7509 + 55.49) = 0.9866$

Notice that the sum of all mole fractions in a mixture is 1. Also, since the mole percent is given by the mole fraction × 100, the sum of all the mole percents is 100:

for NaCl, mole % = 0.01335 × 100 = 1.335 %

for H_2O, mole % = 0.9866 × 100 = 98.66 %

4. $X_{O_2} = P_{O_2}/P_{total}$

X_{O_2} = 116 torr/760 torr = 0.153, and mole % = 15.3 %

5.
$$2000 \text{ g } H_2O \times \frac{0.250 \text{ mol } CH_3OH}{1000 \text{ g } H_2O} \times \frac{32.0 \text{ g } CH_3OH}{1 \text{ mol } CH_3OH} = 16.0 \text{ g } CH_3OH \text{ needed}$$

6. We need to know the number of moles of NaOH and the number of kg of water.

$4.00 \text{ g NaOH} \div 40.0 \text{ g/mol} = 0.100 \text{ mol NaOH}$

$250 \text{ g } H_2O \times 1 \text{ kg}/1000 \text{ g} = 0.250 \text{ kg } H_2O$

The molality is thus given by:

$m = 0.100 \text{ mol}/0.250 \text{ kg} = 0.400 \text{ mol NaOH/kg } H_2O$

7. An HCl solution that is 37 % (w/w) has 37 grams of HCl for every 1.0×10^2 grams of solution.

$$7.5 \text{ g HCl} \times \frac{1.0 \times 10^2 \text{ g HCl solution}}{37 \text{ g HCl}} = 2.0 \times 10^1 \text{ g HCl solution}$$

8. The total mass of the solution is to be 750 g. If the solution is to be 10 % (w/w) NaOH, then the mass of NaOH will be:

$750 \text{ g} \times 10 \text{ g NaOH}/100 \text{ g solution} = 75 \text{ g NaOH}$

We therefore need 75 g of NaOH and (750 - 75) = 675 g H_2O.

9. For every 100 qt of solution, there will be 35 qt of methyl alcohol. The amount of alcohol for 20 qt is, therefore:

$$20 \text{ qt} \times \frac{35 \text{ qt methyl alcohol}}{100 \text{ qt solution}} = 7.0 \text{ qt methyl alcohol}$$

It would thus be necessary to add 7.0 qt of alcohol to enough water to bring the total volume to 20 qt.

10. If a solution is 37.0 % (w/w) HCl, then 37.0 % of the mass of any sample of such a solution is HCl and (100.0 - 37.0) = 63.0 % of the mass is water.

(a) In order to determine the molality of the solution, we can conveniently choose 100.0 g of the solution as a starting point. Then 37.0 g of this solution are HCl and 63.0 g are H_2O. For molality, we need to know the number of moles of HCl and the mass in kg of the solvent:

$37.0 \text{ g HCl} \div 36.46 \text{ g/mol} = 1.01 \text{ mol HCl}$

$63.0 \text{ g } H_2O \times 1 \text{ kg}/1000 \text{ g} = 0.0630 \text{ kg } H_2O$

$\text{molality} = \text{mol HCl/kg } H_2O = 1.01 \text{ mol}/0.0630 = 16.0 \text{ m}$

(b) The conversion to mole % values requires that we know the number of moles of all components in the mixture. From (a) above we know the number of moles of HCl: 1.01 mol HCl. The number of moles of water are:

63.0 g H_2O ÷ 18.02 g/mol = 3.50 mol H_2O

The mole fractions of the two components are:

X_{H_2O} = 3.50/(3.50 + 1.01) = 0.776

X_{HCl} = 1.01/(3.50 + 1.01) = 0.224

and the mole % values are 77.6 % H_2O and 22.4 % HCl

11. If the mole fraction of benzene is 0.450, then the mole fraction of carbon tetrachloride is (1.000 - 0.450) = 0.550. We can convert these mole quantities into mass quantities directly:

For benzene, 0.450 mol × 78.12 g/mol = 35.2 g

For carbon tetrachloride, 0.550 mol × 153.81 = 84.6 g

The total mass of such a sample of the solution is (84.6 + 35.2) = 119.8 g, and the % by weight values are:

For benzene: 35.2 g/119.8 g × 100 = 29.4 %

For carbon tetrachloride: 84.6 g/119.8 g × 100 = 70.6 %

12. We need to know the mole amounts of each component of the mixture, and we work conveniently from an amount of solution that contains 1 kg of water, and hence 4.57 mol of $CaCl_2$:

(1 kg × 1000 g/kg) ÷ 18.02 g/mol = 55.5 mol H_2O

The total number of moles is (55.5 + 4.57) = 60.07 and the mole fractions are:

For $CaCl_2$, 4.57/60.07 = 0.0761

For H_2O, 55.5/60.07 = 0.924

Thus we have 7.61 mol % $CaCl_2$ and 92.4 mol % H_2O

13. We first determine the mass of one L (1000 mL) of this solution, using the density:

1000 mL × 1.38 g/mL = 1.38×10^3 g

Next, we use the fact that 40.0 % of this total mass is due to HBr, and calculate the mass of HBr in the 1000 mL of solution:

$0.400 \times 1.38 \times 10^3 = 552$ g HBr

This is converted to the number of moles of HBr in 552 g:

552 g HBr ÷ 80.91 g/mol = 6.82 mol HBr

Last, the molarity is the number of moles of HBr per liter of solution:

6.82 mol/1 L = 6.82 M

14. The total mass of one liter of this solution is:

$1000 \text{ mL} \times 1.50 \text{ g/mL} = 1.50 \times 10^3 \text{ g}$

and the mass of HI that is present in this much solution is:

5.51 mol/L × 1.00 L × 127.91 g/mol = 705 g HI

The % (w/w) of HI in this solution is:

$705 \text{ g}/1.50 \times 10^3 \text{ g} \times 100 = 47.0$ % HI

15. First determine the number of moles of each component of the solution:

For $C_{16}H_{22}O_4$, 20.0 g/278 g/mol = 0.0719 mol

For C_8H_{18}, 50.0 g/114 g/mol = 0.439 mol

and the mole fraction of solvent is:

0.439 mol/(0.439 mol + 0.0719 mol) = 0.859

Using Raoult's Law, we next find the vapor pressure to expect for the solution, which arises only from the solvent (since the solute is known to be nonvolatile):

$P_{solvent} = X_{solvent} \times P°_{solvent} = 0.859 \times 10.5 \text{ torr} = 9.02 \text{ torr}$

16. $P_{cyclohexane} = X_{cyclohexane} \times P°_{cyclohexane} = 0.500 \times 66.9 \text{ torr} = 33.4 \text{ torr}$

$P_{toluene} = X_{toluene} \times P°_{toluene} = 0.500 \times 21.1 \text{ torr} = 10.6 \text{ torr}$

$P_{total} = P_{cyclohexane} + P_{toluene} = 33.4 \text{ torr} + 10.6 \text{ torr} = 44.0 \text{ torr}$

17. $\Delta T = k_b m = 0.51$ °C kg/mol × 16 mol/kg = 8.2 °C

Since this is the amount by which the boiling point of water is raised due to the presence of the solute, the boiling point of the solution is this much higher than that of pure water: (100.0 °C + 8.2 °C) = 108.2 °C.

18. $\Delta T = k_b m = (63.53 - 61.20) = 2.33$ °C

Hence $m = \Delta T/k_b = 2.33$ °C/(3.63 °C kg mol^{-1}) = 0.642 mol solute/kg solvent

Now we determine the number of moles of solute that were employed:

0.642 mol solute/kg solvent × 0.100 kg solvent = 0.0642 mol solute

and the formula weight is given by:

11.0 g ÷ 0.0642 mol = 171 g/mol

19. From Practice Exercise 17 we know that the molality of the solution is 16 mol/kg. The freezing point depression is calculated as follows, using the freezing point depression constant from Table 13.6 of the text:

$\Delta T = k_f m = 1.86$ °C kg/mol × 16 mol/kg = 30 °C

The freezing point of the solution is then 30 °C less than the freezing point of the pure solvent, namely -30 °C.

20. It is first necessary to obtain the values of the freezing point of pure benzene and the value of k_f for benzene from Table 13.6 of the text. We proceed to determine the number of moles of solute that are present and that have caused this depression in the freezing point:

$\Delta T = k_f m$, $\therefore$ $m = \Delta T/k_f$ = (5.45 °C - 4.13 °C)/(5.07 °C kg mol^{-1}) = 0.260 m

Next use this molality to determine the number of moles of solute that must be present:

0.260 mol solute/kg solvent × 0.0850 kg solvent = 0.0221 mol solute

Last, determine the formula weight of the solute:

3.46 g/0.0221 mol = 157 g/mol

21. Since the number of moles of this solute is known, we can calculate the change in the freezing point directly, using the value of k_f for camphor from Table 13.6:

$\Delta T = k_f m = 37.7$ °C kg/mol × (0.0221 mol/0.0850 kg) = 9.80 °C

$T_f = 178.4 - 9.80 = 168.6$ °C

The change in the freezing point for this solute in camphor is seen to be larger than that in benzene (1.32 °C, Practice Exercise 20 above). It is a general result, as illustrated by this comparison, that the solvent with the largest value for k_f offers the largest ΔT in such determinations, assuming that comparable amounts of solutes are employed.

22. (a) $M = \Pi/RT$, where the osmotic pressure Π is to be expressed in the units atm. The conversion to atmospheres is:

$\Pi = 26.0$ torr/760 torr atm^{-1} = 3.42×10^{-2} atm

and the calculation of molarity is as follows:

$$M = \frac{3.42 \times 10^{-2}\ \text{atm}}{(0.0821\ \text{L atm K}^{-1}\ \text{mol}^{-1})(293\ \text{K})} = 1.42 \times 10^{-3}\ \text{molar}$$

(b) First, determine the number of moles of solute in 100 mL of solution:

1.42×10^{-3} mol/L $\times$ 0.100 L = 1.42×10^{-4} mol

and then determine the molecular weight:

0.122 g/1.42×10^{-4} mol = 859 g/mol

(c) $\Delta T = k_f m$ = (1.86 °C kg mol^{-1})(1.42×10^{-3} mol/kg) = 2.6×10^{-3} °C

The freezing point is thus -0.0026 °C

(d) We can see by a comparison of the osmotic pressure effect in part (a) of this problem and the freeazng point depression calculated in part (c), that the osmotic pressure entails a larger effect. Since this larger effect is more readily measured in the lab, the osmosis technique is the better one for determining formula weights of substances with high molar mass.

23. For the solution as if the solute were 100 % dissociated:

ΔT = (1.86 °C kg mol^{-1})(2 $\times$ 0.237) = 0.882 °C

and the freezing point should be -0.882 °C

For the solution as if the solute were 0 % dissociated:

ΔT = (1.86 °C kg mol^{-1})(1 $\times$ 0.237) = 0.441 °C

and the freezing point should be -0.441 °C.

Chapter 13

REVIEW EXERCISES

Kinds of Mixtures

13.1 The three kinds of mixtures are solutions, colloidal dispersions and suspensions.

13.2 Mixtures, because they can be made in any proportion, do not obey the Law of Definite Proportions. An infinite variety of mixtures is possible. Not only can the components of a mixture be changed, but also the amount of any of the components of a mixture can be changed.

13.3 As shown in Table 13.1, it is the size of the particles that distinguishes a solution, a colloidal dispersion and a suspension. Among other properties listed in Table 13.1, we can see that the homogeneity of a mixture depends on the particle size changes.

Colloidal Dispersions

13.4 The stability of a colloidal dispersion can be influenced by Brownian movement, the attachment of charged particles to the outer surface of the particles, and the use of an emulsifying agent.

13.5 A dispersion of one liquid in another is an emulsion.

13.6 Some useful emulsifying agents are egg yoke protein, natural products from soybeans, and the milk protein casein.

13.7 The Tyndall effect is the light scattering that is caused by the relatively large particles of a colloidal dispersion. The particles of a solution are not large enough to cause this effect.

13.8 Brownian movement by colloidal particles is caused by the continuous, random movement of molecules of solvent, which collide with the colloidal particles and cause them to move erratically.

13.9 Sols are dispersions of solid particles in a liquid medium. They are often stabilized by attracting ions of one charge type to their particle surfaces.

13.10 A Tyndall effect should not be exhibited by a solution, but it should be evident in a colloidal dispersion. Also, solutions should not exhibit Brownian movement.

Chapter 13

Why Solutions Form

13.11 The "like" refers to the properties of solute and solvent, most notably polarity, which govern the tendency of a solute to become soluble in a solvent. Thus, solutes dissolve most readily in solvents having similar (like) polarities .

13.12 Since iodine is nonpolar, it dissolves more readily in the solvent that is also nonpolar, in this case carbon tetrachloride.

13.13 Ethyl alcohol molecules are more nonpolar (less polar) than those of water.

13.14 This event, diagramed in Figure 13.6, is due to the tendency for all systems to proceed spontaneously towards a state with a higher degree of randomness (disorder).

13.15 If these two liquids are miscible in all proportions, then a mixture of one in the other can have any of an infinite variety of proportions. All proportions of one liquid to another are found to give a homogeneous solution. The forces of attraction between acetone and water molecules must be comparable to those in pure water and to those in pure acetone.

13.16 Since water and methyl alcohol both have OH groups, there can be hydrogen bonding between a water molecule and a methyl alcohol molecule. This allows any proportion of methyl alcohol in water to be nearly as stable as either separate water samples or separate methyl alcohol samples.

13.17 Water molecules are tightly linked one to another by hydrogen bonding. In benzene, however, which is a nonpolar organic substance, we have only weak London forces of attraction. This means that benzene as a solute in water offers no advantage in attraction to individual water molecules, and the solvent is not disrupted to allow the solute to dissolve.

13.18 This is very much like the explanation given in Figure 13.8 for NaCl. The dipole moments of water molecules can be oriented so as to stabilize both the dissolved cation and the dissolved anion.

13.19 Hydration of electrolytes is the stabilization of ions in water solution by interaction of the ions with the dipole moments of water molecules, as described in the answer to Review Exercise 13.18.

13.20 There is no solvating force provided by benzene that can overcome and offset the very strong ion-ion forces of the solid KCl sample.

13.21 Hydrophilic compounds are water "loving" and, thus, are soluble in water.

13.22 The formation of micelles is diagramed in Figure 13.10. The driving force for this process is the tendency of the hydrophobic hydrocarbon groups to avoid interaction with water.

13.23 This is diagramed in Figure 13.11. The hydrocarbon groups penetrate the oily material effectively because both are nonpolar. The ionic end of the soap conversely interacts favorably only with the water. This combination of properties allows a soap to disperse the oil.

13.24 The strong ion-ion forces in solid KCl are not offset and compensated by any interaction of the ions with this nonpolar liquid.

Heats of Solution

13.25 The heat of solution is a molar quantity, the enthalpy change associated with dissolving a mole of a solute in a solvent. The net enthalpy change arises from a combination of the energy necessary to dissociate both the solute and the solvent, and the energy associated with solvating the solute.

13.26 This analysis works because it employs Hess's Law. It is convenient to view the process in this two-step fashion because it divides the enthalpy changes into two readily understandable and distinguishable processes, one involving the disruption of the solute's forces (lattice energy), and one involving the solvent-to-solute type forces (solvation energy) that exist once the solution is obtained.

13.27 Since the enthalpy of solution is positive, the process is endothermic. The system thus requires heat for the dissolving process, and the heat flow should cause the temperature to decrease.

13.28 It must be the endothermic step (disruption of the lattice) that is larger.

13.29 This is to be very much like that shown in Figure 13.13:

(a) $KCl(s) \rightarrow K^+(g) + Cl^-(g)$, $\Delta H = +690$ kJ

$K^+(g) + Cl^-(g) \rightarrow K^+(aq) + Cl^-(aq)$, $\Delta H = -686$ kJ

(b) $KCl(s) \rightarrow K^+(aq) + Cl^-(aq)$, $\Delta H = +4$ kJ

13.30 Two percent of 690 kJ/mol is $0.02 \times 690 = 17.8$ kJ, and a value that is 2 % greater than 690 is $690 + 17.8 = 708$ kJ/mol. If this value is used in the calculation of Review Exercise 13.29, the net ΔH is now $708 - 686 = +22$ kJ/mol. A small difference (only 2 % in this case) leads to a large difference in the overall value for the process.

13.31 The Al^{3+} ion, having the greater positive charge, should have the larger hydration energy.

13.32 When a gas dissolves in a liquid, there is no endothermic step analogous to the lattice energy of a solid. The only enthalpy change is the one associated with hydration, and this is always negative.

13.33 This is case (a) in Figure 13.17. There is a greater attraction between water and acetone molecules in the resulting solution than there is among acetone molecules in the starting pure solute.

13.34 This is case (b) in Figure 13.17. The disruption of ethyl alcohol and the disruption of hexane together cost more energy than is gained on formation of the solution. This is because the two liquids are not alike; ethyl alcohol is a polar substance with hydrogen bonding, whereas hexane is a nonpolar liquid having only London forces.

13.35 As shown in Figure 13.19, most substances dissolve in water more extensively at the higher temperatures; in other words, solubility increases as temperature increases. For substances of this class, heat may be regarded as a reactant, and it is written into the left side of the thermochemical equation, meaning that dissolution is endothermic. By the principle of Le Chatelier, the stress that is caused by adding heat to such a system is relieved by the equilibrium's shifting to the right, the result being that more material dissolves.

Temperature and Solubility

13.36 We can estimate from Figure 13.19 that the solubility of NH_4NO_3 in 100 g of H_2O is 500 g at 70 °C and 120 g at 10 °C. The amount of solid that will crystallize is the difference between these two solubilities, namely 500 - 120 = 380 g.

13.37 NaBr is the less soluble, and it should crystallize first.

13.38 (a) For each substance A and B, we must determine the amount that will crystallize from solution when the temperature is brought down from the boiling point of the solvent to 5 °C.

For compound A in water we have the following change in solubility in 100 g of solvent: 35.2 - 1.2 = 34.0 g of A will crystallize on cooling a saturated solution in water.

For compound A in ethyl alcohol, we have the following change in solubility in 100 g of solvent: 36.8 - 15.2 = 21.6 g of A will crystallize on cooling a saturated solution in ethyl alcohol.

For compound B, which is present only to the extent of 1 g according to the problem, all should remain dissolved, even at 5 °C, in either solvent, because the solubility for both solvents is greater than 1 g per 100 g of solvent.

The best solvent to pick is the one that will give the larger yield of compound A upon crystallization, since in either solvent we need not worry about the likely co-crystallization of B. We pick water since it yields the greater return of recrystallized compound A.

(b) The recrystallization goes as follows: Dissolve the combined 31 grams of impure substance in 100 g of water at 100 °C. Slowly cool the resulting solution to give pure crystals of compound A. Filter the crystals, discarding the liquid, which now contains, ideally, all of B and some amount of compound A. Rinse the crystals of compound A with a little cold water, and dry the product.

Pressure and Solubility

13.39 Henry's Law is the statement, applied to the dissolving of a gas in a solvent, that at a given temperature, the concentration (C_g) of the gas in a solution is directly proportional to the partial pressure (P_g) of the gas on the solution, where k in the following equation is the constant of proportionality:

$$C_g = k \times P_g$$

As discussed in the text, an alternate statement expresses the relationship of concentration at one pressure P_1 to the concentration that would exist at some new pressure P_2:

$$C_1/P_1 = C_2/P_2$$

13.40 We can use the alternate statement of Henry's Law stated in the text and in Review Exercise 13.39:

$$C_1/P_1 = C_2/P_2 \quad \therefore \quad C_2 = (C_1 \times P_2)/P_1$$
$$= (0.025 \text{ g/L} \times 1.40 \text{ atm})/1.0 \text{ atm}$$
$$= 0.035 \text{ g/L}$$

13.41 We can compare the solubility that is actually observed with the predicted solubility based on Henry's Law. If the actual and the predicted solubilities are the same, we conclude that the gas obeys Henry's Law. We proceed as in Review Exercise 13.40:

$$C_1/P_1 = C_2/P_2 \quad \therefore \quad C_2 = (C_1 \times P_2)/P_1$$
$$C_2 = (0.018 \text{ g/L} \times 620 \text{ torr})/740 \text{ torr}$$
$$C_2 = 0.015$$

The calculated value of C_2 is the same as the observed value, and we conclude that over this pressure range, nitrogen does obey Henry's Law.

13.42 Ammonia is more soluble in water than nitrogen because ammonia is able to enter into hydrogen bonding with solvent molecules, whereas nitrogen is not. Nitrogen is a nonpolar molecular substance, whereas ammonia is a polar substance that is capable of hydrogen bonding. Also, ammonia reacts with water to form nonvolatile ions:

$$NH_3(g) + H_2O(\ell) \rightleftharpoons NH_4^+(aq) + OH^-(aq)$$

13.43 Although carbon dioxide and oxygen are both nonpolar molecular substances, carbon dioxide reacts in water so that its solubility is enhanced:

$$CO_2(g) + H_2O(\ell) \rightleftharpoons H^+(aq) + HCO_3^-(aq)$$

13.44 Sulfur dioxide reacts with water to give an aqueous solution of the acid H_2SO_3:

$$SO_2(g) + H_2O(\ell) \rightleftharpoons H^+(aq) + HSO_3^-(aq)$$

13.45 When it dissolves in water, ammonia reacts to give the equilibrium:

$$NH_3(g) + H_2O(\ell) \rightleftharpoons NH_4^+(aq) + OH^-(aq)$$

Only a small amount of what is present in such a solution is actually called ammonium hydroxide.

Expressions of Concentration

13.46 First determine the number of moles of each component:

For C_7H_8, 60 g ÷ 92 g/mol = 0.65 mol

For C_6H_5Cl, 60 g ÷ 113 g/mol = 0.53 mol

The total number of moles is (0.65 + 0.53) = 1.18 mol, and the mole fractions are determined as follows:

$X_{C_7H_8} = 0.65/1.18 = 0.55$ and $X_{C_6H_5Cl} = 0.53/1.18 = 0.45$

The sum of the mole fractions is 1, and the sum of the mole percents ($X \times 100$) is 100:

mol % $C_7H_8 = 0.55 \times 100 = 55$ %

mol % $C_6H_5Cl = 0.45 \times 100 = 45$ %

13.47 Determine the number of moles of each component of the mixture:

For methyl alcohol:

$$2.50 \times 10^3 \text{ mL} \times \frac{0.780 \text{ g}}{1 \text{ mL}} \times \frac{1 \text{ mol}}{32.0 \text{ g}} = 60.9 \text{ mol}$$

For water:

$$2.50 \times 10^3 \text{ mL} \times \frac{1.00 \text{ g}}{1 \text{ mL}} \times \frac{1 \text{ mol}}{18.0 \text{ g}} = 139 \text{ mol}$$

The total number of moles is (60.9 + 139) = 200 mol

and the mole fractions are:

$X_{CH_3OH} = 60.9/200 = 0.305$ mol, and $X_{H_2O} = 139/200 = 0.695$

The mole percents are 30.5 and 69.5 % respectively.

13.48 The sum of the individual values for mole % should be 100. The mole percent water is thus 100 - (33 + 25) = 42 %.

13.49 The mole fraction of each constituent is given by the ratio of the partial pressure of that constituent to the total pressure:

$X_i = P_i/P_{total}$

$X_{N_2} = 570/760 = 0.750$, $X_{O_2} = 103/760 = 0.14$

$X_{CO_2} = 40/760 = 0.053$, $X_{H_2O} = 47/760 = 0.062$

The mole percent values are obtained by multiplying each mole fraction by 100:

% $N_2 = 75$, % $O_2 = 14$, % $CO_2 = 5.3$, % $H_2O = 6.2$

and the total of these values should be close to 100 since these are the major constituents of air.

13.50 If no other components are present in significant amounts, then the sum of these two partial pressures only is the total pressure with which we are to work:

(a) P_{total} = 197 + 53 = 250 torr and

X_{O_2} = 53 torr/250 torr = 0.21; mol % = 21

X_{N_2} = 197 torr/250 torr = 0.79; mol % = 79

(b) P_{total} = 760 torr

X_{O_2} = 160 torr/760 torr = 0.21; mol % = 21

X_{N_2} = 600 torr/760 torr = 0.79; mol % = 79

(c) The point here is that although the relative proportions of O_2 and N_2 do not change with the altitude, the amounts of oxygen at the higher altitude are still diminished compared to the amount that is available at a lower altitude.

13.51 (a) 18.0 g glucose ÷ 180 g/mol = 0.100 mol glucose

molality = 0.100 g glucose/1.00 kg solvent = 0.100 molal

(b) In an amount of this solution that contains 1 kg of solvent, there are 0.100 mol of glucose, and the following number of moles of water:

1.00×10^3 g × 1 mol/18.0 g = 55.6 mol H_2O

The total number of moles is (55.6 + 0.100) = 55.7 moles, and the mole fraction of glucose is:

$X_{glucose} = 0.100/55.7 = 1.80 \times 10^{-3}$

(c) $X_{H_2O} = 1 - 1.80 \times 10^{-3} = 0.998$

13.52

$$10.0 \text{ g NaCl} \times \frac{1 \text{ mol NaCl}}{58.5 \text{ g NaCL}} = 0.171 \text{ mol NaCL}$$

$$\text{molality} = \frac{0.171 \text{ mol NaCl}}{1.00 \text{ kg } H_2O} = 0.171 \text{ m}$$

Since the density of water is 1.00 g/mL, the volume of 1 kg is 1 L. Thus the molarity is:

0.171 mol/1.00 L = 0.171 M

A solvent must have a density close to 1.0 for this to happen. Also the volume of the solvent must not change appreciably on addition of the solute.

13.53 We need to know the mole amounts of both components of the mixture. It is convenient to work from an amount of solution that contains 1.50 mol of methyl alcohol and, therefore, 1.00 kg of solvent.

The number of moles of water in such an amount of solution is:

1.00×10^3 g H_2 ÷ 18.0 g/mol = 55.6 mol

(a) The mole fractions are :

X_{CH_3OH} = 1.50/57.1 = 0.0263 and X_{H_2O} = 55.6/57.1 = 0.974

(b) Convert the mole fractions into mass amounts as follows:

For CH_3OH, 0.0263 mol × 32.0 g/mol = 0.842 g

For H_2O, 0.974 g × 18.0 g/mol = 17.5 g

Thus, the total mass is (17.5 + 0.842) = 18.3 g, and the mol % of alcohol is given by:

0.842/18.3 × 100 = 4.60 %

13.54 If we choose, for convenience, an amount of solution that contains 1 kg of solvent, then it also contains 0.370 moles of Na_2SO_4. The number of moles of solvent is:

(a) 1.00×10^3 g ÷ 18.0 g/mol = 55.6 mol H_2O

The mole fractions are thus:

$X_{Na_2SO_4}$ = 0.370/(55.6 + 0.370) = 6.61×10^{-3}

X_{H_2O} = 1 - 6.61×10^{-3} = 0.9934

(b) If we use the above mole fractions, we can convert them into mass amounts as follows:

For Na_2SO_4, 6.61×10^{-3} mol × 142 g/mol = 0.939 g

For H_2O, 0.9934 × 18.02 g/mol = 17.90 g

and the percent (w/w) values are:

% Na_2SO_4 = 0.939 g/18.84 g × 100 = 4.98 %

% H_2O = 17.90 g/18.84 g × 100 = 95.01 %

(c) First determine the mass of 1.00 L of this solution, using the known density:

1000 mL × 1.0436 g/mL = 1043.6 g

Next, we use the fact that 4.98 % of any sample of this solution is Na_2SO_4, and calculate the mass of Na_2SO_4 that is contained in 1000 mL of the solution:

0.0498 × 1043.6 g = 52.0 g Na_2SO_4

The number of moles of Na_2SO_4 is given by:

52.0 g ÷ 142 g/mol = 0.366 mol Na_2SO_4

The molarity is the number of moles of Na_2SO_4 per liter of solution:

0.366 mol/1.00 L = 0.366 M

13.55 (a) If the sample is 2.40 mol % H_2SO_4, then an amount of the solution that contains 2.40 mol of H_2SO_4 also contains (100 - 2.42) = 97.60 mol % water. We can calculate the molality if we know the number of moles of H_2SO_4 and the number of kg of solvent. The latter is determined as follows:

97.60 mol H_2O × 18.02 g/mol × 1 kg/1000 g = 1.759 kg H_2O

molality = 2.40 mol H_2SO_4/1.759 kg H_2O = 1.36 m H_2SO_4

(b) The mass of H_2SO_4 in the above sample is:

2.40 mol × 98.1 g/mol = 235 g H_2SO_4

The total mass of the solution is then equal to

(235 g + 1.759 × 10^3 g) = 1.99 × 10^3 g, and the % (w/w) values are:

for H_2SO_4, 235 g/1.99 × 10^3 g × 100 = 11.8 %.

for H_2O, 1.759 × 10^3 g/1.99 × 10^3 g × 100 = 88.4 %.

(c) If we have on hand 100 mL (0.100 L) of this solution, it will have a mass that can be determined using its known density:

mass = 1.079 g/mL × 100.0 mL = 107.9 g of solution

Since this solution has 11.8 % (w/w) H_2SO_4, the mass of H_2SO_4 in 0.100 L of the solution is:

107.9 g × 0.118 = 12.7 g H_2SO_4

The number of moles of H_2SO_4 is thus:

12.7 g ÷ 98.1 g/mol = 0.129 mol H_2SO_4

The molarity is the number of moles of H_2SO_4 divided by the volume of solution:

0.129 mol/0.100 L = 1.29 M H_2SO_4

13.56 (a) If the solution has 3.28 mol % KNO_3, then it also has (100 - 3.28) = 96.72 mol % H_2O. An amount of this solution that contains 3.28 mol of KNO_3 then also contains:

96.72 mol H_2O × 18.02 g/mol × 1 kg/1000 g = 1.743 kg H_2O

The molality is calculated as follows:

3.28 mol KNO_3/1.743 kg H_2O = 1.88 m KNO_3

(b) If there are 1.743 kf (= 1.743×10^3 g) of water, and if there are simultaneously the following number of grams of KNO_3:

3.28 mol KNO_3 × 101 g/mol = 331 g KNO_3

then the % (w/w) values are the following:

For KNO_3, 331 g/(331 g + 1.743×10^3 g) × 100 = 16.0 % KNO_3

For H_2O, 1.743×10^3 g/(331 g + 1.743×10^3 g) × 100 = 84.0 % H_2O

(c) In 100 mL (0.100 L) of this solution we will have the following mass:

100 mL × 1.1039 g/mL = 110.39 g

Since we know the % (w/w) of KNO_3 in this solution, we can determine the mass of KNO_3 that is present:

110.39 g × 0.160 = 17.7 g KNO_3

The number of moles of KNO_3 in this much solution is thus:

17.7 g/101.1 g/mol = 0.175 mol KNO_3

and the molarity is:

0.175 mol/0.100 L = 1.75 M KNO_3

13.57 (a) If we have 100.0 g of the solution, then 22.0 g is NaCl and the remainder, 78.0 g, is water. We need to know the number of moles of NaCl and the number of kg of water:

22.0 g NaCl ÷ 58.5 g/mol = 0.376 mol NaCl

78.0 g ÷ 1000 g/kg = 7.80×10^{-2} kg H_2O

molality = 0.376 mol/7.80×10^{-2} kg = 4.82 m NaCl

(b) We can work conveniently from an amount of solution that contains one kg of water:

(1 kg × 1000 g/kg) ÷ 18.02 g/mol = 55.5 mol H_2O

This is also an amount of solution, which by the very definition of molality, contains 4.82 mol of NaCl. The total number of moles is thus 55.5 + 4.82 = 60.32 mol, and the mole fractions are the following:

X_{NaCl} = 4.82/60.32 = 0.0799

X_{H_2O} = 55.5/60.32 = 0.9201

13.58 If a solution is 10.00 % (w/w) HNO_3, then 10.00 % of the mass of any sample of the solution is HNO_3, and (100.00 - 10.00) = 90.00 % of the mass is water.

(a) We work conveniently by choosing 100.00 g of this solution, which contains 10.00 g HNO_3 and 90.00 g H_2O. The molality can be determined if we know the number of moles of HNO_3 and the number of kg of H_2O:

mol HNO_3 = 10.00 g ÷ 63.02 g/mol = 0.1587 mol HNO_3

mass of water = 90.00 g ÷ 1000 g/kg = 9.000×10^{-2} kg

molality = 0.1587 mol/9.000×10^{-2} kg = 1.763 m HNO_3

(b) We need only to obtain the number of moles of water to proceed:

mol H_2O = 90.00 g ÷ 18.02 g/mol = 4.994 mol H_2O

Thus the mole % HNO_3 is:

0.1587 mol HNO_3/(0.1587 + 4.994) × 100 = 3.080 % HNO_3

(c) We first determine the mass of one L (1000 mL) of this solution using density:

1000 mL × 1.0543 g/mL = 1054.3 g solution

The number of moles of HNO_3 in this much solution is:

1054.3 g × 0.1000 ÷ 63.02 g/mol = 1.673 mol HNO_3

The molarity is: 1.673 mol/1.000 L = 1.673 M HNO_3

13.59 (a) If the molar concentration is 0.733 M, then 1.00 L of this solution contains 0.733 mol $NaNO_3$. This amount of solution also weighs:

1000 mL × 1.0392 g/mL = 1039.2 g solution

The mass of solute in this much solution is:

0.733 mol × 85.01 g/mol = 62.3 g $NaNO_3$

Thus the mass of solvent is (1039.2 - 62.3) = 976.9 g H_2O

The molality is the number of moles of solute divided by the number of kg of solvent:

0.733 mol/0.9769 kg = 0.750 m KNO_3

(b) The number of moles of water is:

976.9 g ÷ 18.02 g/mol = 54.21 mol H_2O

mole % $NaNO_3$ = 0.733 mol/(54.21 mol + 0.733 mol) × 100 = 1.33 %

(c) % (w/w) $NaNO_3$ = 62.3 g/(62.3 g + 976.9 g) × 100 = 5.99 %

13.60 (a) If the molarity of the solution is 0.275 M, then 1.000 L of this solution contains 0.275 mol HCl. The mass of this much solution is:

1000 mL × 1.0031 g/mL = 1003.1 g

of which the following is HCl:

0.275 mol × 36.46 g/mol = 10.03 g HCl

The mass of water is thus (1003.1 - 10.03) = 993.1 g H_2O, and the molality is:

0.275 mol HCl/0.9931 kg H_2O = 0.277 m HCl

The molality is numerically close to the molarity because the solution is dilute and the density is close to 1 g/mL.

(b) % (w/w) HCl = 10.03 g/(10.03 g + 993.1 g) × 100 = 1.000 % HCl

(c) The number of moles of water is:

993.1 g ÷ 18.02 g/mol = 55.11 mol H_2O

mol % HCl = 0.275 mol/(0.275 mol + 55.11 mol) × 100 = 0.495 % HCl

13.61 (a) Since the molarity of the solution is 6.211 mol/L, then one L of this solution contains:

6.211 mol × 46.08 g/mol = 286.2 g C_2H_5OH

The mass of the total 1 L of solution is:

1000 mL × 0.9539 g/mL = 953.9 g

The mass of water is thus 953.9 g - 286.2 g = 667.7 g H_2O, and the molality is:

6.211 mol C_2H_5OH/0.6677 kg H_2O = 9.302 m

(b) % (w/w) C_2H_5OH = 286.2 g/953.9 g × 100 = 30.00 %

(c) mol % C_2H_5OH = 6.211 mol/(37.05 mol + 6.211 mol) × 100 = 14.36 %

(d) Vol of ethyl alcohol = 286.2 g ÷ 0.7893 g/mL = 362.6 mL

Vol of H_2O = 667.7 g ÷ 0.99823 g/mL = 668.9 mL

% (v/v) C_2H_5OH = 362.6 mL/(362.6 mL + 668.9 mL) × 100 = 35.15 %

13.62 (a) The mass of one L of this solution is:

$1000 \text{ mL} \times 0.9938 \text{ g/mL} = 993.8 \text{ g}$

The mass of HCl in this much solution is:

$993.8 \times 0.0100 = 9.938 \text{ g HCl}$

$9.938 \text{ g HCl} \div 36.46 \text{ g/mol} = 0.2726 \text{ mol HCl}$

molarity = 0.2726 mol/1.000 L = 0.2726 M HCl

(b) The mass of water in one L of this solution is:

$(993.8 \text{ g} - 9.938 \text{ g}) \times 1 \text{ kg}/1000 \text{ g} = 0.9839 \text{ kg } H_2O$

molality = 0.2726 mol HCl/0.9839 kg H_2O = 0.2771 m

(c) The number of moles of water is:

$983.9 \text{ g} \div 18.02 \text{ g/mol} = 54.60 \text{ mol } H_2O$

mol % HCl = 0.2726 mol/(0.2726 mol + 54.60 mol) × 100 = 0.4968 %

Dilutions Revisited

13.63 (a) The amount of solute will not change on dilution of a given volume of a concentrated solution to make a dilute solution. If the mass of the initial solution is $mass_1$, then the amount of solute in this much solution is:

$$mass_1 \times \frac{[\% \text{ (w/w) solute}]_1}{(100)}$$

Similarly, on dilution, we have the same expression for the amount of solute in the new solution:

$$mass_2 \times \frac{[\% \text{ (w/w) solute}]_2}{(100)}$$

Now we set these two expressions equal to one another:

$$mass_1 \times (\%)_1 = mass_2 \times (\%)_2$$

(b) The amount of solute will not change on dilution of a certain volume of a concentrated solution, but the overall volume will change. If V_1 and V_2 are the starting and the final volumes, respectively, then the amount of solute in each case is given by the expression:

$$V \times \frac{[\ \%\ (v/v)]}{(100)}$$

Setting these two expressions equal to one another, we have:

$$V_1 \times (\%)_1 = V_2 \times (\%)_2$$

13.64 Use the approach developed in Review Exercise 13.63:

$$10.0\ \% \times \text{mass}_1 = 1.50\ \% \times 250\ g$$

$$\text{mass}_1 = 37.5\ g$$

Therefore, add sufficient water to 37.5 g of the concentrated solution to bring the total mass (i.e. the mass of the diluted solution) up to the value 250 g.

13.65 Use the equation developed in Review Exercise 13.63:

$$1.00\ \% \times \text{mass}_1 = 0.200\ \% \times 50.0\ g$$

$$\text{mass}_1 = 10.0\ g$$

Therefore, add water to 10.0 of the original solution until the total mass of the diluted solution equals 50.0 g.

13.66 Use the approach outlined in Review Exercise 13.63:

$$25\ \% \times V_1 = 15\ \% \times 8\ \text{quarts}$$

$V_1 = 5$ quarts, when rounded to one significant figure

Thus, add water to 5 quarts of the concentrated solution until the final volume of the diluted solution equals 8 quarts.

13.67 Use the method of Review Exercise 13.63:

$$95.5\ \% \times V_1 = 50.0\ \% \times 2.00\ L$$

$$V_1 = 1.05\ L$$

Thus, add water to 1.05 L of the original alcohol until the total volume equals 2.00 L.

Lowering of the Vapor Pressure

13.68 A colligative property of a solution is one that depends only on the molal concentration of the solute particles, and particularly not on the identity of the solute.

13.69 The colligative properties of solutions, as studied in this chapter, are vapor pressure lowering, freezing point depression, boiling point elevation, and osmotic pressure of solutions.

13.70 $P_{solution} = X_{solvent} \times P°_{solvent}$

13.71 The vapor pressure of a solvent is lowered by the presence of a solute because the statistical opportunity for solvent molecules to reach the surface of the solution and evaporate is decreased. In other words, the solute molecules interfere with the evaporation of the solvent. This is illustrated in Figure 13.23.

13.72 We use Raoult's Law to calculate the individual partial pressures for such a solution, and the sum of the partial pressures is taken to be the total vapor pressure of the solution.

13.73 A solution is ideal if the sum of the partial pressures of the components of the solution equals the observed vapor pressure of the solution, i.e. if the solution obeys Raoult's Law. Also, it should be true that the heat of solution is nearly zero.

13.74 $P_{solution} = P°_{solvent} \times X_{solvent}$

$= 23.8 \text{ torr} \times 0.998 = 23.8 \text{ torr}$

13.75 In 100 g of the mixture we have the following mole amounts:

$80.0 \text{ g } H_2O \div 18.02 \text{ g/mol} = 4.44 \text{ mol } H_2O$

$20.0 \text{ g } C_2H_6O_2 \div 62.08 \text{ g/mol} = 0.322 \text{ mol ethylene glycol}$

$X_{H_2O} = 4.44/(4.44 + 0.322) = 0.932$

$P_{solution} = P°_{solvent} \times X_{solvent} = 17.5 \text{ torr} \times 0.932 = 16.3 \text{ torr}$

13.76 The mole fraction values are:

$X_{pentane} = 30.0\ \%/100 = 0.300$ and $X_{heptane} = 70.0\ \%/100 = 0.700$

$P_{pentane} = X_{pentane} \times P°_{pentane} = 0.300 \times 420 \text{ torr} = 126 \text{ torr}$

$P_{heptane} = X_{heptane} \times P°_{heptane} = 0.700 \times 36 \text{ torr} = 25 \text{ torr}$

$P_{total} = P_{pentane} + P_{heptane} = (126 + 25)$ torr = 151 torr

13.77 The following relationships are to be established:

$P_{total} = 96 \text{ torr} = P°_{benzene} \times X_{benzene} + P°_{toluene} \times X_{toluene}$

The relationship between the two mole fractions is:

$X_{benzene} = 1 - X_{toluene}$

since the sum of the two mole fractions is one. Substituting this expression for $X_{benzene}$ into the first equation gives:

$96 \text{ torr} = P°_{benzene} \times (1 - X_{toluene}) + P°_{toluene} \times X_{toluene}$

$96 \text{ torr} = 180 \text{ torr} \times (1 - X_{toluene}) + 60 \text{ torr} \times X_{toluene}$

Solving for $X_{toluene}$ we get:

$120 \times X_{toluene} = 84$

$\therefore X_{toluene} = 0.70$ and we also then know that $X_{benzene} = 0.30$

The mole % values are to be 70 % toluene and 30 % benzene.

13.78 (a) $P_{solvent} = X_{solvent} \times P°_{solvent}$

$336.0 \text{ torr} = X_{solvent} \times 400.0 \text{ torr} \quad \therefore \quad X_{solvent} = 0.8400$

$X_{solute} = 1 - 0.8400 = 0.1600$

(b) The number of moles of solvent is:

34.88 g ÷ 109.0 g/mol = 0.3200 mol

and the following expression for mole fraction of solvent can be solved to determine the number of moles of solute:

$$0.8400 = \frac{0.3200}{\text{mol solute} + 0.3200} \qquad \therefore \text{mol solute} = 0.0610$$

(c) 19.35 g/0.0160 mol = 317 g/mol

13.79 As discussed on pages 527 - 528 of the text, a positive deviation from Raoult's Law is displayed by the data plotted in the left-hand side of Figure 13.25. In such a mixture, there are less strong intermolecular forces than is present in either of the two pure components. The mixture thus has a greater vapor pressure than might ideally be expected because the molecules of the mixture are more free to evaporate than are those of either of the separate pure components of the mixture.

Freezing Point Depression and Boiling Point Elevation

13.80 The crystal structure of ice accommodates only water molecules, not ions of a solute.

13.81 (a) The solute particles decrease the vapor pressure, so to raise the vapor pressure to be equal to atmospheric pressure (the definition of the boiling point), a higher temperature is required.

(b) The solute particles interfere with the crystallization of the solvent, i.e. with the transition of solvent molecules from the liquid to crystalline state. Lowering the temperature aids in the crystallization.

13.82 (a) Since -40 °F is also equal to -40 °C, the following expression applies:

$$\Delta T = k_f\, m, \text{ so } 40\ °C = (1.86\ °C\ kg\ mol^{-1}) \times m$$

$$\therefore\ m = 40/1.86\ mol/kg = 22\ molal$$

(b)

$$22\ mol \times \frac{62.1\ g}{1\ mol} \times \frac{1.00\ mL}{1.11\ g} = 1.2 \times 10^3\ mL$$

(c) There are 946 mL in one quart. Thus for 1 qt of water we are to have 946 mL, and the required number of quarts of ethylene glycol is:

$$946\ mL\ H_2O \times \frac{1.00\ g\ H_2O}{1.00\ mL\ H_2O} \times \frac{1.2 \times 10^3\ mL\ C_2H_6O_2}{1000\ g\ H_2O} \times \frac{1\ qt\ C_2H_6O_2}{946\ mL\ C_2H_6O_2}$$

$$= 1.2\ qt\ C_2H_6O_2$$

The proper ratio of ethylene glycol to water is 1.2 qt to 1 qt.

13.83 $\Delta T = K_b m = 0.51\ °C\ kg\ mol^{-1} \times 2.0\ mol\ kg^{-1} = 1.0\ °C$

$\therefore\ T_b = 100.0 + 1.0 = 101\ °C$

$\Delta T = k_f m = 1.86\ °C\ kg\ mol^{-1} \times 2.00\ mol\ kg^{-1} = 3.72\ °C$

$\therefore\ T_f = 0.0 - 3.72 = -3.72\ °C$

13.84 The number of moles of glycerol is:

46.0 g ÷ 92 g/mol = 0.50 mol

and the molality of this solution is 0.50 mol/0.250 kg = 2.0 m

(a) $\Delta T = k_b \times m = 0.51\ °C\ kg\ mol^{-1} \times 2.0\ mol\ kg^{-1} = 1.0\ °C$

and the boiling point is 100.0 + 1.0 = 101 °C

(b) $\Delta T = k_f \times m = 1.86\ °C\ kg\ mol^{-1} \times 2.0\ mol\ kg^{-1} = 3.72\ °C$

and the freezing point is 0.0 - 3.72 = -3.72 °C

(c) $P_{solution} = P°_{solvent} \times X_{solvent}$

Now the number of moles of solute is:

46.0 g ÷ 92.1 g/mol = 0.499 mol $C_3H_8O_3$

The number of moles of solvent is:

250 g H_2O ÷ 18.02 g/mol = 13.9 mol H_2O

and the mole fraction of solvent is:

$X_{H_2O} = 13.9/(13.9 + 0.499) = 0.965$

Thus, $P_{solution}$ = 23.8 torr × 0.965 = 23.0 torr

13.85 $\Delta T = (5.45 - 3.45) = 2.00\ °C = k_f \times m = 5.07\ °C\ kg\ mol^{-1} \times m$

∴ m = 0.394 mol solute/kg solvent

0.394 mol/kg benzene × 0.200 kg benzene = 0.0788 mol solute

and the molecular weight is: 12.00 g/0.0788 mol = 152 g/mol

13.86 $\Delta T = k_b \times m = (81.7 - 80.2)$, ∴ m = 0.59 mol solute/kg benzene

0.59 mol/kg benzene × 1.0 kg benzene = 0.59 mol solute

and the formula weight is: 14 g/0.59 mol = 24 g/mol

13.87 (a) For convenience we choose to work with 100 g of the compound, and then to convert the mass amounts of each element found in this compound into mole amounts:

for C, 42.86 g ÷ 12.01 g/mol = 3.569 mol C
for H, 2.40 g ÷ 1.01 g/mol = 2.37 mol H
for N, 16.67 g ÷ 14.01 g/mol = 1.190 mol N

for O, 38.07 g ÷ 16.00 g/mol = 2.379 mol O

The relative mole amounts that represent the empirical formula are determined by dividing the above mole amounts each by the smallest mole amount:

for C: 3.569 ÷ 1.190 = 2.999
for H: 2.37 ÷ 1.190 = 1.99
for N: 1.190 ÷ 1.190 = 1.000
for O: 2.379 ÷ 1.190 = 1.999

and the empirical formula is $C_3H_2NO_2$

(b) $\Delta T = 1.84\ °C = k_b \times m = 2.53\ °C\ kg\ mol^{-1} \times m$

$\therefore m = 0.727$ mol solute/kg benzene

The number of moles of solute is:

0.727 mol/kg benzene × 0.045 kg benzene = 0.0327 mol

and the formula weight is: 5.5g/0.0327 mol = 168 g/mol.

Since the weight of the empirical unit is 84, the molecular formula must be twice the empirical formula, namely $C_6H_4N_2O_4$.

13.88 (a) $6e^- + 14H^+ + Cr_2O_7^{2-} \rightarrow 2Cr^{3+} + 7H_2O$

$3 \times (C_3H_8O_2 \rightarrow C_3H_6O + 2H^+ + 2e^-)$

$8H^+ + Cr_2O_7^{2-} + 3C_3H_8O \rightarrow 2Cr^{3+} + 7H_2O + 3C_3H_6O$

(b) The formula weights are $Na_2Cr_2O_7 \cdot 2H_2O$: 298 g/mol,

C_3H_8O: 60.1 g/mol, and C_3H_6O: 58.1 g/mol.

$$25.0\ g\ C_3H_8O \times \frac{1\ mol\ C_3H_8O}{60.1\ g\ C_3H_8O} \times \frac{1\ mol\ Na_2Cr_2O_7 \cdot 2H_2O}{3\ mol\ C_3H_8O}$$

$$\times \frac{298\ g\ Na_2Cr_2O_7 \cdot 2H_2O}{1\ mol\ Na_2Cr_2O_7 \cdot 2H_2O} = 41.3\ g\ Na_2Cr_2O_7 \cdot 2H_2O$$

(c)

$$25.0\ g\ C_3H_8O \times \frac{1\ mol\ C_3H_8O}{60.1\ mol\ C_3H_8O} \times \frac{3\ mol\ C_3H_6O}{3\ mol\ C_3H_8O} \times \frac{58.1\ g\ C_3H_6O}{1\ mol\ C_3H_6O}$$

$$= 24.2\ g\ C_3H_6O$$

(d) First, we determine the number of grams of C, H, and O that are found in the products, and then the % by weight of C, H, and O that were present in the sample that was analyzed by combustion, i.e. the by-product:

For C,

$$22.365 \times 10^{-3} \text{ g } CO_2 \times \frac{12.01 \text{ g C}}{44.01 \text{ g } CO_2} = 6.103 \times 10^{-3} \text{ g C}$$

and the % C is: 6.103×10^{-3} g/8.654×10^{-3} g $\times$ 100 = 70.52 % C

For H,

$$10.655 \times 10^{-3} \text{ g } H_2O \times \frac{2.016 \text{ g H}}{18.02 \text{ g } H_2O} = 1.192 \times 10^{-3} \text{ g H}$$

and the % H is: 1.192×10^{-3} g H/8.654×10^{-3} g $\times$ 100 = 13.77 % H

For O, the mass is the total mass minus that of C and H in the sample that was analyzed:

$$8.654 \times 10^{-3} \text{ g total} - (6.103 \times 10^{-3} \text{ g C} + 1.192 \times 10^{-3} \text{ g H})$$

$$= 1.359 \times 10^{-3} \text{ g O}$$

and the % O is: 1.359×10^{-3} g/8.654×10^{-3} g $\times$ 100 = 15.70 % O

Alternatively, we could have determined the amount of oxygen by using the mass % values, realizing that the sum of the weight percent values should be 100.

Next, we convert these mass amounts for C, H, and O into mole amounts by dividing the amount of each element by the atomic weight of each element:

For C, 6.103×10^{-3} g C $\div$ 12.01 g/mol = 0.5082×10^{-3} mol C

For H, 1.192×10^{-3} g H $\div$ 1.008 g/mol = 1.183×10^{-3} mol H

For O, 1.359×10^{-3} g O $\div$ 16.00 g/mol = 0.08494×10^{-3} mol O

Lastly, these are converted to relative mole amounts by dividing each of the above mole amounts by the smallest of the three:

For C, 0.5082/0.08494 = 5.983
For H, 1.183/0.08494 = 13.93
For O, 0.08494/0.08494 = 1.000

and the empirical formula is given by this ratio of relative mole amounts, namely $C_6H_{14}O$.

(e) $\Delta T_f = k_f m$

$(5.45\ °C - 4.87\ °C) = (5.07\ °C/m) \times m \quad \therefore m = 0.114$ molal

and there are 0.114 moles of solute dissolved in each kg of solvent. Thus, the number of moles of solute that have been used here is:

$0.114\ mol/kg \times 0.1150\ kg = 1.31 \times 10^{-2}$ mol solute

The formula weight is thus: 1.338 g/0.0131 mol = 102 g/mol

Since the empirical formula has this same mass, we conclude that the molecular formula is the same as the empirical formula, i.e. $C_6H_{14}O$.

Dialysis and Osmosis

13.89 Osmosis is a process in which solvent flows spontaneously across a membrane from a solution of low solute concentration to one of high solute concentration. So long as the two solutions that are separated by an osmotic membrane have differing concentrations, the process of osmosis brings about a decrease in concentration of solute for the more concentrated of the two solutions. This is illustrated in Figure 13.28.

13.90 In osmosis, only solvent is passed across the membrane. In dialysis, solvent and certain solute particles pass selectively across the membrane. See Figure 13.31.

13.91 An osmotic membrane allows only solvent to pass, whereas a dialyzing membrane allows solvated ions of a certain minimum size to pass as well as solvent molecules. A dialyzing membrane prevents the passage of only certain solute particles, usually those of large size, such as colloid particles.

13.92 This is a statistical effect. There are more solvent molecules on the side of the membrane having the lower solute concentration. This means that there is less interference to migration on the side of the membrane with the lower solute concentration, whereas on the side of the membrane having the higher solute concentration, there is greater interference with a solvent molecule's chance to approach and pass through the membrane.

13.93 The solution that loses solvent into the other solution is the one with the lower molarity.

13.94 The answer is the same as for Review Exercise 13.93.

13.95 In each case, the osmotic pressure Π is given by the equation:

$$\Pi = M \times R \times T$$

Since we do not know either the density of the solution or the volume of the solution, we cannot convert values for % by weight into molarities. However, we do know that glucose, having the smaller molecular weight, has the higher molarity, and we conclude that it will have the larger osmotic pressure.

13.96 $\Pi = MRT = (0.0100 \text{ mol/L})(0.0821 \text{ L atm K}^{-1} \text{ mol}^{-1})(298 \text{ K})$

$= 0.24 \text{ atm}$

$0.24 \text{ atm} \times 760 \text{ torr/atm} = 1.8 \times 10^2 \text{ torr}$

13.97 (a) If the equation is correct, the units on both sides of the equation should be g/mol. The units of the right side of this equation are:

$$\frac{(\text{g}) \times (\text{L atm K}^{-1} \text{ mol}^{-1}) \times \text{K}}{\text{atm} \times \text{L}} = \text{g/mol}$$

which is correct.

(b) $\Pi = MRT = (n/V)RT$

$\therefore n = \Pi V/RT$

This means that we can calculate the number of moles of solute in one L of solution, as follows:

$$n = \frac{(0.021 \text{ torr} \times 1 \text{ atm/760 torr})(1.00 \text{ L})}{(0.0821 \text{ l atm K}^{-1} \text{ mol}^{-1})(298 \text{ K})} = 1.1 \times 10^{-6} \text{ mol}$$

The formula weight is the mass in 1 L divided by the number of moles in 1 L:

$2.0 \text{ g}/1.1 \times 10^{-6} \text{ mol} = 1.8 \times 10^{6} \text{ g/mol}$

13.98 By the "association of solute particles" we mean that some particles are attracted to others, or that solvent does not perfectly insulate solute particles from attachment to one another. This is another way of saying that there is something less than 100 % dissociation of solute in solution.

13.99 Ionic compounds characteristically dissociate in solution to produce greater numbers of solute particles than formula units that have dissolved.

13.100 The dissociation of sodium chloride is represented as follows:

$NaCl(s) \rightarrow Na^+(aq) + Cl^-(aq)$

and the molality of dissolved and dissociated solute particles is:

0.171 mol NaCl/kg solvent × 2 mol particles/mol NaCl = 0.342 molal

The number of moles of solvent in 1000 g (1 kg) is:

1000 g H_2O/18.02 g/mol = 55.6 mol H_2O

The solvent mole fraction is thus:

$X_{H_2O} = 55.6/(55.6 + 0.342) = 0.994$

and the vapor pressure is:

$P_{solution} = P°_{H_2O} \times X_{H_2O} = 17.5 \text{ torr} \times 0.994 = 17.4 \text{ torr}$

13.101 The solute NaI has the larger formula weight, and it thus has the solution with the smaller number of moles per kg of solvent. Thus, the NaI solution should have the smaller depression in the freezing point. It is, therefore, the NaCl solution that should have the lower freezing point.

13.102 The solute with the larger formula weight is Na_2CO_3. It is therefore the solute that gives the solution with the smaller molality. The Na_2CO_3 solution thus gives the smaller elevation of the boiling point, and the lower boiling point overall.

13.103 Since glucose has the larger molecular weight, it's solution has the smaller molarity. The solution with the smaller molarity has the lower osmotic pressure.

13.104 The van't Hoff factor is the ratio of the value for a colligative property as actually measured to that value of the colligative property that is expected in the complete absence of any solute dissociation. Some typical values are given in Table 13.7. For all molecular solutes, the van't Hoff factor should be 1. For solutes such as Na_2SO_4, which dissociate to give 3 ions in solution, the factor should be 3.

13.105 Any electrolyte such as $NiSO_4$, that dissociated to give 2 ions, if fully dissociated should have a van't Hoff factor of 2.

13.106 (a) $\Delta T_f = i \times k_f \times m = 1.89 \times 0.118 \times 1.86\ °C/m = 0.415\ °C$

$\therefore T_f = -0.415\ °C$

(b) There is more effective dissociation of NaCl than of $NiSO_4$ at this concentration.

13.107 (a) $\Delta T_b = k_b \times m = 0.51\ °C/m \times 1.00 = 0.51\ °C \quad \therefore T_b = 100.51\ °C$

(b) $\Delta T_b = k_b \times (m \times 4) = 0.51\ °C/m(4 \times 1.00\ m) = 2.04\ °C$

$\therefore T_b = 102.04\ °C$

(c) $i = 1.183/0.51 = 2.3$

13.108 The freezing point depression that is expected from this solution if HF behaves as a nonelectrolyte is:

$\Delta T_f = 1.86\ °C/m \times 1.00\ m = 1.86\ °C$

The freezing point that is expected upon complete dissociation of HF is:

$\Delta T_f = k_f \times (2 \times m) = 3.72\ °C$

The observed freezing point depression is 1.91 °C, and the apparent molality is:

$m = 1.91\ °C/k_f = 1.03$ mol solute particles per kg of solvent

This represents a mole excess of 3 solute particles per mol of HF, and we conclude that the percent ionization is 3 %.

13.109 $\Delta T_f = 0.261\ °C = k_f \times$ (apparent molality)

Thus the apparent molality of solute particles is:

$m = 0.261\ °C/(1.86\ °C\ \text{molal}^{-1}) = 0.140$ molal

If the solute were dissolved as a nonelectrolyte, the apparent molality would be 0.125. The excess apparent molality arises from dissociation of the solute, and the amount (0.140 - 0.125) = 0.015 is the excess molality due to dissociation in this case. That is, 0.015 mol of solute per kg of solvent have been generated by dissociation of some certain % of the solute.

% ionization = 0.015/0.125 × 100 = 12 %

CHAPTER FOURTEEN

PRACTICE EXERCISES

1.

$$K_c = \frac{[H_2O(g)]^2}{[H_2(g)]^2[O_2(g)]}$$

$$K_c = \frac{[NH_4^+(aq)][OH^-(aq)]}{[NH_3(aq)][H_2O(\ell)]}$$

2.

$$K_p = \frac{P_{HI}^2}{P_{H_2} \times P_{I_2}}$$

3.

$$K_p = K_c(RT)^{\Delta n_g}$$

In this case, $\Delta n_g = (1 - 3) = -2$, and we have:

$3.8 \times 10^{-2} = K_c[(0.0821 \text{ L atm K}^{-1} \text{ mol}^{-1})(473 \text{ K})]^{-2}$

Thus $3.8 \times 10^{-2} = K_c(6.6 \times 10^{-4})$, and $K_c = 57$

4.

$$K_p = K_c(RT)^{\Delta n_g}$$

In this reaction, $\Delta n_g = 3 - 2 = 1$

$K_p = (7.3 \times 10^{34})[(0.0821 \text{ L atm K}^{-1} \text{ mol}^{-1})(298 \text{ K})]^1$, and $K_p = 1.8 \times 10^{36}$

5. Since reaction (2) has the largest value of K_c, it is the reaction that, at equilibrium, has proceeded farthest towards completion, i.e. farthest towards the formation of products.

6. (a) The equilibrium will shift to the right, decreasing the concentration of Cl_2 at equilibrium, and consuming some of the added PCl_3. The value of K_p will be unchanged.

(b) The equilibrium will shift to the left, consuming some of the added PCl_5 and increasing the amount of Cl_2 at equilibrium. The value of K_p will be unchanged.

(c) For any exothermic equilibrium, an increase in temperature causes the equilibrium to shift to the left, in order to remove energy in response to the stress. This equilibrium is shifted to the left, making more Cl_2 and more PCl_3 at the new equilibrium. The value of K_p is given by the following:

$$K_p = \frac{P_{PCl_5}}{P_{PCl_3} \times P_{Cl_2}}$$

In this system, an increase in temperature (which causes an increase in the equilibrium concentrations of both PCl_3 and Cl_2 and a decrease in the equilibrium concentration of PCl_5) causes an increase in the denominator of the above expression as well as a decrease in the numerator of the above expression. Both of these changes serve to cause a decrease in the value of K_p.

(d) Decreasing the container volume for a gaseous system will produce an increase in partial pressures for all gaseous reactants and products. In order to lower the increase in partial pressures, the equilibrium will shift so as to favor the reaction side having the smaller number of gaseous molecules, in this case to the right. This shift will decrease the amount of Cl_2 and PCl_3 at equilibrium, and it will increase the amount of PCl_5 at equilibrium. This increases the size of the numerator and decreases the size of the denominator in the above expression for K_p, causing the value of K_p to increase.

7. $2CO \quad + \quad O_2 \quad \rightleftharpoons \quad 2CO_2$

Changes in concentrations for the above materials are:

-2 × (0.030 mol/L) -1 × (0.030 mol/L) +2 × (0.030 mol/L)

In other words, the changes in the concentrations of the reactants and products are governed by the stoichiometry of the reaction:

[CO] decreases by 0.060 mol/L and $[CO_2]$ increases by 0.060 mol/L

8.

$$K_c = \frac{[H_2][CO_2]}{[H_2O][CO]} = \frac{(0.200\ M)(0.150\ M)}{(0.0411\ M)(0.180\ M)} = 4.06$$

9. (a) The initial concentrations were:

$[PCl_3] = 0.20$ mol/1.00 L = 0.20 M; $[Cl_2] = 0.10$ mol/1.00 L = 0.10 M

$[PCl_5] = 0.00$ mol/1.00 L = 0.00 M

(b) The change in concentration of PCl_3 was (0.20 - 0.12) M = 0.08 mol/L.

The other materials must have undergone changes in concentration that are dictated by the coefficients of the balanced chemical equation, namely:

PCl_3	+	Cl_2	⇌	PCl_5
-1 × (0.08 M)		-1 × (0.08 M)		+1 × (0.08 M)

or both PCl_3 and Cl_2 have decreased by 0.08 M and PCl_5 has increased by 0.08 M.

(c)

	PCl_3	+	Cl_2	⇌	PCl_5
initial conc.	0.20 M		0.10 M		0.00 M
change in conc.	-0.08 M		-0.08 M		+0.08 M
equilibrium conc.	0.12 M		0.02 M		0.08 M

(d)

$$K_c = \frac{[PCl_5]}{[PCl_3][Cl_2]} = \frac{(0.08)}{(0.12)(0.02)} = 33 = 3 \times 10^1$$ if we use the proper number of sig. figs.

10.

$$K_c = \frac{[H_2O][CH_3CO_2C_2H_5]}{[CH_3CO_2H][C_2H_5OH]} = 4.10$$

Solving for $[C_2H_5OH]$, we get:

$$[C_2H_5OH] = \frac{[H_2O][CH_3CO_2C_2H_5]}{[CH_3CO_2H][4.10]} = \frac{(0.00850)(0.910)}{(0.210)(4.10)}$$

$$[C_2H_5OH] = 8.98 \times 10^{-3}\ M$$

11.

	$H_2(g)$	+	$I_2(g)$	$\rightleftharpoons$	$2HI(g)$
initial conc. (M)	0.200 mol/1.00 L		0.200 mol/1.00 L		0
change in conc. (M)	-x		-x		+2x
equilibrium conc. (M)	0.200 - x		0.200 - x		2x

Substitution of the above equilibrium values into the mass action expression gives the following:

$$K_c = \frac{[HI]^2}{[H_2][I_2]} = 49.5 = \frac{(2x)^2}{(0.200 - x)(0.200 - x)}$$

Taking the square root of both sides of the above equation we have:

$\frac{2x}{(0.200 - x)} = 7.04$, and on solving for x we get:

$9.04x = (0.200)(7.04)$, or $x = 0.156$ M

The equilibrium concentrations are thus:

$[HI] = 2x = 2(0.156) = 0.312$ M; $[H_2] = [I_2] = 0.200 - 0.156 = 0.044$ M

12.

	$N_2(g)$	+	$O_2(g)$	$\rightleftharpoons$	$2NO(g)$
initial conc. (M)	0.033		0.00810		0
change in conc. (M)	-x		-x		+2x
equilibrium conc. (M)	(0.033 - x)		(0.00810 - x)		2x

Substituting the above values for equilibrium concentrations into the mass action expression gives:

$$K_c = \frac{[NO]^2}{[N_2][O_2]} = 4.8 \times 10^{-31} = \frac{(2x)^2}{(0.033 - x)(0.00810 - x)}$$

Because 0.033 and 0.00810 are much larger than $1000 \times K_c$, we can make the simplifying assumption that:

$(0.033 - x) \approx 0.033$ and $(0.00810 - x) \approx 0.00810$

which allows us to write:

$4.8 \times 10^{-31} = \frac{(2x)^2}{(0.033)(0.00810)}$ which is solved for x to give:

$4x^2 = (4.8 \times 10^{-31})(8.10 \times 10^{-3})(3.3 \times 10^{-2})$ or $x = 5.7 \times 10^{-18}$ M and we see that the simplifying assumption was valid.

At equilibrium, $[NO] = 2x = 1.1 \times 10^{-17}$ M

13. (a)

$$K_c = \frac{1}{[Cl_2(g)]}$$

(b)

$$K_c = \frac{1}{[NH_3(g)][HCl(g)]}$$

14. (a) $K_{sp} = [Ba^{2+}][CrO_4^{2-}]$ (b) $K_{sp} = [Ag^+]^3[PO_4^{3-}]$

15.

	$PbF_2(s)$ ⇌	$Pb^{2+}(aq)$	+	$2F^-(aq)$
initial conc. (M)		0.0		0.0
change in conc. (M)		2.15×10^{-3}		$2 \times (2.15 \times 10^{-3})$
equilibrium conc. (M)		2.15×10^{-3}		4.30×10^{-3}

Substituting the above values for equilibrium concentrations into the expression for K_{sp} gives:

$K_{sp} = [Pb^{2+}][F^-]^2 = (2.15 \times 10^{-3})(4.30 \times 10^{-3})^2 = 3.98 \times 10^{-8}$

16.

	$CoCO_3$ ⇌	Co^{2+}	+	CO_3^{2-}
initial conc. (M)		0.0		0.100
change in conc. (M)		1.0×10^{-9}		1.0×10^{-9}
equilibrium conc. (M)		1.0×10^{-9}		$(0.100 + 1.0 \times 10^{-9})$

Substituting the above values for equilibrium concentrations into the expression for K_{sp} gives:

$K_{sp} = [Co^{2+}][CO_3^{2-}] = (1.0 \times 10^{-9})(0.10 + 1.0 \times 10^{-9}) = 1.0 \times 10^{-10}$

17.

	$PbF_2(s)$ ⇌	Pb^{2+}	+	$2F^-$
initial conc. (M)		0.10		0.0
change in conc. (M)		3.1×10^{-4}		$2(3.1 \times 10^{-4})$
equilibrium conc. (M)		$0.10 + 3.1 \times 10^{-4}$		6.2×10^{-4}

Substituting the above values for equilibrium concentrations into the expression for K_{sp} gives:

$K_{sp} = [Pb^{2+}][F^-]^2 = [0.10 + 3.1 \times 10^{-4}][6.2 \times 10^{-4}]^2$

Now $(0.10 + 3.1 \times 10^{-4})$ is also ≈ 0.10:

Hence, $K_{sp} = (0.10)(6.2 \times 10^{-4})^2 = 3.8 \times 10^{-8}$

18.

	$AgBr(s)$	$\rightleftharpoons$	Ag^+	+	Br^-
initial conc. (M)			0.0		0.0
change in conc. (M)			+x		+x
equilibrium conc. (M)			x		x

Substituting the above values for equilibrium concentrations into the expression for K_{sp} gives:

$$K_{sp} = 5.0 \times 10^{-13} = [Ag^+][Br^-] = (x)(x)$$

$$x = \sqrt{5.0 \times 10^{-13}} = 7.1 \times 10^{-7} \text{ mol/L}$$

Thus the solubility is 7.1×10^{-7} M AgBr.

19.

	$Ag_2CO_3(s)$	$\rightleftharpoons$	$2Ag^+$	+	CO_3^{2-}
initial conc. (M)			0.0		0.0
change in conc. (M)			+2x		+x
equilibrium conc. (M)			2x		x

Substituting the above values for equilibrium concentrations into the expression for K_{sp} gives:

$$K_{sp} = 8.1 \times 10^{-12} = [Ag^+]^2[CO_3^{2-}] = (2x)^2(x)$$

and $4x^3 = 8.1 \times 10^{-12}$

$$x = \sqrt[3]{\frac{8.1 \times 10^{-12}}{4}} = 1.3 \times 10^{-4} \text{ mol/L}$$

20.

	$AgI(s)$	$\rightleftharpoons$	Ag^+	+	I^-
initial conc. (M)			0.0		0.20
change in conc. (M)			+x		+x
equilibrium conc. (M)			x		0.20 + x

Substituting the above values for equilibrium concentrations into the expression for K_{sp} gives:

$$K_{sp} = 8.3 \times 10^{-17} = [Ag^+][I^-] = (x)(0.20 + x)$$

We know that the value of K_{sp} is very small, and it suggests the simplifying assumption that $(0.20 + x) \approx 0.20$:

Hence, $8.3 \times 10^{-17} \approx (0.20)x$, and $x = 4.2 \times 10^{-16}$. The assumption that $(0.20 + x) \approx 0.20$ is seen to be valid indeed.

Thus 4.2×10^{-16} mol of AgI will dissolve in 1.0 L of 0.20 M NaI solution.

21.

	$Fe(OH)_3(s)$	$\rightleftharpoons$	Fe^{3+}	+	$3OH^-$
initial conc. (M)			0.0		5.0×10^{-2}
change in conc. (M)			+x		+3x
equilibrium conc. (M)			x		$5.0 \times 10^{-2} + 3x$

Substituting the above values for equilibrium concentrations into the expression for K_{sp} gives:

$K_{sp} = 1.6 \times 10^{-39} = [Fe^{3+}][OH^-]^3 = (x)[5.0 \times 10^{-2} + 3x]^3$

We try to simplify by making the approximation that $(5.0 \times 10^{-2} + 3x) \approx 5.0 \times 10^{-2}$:

$1.6 \times 10^{-39} = (x)(5.0 \times 10^{-2})^3$ or $x = 1.3 \times 10^{-35}$

Clearly the assumption that $(5.0 \times 10^{-2} + 3x) \approx 5.0 \times 10^{-2}$ is justified.

Thus 1.3×10^{-35} mol of $Fe(OH)_3$ will dissolve in 1.0 L of 0.050 M sodium hydroxide solution.

22. The expression for K_{sp} is $K_{sp} = [Ca^{2+}][SO_4^{2-}] = 2.4 \times 10^{-5}$

and the ion product for this solution would be:

$[Ca^{2+}][SO_4^{2-}] = (2.5 \times 10^{-3})(3.0 \times 10^{-2}) = 7.5 \times 10^{-5}$

Since the ion product is larger than the value of K_{sp}, a precipitate is expected to form.

23. The solubility product constant is $K_{sp} = [Ag^+]^2[CrO_4^{2-}] = 1.2 \times 10^{-12}$

and the ion product for this solution would be:

$[Ag^+]^2[CrO_4^{2-}] = (4.8 \times 10^{-5})^2(3.4 \times 10^{-4}) = 7.8 \times 10^{-13}$

Since the ion product is smaller than the value of K_{sp}, we do not expect a precipitate to form.

24. Because two solutions are to be mixed together, there will be a dilution of the concentrations of the various ions, and the diluted ion concentrations

must be used. In general, on dilution, the following relationship is found for the concentrations of the initial solution (M_i) and the concentration of the final solution (M_f):

$$M_i \times V_i = M_f \times V_f$$

Thus the final or diluted concentrations are:

$$[Pb^{2+}]_f = 1.0 \times 10^{-3}\ M \times \frac{0.100\ mL}{0.200\ L} = 5.0 \times 10^{-4}\ M$$

and

$$[SO_4^{\ 2-}]_f = 2.0 \times 10^{-3}\ M \times \frac{0.100\ L}{0.200\ L} = 1.0 \times 10^{-3}\ M$$

The value of the ion product for the final or diluted solution is:

$$[Pb^{2+}][SO_4^{\ 2-}] = (5.0 \times 10^{-4})(1.0 \times 10^{-3}) = 5.0 \times 10^{-7}$$

Since this is smaller than the value of K_{sp} (6.3×10^{-7}), a precipitate of $PbSO_4$ is not expected.

25. We proceed as in Practice Exercise 24.

$$M_i \times V_i = M_f \times V_f$$

$$[Pb^{2+}]_f = 0.10\ M \times \frac{50.0\ mL}{70.0\ mL} = 7.1 \times 10^{-2}\ M$$

and

$$[Cl^-]_f = 0.040\ M \times \frac{20.0\ mL}{70.0\ mL} = 1.1 \times 10^{-2}\ M$$

The value of the ion product for such a solution would be:

$$[Pb^{2+}][Cl^-]^2 = (7.1 \times 10^{-2})(1.1 \times 10^{-2})^2 = 8.6 \times 10^{-6}$$

Since the ion product is smaller than K_{sp}, we expect that no precipitate of $PbCl_2$ can form.

26. As developed in Example 14.20 on page 586 of the text, the product of the values for K_{sp} and $K_{formation}$ is the correct equilibrium constant for the net reaction that we are interested in:

$$AgCl(s) + 2NH_3(aq) \rightleftharpoons Ag(NH_3)_2^{\ +}(aq) + Cl^-(aq)$$

for which $K_c = K_{sp} \times K_{formation} = (1.6 \times 10^7)(1.8 \times 10^{-8}) = 2.9 \times 10^{-3}$

and for which K_c is also given by the usual mass action expression:

$$K_c = \frac{[Ag(NH_3)_2^+][Cl^-]}{[NH_3]^2} = 2.9 \times 10^{-3}$$

Next we set up the concentration table as usual:

	AgCl(s) +	$2NH_3(aq)$ ⇌	$Ag(NH_3)_2^+(aq)$ +	$Cl^-(aq)$
initial conc. (M)		0.10	0.0	0.0
change in conc. (M)		-2x	+x	+x
equilibrium conc. (M)		(0.10 - 2x)	x	x

Substituting the above values for equilibrium concentrations into the mass action expression gives:

$$K_c = 2.9 \times 10^{-3} = \frac{(x)(x)}{(0.10 - 2x)^2}$$

Finally we can solve for x by taking the square root of both sides of the above equation:

$5.4 \times 10^{-2} = x/(0.10 - 2x)$ which simplifies to $1.1(x) = 5.4 \times 10^{-3}$

and $x = 4.9 \times 10^{-3}$. In other words, 4.9×10^{-3} mol of AgCl(s) dissolves in 1.0 L of 0.10 M NH_3 solution.

Now we can proceed to determine the amount of AgCl(s) that would dissolve on the other hand in water:

	AgCl(s) ⇌	$Ag^+(aq)$	+ $Cl^-(aq)$
initial conc. (M)		0.0	0.0
change in conc. (M)		+x	+x
equilibrium conc. (M)		x	x

Substituting the above values for equilibrium concentrations into the mass action expression gives:

$K_{sp} = [Ag^+][Cl^-] = 1.8 \times 10^{-10} = (x)(x)$ from which we have $x = 1.3 \times 10^{-5}$ M

This means that, in water only, the solubility of AgCl(s) is 1.3×10^{-5} mol/L, or about 380 times less soluble than is true in 0.10 M NH_3 solution.

27. This is the same equilibrium that was considered in Practice Exercise 26.

$$AgCl(s) + 2NH_3(aq) \rightleftharpoons Ag(NH_3)_2^+(aq) + Cl^-(aq)$$

for which the mass action expression is:

$$K_c = \frac{[Ag(NH_3)_2^+][Cl^-]}{[NH_3]^2} = 2.9 \times 10^{-3}$$

The stoichiometry of the above equilibrium indicates that in order to form 0.20 mol of $Ag(NH_3)_2^+$ from 0.20 mol of AgCl(s) requires the use of:

$$0.20 \text{ mol } Ag(NH_3)_2^+ \times \frac{2 \text{ mol } NH_3}{1 \text{ mol } Ag(NH_3)_2^+} = 0.40 \text{ mol } NH_3$$

and it will simultaneously produce 0.20 mol of Cl^-(aq). There will also need to be some certain additional amount of NH_3(aq) that is present at equilibrium in order to satisfy the mass action expression for this system, i.e. NH_3(aq) cannot have a zero value. This latter amount, plus the 0.40 mol that are required for formation of the $Ag(NH_3)_2^+$ ion, is the total amount of ammonia that is required for this dissolving process. The amount of ammonia that will satisfy the mass action expression is:

$$2.9 \times 10^{-3} = \frac{[0.20 \text{ M}][0.20 \text{ M}]}{[NH_3]^2}$$

Taking the square root of both sides of the above equation allows us to solve for the concentration of ammonia:

$[NH_3]$ = 3.7 mol/L

To dissolve 0.20 mol of AgCl(s) in 1.0 L will thus require the use of (0.40 + 3.7) = 4.1 mol NH_3.

Chapter 14

REVIEW EXERCISES

14.1 At equilibrium, both the forward and the reverse reactions constantly proceed at identical rates.

14.2 See Figure 14.1.

14.3 By reversibility we mean first that the reaction can proceed in either of the two possible directions, and, secondly, that for a given overall composition, the same equilibrium mixture can be attained from either the forward or the reverse directions.

Mass Action Expressions, K_p and K_c

14.4 The coefficients become the exponents in the mass action expression. This is shown in general on page 555.

14.5 The mass action expression becomes numerically equal to the value of K_c once the system has reached the state of equilibrium.

14.6 The units are typically mol/L, i.e. molar concentrations.

14.7 The reaction quotient is the numerical value of the mass action expression.

14.8 The reaction quotient becomes equal to the value of K_c once equilibrium has been attained.

14.9 An equilibrium law is the statement that the reaction quotient (i.e. the value of K_c) must be equal to the numerical value of the mass action expression once equilibrium is attained.

14.10 (a)

$$K_c = \frac{[POCl_3]^2}{[PCl_3]^2[O_2]}$$

(b)

$$K_c = \frac{[SO_2]^2[O_2]}{[SO_3]^2}$$

(c)

$$K_c = \frac{[NO]^2[H_2O]^2}{[N_2H_4][O_2]^2}$$

(d)

$$K_c = \frac{[NO_2]^2[H_2O]^8}{[N_2H_4][H_2O_2]^6}$$

14.11 (a)

$$K_c = \frac{[HCl]^2}{[H_2][Cl_2]}$$

(b)

$$K_c = \frac{[HCl]}{[H_2]^{\frac{1}{2}}\,[Cl_2]^{\frac{1}{2}}}$$

K_c for (a) is the square of K_c for (b).

14.12

$$K_c = \frac{[H_2][Cl_2]}{[HCl]^2}$$

This is equal to $1/K_c$ for Review Exercise 14.11 (a).

14.13 (a)

$$K_p = \frac{P^2_{POCl_3}}{P^2_{PCl_3} \times P_{O_2}}$$

(c)

$$K_p = \frac{P^2_{NO} \times P^2_{H_2O}}{P_{N_2H_4} \times P^2_{O_2}}$$

(b)

$$K_p = \frac{P^2_{SO_2} \times P_{O_2}}{P^2_{SO_3}}$$

(d)

$$K_p = \frac{P^2_{NO_2} \times P^8_{H_2O}}{P_{N_2H_4} \times P^6_{H_2O_2}}$$

14.14 By convention, the products are always written into the numerator and the reactants are written into the denominator of the mass action expression.

14.15 This equilibrium constant is small, and we do not expect the equilibrium to favor products.

14.16 (a) < (c) < (b), based on the relative magnitudes of K_c.

Converting Between K_p and K_c

14.17

$$K_p = K_c(RT)^{\Delta n_g}$$

where Δn_g is equal to the change in the number of moles of gaseous material on going from reactants to products.

14.18 First we examine the units that characterize the K_c and K_p mass action expressions:

$$K_c = \frac{[N_2O_4(g)]}{[NO_2(g)]^2} = \frac{M}{M^2} = M^{-1} = \frac{L}{mol}$$

$$K_p = \frac{P_{N_2O_4(g)}}{P^2_{NO_2(g)}} = \frac{atm}{atm^2} = atm^{-1} = \frac{1}{atm}$$

We next substitute these units into the following relationship between K_p and K_c:

$$K_p = K_c(RT)^{\Delta n_g}$$

and we have

$$\frac{1}{atm} = \frac{L}{mol} \times (RT)^{\Delta n_g}$$ and since $\Delta n_g = (1 - 2) = -1$, we arrive at:

$$\frac{1}{atm} = \frac{L}{mol} \times \frac{1}{RT}$$

This last equality is valid only if the units of the right side of the equation can be made to be equal to the units of the left side of the equation. This will happen only if R has the specified units, i.e.

L atm K^{-1} mol^{-1}, and we conclude that R must be 0.0821 L atm K^{-1} mol^{-1}, the familiar gas law constant.

14.19

$$K_p = K_c(RT)^{\Delta n_g}$$

If the value of $(RT)^{\Delta n_g}$ is unity, then the left and right sides of this equation become equal to one another. This can happen only if there is no change in the total number of gaseous substances on going from reactants to products. This happens in reactions (a) and (d) only.

14.20

$$K_p = K_c(RT)^{\Delta n_g}$$

$$6.3 \times 10^{-3} = K_c[(0.0821 \text{ L atm K}^{-1} \text{ mol}^{-1})(498 \text{ K})]^{-2} = 5.98 \times 10^{-4} \times K_c$$

$K_c = 11$

14.21

$K_p = K_c(RT)^{\Delta n_g}$

$K_p = 2.2 \times 10^{59} \times [(0.0821 \text{ L atm K}^{-1} \text{ mol}^{-1})(573 \text{ K})]^{-1} = 4.7 \times 10^{57}$

14.22

$K_p = K_c(RT)^{\Delta n_g}$

$K_p = (0.40)[(0.0821 \text{ L atm K}^{-1} \text{ mol}^{-1})(1046 \text{ K})]^{-2} = 5.4 \times 10^{-5}$

14.23

$K_p = K_c(RT)^{\Delta n_g}$

$4.6 \times 10^{-2} = K_c[(0.0821 \text{ L atm K}^{-1} \text{ mol}^{-1})(668 \text{ K})]^{1} = 54.8 \times K_c$

$K_c = 8.4 \times 10^{-4}$

Le Chatelier's Principle

14.24 When a system at equilibrium is disturbed so that the equilibrium is upset, the system changes in a way that opposes the disturbance and returns the system to a new state of equilibrium.

14.25 (a) The system shifts to the right to consume some of the added methane.
(b) The system shifts to the left to consume some of the added hydrogen.
(c) The system shifts to the right to make some more carbon disulfide.
(d) The system shifts to the left to decrease the amount of gas.
(e) The system shifts to the right to absorb some of the added heat.

14.26 (a) increase (b) decrease (c) increase (d) no change (e) decrease

14.27 The value of K_c can be changed only by a change in temperature.

14.28 (a) increase (b) increase (c) increase (d) decrease

14.29 The ones not affected are those with no net change in the number of moles of gaseous materials, i.e. (a) and (d).

14.30 $\Delta H° = 0$

Equilibrium Calculations

14.31

$$K_c = \frac{[CH_3OH]}{[CO][H_2]^2} = \frac{(0.00261)}{(0.105)(0.250)^2} = 0.398$$

14.32 The mass action expression for this equilibrium is:

$$K_c = \frac{[PCl_5]}{[PCl_3][Cl_2]} = 0.18$$

and the value of the ion product for this system is:

$$\frac{(0.00600)}{(0.0520)(0.0140)} = 8.24$$

(a) This is not the value of the equilibrium constant, and we conclude that the system is not at equilibrium.

(b) Since the value of the ion product for this system is larger than that of the equilibrium constant, the system must shift to the left to reach equilibrium.

14.33 The mass action expression for the system is:

$$K_c = \frac{[NO][SO_3]}{[SO_2][NO_2]} = 85.0$$

and the ion product for the system is:

$$\frac{(0.0100)(0.0400)}{(0.00150)(0.00300)} = 88.9$$

(a) The system is not at equilibrium.

(b) Since the value of the ion product is larger than the value of the equilibrium constant, the system must shift to the left in order to reach equilibrium.

14.34 (a) The mass action expression is:

$$K_p = \frac{P_{NO_2}^2}{P_{N_2O_4}} = 0.140$$

Solving the above expression for the partial pressure of NO_2, we get:

$$P_{NO_2} = \sqrt{K_P \times P_{N_2O_4}} = \sqrt{(0.140)(0.300)} = 0.205$$

(b) $P_{total} = P_{NO_2} + P_{N_2O_4} = 0.205 + 0.300 = 0.505$ atm

14.35

$$K_c = \frac{[C_2H_5OH]}{[C_2H_4][H_2O]} = \frac{(0.150)}{(0.0222)(0.0225)} = 3.00 \times 10^2$$

14.36 The mass action expression is:

$$K_c = \frac{[CH_3OH]}{[CO][H_2]^2} = 0.500$$

Solving the above expression for $[CH_3OH]$ gives:

$$[CH_3OH] = K_c \times [CO] \times [H_2]^2 = (0.500)(0.210)(0.100)^2 = 1.05 \times 10^{-3}\ M$$

14.37 The mass action expression is:

$$K_c = \frac{[NH_3]^2}{[N_2][H_2]^3} = 64$$

Solving for $[H_2]$ gives:

$$[H_2] = \sqrt[3]{\frac{(0.280)^2}{(0.00840)(64)}} = 0.53\ mol/L$$

14.38

	$2HBr$	$\rightleftharpoons$	H_2	+	Br_2
initial conc. (M)	0.500		0.0		0.0
change in conc. (M)	-2(0.130)		+1(0.130)		+1(0.130)
equilibrium conc. (M)	0.240		0.130		0.130

Substituting the above values for equilibrium concentrations into the mass action expression gives:

$$K_c = \frac{[Br_2][H_2]}{[HBr]^2} = \frac{(0.130)(0.130)}{(0.240)^2} = 0.293$$

14.39

	CH_2O	$\rightleftharpoons$	H_2	+	CO
initial conc. (M)	0.100		0.0		0.0
change in conc. (M)	-(0.020)		+1(0.020)		+1(0.020)
equilibrium conc. (M)	0.080		0.020		0.020

Substituting the above values for equilibrium concentrations into the mass action expression gives:

$$K_c = \frac{[CO][H_2]}{[CH_2O]} = \frac{(0.020)(0.020)}{(0.080)} = 5.0 \times 10^{-3}$$

14.40

The initial concentrations are each 0.300 mol/2.00 L = 0.150 M.

	SO_3	+	NO	$\rightleftharpoons$	NO_2	+	SO_2
initial conc. (M)	0.150		0.150		0.0		0.0
change in conc. (M)	-x		-x		+x		+x
equilibrium conc. (M)	0.150 - x		0.150 - x		x		x

Substituting the above values for equilibrium concentrations into the mass action expression gives:

$$K_c = \frac{[NO_2][SO_2]}{[NO][SO_3]} = \frac{(x)(x)}{(0.150 - x)(0.150 - x)} = 0.500$$

$$0.500 = \frac{x^2}{(0.150 - x)^2}$$

Taking the square root of both sides of this equation gives:

$0.707 = x/(0.150 - x)$

Solving for x we have:

$1.707(x) = 0.106 \quad \therefore \quad x = 0.0621 \text{ mol/L} = [NO_2] = [SO_2]$

$[NO] = [SO_3] = 0.150 - x = 0.088 \text{ mol/L}$

14.41 The initial concentrations are all 1.00 mol/100 L = 0.0100 M. Since the initial concentrations are all the same, the ion product is equal to 1.0, and we conclude that the system must shift to the left to reach equilibrium.

	CO	+	H_2O	$\rightleftharpoons$	CO_2	+	H_2
initial conc. (M)	0.0100		0.0100		0.0100		0.0100
change in conc. (M)	+x		+x		-x		-x
equilibrium conc. (M)	(0.0100 + x)		(0.0100 + x)		(0.0100 - x)		(0.0100 - x)

Substituting the above values for equilibrium concentrations into the mass action expression gives:

$$K_c = \frac{[H_2][CO_2]}{[CO][H_2O]} = \frac{(0.0100 - x)(0.0100 - x)}{(0.0100 + x)(0.0100 + x)} = 0.400$$

We take the square root of both sides of the above equation:

$$0.632 = \frac{(0.0100 - x)}{(0.0100 + x)}$$

and $(0.632)(0.0100 + x) = 0.0100 - x$

$(1.632)x = 3.68 \times 10^{-3}$, or $x = 2.25 \times 10^{-3}$ mol/L

The equilibrium concentrations are then:

$[H_2] = [CO_2] = (0.0100 - 2.25 \times 10^{-3}) = 7.8 \times 10^{-3}$ M

$[CO] = [H_2O] = (0.0100 + 2.25 \times 10^{-3}) = 0.0123$ M

14.42

	$2HCl$	$\rightleftharpoons$	H_2	+	Cl_2
initial conc. (M)	2.00		0.0		0.0
change in conc. (M)	2.00 -2x		+x		+x
equilibrium conc. (M)	2.00 -2x		x		x

Substituting the above values for equilibrium concentrations into the mass action expression gives:

$$K_c = \frac{[H_2][Cl_2]}{[HCl]^2} = 3.2 \times 10^{-34} = \frac{(x)(x)}{(2.00 - 2x)^2}$$

Because K_c is so exceedingly small, we can make the simplifying assumption that x is also small enough to make $(2.00 - 2x) \approx 2.00$. Thus we have:

$3.2 \times 10^{-34} = (x)^2/(2.00)^2$ Taking the square root of both sides, and solving for the value of x gives:

$x = 3.6 \times 10^{-17}$ M = $[H_2]$ = $[Cl_2]$

$[HCl] = (2.00 - x) \approx 2.00$ mol/L

14.43 Because of the very large value of K_c, we start by realizing that nearly all of the 0.100 moles of H_2 will react with nearly all of the 0.100 moles of Br_2 to form 0.200 moles of HBr. This brings us to an initial condition that is more realistically close to the true equilibrium condition, i.e. (0.200 mol/10.0 L) = 0.0200 M Br_2. Next we proceed in the normal fashion,

allowing 2x mol/L of HBr to disappear making x mol/L each of H_2 and Br_2 in order to reach equilibrium:

	H_2	+	Br_2	$\rightleftharpoons$	2HBr
initial conc. (M)	0.0		0.0		0.0200
change in conc. (M)	+x		+x		0.0200 - 2x
equilibrium conc. (M)	x		x		0.0200 - 2x

Substituting the above values for equilibrium concentrations into the mass action expression gives:

$$K_c = \frac{[HBr]^2}{[H_2][Br_2]} = 2.0 \times 10^9 = \frac{(0.0200 - 2x)^2}{(x)(x)}$$

We next make the assumption that $(0.0200 - 2x) \approx 0.0200$, giving

$2.0 \times 10^9 = (0.0200)^2/x^2$

Taking the square root of both sides of the above equation gives:

$4.47 \times 10^4 = 0.0200/x$ and $x = 4.47 \times 10^{-7}$ M = $[H_2]$ = $[Br_2]$

The small size of x demonstrates that the assumption made above was justified.

$[HBr] = 0.0200 - 2x \approx 0.0200$ M

14.44

	SO_3	+	NO	⇌	NO_2	+	SO_2
initial conc. (M)	0.0500		0.100		0.0		0.0
change in conc. (M)	-x		-x		+x		+x
equilibrium conc. (M)	0.0500 - x		0.100 - x		x		x

Substituting the above values for equilibrium concentrations into the mass action expression gives:

$$K_c = \frac{[NO_2][SO_2]}{[SO_3][NO]} = 0.500 = \frac{(x)(x)}{(0.0500 - x)(0.100 - x)}$$

Since the equilibrium constant is not much larger than either of the values 0.0500 or 0.100, we cannot neglect the sixe of x in the above expression. A simplifying assumption is not therefore possible, and we must solve for the value of x using the quadratic equation. Multiplying out the above denominator, collecting like terms, and putting the result into the standard quadratic form gives:

$$0.500x^2 + (7.5 \times 10^{-2})x - (2.5 \times 10^{-3}) = 0$$

Dividing each side by 2.5×10^{-3} gives:

$$200x^2 + 30x - 1 = 0$$

Applying the quadratic formula gives:

$$x = \frac{-30 +/- \sqrt{900 - (4)((200)(-1)}}{400} = 0.028 \text{ mol/L, using the (+) root.}$$

$$[NO_2] = [SO_2] = 0.028 \text{ M}$$

Chapter 14

Heterogeneous Equilibria

14.45 In a homogeneous equilibrium, all of the reactants and products are in the same phase. In heterogeneous equilibria, at least two different phases are found among the reactants and products. The mass action expression for a homogeneous equilibrium includes a term for every reactant and product, whereas the mass action expression for a heterogeneous system does not include a concentration term for every reactant and product.

14.46 (a) $\frac{[CO]^2}{[O_2]} = K_c$ (b) $[SO_2][H_2O] = K_c$

(c) $\frac{[CH_4][CO_2]}{[H_2O]^2} = K_c$ (d) $\frac{[H_2O][CO_2]}{[HF]^2} = K_c$

14.47 This is possible because their concentrations are constants that are incorporated into the numerical values of equilibrium constants.

14.48

	$2HCl(g)$	+	$I_2(s)$	⇌	$2HI(g)$	+	$Cl_2(g)$
initial conc. (M)	1.00		solid		0.0		0.0
change in conc. (M)	-2x				+2x		+x
equilibrium conc. (M)	1.00 - 2x		solid		2x		x

Substituting the above values for equilibrium concentrations into the mass action expression gives:

$$K_c = \frac{[HI]^2[Cl_2]}{[HCl]^2} = 1.6 \times 10^{-34} = \frac{(2x)^2(x)}{(1.00 - 2x)^2}$$

Because the value of K_c is so small, we make the simplifying assumption that $(1.00 - 2x) \approx 1.00$, and the above equation becomes:

$$1.6 \times 10^{-34} = \frac{(2x)^2(x)}{(1.00)^2}$$

$4x^3 = 1.6 \times 10^{-34}$; $\therefore$ $x = 3.4 \times 10^{-12}$, and the above assumption is seen to have been valid.

$[HI] = 2x = 6.8 \times 10^{-12}$ M

$[Cl_2] = x = 3.4 \times 10^{-12}$ M

$[HCl] = (1.00 - 2x) \approx 1.00$ M

14.49 In each case we get 55.6 M:

(a)

$$18.0 \text{ mL} \times \frac{1 \text{ g}}{1 \text{ mL}} \times \frac{1 \text{ mol}}{18.0 \text{ g}} = 1.00 \text{ mol } H_2O$$

$$\frac{1.00 \text{ mol}}{0.0180 \text{ L}} = 55.6 \text{ M}$$

(b)

$$100.0 \text{ mL} \times \frac{1 \text{ g}}{1 \text{ mL}} \times \frac{1 \text{ mol}}{18.0 \text{ g}} = 5.56 \text{ mol } H_2O$$

$$\frac{5.56 \text{ mol}}{0.100 \text{ L}} = 55.6 \text{ M}$$

(c)

$$1.00 \text{ L} \times \frac{1000 \text{ mL}}{1 \text{ L}} \times \frac{1 \text{ g}}{1 \text{ mL}} \times \frac{1 \text{ mol}}{18.0 \text{ g}} = 55.6 \text{ mol } H_2O$$

$$\frac{55.6 \text{ mol}}{1 \text{ L}} = 55.6 \text{ M}$$

14.50 The density is 2.165 g/mL.

$$\frac{2.165 \text{ g}}{\text{mL}} \times \frac{1 \text{ mol}}{58.44 \text{ g}} \times \frac{1000 \text{ mL}}{1 \text{ L}} = 37.05 \text{ mol/L}$$

Solubility Products

14.51 The ion product is the expression (obtained from the mass action expression) of the product of the ion concentrations, each raised to the power of the ion's stoichiometry for the reaction. The ion product constant is the numerical value of the ion product (mass action expression) for a saturated solution. Any solution can have an ion product. The question is, is the ion product equal numerically to the value that typifies a saturated solution, i.e. equal to the ion product constant?

14.52 (a)

$$K_c = \frac{[Bi^{3+}(aq)]^2[S^{2-}(aq)]^3}{[Bi_2S_3(s)]}$$

(b) Since both the density and the molarity of a pure solid or a pure liquid are constant, we can multiply both sides of the above expression by the value of this constant:

$K_c \times [Bi_2S_3(s)] = [Bi^{3+}(aq)]^2[S^{2-}(aq)]^3$

The left side of the above expression, being the product of two constants, is itself a constant - which we define to be K_{sp} for $Bi_2S_3(s)$.

(c) $K_{sp} = [Bi^{3+}(aq)]^2[S^{2-}(aq)]^3$

14.53 (a) for CaF_2, $K_{sp} = [Ca^{2+}][F^-]^2$

(b) for Ag_2CO_3, $K_{sp} = [Ag^+]^2[CO_3{}^{2-}]$

(c) for $PbSO_4$, $K_{sp} = [Pb^{2+}][SO_4{}^{2-}]$

(d) for $Fe(OH)_3$, $K_{sp} = [Fe^{3+}][OH^-]^3$

(e) for PbI_2, $K_{sp} = [Pb^{2+}][I^-]^2$

(f) for $Cu(OH)_2$, $K_{sp} = [Cu^{2+}][OH^-]^2$

14.54 (a) for AgI, $K_{sp} = [Ag^+][I^-]$

(b) for Ag_3PO_4, $K_{sp} = [Ag^+]^3[PO_4{}^{3-}]$

(c) for $PbCrO_4$, $K_{sp} = [Pb^{2+}][CrO_4{}^{2-}]$

(d) for $Al(OH)_3$, $K_{sp} = [Al^{3+}][OH^-]^3$

(e) for $ZnCO_3$, $K_{sp} = [Zn^{2+}][CO_3{}^{2-}]$

(f) for $Zn(OH)_2$, $K_{sp} = [Zn^{2+}][OH^-]^2$

14.55 The common ion effect operates to make a salt less soluble in a solution that contains an additional source of one of its ions than it is in pure water. According to Le Chatelier's principle, the common ion shifts the position of the equilibrium in the direction of the solid, so that the solubility of the solid is reduced.

14.56 Precipitation can occur only if an ion product is larger than the value of K_{sp} for a solid.

14.57 (a) $0.00245\ g \div 233\ g/mol = 1.05 \times 10^{-5}$ mol $BaSo_4$

molarity = 1.05×10^{-5} mol/L

(b) These are the concentrations generated in the solution from part (a), upon complete ionization of $BaSO_4$:

$[Ba^{2+}] = 1.05 \times 10^{-5}$ M

$[SO_4^{2-}] = 1.05 \times 10^{-5}$ M

(c) $K_{sp} = [Ba^{2+}][SO_4^{2-}] = (1.05 \times 10^{-5})(1.05 \times 10^{-5})$

$K_{sp} = 1.10 \times 10^{-10}$

14.58 (a) 7.05×10^{-3} g ÷ 58.3 g/mol = 1.21×10^{-4} mol $Mg(OH)_2$

molarity = 1.21×10^{-4} M

(b) $Mg(OH)_2 = Mg^{2+} + 2OH^-$

$[Mg^{2+}] = 1.21 \times 10^{-4}$ M

$[OH^-] = 2 \times [Mg^{2+}] = 2.42 \times 10^{-4}$ M

(c) $K_{sp} = [Mg^{2+}][OH^-]^2 = (1.21 \times 10^{-4})(2.42 \times 10^{-4})^2$

$K_{sp} = 7.09 \times 10^{-12}$

14.59

	Ag_3PO_4 ⇌	$3Ag^+(aq)$	+	$PO_4^{3-}(aq)$
initial conc. (M)		0.0		0.0
change in conc. (M)		$3(1.8 \times 10^{-5})$		1.8×10^{-5}
equilibrium conc. (M)		5.4×10^{-5}		1.8×10^{-5}

Substituting the above values for equilibrium concentrations into the expression for K_{sp} gives:

$K_{sp} = [Ag^+]^3[PO_4^{3-}] = (5.4 \times 10^{-5})^3(1.8 \times 10^{-5}) = 2.8 \times 10^{-18}$

14.60

	$Ba_3(PO_4)_2(s)$ ⇌	$3Ba^{2+}(aq)$	+	$2PO_4^{3-}(aq)$
initial conc. (M)		0.0		0.0
change in conc. (M)		$3(1.4 \times 10^{-8})$		$2(1.4 \times 10^{-8})$
equilibrium conc. (M)		4.2×10^{-8}		2.8×10^{-8}

Substituting the above values for equilibrium concentrations into the expression for K_{sp} gives:

$K_{sp} = [Ba^{2+}]^3[PO_4^{3-}]^2 = (4.2 \times 10^{-8})^3(2.8 \times 10^{-8})^2 = 5.8 \times 10^{-38}$

14.61

	$BaSO_3$ ⇌	$Ba^{2+}(aq)$	+	$SO_3^{2-}(aq)$
initial conc. (M)		0.10		0.0
change in conc. (M)		8.0×10^{-6}		8.0×10^{-6}
equilibrium conc. (M)		$0.10 + 8.0 \times 10^{-6}$		8.0×10^{-6}

Substituting the above values for equilibrium concentrations into the expression for K_{sp} gives:

$K_{sp} = [Ba^{2+}][SO_3^{2-}] = (0.10 + 8.0 \times 10^{-6})(8.0 \times 10^{-6}) = 8.0 \times 10^{-7}$

14.62 The number of moles of material that was dissolved in this amount of solution is:

$0.416\ g \div 156\ g/mol = 2.67 \times 10^{-3}\ mol$

and the molarity of the solution must have been:

$2.67 \times 10^{-3}\ mol/0.100\ L = 0.0267\ M$, which represents the concentration of a saturated solution, i.e. the molar solubility of the material.

	$CaCrO_4(s)$	$\rightleftharpoons$	$Ca^{2+}(aq)$	+	$CrO_4^{2-}(aq)$
initial conc. (M)			0.0		0.0
change in conc. (M)			0.0267		0.0267
equilibrium conc. (M)			0.0267		0.0267

Substituting the above values for equilibrium concentrations into the expression for K_{sp} gives:

$K_{sp} = [Ca^{2+}][CrO_4^{2-}] = (0.0267)(0.0267) = 7.13 \times 10^{-4}$

14.63 (a)

	$CuCl(s)$	$\rightleftharpoons$	$Cu^{+}(aq)$	+	$Cl^{-}(aq)$
initial conc. (M)			0.0		0.0
change in conc. (M)			x		x
equilibrium conc. (M)			x		x

Substituting the above values for equilibrium concentrations into the mass action expression gives:

$K_{sp} = [Cu^{+}][Cl^{-}] = (x)(x) = 1.9 \times 10^{-7}$

$\therefore$ x = molar solubility = 4.4×10^{-4}

(b)

	$CuCl(s)$	$\rightleftharpoons$	$Cu^{+}(aq)$	+	$Cl^{-}(aq)$
initial conc. (M)			0.0		0.010
change in conc. (M)			x		x
equilibrium conc. (M)			x		0.010 + x

Substituting the above values for equilibrium concentrations into the mass action expression gives:

$K_{sp} = [Cu^+][Cl^-] = (x)(0.010 + x) = 1.9 \times 10^{-7}$

$\therefore$ x = molar solubility = 1.9×10^{-5}

(c)

	$CuCl(s)$ ⇌	$Cu^+(aq)$	+	$Cl^-(aq)$
initial conc. (M)		0.0		0.100
change in conc. (M)		x		x
equilibrium conc. (M)		x		0.100 + x

Substituting the above values for equilibrium concentrations into the mass action expression gives:

$K_{sp} = [Cu^+][Cl^-] = (x)(0.100 + x) = 1.9 \times 10^{-7}$

$\therefore$ x = molar solubility = 1.9×10^{-6}

(d)

	$CuCl(s)$ ⇌	$Cu^+(aq)$	+	$Cl^-(aq)$
initial conc. (M)		0.0		0.200
change in conc. (M)		x		x
equilibrium conc. (M)		x		0.200 + x

Substituting the above values for equilibrium concentrations into the mass action expression gives:

$K_{sp} = [Cu^+][Cl^-] = (x)(0.200 + x) = 1.9 \times 10^{-7}$

$\therefore$ x = molar solubility = 9.5×10^{-7}

14.64 (a)

	$AuCl_3(s)$ ⇌	$Au^{3+}(aq)$	+	$3Cl^-(aq)$
initial conc. (M)		0.0		0.0
change in conc. (M)		x		3x
equilibrium conc. (M)		x		3x

Substituting the above values for equilibrium concentrations into the mass action expression gives:

$K_{sp} = [Au^{3+}][Cl^-]^3 = 3.2 \times 10^{-25} = (x)(3x)^3 = 27(x)^4$

$\therefore$ x = molar solubility = 3.3×10^{-7}

(b)

	$AuCl_3(s)$	$\rightleftharpoons$	$Au^{3+}(aq)$	+	$3Cl^-(aq)$
initial conc. (M)			0.0		0.020
change in conc. (M)			x		3x
equilibrium conc. (M)			x		0.020 + 3x

Substituting the above values for equilibrium concentrations into the mass action expression gives:

$$K_{sp} = [Au^{3+}][Cl^-]^3 = 3.2 \times 10^{-25} = (x)(0.020 + 3x)^3 \approx (x)(0.020)^3$$

$\therefore$ x = molar solubility = 4.0×10^{-20}

(c)

	$AuCl_3(s)$	$\rightleftharpoons$	$Au^{3+}(aq)$	+	$3Cl^-(aq)$
initial conc. (M)			0.0		0.040
change in conc. (M)			x		3x
equilibrium conc. (M)			x		0.040 + 3x

Substituting the above values for equilibrium concentrations into the mass action expression gives:

$$K_{sp} = [Au^{3+}][Cl^-]^3 = 3.2 \times 10^{-25} = (x)(0.040 + 3x)^3 \approx (x)(0.040)^3$$

$\therefore$ x = molar solubility = 5.0×10^{-21}

(d)

	$AuCl_3(s)$	$\rightleftharpoons$	$Au^{3+}(aq)$	+	$3Cl^-(aq)$
initial conc. (M)			0.020		0.0
change in conc. (M)			x		3x
equilibrium conc. (M)			0.020 + x		3x

Substituting the above values for equilibrium concentrations into the mass action expression gives:

$$K_{sp} = [Au^{3+}][Cl^-]^3 = 3.2 \times 10^{-25} = (0.020 + x)(3x)^3 \approx (0.020)(3x)^3$$

$(3x)^3 = 1.6 \times 10^{-23}$ and taking the cube root of both sides of this equation gives:

$3x = 2.5 \times 10^{-8}$, or $x = 8.4 \times 10^{-9}$ = molar solubility

14.65 Since the mathematical form of the K_{sp} expressions for LiF and BaF_2 is not the same, it would be inappropriate to choose the more soluble substance based simply on a comparison of the magnitudes of the values for K_{sp}. Instead, we must proceed to determine a molar solubility for each substance in the usual manner.

For LiF:

	$LiF(s)$	$\rightleftharpoons$	$Li^+(aq)$	+	$F^-(aq)$
initial conc. (M)			0.0		0.0
change in conc. (M)			x		x
equilibrium conc. (M)			x		x

Substituting the above values for equilibrium concentrations into the expression for K_{sp} gives:

$K_{sp} = [Li^+][F^-] = (x)(x) = 1.7 \times 10^{-3}$ $\therefore$ $x = 4.1 \times 10^{-2}$ mol/L, which is the molar solubility of LiF.

For BaF_2:

	$BaF_2(s)$	$\rightleftharpoons$	$Ba^{2+}(aq)$	+	$2F^-(aq)$
initial conc. (M)			0.0		0.0
change in conc. (M)			x		2x
equilibrium conc. (M)			x		2x

Substituting the above values for equilibrium concentrations into the expression for K_{sp} gives:

$K_{sp} = [Ba^{2+}][F^-]^2 = (x)(2x)^2 = 1.7 \times 10^{-6}$

$4x^3 = 1.7 \times 10^{-6}$ $\therefore$ $x = \text{molar solubility} = 7.5 \times 10^{-3}$

We have found that LiF has the larger molar solubility.

14.66 In this calculation, we must consider both the different forms for the K_{sp} expressions and the different formula weights.

For AgCN:

	$AgCN(s)$	$\rightleftharpoons$	$Ag^+(aq)$	+	$CN^-(aq)$
initial conc. (M)			0.0		0.0
change in conc. (M)			x		x
equilibrium conc. (M)			x		x

Substituting the above values for equilibrium concentrations into the mass action expression gives:

$K_{sp} = [Ag^+][CN^-] = (x)(x) = 2.2 \times 10^{-16}$ $\therefore$ $x = 1.5 \times 10^{-8}$ mol/L

The solubility expressed as the number of grams per 100 mL is:

1.5×10^{-8} mol/1000 mL $\times$ 134 g/mol $\times$ 100 mL = 2.0×10^{-7} g AgCN

For $Zn(CN)_2$:

	$Zn(CN)_2(s)$ $\rightleftharpoons$	$Zn^{2+}(aq)$	+	$2CN^-(aq)$
initial conc. (M)		0.0		0.0
change in conc. (M)		x		2x
equilibrium conc. (M)		x		2x

Substituting the above values for equilibrium concentrations into the mass action expression gives:

$K_{sp} = [Zn^{2+}][CN^-]^2 = (x)(2x)^2 = 3 \times 10^{-16}$ or $x^3 = 7.5 \times 10^{-17}$

$x = 4.2 \times 10^{-6}$ mol/L

The solubility expressed as the number of grams per 100 mL is:

4.2×10^{-6} mol/1000 mL $\times$ 117 g/mol $\times$ 100 mL = 5×10^{-5} g $Zn(CN)_2$

This means that $Zn(CN)_2$ has the larger solubility when expressed as the number of grams per 100 mL of solution.

14.67 First determine the molar solubility of the MX salt:

	$MX(s)$ $\rightleftharpoons$	$M^+(aq)$	+	$X^-(aq)$
initial conc. (M)		0.0		0.0
change in conc. (M)		x		x
equilibrium conc. (M)		x		x

Substituting the above values for equilibrium concentrations into the mass action expression gives:

$K_{sp} = x^2 = 2.0 \times 10^{-10}$ $\therefore$ $x = 1.4 \times 10^{-5}$

We use this value for x in the following equilibrium for MX_3:

	$MX_3(s)$	$\rightleftharpoons$	$M^{3+}(aq)$	+	$3X^-(aq)$
initial conc. (M)			0.0		0.0
change in conc. (M)			x		3x
equilibrium conc. (M)			x		3x

Substituting the above values for equilibrium concentrations into the mass action expression gives:

$K_{sp} = [M^{3+}][X^-]^3 = (x)(3x)^3 = 27x^4$

We now use the value of x determined above for the MX substance, namely:

1.4×10^{-5} M

$K_{sp} = 27(1.4 \times 10^{-5})^4 = 1.0 \times 10^{-18}$

14.68 We must solve the typical mass action equation for the M_2X_3 salt, and then use the solubility that results, times 2, in the next equilibrium calculation.

For M_2X_3:

	$M_2X_3(s)$	$\rightleftharpoons$	$2M^{3+}(aq)$	+	$3X^{2-}(aq)$
initial conc. (M)			0.0		0.0
change in conc. (M)			2x		3x
equilibrium conc. (M)			2x		3x

Substituting the above values for equilibrium concentrations into the mass action expression gives:

$K_{sp} = [M^{3+}]^2[X^{2-}]^3 = (2x)^2(3x)^3 = 1.0 \times 10^{-20}$

$108(x)^5 = 1.0 \times 10^{-20}$ and $x = 3.9 \times 10^{-5}$ M, the molar solubility of M_2X_3.

This means that we are to have a molar solubility of $2(3.9 \times 10^{-5}\text{ M}) = 7.8 \times 10^{-5}$ M for the other substance, M_2X. That is, the value of x in the following equilibrium is to be 7.8×10^{-5} M.

For M_2X:

	$M_2X(s)$	⇌	$2M^+(aq)$	+	$X^{2-}(aq)$
initial conc. (M)			0		0
change in conc. (M)			2x		x
equilibrium conc. (M)			2x		x

Substituting the above values for equilibrium concentrations into the mass action expression gives:

$$K_{sp} = [M^+]^2[X^{2-}] = (2x)^2(x) = (4)(7.8 \times 10^{-5})^2(7.8 \times 10^{-5}) = 1.9 \times 10^{-12}$$

14.69

	$CaSO_4(s)$	⇌	$Ca^{2+}(aq)$	+	$SO_4^{2-}(aq)$
initial conc. (M)			0.0		0.0
change in conc. (M)			x		x
equilibrium conc. (M)			x		x

Substituting the above values for equilibrium concentrations into the expression for K_{sp} gives:

$$K_{sp} = [Ca^{2+}][SO_4^{2-}] = (x)(x) = 2.4 \times 10^{-5}, \quad \therefore x = 4.9 \times 10^{-3} \text{ M } CaSO_4$$

14.70

	$CaCO_3(s)$	⇌	$Ca^{2+}(aq)$	+	$CO_3^{2-}(aq)$
initial conc. (M)			0.0		0.0
change in conc. (M)			x		x
equilibrium conc. (M)			x		x

Substituting the above values for equilibrium concentrations into the expression for K_{sp} gives:

$$K_{sp} = [Ca^{2+}][CO_3^{2-}] = (x)(x) = 4.5 \times 10^{-9}, \quad \therefore x = 6.7 \times 10^{-5} \text{ M } CaCO_3$$

The number of grams that dissolve in 100 mL is:

$$6.7 \times 10^{-5} \text{ mol/1000 mL} \times 100 \text{ g/mol} \times 100 \text{ mL} = 6.7 \times 10^{-4} \text{ g } CaCO_3$$

14.71

	$PbI_2(s)$	⇌	$Pb^{2+}(aq)$	+	$2I^-(aq)$
initial conc. (M)			0.0		0.0
change in conc. (M)			x		2x
equilibrium conc. (M)			x		2x

Substituting the above values for equilibrium concentrations into the expression for K_{sp} gives:

$K_{sp} = [Pb^{2+}][I^-]^2 = (x)(2x)^2 = 4x^3 = 7.9 \times 10^{-9}, \quad \therefore x = 1.3 \times 10^{-3}$ M

14.72

	$Ag_2CO_3(s) \rightleftharpoons$	$2Ag^+(aq)$	+	$CO_3^{2-}(aq)$
initial conc. (M)		0.0		0.0
change in conc. (M)		2x		x
equilibrium conc. (M)		2x		x

Substituting the above values for equilibrium concentrations into the mass action expression gives:

$K_{sp} = [Ag^+]^2[CO_3^{2-}] = (2x)^2(x) = 4x^3 = 8.1 \times 10^{-12}, \therefore x = 1.3 \times 10^{-4}$ M

14.73

	$Ag_2CrO_4(s) \rightleftharpoons$	$2Ag^+(aq)$	+	$CrO_4^{2-}(aq)$
initial conc. (M)		0.10		0.0
change in conc. (M)		2x		x
equilibrium conc. (M)		0.10 + 2x		x

Substituting the above values for equilibrium concentrations into the mass action expression gives:

$K_{sp} = [Ag^+]^2[CrO_4^{2-}] = (0.10 + 2x)^2(x) = 1.2 \times 10^{-12}$

We make the assumption that $(0.10 + 2x) \approx 0.10$ and solve the above expression for the value of x:

$1.2 \times 10^{-12} = (0.10)^2x, \quad \therefore \quad x = 1.2 \times 10^{-10}$ and the assumption is seen to have been valid.

The molar solubility is thus 1.2×10^{-10} M.

14.74

	$Mg(OH)_2(s) \rightleftharpoons$	$Mg^{2+}(s)$	+	$2OH^-(aq)$
initial conc. (M)		0.0		0.10
change in conc. (M)		x		2x
equilibrium conc. (M)		x		0.10 + 2x

Substituting the above values for equilibrium concentrations into the mass action expression gives:

$K_{sp} = [Mg^{2+}][OH^-]^2 = (x)(0.10 + 2x)^2$

We simplify the above expression by assuming that $(0.10 + 2x) \approx 0.10$:

$(x)(0.10)^2 = 7.1 \times 10^{-12}, \quad \therefore \quad x = 7.1 \times 10^{-10}$ M

The molar solubility of $Mg(OH)_2$ in 0.10 M NaOH solution is 7.1×10^{-10} M.

14.75 The precipitate will form only if the value of the ion product $[Pb^{2+}][Cl^-]^2$ is greater than the value of the solubility product constant, K_{sp}. The apparent molar concentrations of the pertinent ions would be:

$[Pb^{2+}]$ = 0.0100 mol/1.00 L = 0.0100 M

$[Cl^-]$ = 0.0100 mol/1.00 L = 0.0100 M

and the value of the ion product is:

$(0.0100)(0.0100)^2 = 1.0 \times 10^{-6}$

Since the ion product is smaller than the value of K_{sp}, no precipitate of $PbCl_2$ will form.

14.76 The expression for K_{sp} is:

$K_{sp} = [Ag^+][C_2H_3O_2^-] = 4 \times 10^{-3}$

The concentrations of the ions would be:

$[Ag^+]$ = 0.010 mol/L

$[C_2H_3O_2^-]$ = 0.600 mol/L

The value of the ion product for such a solution would be:

$(0.010)(0.600) = 6.0 \times 10^{-3}$, and since the value of the ion product is greater than the value of K_{sp} for $AgC_2H_3O_2$, we expect that a precipitate will form from such a mixture.

14.77 The expression for K_{sp} is:

$K_{sp} = [Pb^{2+}][Br^-]^2$

(a) The concentrations of each of the ions can be calculated from the following general equation used to determine the molarity (M_f) of a diluted solution prepared from a more concentrated solution (M_i):

$M_f = M_i \times (V_i/V_f)$

$[Pb^{2+}]$ = 0.010 M × 50.0 mL/100.0 mL = 5.0×10^{-3} M

$[Br^-]$ = 0.010 M × 50.0 mL/100.0 mL = 5.0×10^{-3} M and the value of the ion product is $(5.0 \times 10^{-3})(5.0 \times 10^{-3})^2 = 1.3 \times 10^{-7}$. Since this is smaller than the value of K_{sp} for $PbBr_2$, we do not expect that a precipitate will form on mixing these two solutions.

(b) Proceeding as in (a) above:

$[Pb^{2+}] = 0.010\ M \times 50.0\ mL/100.0\ mL = 5.0 \times 10^{-3}\ M$

$[Br^-] = 0.10\ M \times 50.0\ mL/100.0\ mL = 5.0 \times 10^{-2}\ M$

The ion product is thus $(5.0 \times 10^{-3})(5.0 \times 10^{-2})^2 = 1.3 \times 10^{-5}$. Since this is larger than the value of K_{sp} for $PbBr_2$, we do expect a precipitate of $PbBr_2$ on mixing these two solutions.

14.78 We can again use the equation set down in Review Exercise 14.77 in order to determine the diluted molarities of the two pertinent ions:

$M_i \times V_i = M_f \times V_f$

$[Ag^+] = 0.10\ M \times 18.0\ mL/58.0\ mL = 0.031\ M$

$[C_2H_3O_2^-] = 0.024\ M \times 40.0\ mL/58.0\ mL = 0.017\ M$

The value of the ion product is therefore:

$[Ag^+][C_2H_3O_2^-] = (0.031)(0.017) = 5.3 \times 10^{-4}$

Since the value of the ion product is smaller than the value of K_{sp} for $AgC_2H_3O_2$, we do not expect that a precipitate of $AgC_2H_3O_2$ will form on mixing these two solutions.

14.79 The gradual addition of $AgNO_3(s)$ will cause the selective precipitation of the much less soluble AgI, thereby reducing the concentration of I^- in solution. The important phrase in this problem is "when AgCl first begins to precipitate." This may be restated as: "when the ion product for AgCl just becomes as large as K_{sp} for AgCl." When this happens, a certain amount of I^- has already been removed from solution as AgI(s), and the ion product $[Ag^+][Cl^-]$ has become numerically equal to $K_{sp} = 1.8 \times 10^{-10}$. Since we are told that the solution contains $[Cl^-] = 0.10\ M$, then we can solve the mass action expression for the value of $[Ag^+]$:

$[Ag^+] = K_{sp} \div [Cl^-] = 1.8 \times 10^{-10}/0.10 = 1.8 \times 10^{-9}$

Additionally, this must be a solution that is now saturated in AgI(s), and the mass action expression for AgI must simultaneously be satisfied. Thus it must be true that the concentration of I^- is that governed by the solubility of AgI(s) in this solution:

K_{sp} for AgI $= 8.3 \times 10^{-17} = [Ag^+][I^-]$

Since we know the concentration of Ag^+, we can solve the last expression for $[I^-]$:

$[I^-] = K_{sp}/[Ag^+] = 8.3 \times 10^{-17}/1.8 \times 10^{-9} = 4.6 \times 10^{-8}$ M

14.80 The number of moles of the two reactants is:

0.10 M $Ag^+ \times 0.050$ L $= 5.0 \times 10^{-3}$ mol Ag^+

0.050 M $Cl^- \times 0.050$ L $= 2.5 \times 10^{-3}$ mol Cl^-

(a)

The precipitation of AgCl proceeds according to the following stoichiometry: $Ag^+ + Cl^- \rightarrow AgCl$ and we conclude that if the product were completely insoluble in the resulting solution, that 2.5×10^{-3} mol of AgCl can be formed, because the limiting reagent is Cl^-, and there is a 1 to 1 stoichiometry between the number of moles of Cl^- that are available and the number of moles of AgCl that can be formed.

Now we also know that not all of the AgCl that can form will remain insoluble in the resulting solution. That is, AgCl does have some finite solubility in the resulting solution. Also we note that the resulting solution still contains $5.0 \times 10^{-3} - 2.5 \times 10^{-3} = 2.5 \times 10^{-3}$ mol of Ag^+ that were present in excess. The problem can then be solved by asking what is the solubility of AgCl in a solution that additionally has a silver ion concentration of 2.5×10^{-3} mol/L. We proceed by constructing the usual equilibrium table:

	$AgCl(s)$	$\rightleftharpoons$	$Ag^+(aq)$	+	$Cl^-(aq)$
initial conc. (M)			0.0250		0.0
change in conc. (M)			x		x
equilibrium conc. (M)			0.0250 + x		x

Substituting the above values for equilibrium concentrations into the mass action expression gives:

$K_{sp} = [Ag^+][Cl^-] = (0.025 + x)(x) = 1.8 \times 10^{-10}$

Solving for x we have: $x = 7.2 \times 10^{-9}$ M AgCl, the solubility of AgCl in the presence of 0.0250 M Ag^+. In this problem, we have only 100.0 mL of solution, so the amount of AgCl that would dissolve is:

$7.2 \times 10^{-9}\ M \times 0.100\ L = 7.2 \times 10^{-10}$ mol of AgCl

The mass of AgCl that would remain in solution is seen to be negligibly small, and the mass of AgCl that is expected to be formed as the solid is:

$2.5 \times 10^{-3}\ mol \times 143\ g/mol = 0.36\ g$

(b)

It was shown in part (a) that the molar solubility of AgCl is 7.2×10^{-9} mol/L. The concentrations in the final solutions are thus:

$[Ag^+] = (2.5 \times 10^{-2} - 7.2 \times 10^{-9})\ mol/1.00\ L = 2.5 \times 10^{-2}\ M$

$[Cl^-] = 7.2 \times 10^{-9}\ M$

For NO_3^- we have: $0.10\ M \times 50.0\ mL/100.0\ mL = 5.0 \times 10^{-2}\ M$

For Na^+ we have: $0.050\ M \times 50.0\ mL/100.0\ mL = 2.5 \times 10^{-2}\ M$

14.81 In this problem, the usual assumption that simplifies the arithmetic is not a valid assumption, and we must use the quadratic equation to solve for the value of x. This happens because the equilibrium constant is not an especially small number.

	$CaSO_4$	$\rightleftharpoons$	$Ca^{2+}(aq)$	+	$SO_4^{2-}(aq)$
initial conc. (M)			0.010		0.0
change in conc. (M)			x		x
equilibrium conc. (M)			0.010 - x		x

Substituting the above values for equilibrium concentrations into the mass action expression gives:

$K_{sp} = 2.4 \times 10^{-5} = [Ca^{2+}][SO_4^{2-}] = (0.010 - x)(x)$

$2.4 \times 10^{-5} = (0.010)x + x^2$ which, in the standard quadratic form becomes:

$x^2 + (0.010)x - 2.4 \times 10^{-5} = 0$, or $ax^2 + bx + c = 0$

and we see that a = 1, b = 0.010, and c = -2.4×10^{-5}

The quadratic formula in general is:

$$x = \frac{-b +/- \sqrt{b^2 - 4ac}}{2a}$$

and in this case we have in particular:

$$x = \frac{-(0.010) +/- \sqrt{(1.0 \times 10^{-2})^2 - 4(1)(-2.4 \times 10^{-5})}}{2(1)}$$

$$x = \frac{-(0.010) + (1.4 \times 10^{-2})}{2} = 2.0 \times 10^{-3}$$

14.82 We can proceed as in Review Exercise 14.79. The precipitation of $CaSO_4$ requires that we have just reached the concentration of $[SO_4^{2-}]$ that satisfies the mass action expression, namely K_{sp} for $CaSO_4(s)$:

$K_{sp} = 2.4 \times 10^{-5} = [Ca^{2+}][SO_4^{2-}] = (0.10)[SO_4^{2-}]$

$[SO_4^{2-}] = 2.4 \times 10^{-5}/0.10 = 2.4 \times 10^{-4}$ M

When the sulfate ion concentration has reached the value 2.4×10^{-4} M, there will just begin to be the precipitation of $CaSO_4$, and the solution will simultaneously be saturated in $SrSO_4$, which, being less soluble, had begun to precipitate earlier in the addition of sodium sulfate. Now we simply solve the equilibrium expression for the dissolving of $CaSO_4$ in the presence of the known amount of SO_4^{2-}:

	$SrSO_4(s)$ ⇌	$Sr^{2+}(aq)$	+	$SO_4^{2-}(aq)$
initial conc. (M)		0.0		2.4×10^{-4}
change in conc. (M)		x		x
equilibrium conc. (M)		x		$2.4 \times 10^{-4} + x$

Substituting the above values for equilibrium concentrations into the mass action expression gives:

$$K_{sp} = [Sr^{2+}][SO_4^{2-}] = (x)(2.4 \times 10^{-4} + x) = 3.2 \times 10^{-7}$$

Rearranging to the standard quadratic form gives:

$$x^2 + (2.4 \times 10^{-4})x - 3.2 \times 10^{-7} = 0$$

This means that a = 1, b = 2.4×10^{-4}, and c = -3.2×10^{-7}

The quadratic formula in general is:

$$x = \frac{-b +/- \sqrt{b^2 - 4ac}}{2a}$$

and in this case we have in particular:

$$x = \frac{-(2.4 \times 10^{-4}) +/- \sqrt{(2.4 \times 10^{-4})^2 - 4(1)(-3.2 \times 10^{-7})}}{2(1)}$$

$$x = \frac{-2.4 \times 10^{-4} + (1.2 \times 10^{-3})}{2} = 4.8 \times 10^{-4} \text{ M}$$

Complex Ion Equilibria

14.83 The three equilibria that operate here are:

$$AgCl(s) \rightleftharpoons Ag^+(aq) + Cl^-(aq)$$

$$Ag^+(aq) + 2NH_3(aq) \rightleftharpoons [Ag(NH_3)_2]^+(aq)$$

$$HNO_3(aq) + NH_3(aq) \rightleftharpoons NH_4^+(aq) + NO_3^-(aq)$$

The addition of NH_3 shifts the second equilibrium to the right, resulting in a decrease in $[Ag^+]$. The response from the first equilibrium is to restore some of the $Ag^+(aq)$, and the solubility of AgCl(s) is increased by the additon of ammonia. The addition of HNO_3 reverses this effect by shifting the third equilibrium so as to remove some $NH_3(aq)$. This causes the second equilibrium to shift to the left, making more NH_3 as well as more Ag^+. The latter enters the first equilibrium, causing it to shift to the left, with a corresponding decrease in solubility for AgCl(s).

14.84 $PbCl_2(s) \rightleftharpoons Pb^{2+}(aq) + 2Cl^-(aq)$

$Pb^{2+}(aq) + 3Cl^-(aq) \rightleftharpoons PbCl_3^-(aq)$

$$K_{form} = \frac{[PbCl_3^-]}{[Pb^{2+}][Cl^-]^3} = \frac{(\text{mol } PbCl_3^- / \text{ volume})}{(\text{mol } Pb^{2+} / \text{ volume})(\text{mol } Cl^- / \text{volume})^3}$$

$$K_{form} = (\text{volume})^3 \times \frac{(\text{mol } PbCl_3^-)}{(\text{mol } Pb^{2+})(\text{mol } Cl^-)^3}$$

Notice that the above expression is the product of a ratio of mole amounts and a volume^3 term. Since the constant K_{form} does not change on dilution, and if the volume term is changed by dilution, then the ratio of moles term in the above expression must change on dilution in order to hold the product constant. The ratio of moles would have to become smaller by a factor of 8 in order for the entire argument to have a constant value, i.e in order for K_{form} to remain constant on dilution.

This means that the concentrations of Pb^{2+} and Cl^- would increase as the solution containing the complex ion is diluted. Eventually the ion product for $PbCl_2$ will exceed the value of K_{sp} for $PbCl_2$, resulting in its precipitation.

14.85 (a) $Co(NH_3)_6^{3+}(aq) \rightleftharpoons Co^{3+}(aq) + 6NH_3(aq)$

$K_{inst} = 1/k_{form} = [Co^{3+}][NH_3]^6/[Co(NH_3)_6^{3+}]$

(b) $HgI_4^{2-}(aq) \rightleftharpoons Hg^{2+}(aq) + 4I^-(aq)$

$K_{inst} = 1/K_{form} = [Hg^{2+}][I^-]^4/[HgI_4^{2-}]$

(c) $Fe(CN)_6^{4-}(aq) \rightleftharpoons Fe^{2+}(aq) + 6CN^-(aq)$

$K_{inst} = 1/K_{form} = [Fe^{2+}][CN^-]^6/[Fe(CN)_6^{4-}]$

14.86 (a) $4NH_3(aq) + Hg^{2+}(aq) \rightleftharpoons Hg(NH_3)_4^{2+}(aq)$

$K_{form} = [Hg(NH_3)_4^{2+}]/[NH_3]^4[Hg^{2+}]$

(b) $6F^-(aq) + Sn^{4+}(aq) \rightleftharpoons SnF_6^{2-}(aq)$

$K_{form} = [SnF_6^{2-}]/[F^-]^6[Sn^{4+}]$

(c) $6CN^-(aq) \; + \; Fe^{3+}(aq) \; \rightleftharpoons \; Fe(CN)_6{}^{3-}(aq)$

$K_{form} = [Fe(CN)_6{}^{3-}]/[CN^-]^6[Fe^{3+}]$

14.87 (a) $K_{form} = 1/K_{inst} = 1/5.6 \times 10^{-2} = 18$

(b) The values of K_{form} in Table 14.3 are all much larger than 18, and we conclude that this complex is less stable than the others.

14.88 There are two events in this net process: one is the formation of a complex ion (an equilibrium which has an appropriate value for K_{form}), and the other is the dissolving of $Fe(OH)_3$, which is governed by K_{sp} for the solid. The net process is:

$$Fe(OH)_3(s) + 6CN^-(aq) \rightleftharpoons Fe(CN)_6{}^{3-}(aq) + 3OH^-(aq)$$

The equilibrium constant for this process should be:

$$K_c = \frac{[Fe(CN)_6{}^{3-}][OH^-]^3}{[CN^-]^6}$$

The numerical value for the above K_c is equal to the product of K_{sp} for $Fe(OH)_3(s)$ and K_{form} for $Fe(CN)_6{}^{3-}$, as can be seen by multiplying the mass action expressions for these two equilibria:

$$K_c = K_{form} \times K_{sp} = 1.6 \times 10^{-8}$$

Because K_{form} is so very large, we can assume that all of the dissolved iron ion is present in solution as the complex, i.e. $[Fe(CN)_6{}^{3-}]$ = 0.10 M. Also the reaction stoichiometry shows that each iron ion that dissolves gives 3 OH^- ions in solution, and we have $[OH^-]$ = 0.30 M. Substituting these values into the K_c expression gives:

$$K_c = \frac{(0.10)(0.30)^3}{[CN^-]^6} = 1.6 \times 10^{-8}$$

Thus we arrive at the concentration of cyanide ion that is required in order to satisfy the mass action requirements of the equilibrium:

$[CN^-]^6 = 1.69 \times 10^5$ and $[CN^-] = 7.4$ mol/L

Additionally, we require enough cyanide to enter into the formation of the complex ion. This is 0.10 × 6 = 0.60 mol, as indicated by the stoichiometry of the reaction, bringing the total required cyanide to (7.4 + 0.6) = 8.0 mol.

8.0 mol × 49.0 g/mol = 3.9×10^2 g NaCN are required.

14.89 $CoCO_3(s) + 6NH_3(aq) \rightleftharpoons Co(NH_3)_6{}^{2+} + CO_3{}^{2-}(aq)$

$K_c = [Co(NH_3)_6{}^{2+}][CO_3{}^{2-}]/[NH_3]^6 = ?$

The numerical value of K_c is equal to the product of K_{form} and K_{sp} as follows:

$K_{form} = [Co(NH_3)_6{}^{2+}]/[Co^{2+}][NH_3]^6 = 5.0 \times 10^4$

$K_{sp} = [Co^{2+}][CO_3{}^{2-}] = 1.0 \times 10^{-10}$

$K_c = K_{form} \times K_{sp} = 5.0 \times 10^{-6}$

We assume that nearly all of the Co^{2+} that dissolves becomes complexed by ammonia in solution, i.e. that $[Co(NH_3)_6{}^{2+}] = 1.0 \times 10^{-4}$ M. Also we know that 1 mol of $CO_3{}^{2-}$ is produced for each mol of complex that forms, such that $[CO_3{}^{2-}] = 1.0 \times 10^{-4}$ M. Substituting these values into the mass action expression gives:

$$K_c = \frac{[Co(NH_3)_6{}^{2+}][CO_3{}^{2-}]}{[NH_3]^6} = 5.0 \times 10^{-6} = \frac{(1.0 \times 10^{-4})(1.0 \times 10^{-4})}{[NH_3]^6}$$

$[NH_3]^6 = (1.0 \times 10^{-4})^2/5.0 \times 10^{-6}$ and solving for $[NH_3]$ gives:

$[NH_3] = 0.35$ mol/L, the concentration required in order to satisfy the mass action expression for the equilibrium. Additionally we require the amount of NH_3 that is stoichiometrically necessary, i.e. an extra $(1.0 \times 10^{-4}) \times 6 = 6.0 \times 10^{-4}$ mol of NH_3.

The total amount of NH_3 is thus $(0.35 + 6.0 \times 10^{-4}) = 0.35$ mol NH_3.

0.35 mol × 22.4 L/mol = 7.8 L = 7.8×10^3 mL NH_3 gas in one L of water.

14.90 The two important equilibria are:

$Ag^+ + 2S_2O_3^{2-} \rightleftharpoons Ag(S_2O_3)_2^{3-}$, $K_{form} = 2.0 \times 10^{13}$

$AgBr(s) \rightleftharpoons Ag^+(aq) + Br^-(aq)$, $K_{sp} = 5.0 \times 10^{-13}$

Adding these two equilibria together gives us the desired net reaction, for which the equilibrium constant, K_c, is given by the product of K_{form} and K_{sp}:

$K_c = (2.0 \times 10^{13})(5.0 \times 10^{-13}) = 10$

	$AgBr(s) + 2S_2O_3^{2-}(aq)$	$\rightleftharpoons Ag(S_2O_3)_2^{3-}(aq)$	$+ Br^-(aq)$
initial conc. (M)	2.0	0.0	0.0
change in conc. (M)	-2x	x	x
equilibrium conc. (M)	2.0 - 2x	x	x

$$K_c = \frac{[Ag(S_2O_3)_2^{3-}][Br^-]}{[S_2O_3^{2-}]^2} = \frac{x^2}{(2.0 - 2x)^2} = 10$$

$x^2 = 10(2.0 - 2x)^2$

$x^2 = 10[4.0 - (4.0)x + 4.0x^2] = 40 - 40x + 40x^2$

$39x^2 - 40x + 40 = 0$, which is in the form $ax^2 + bx + c = 0$

and we have the quadratic equation where a = 39, b = -40, and c = 40.

The quadratic formula in general is:

$$x = \frac{-b +/- \sqrt{b^2 - 4ac}}{2a}$$

and in this case we have in particular:

$$x = \frac{-(-40) +/- \sqrt{(-40)^2 - 4(39)(40)}}{2(39)}$$

x = (40 + 68)/78 = 1.4 mol/L

Thus the concentration of Br^- in 2.0 M $S_2O_3^{2-}$ solution would be 1.4 M.

Since we have only 500 mL of solution, the amount of Br^- (and hence the dissolved amount of AgBr) would be:

1.4 mol/L × 0.500 L × 188 g/mol = 132 g AgBr dissolve.

CHAPTER FIFTEEN

PRACTICE EXERCISES

1. $K_w = 1.0 \times 10^{-14} = [H^+][OH^-]$

   $$[H^+] = \frac{1.0 \times 10^{-14}}{[OH^-]} = \frac{1.0 \times 10^{-14}}{7.8 \times 10^{-6}} = 1.3 \times 10^{-9}\ M$$

   Since $[OH^-] > [H^+]$, the solution is basic.

2. (a) $pH = -\log[H^+] = -\log(0.020) = 1.70$

   $pOH = 14.00 - pH = 14.00 - 1.70 = 12.30$

   (b) $pOH = -\log[OH^-] = -\log(5.0 \times 10^{-3}) = 2.30$

   $pH = 14.00 - pOH = 14.00 - 2.30 = 11.70$

   (c) $pH = -\log[H^+] = -\log(7.2 \times 10^{-8}) = 7.14$

   $pOH = 14.00 - pH = 14.00 - 7.14 = 6.86$

   This is a slightly basic solution since pH > 7.0.

   (d) $[OH^-] = 3.5 \times 10^{-4}$ mol $Ba(OH)_2$/L $\times$ 2 mol OH^-/1 mol $Ba(OH)_2$

   $= 7.0 \times 10^{-4}$ M

   $pOH = -\log[OH^-] = -\log(7.0 \times 10^{-4}) = 3.15$

   $pH = 14.00 - pOH = 14.00 - 3.15 = 10.85$

3. In general, we have the following relationships between pH and $[H^+]$:

   $pH = -\log[H^+]$ and $[H^+] = 10^{-pH}$ or $[H^+] = \text{antilog}(-pH)$

   and the following relationships between $[OH^-]$ and pOH:

   $pOH = -\log[OH^-]$ and $[OH^-] = 10^{-pOH}$ or $[OH^-] = \text{anitlog}(-pOH)$

   Also, we know that $K_w = 1.0 \times 10^{-14} = [H^+][OH^-]$, and as a check of the answers, we can multiply $[OH^-] \times [H^+]$ to see if the product is actually equal to K_w.

   (a) $[H^+] = \text{antilog}(-2.9) = 1.3 \times 10^{-3}$ M

$pOH = 14.00 - pH = 14.00 - 2.90 = 11.10$

$[OH^-] = antilog(-11.10) = 7.9 \times 10^{-12}$ M

(b) $[H^+] = antilog(-3.85) = 1.4 \times 10^{-4}$ M

$pOH = 14.00 - pH = 14.00 - 3.85 = 10.15$

$[OH^-] = antilog(-10.15) = 7.1 \times 10^{-11}$

(c) $[H^+] = antilog(-10.81) = 1.5 \times 10^{-11}$ M

$pOH = 14.00 - pH = 14.00 - 10.81 = 3.19$

$[OH^-] = antilog(-3.19) = 6.5 \times 10^{-4}$ M

(d) $[H^+] = antilog(-4.11) = 7.8 \times 10^{-5}$ M

$pOH = 14.00 - pH = 14.00 - 4.11 = 9.89$

$[OH^-] = antilog(-9.89) = 1.3 \times 10^{-10}$

(e) $[H^+] = antilog(-11.61) = 2.5 \times 10^{-12}$ M

$pOH = 14.00 - pH = 14.00 - 11.61 = 2.39$

$[OH^-] = antilog(-2.39) = 4.1 \times 10^{-3}$

4. Solutions that are acidic are those for which $[H^+] > 1.0 \times 10^{-7}$ and $pH < 7.0$. Solutions that are basic are those for which $[H^+] < 1.0 \times 10^{-7}$ and $pH > 7.0$.

(a) acidic (b) acidic (c) basic (d) acidic (e) basic

5. $HNO_2 \rightleftharpoons H^+ + NO_2^-$

$$K_a = \frac{[H^+][NO_2^-]}{[HNO_2]}$$

6. $HPO_4^{2-} \rightleftharpoons H^+ + PO_4^{3-}$

$$K_a = \frac{[H^+][PO_4^{3-}]}{[HPO_4^{2-}]}$$

7. For water:

$$H_2O \rightleftharpoons H^+ + OH^-$$

$$K_a = \frac{[H^+][OH^-]}{[H_2O]}$$

For H_3O^+:

$$H_3O^+ \rightleftharpoons H^+ + H_2O$$

$$K_a = \frac{[H_2O][H^+]}{[H_3O^+]}$$

8. $HBu \rightleftharpoons H^+ + Bu^-$

$$K_a = \frac{[H^+][Bu^-]}{[HBu]}$$

Since pH = 3.40, we can calculate $[H^+]$ and $[Bu^-]$ as follows:

$$[H^+] = \text{antilog}(-pH) = \text{antilog}(-3.40) = 4.0 \times 10^{-4}\ M$$

This is also the equilibrium concentration of $[Bu^-]$, since the stoichiometry of the ionization equilibrium is one to one.

	HBu	⇌	H^+	+	Bu^-
initial conc. (M)	0.0100		0.0		0.0
change in conc. (M)	-4.0×10^{-4}		$+4.0 \times 10^{-4}$		$+ 4.0 \times 10^{-4}$
equilibrium conc. (M)	0.0096		4.0×10^{-4}		4.0×10^{-4}

Substituting the above values for equilibrium concentrations into the expression for the acid ionization gives:

$$K_a = \frac{[H^+][Bu^-]}{[HBU]} = \frac{(4.0 \times 10^{-4})(4.0 \times 10^{-4})}{(0.0096)} = 1.7 \times 10^{-5}$$

9. $HBu \rightleftharpoons H^+ + Bu^-$

$$K_a = \frac{[H^+][Bu^-]}{[HBu]}$$

Since pH = 2.98, we can calculate $[H^+]$ and $[Bu^-]$ as follows:

$$[H^+] = \text{antilog}(-pH) = \text{antilog}(-2.98) = 1.0 \times 10^{-3}\ M$$

This is also the equilibrium concentration of $[Bu^-]$, since the stoichiometry of the ionization equilibrium is one to one.

	HBu	⇌	H^+	+	Bu^-
initial conc. (M)	0.0100		0.0		0.0
change in conc. (M)	-1.0×10^{-3}		$+1.0 \times 10^{-3}$		$+ 1.0 \times 10^{-3}$
equilibrium conc. (M)	0.0090		1.0×10^{-3}		1.0×10^{-3}

Substituting the above values for equilibrium concentrations into the expression for the acid ionization gives:

$$K_a = \frac{[H^+][Bu^-]}{[HBu]} = \frac{(1.0 \times 10^{-3})(1.0 \times 10^{-3})}{(0.0090)} = 1.1 \times 10^{-4}$$

10.

	$HC_2H_4NO_2$	⇌	H^+	+	$C_2H_4NO_2^-$
initial conc. (M)	0.010		0.0		0.0
change in conc. (M)	-x		+x		+x
equilibrium conc. (M)	0.010 - x (≈ 0.010)		x		x

Substituting the above values for equilibrium concentrations into the mass action expression gives:

$$K_a = \frac{[H^+][C_2H_4NO_2^-]}{[HC_2H_4NO_2]} = \frac{(x)(x)}{(0.010)} = 1.4 \times 10^{-5}$$

$x^2 = 1.4 \times 10^{-7}$ $\therefore$ $x = 3.7 \times 10^{-4}$ M = $[H^+]$

$pH = -\log[H^+] = -\log(3.7 \times 10^{-4}) = 3.43$

11.

(a)

	$HC_2H_3O_2$	⇌	H^+	+	$C_2H_3O_2^-$
initial conc. (M)	0.010		0.0		0.0
change in conc. (M)	-x		+x		+x
equilibrium conc. (M)	0.010 - x (≈ 0.010)		x		x

Substituting the above values for equilibrium concentrations into the mass action expression gives:

$$K_a = \frac{[H^+][C_2H_3O_2^-]}{[HC_2H_3O_2]} = \frac{(x)(x)}{(0.010)} = 1.8 \times 10^{-5}$$

$x^2 = 1.8 \times 10^{-7}$ $\therefore x = 4.2 \times 10^{-4}$ M

% ionization = $4.2 \times 10^{-4}/0.010 \times 100 = 4.2$ %

(b)

	$HC_2H_3O_2$	⇌	H^+	+	$C_2H_3O_2^-$
initial conc. (M)	0.0010		0.0		0.0
change in conc. (M)	-x		+x		+x
equilibrium conc. (M)	0.0010 - x (≈ 0.0010)		x		x

Substituting the above values for equilibrium concentrations into the mass action expression gives:

$$K_a = \frac{[H^+][C_2H_3O_2^-]}{[HC_2H_3O_2]} = \frac{(x)(x)}{(0.0010)} = 1.8 \times 10^{-5}$$

$x^2 = 1.8 \times 10^{-8}$ $\therefore x = 1.3 \times 10^{-4}$ M

Notice that the assumption made above that (0.0010 - x) ≈ (0.0010) is close to being invalid. This can happen in very dilute solutions.

% ionization = $1.3 \times 10^{-4}/0.0010 \times 100 = 13$ %

12. (a)

$$K_b = \frac{[HCN][OH^-]}{[CN^-]}$$

(b)

$$K_b = \frac{[HC_2H_3O_2][OH^-]}{[C_2H_3O_2^-]}$$

(c)

$$K_b = \frac{[C_6H_5NH_3^+][OH^-]}{[C_6H_5NH_2]}$$

(d)

$$2H_2O \rightleftharpoons H_3O^+ + OH^-$$

$$K_b = [H_3O^+][OH^-]$$

13. morphine + H_2O $\rightleftharpoons$ morphineH^+ + OH^-

K_b = [morphineH^+][OH^-]/[morphine]

pOH = 14.00 - pH = 14.00 - 10.10 = 3.90

[OH^-] = antilog(-pOH) = antilog(-3.90) = 1.3×10^{-4} M. This is also the concentration of the protonated morphine cation, morphineH^+, since this and OH^- are formed on a one to one mole basis.

	H_2O + morphine	$\rightleftharpoons$ morphineH^+	+ OH^-
initial conc. (M)	0.010	0.0	0.0
change in conc. (M)	-1.3×10^{-4}	$+1.3 \times 10^{-4}$	$+1.3 \times 10^{-4}$
equilibrium conc. (M)	0.00987	1.3×10^{-4}	1.3×10^{-4}

Substituting the above values for equilibrium concentrations into the mass action expression gives:

$$K_b = \frac{[\text{morphineH}^+][\text{OH}^-]}{[\text{morphine}]} = \frac{(1.3 \times 10^{-4})(1.3 \times 10^{-4})}{(0.000987)} = 1.7 \times 10^{-6}$$

14.

	H_2O + NH_3	$\rightleftharpoons$ NH_4^+	+ OH^-
initial conc. (M)	0.020	0.0	0.0
change in conc. (M)	-x	x	x
equilibrium conc. (M)	0.020 - x ($\approx$ 0.020)	x	x

Substituting the above values for equilibrium concentrations into the mass action expression gives:

$$K_b = \frac{[NH_4^+][OH^-]}{[NH_3]} = \frac{(x)(x)}{(0.020)} = 1.8 \times 10^{-5}$$

$x^2 = 3.6 \times 10^{-7}$ $\quad\therefore$ $x = 6.0 \times 10^{-4}$ M = [OH^-] = [NH_4^+]

pOH = -log[OH^-] = -log(6.0×10^{-4}) = 3.22

pH = 14.00 - pOH = 14.00 - 3.22 = 10.78

15. pK_a = -logK_a = -log(7.1×10^{-4}) = 3.15

16. In general, the smaller the value of pK_a, the stronger the acid. Hence,

acetic acid (pK_a = 4.76) is stronger than hydrocyanic acid (pK_a = 9.21).

17. $pK_b = -\log K_b = -\log(6.41 \times 10^{-4}) = 3.193$

18. $pK_a + pK_b = 14.00$

$\therefore pK_b = 14.00 - pK_a = 14.00 - 9.2 = 4.8$

$CN^- + H_2O \rightleftharpoons HCN + OH^-$

$$K_b = \frac{[HCN][OH^-]}{[CN^-]}$$

19.

	$HC_2H_3O_2$	$\rightleftharpoons$ H^+	+ $C_2H_3O_2^-$
initial conc. (M)	0.090	0.0	0.0
change in conc. (M)	-x	x	x
equilibrium conc. (M)	0.090 - x (≈ 0.090)	x	x

Substituting the above values for equilibrium concentrations into the expression for K_a of acetic acid gives:

$$K_a = \frac{[H^+][C_2H_3O_2^-]}{[HC_2H_3O_2]} = \frac{(x)(x)}{(0.090)} = 1.8 \times 10^{-5}$$

$x^2 = 1.6 \times 10^{-6} \quad \therefore \quad x = 1.3 \times 10^{-3} = [H^+]$

$pH = -\log[H^+] = -\log(1.3 \times 10^{-3}) = 2.89$

20. In general, for an acid HA which ionizes to give H^+ and A^-, we have:

$$pH = pK_a + \log\frac{[\text{anion}]}{[\text{acid}]} = pK_a + \log\frac{[A^-]}{[HA]}$$

(a) $pH = pK_a + \log(10) = pK_a + 1$

(b) $pH = pK_a + \log(1/10) = pK_a - 1$

21.

$$pH = pK_a + \log\frac{[\text{anion}]}{[\text{acid}]} = pK_a + \log\frac{[A^-]}{[HA]}$$

$pH = 4.74 + \log(0.015/0.10) = 4.74 + (-0.82) = 3.92$

22. In general, one chooses a weak acid that has a value of pK_a close to the

desired pH. Formic acid will do nicely, because it has $K_a = 1.8 \times 10^{-4}$ and $pK_a = 3.74$.

$$pH = pK_a + \log \frac{[anion]}{[acid]} = pK_a + \log \frac{[A^-]}{[HA]}$$

$$pH = 3.74 + \log([anion]/[acid]) = 3.90$$

$$\log([anion]/[acid]) = 0.15$$

Taking the log of both sides of the above equation gives:

$$([anion]/[acid]) = 1.4$$

Thus we know that the ratio of the concentration of the anion to that of the acid is to be 1.4 to 1, if formic acid, together with its anion is to give a solution having a pH of 3.90.

23. First determine the concentration of the ammonium cation that results:

$$2.5\ g\ NH_4Cl \times \frac{1\ mol\ NH_4Cl}{53.5\ g\ NH_4Cl} \times \frac{1}{0.125\ L} = 0.37\ M\ NH_4Cl$$

$$pOH = pK_b + \log \frac{[cation]}{[base]} = 4.74 + \log \frac{0.37}{0.24} = 4.93$$

24.

	$H_2PO_4^-$	⇌	HPO_4^-	+	H^+
initial conc. (M)	0.10		0.0		0.0
change in conc. (M)	-x		x + y		x - y
equilibrium conc. (M)	0.10 - x (≈ 0.10)		x + y (≈ x)		x - y (≈ x)

Substituting the above values for equilibrium concentrations into the mass action expression that is derived from K_{a_2} for H_3PO_4 gives:

$$K_{a_2} = \frac{[H^+][HPO_4^-]}{[H_2PO_4^-]} = \frac{x^2}{0.10} = 6.3 \times 10^{-8}$$

$$x = 7.9 \times 10^{-5} = [H^+]$$

$$pH = -\log[H^+] = -\log(7.9 \times 10^{-5}) = 4.10$$

For the concentration of PO_4^{3-}, we solve the usual mass action expression for the ionization of the weak acid HPO_4^{2-}:

$$K_{a_3} = \frac{[H^+][PO_4{}^{3-}]}{[HPO_4{}^{2-}]} = \frac{x}{x} \times [PO_4{}^{3-}]$$

We can see from the form of the above equation that K_{a_3} for phosphoric acid is equal to the concentration of the anion $PO_4{}^{3-}$:

$[PO_4{}^{3-}] = 4.5 \times 10^{-13}$

25. The ions that should hydrolyze in water to produce basic solutions are those that are the anions (conjugate bases) of weak acids:

$CO_3{}^{2-}$, S^{2-}, $HPO_4{}^{2-}$, $NO_2{}^-$, and F^-.

The cations that should hydrolyze to give acidic solutions are those of the transition elements and Be. From this list we have only Cu^{2+} in this category.

26. This salt contains $C_2H_3O_2{}^-$, the conjugate base of the weak acid acetic acid. The anion should hydrolyze in water to give a basic solution:

$C_2H_3O_2{}^- + H_2O \rightleftharpoons HC_2H_3O_2 + OH^-$

27. Since the anion here is derived from the strong acid HNO_3, it does not hydrolyze in water. The cation, however, is a transition metal ion, and it hydrolyzes to give an acidic solution.

28. Since the anion here is the anion of the strong acid HNO_3, it is not hydrolyzed in water. The cation, however, is hydrolyzed by water, making an acidic solution (i.e. the pH is lowered):

$NH_4{}^+ + H_2O \rightleftharpoons NH_3 + H_3O^+$

29. Only the hydrolysis of the anion needs to be considered, because the cation is from Group IA.

$K_b = K_w/K_a = 1.0 \times 10^{-14}/1.4 \times 10^{-4} = 7.1 \times 10^{-11}$

	$C_3H_5O_3{}^-$	+ H_2O $\rightleftharpoons$	$HC_3H_5O_3$	+ OH^-
initial conc. (M)	0.15		0.0	0.0
change in conc. (M)	-x		x	x
equilibrium conc. (M)	(0.15 - x) (≈ 0.15)		x	x

Substituting the above values for equilibrium concentrations into the mass action expression gives:

$$K_b = \frac{[HC_3H_5O_3][OH^-]}{[C_3H_5O_3^-]} = \frac{(x)(x)}{(0.15 - x)} = 7.1 \times 10^{-11}$$

$x^2 = 1.065 \times 10^{-11}$ $\therefore$ $x = 3.3 \times 10^{-6}$ M = $[OH^-]$

pOH = $-\log[OH^-] = -\log(3.3 \times 10^{-6}) = 5.48$

pH = 14.00 - pOH = 14.00 - 5.48 = 8.52

30. $SO_3^{2-} + H_2O \rightleftharpoons HSO_3^- + OH^-$

$K_b = K_w/K_{a_2} = 1.0 \times 10^{-14}/6.6 \times 10^{-8} = 1.5 \times 10^{-7}$

We can ignore any subsequent hydrolysis of HSO_3^-, whose concentration will be quite low in this solution. Considering then only the first of two hydrolysis equilibria:

	SO_3^{2-}	+	H_2O	$\rightleftharpoons$	HSO_3^-	+	OH^-
initial conc. (M)	0.20				0.0		0.0
change in conc. (M)	-x				x		x
equilibrium conc. (M)	0.20 - x (≈ 0.20)				x		x

Substituting the above values for equilibrium concentrations into the mass action expression for the equilibrium shown above gives:

$$K_b = \frac{[HSO_3^-][OH^-]}{[SO_3^{2-}]} = \frac{(x)(x)}{0.20} = 1.5 \times 10^{-7}$$

$x^2 = 3.0 \times 10^{-8}$ $\therefore$ $x = 1.7 \times 10^{-4}$ M = $[OH^-]$

pOH = $-\log[OH^-] = -\log(1.7 \times 10^{-4}) = 3.77$

pH = 14.00 - pOH = 14.00 - 3.77 = 10.23

31. The initial number of moles of NaOH and of $HC_2H_3O_2$ are:

0.20 M NaOH × 0.01500 L = 3.0×10^{-3} mol NaOH

0.20 M $HC_2H_3O_2$ × 0.02500 L = 5.0×10^{-3} M $HC_2H_3O_2$

The amount of acetic acid that is not neutralized is thus 2.0×10^{-3} moles, and the concentrations of acid and anion in the resulting solution are:

For $HC_2H_3O_2$, 2.0×10^{-3} mol/0.0400 L = 5.0×10^{-2} M

For $C_2H_3O_2^-$, 3.0×10^{-3} mol/0.0400 L = 7.5×10^{-2} M

$$pH = pK_a + \log \frac{[\text{anion}]}{[\text{acid}]} = pK_a + \log \frac{[C_2H_3O_2^-]}{[HC_2H_3O_2]}$$

$$pH = 4.74 + \log(7.5 \times 10^{-2}/5.0 \times 10^{-2}) = 4.74 + 0.18 = 4.92$$

REVIEW EXERCISES

Self-Ionization of Water and pH

15.1 The true equilibrium constant expression for the self ionization of water has the term for the concentration of water in its denominator. The simple ion product constant expression treats the concentration of water as being a constant that is incorporated into the numerical value for the product. These differences are illustrated below:

$$K_{eq} = \frac{[H_3O^+][OH^-]}{[H_2O]^2}$$

$$K_w = K_{eq} \times [H_2O]^2 = [H_3O^+][OH^-]$$

15.2 (a) acidic: $[H^+] > [OH^-]$

basic: $[OH^-] > [H^+]$

neutral: $[OH^-] = [H^+]$

(b) acidic: $pH < 7.0$

basic: $pH > 7.0$

neutral: $pH = 7.0$

15.3 (a) The pH should decrease with increasing temperature.

(b) Water should remain neutral at all temperatures at which it is thermally stable.

(c) If the value of K_w increases, then so too must the product of $[OH^-]$ and $[H^+]$ increase, since $K_w = [H^+] \times [OH^-]$. Also, each of $[H^+]$ and $[OH^-]$ must undergo the same increase, since the stoichiometry of their formation is one to one. Thus, water should remain neutral, because at all temperatures, $[H^+] = [OH^-]$. Nevertheless, since $[H^+]$ increases, so too must pH decrease, and so must pOH decrease. There is more on this in the answer to Review Exercise 15.4 below.

15.4 It should be stated from the outset that water at this temperature is neutral by definition, since $[H^+] = [OH^-]$, because the self-ionization of water is still on a one to one mole basis:

$H_2O \rightleftharpoons H^+ + OH^-$, $K_w = 2.4 \times 10^{-14} = [H^+] \times [OH^-]$

Since $[H^+] = [OH^-]$, we can rewrite the above relationship:

$2.4 \times 10^{-14} = ([H^+])^2$, $\therefore [H^+] = [OH^-] = 1.5 \times 10^{-7}$ M

$pH = -\log[H^+] = -\log(1.5 \times 10^{-7}) = 6.82$

$pOH = -\log[OH^-] = -\log(1.5 \times 10^{-7}) = 6.82$

$pK_w = pH + pOH = 6.82 + 6.82 = 13.64$

Alternatively, for the last calculation we can write:

$pK_w = -\log(K_w) = -\log(2.4 \times 10^{-14}) = 13.62$

15.5 $D_2O \rightleftharpoons D^+ + OD^-$, $K_w = [D^+] \times [OD^-] = 8.9 \times 10^{-16}$

Since $[D^+] = [OD^-]$, we can rewrite the above expression to give:

$8.9 \times 10^{-16} = ([D^+])^2$, $\therefore [D^+] = 3.0 \times 10^{-8}$ M $= [OD^-]$

$pD = -\log[D^+] = -\log(3.0 \times 10^{-8}) = 7.52$

$pOD = -\log[OD^-] = -\log(3.0 \times 10^{-8}) = 7.52$

$pK_w = pD + pOd = 15.04$

Alternatively, we can calculate:

$pK_w = -\log(K_w) = -\log(8.9 \times 10^{-16}) = 15.05$

15.6 $pH = -\log[H^+]$

$pH = -\log(0.010) = 2.00$

15.7 $pH = -\log[H^+]$

$pH = -\log(5.0 \times 10^{-3}) = 2.30$

15.8 $6.0\ g\ NaOH \times 1\ mol/40.0\ g = 0.15\ M\ NaOH = 0.15\ M\ OH^-$

$pOH = -\log[OH^-] = -\log(0.15) = 0.82$

$pH = 14.00 - pOH = 14.00 - 0.82 = 13.18$

15.9

$$0.837\ g\ Ba(OH)_2 \times \frac{1\ mol\ Ba(OH)_2}{171\ g\ Ba(OH)_2} \times \frac{2\ mol\ OH^-}{1\ mol\ Ba(OH)_2} \times \frac{1}{0.100\ L}$$

$$= 9.79 \times 10^{-2}\ M\ OH^-$$

$pOH = -\log[OH^-] = -\log(9.79 \times 10^{-2}) = 1.01$

$pH = 14.00 - pOH = 14.00 - 1.01 = 12.99$

15.10 $pH = -\log[H^+] = -\log(1.9 \times 10^{-5}) = 4.72$

15.11 $pH = -\log[H^+] = -\log(1.4 \times 10^{-5}) = 4.85$

Acid and Base Ionization Constants - K_a, K_b, pK_a, and pK_b

15.12 $C_6H_5CO_2H \rightleftharpoons H^+ + C_6H_5CO_2^-$

$K_a = [H^+][C_6H_5CO_2^-]/[C_6H_5CO_2H]$

15.13 $C_6H_5CO_2^- + H_2O \rightleftharpoons C_6H_5CO_2H + OH^-$

$K_b = [HC_6H_5CO_2][OH^-]/[C_6H_5CO_2^-]$

15.14 In general, there is an inverse relationship between the strengths of an acid and its conjugate base; the stronger the acid, the weaker is its conjugate base. Since HCN has the larger value for pK_a, we know that it is a weaker acid than is HF. Accordingly, CN^- is a stronger base than is F^-.

15.15

	HF	$\rightleftharpoons$	H^+	+	F^-
initial conc. (M)	0.15		0.0		0.0
change in conc. (M)	-x		x		x
equilibrium conc. (M)	0.15 - x (≈ 0.15)		x		x

Substituting the above values for equilibrium concentrations into the mass action expression gives:

$K_a = [H^+][F^-]/[HF] = x^2/0.15 = 6.8 \times 10^{-4}$

$x^2 = 1.0 \times 10^{-4}$, $\therefore x = 1.0 \times 10^{-2}$ M = $[H^+]$

pH = $-\log[H^+] = -\log(0.010) = 2.00$

% ionization = $0.010/0.15 \times 100 = 6.7$ %

15.16

	$HC_2H_3O_2$	$\rightleftharpoons$	H^+	+	$C_2H_3O_2^-$
initial conc. (M)	1.0		0.0		0.0
change in conc. (M)	-x		x		x
equilibrium conc. (M)	1.0 - x (≈ 1.0)		x		x

Substituting the above values for equilibrium concentrations into the expression for K_a for acetic acid gives:

$K_a = [H^+][C_2H_3O_2^-]/[HC_2H_3O_2] = 1.8 \times 10^{-5} = x^2/1.0$

$x^2 = 1.8 \times 10^{-5}$, $\therefore x = 4.2 \times 10^{-3}$ M

pH = $-\log[H^+] = -\log(4.2 \times 10^{-3}) = 2.38$

% ionization = $4.2 \times 10^{-3}/1.0 \times 100 = 0.42$ %

15.17 (a) $HSO_4^- \rightleftharpoons H^+ + SO_4^{2-}$

$K_a = [H^+][SO_4^{2-}]/[HSO_4^-] = 1.0 \times 10^{-2}$

(b)

	HSO_4^-	$\rightleftharpoons$	H^+	+	SO_4^{2-}
initial conc. (M)	0.010		0.0		0.0
change in conc. (M)	-x		x		x
equilibrium conc. (M)	0.010 - x		x		x

Substituting the above values for equilibrium concentrations into the mass action expression written in part (a) gives:

$K_a = 1.0 \times 10^{-2} = x^2/(1.0 \times 10^{-2} - x)$

$x^2 + (1.0 \times 10^{-2})x + (-1.0 \times 10^{-4}) = 0$, which is in the form of the quadratic equation, where a = 1.0, b = 1.0×10^{-2}, and c = -1.0×10^{-4}.

Solving for x we get:

$$x = \frac{-b +/- \sqrt{b^2 - 4ac}}{2a}$$

$$x = \frac{-(1.0 \times 10^{-2}) +/- \sqrt{(1.0 \times 10^{-2})^2 - 4(1)(-1.0 \times 10^{-4})}}{2}$$

$x = 6.0 \times 10^{-3}$ M = $[H^+]$

(c) The simplifying assumption would be, for convenience, to set the value of the equilibrium concentration of HSO_4^- to be $(0.010 - x) \approx 0.010$ M.

Thus $x^2 = 1.0 \times 10^{-4}$ M, and $x = 1.0 \times 10^{-2}$ M, which is clearly a much different result than was obtained in part (b).

(d) % ionization = $6.0 \times 10^{-3}/0.010 \times 100 = 60$ %

(e) % ionization = $1.0 \times 10^{-2}/0.010 \times 100 = 100$ %

This example nicely illustrates the error that can be introduced into this sort of calculation when wrongly making a simplifying assumption, especially in systems where the equilibrium constant for ionization is large, i.e. where there is a high % ionization. In these cases, the value of x is not negligible, and ignoring it in the calculation leads to significant error. The full quadratic equation must be employed in order to solve for the value of x.

15.18

	HIO_4	$\rightleftharpoons$	H^+	+	IO_4^-
initial conc. (M)	0.10		0.0		0.0
change in conc. (M)	-0.038		0.038		0.038
equilibrium conc. (M)	0.10 - 0.038 (= 0.06)		0.038		0.038

Substituting the above values for equilibrium concentrations into the mass action expression gives:

$K_a = [H^+][IO_4^-]/[HIO_4] = (0.038)(0.038)/(0.06) = 2 \times 10^{-2}$

$pK_a = -\log K_a = -\log(2 \times 10^{-2}) = 1.7$

15.19 $[H^+] = \text{antilog}(-pH) = \text{antilog}(-1.96) = 0.011$ M

	$ClCH_2CO_2H$	$\rightleftharpoons$	$ClCH_2CO_2^-$	+	H^+
initial conc. (M)	0.10		0.0		0.0
change in conc. (M)	-0.011		0.011		0.011
equilibrium conc. (M)	0.09		0.011		0.011

Substituting the above values for equilibrium concentrations into the mass action expression gives:

$K_a = [ClCH_2CO_2^-][H^+]/[ClCH_2CO_2H] = (0.011)(0.011)/0.09 = 1 \times 10^{-3}$

15.20 $K_a = \text{antilog}(-K_a) = \text{antilog}(-4.92) = 1.2 \times 10^{-5}$

	HPaba	$\rightleftharpoons$	$Paba^-$	+	H^+
initial conc. (M)	0.030		0.0		0.0
change in conc. (M)	-x		x		x
equilibrium conc. (M)	0.030 - x (≈ 0.030)		x		x

Substituting the above values for equilibrium concentrations into the mass action expression gives:

$K_a = [H^+][Paba^-]/[HPaba] = x^2/0.030 = 1.2 \times 10^{-5}$

$x^2 = 3.6 \times 10^{-7}, \quad \therefore x = 6.0 \times 10^{-4} \text{ M} = [H^+]$

$pH = -\log[H^+] = -\log(6.0 \times 10^{-4}) = 3.22$

15.21 $K_a = \text{antilog}(-K_a) = \text{antilog}(-4.01) = 9.8 \times 10^{-5}$

	HBar	⇌	Bar^-	+	H^+
initial conc. (M)	0.050		0.0		0.0
change in conc. (M)	-x		x		x
equilibrium conc. (M)	0.050 - x (≈ 0.050)		x		x

Substituting the above values for equilibrium concentrations into the mass action expression gives:

$K_a = [H^+][Bar^-]/[HBar] = x^2/0.050 = 9.8 \times 10^{-5}$

$x^2 = 4.9 \times 10^{-6}, \quad \therefore x = 2.2 \times 10^{-3}\ M = [H^+]$

$pH = -\log[H^+] = -\log(2.2 \times 10^{-3}) = 2.66$

15.22 $pOH = 14.00 - pH = 14.00 - 11.86 = 2.14$

$[OH^-] = \text{antilog}(-pOH) = \text{antilog}(-2.14) = 7.2 \times 10^{-3}\ M$

	$CH_3CH_2NH_2$	+	H_2O	⇌	$CH_3CH_2NH_3^+$	+	OH^-
initial conc. (M)	0.10				0.0		0.0
change in conc. (M)	-7.2×10^{-3}				$+7.2 \times 10^{-3}$		$+7.2 \times 10^{-3}$
equilibrium conc. (M)	9×10^{-2}				7.2×10^{-3}		7.2×10^{-3}

Substituting the above values for equilibrium concentrations into the mass action expression gives:

$K_b = [OH^-][CH_3CH_2NH_3^+]/[CH_3CH_2NH_2] = (7.2 \times 10^{-3})(7.2 \times 10^{-3})/(9 \times 10^{-2})$

$K_b = 6 \times 10^{-4}$

$pK_b = -\log K_b = -\log(6 \times 10^{-4}) = 3.2$

For any conjugate acid-base pair, we have the general relationship:

$pK_a + pK_b = 14.00$

Therefore: pK_a for the acid $CH_3CH_2NH_3^+$ is:

$pK_a = 14.00 - 3.2 = 10.8$

15.23 $pOH = 14.00 - pH = 14.00 - 10.70 = 3.30$

$[OH^-] = \text{antilog}(-pOH) = \text{antilog}(-3.30) = 5.0 \times 10^{-4}\ M$

	N_2H_4 + H_2O	⇌	$N_2H_5^+$	+	OH^-
initial conc. (M)	0.15		0.0		0.0
change in conc. (M)	-5.0×10^{-4}		$+5.0 \times 10^{-4}$		$+5.0 \times 10^{-4}$
equilibrium conc. (M)	0.15		5.0×10^{-4}		5.0×10^{-4}

Substituting the above values for equilibrium concentrations into the mass action expression gives:

$K_b = [OH^-][N_2H_5^+]/[N_2H_4] = (5.0 \times 10^{-4})(5.0 \times 10^{-4})/(0.15)$

$K_b = 1.7 \times 10^{-6}$

$pK_b = -\log K_b = -\log(1.7 \times 10^{-6}) = 5.77$

For any conjugate acid-base pair, we have the general relationship:

$pK_a + pK_b = 14.00$

Therefore: pK_a for the acid $N_2H_5^+$ is:

$pK_a = 14.00 - 5.77 = 8.23$

15.24 $K_b = \text{antilog}(-K_b) = \text{antilog}(-5.79) = 1.6 \times 10^{-6}$

	Cod + H_2O	⇌	$CodH^+$	+	OH^-
initial conc. (M)	0.020		0.0		0.0
change in conc. (M)	-x		x		x
equilibrium conc. (M)	0.020 - x (≈ 0.020)		x		x

Substituting the above values for equilibrium concentrations into the mass action expression for K_b of codeine gives:

$K_b = [CodH^+][OH^-]/[Cod] = x^2/0.020 = 1.6 \times 10^{-6}$

$x^2 = 3.2 \times 10^{-8}$, $\therefore x = 1.8 \times 10^{-4}$ M $= [OH^-]$

$pOH = -\log[OH^-] = -\log(1.8 \times 10^{-4}) = 3.74$

$pH = 14.00 - pOH = 14.00 - 3.74 = 10.26$

15.25 K_b = antilog($-K_b$) = antilog(-8.04) = 9.1×10^{-9}

	NH_2OH + H_2O	⇌	NH_3OH^+	+	OH^-
initial conc. (M)	0.25		0.0		0.0
change in conc. (M)	-x		x		x
equilibrium conc. (M)	0.25 - x (≈ 0.25)		x		x

Substituting the above values for equilibrium concentrations into the mass action expression for K_b of hydroxyl amine gives:

$K_b = [NH_3OH^+][OH^-]/[NH_2OH] = x^2/0.25 = 9.1 \times 10^{-9}$

$x^2 = 2.3 \times 10^{-9}$, $\therefore x = 4.8 \times 10^{-5}$ M = $[OH^-]$

pOH = $-\log[OH^-] = -\log(4.8 \times 10^{-5}) = 4.32$

pH = 14.00 - pOH = 14.00 - 4.32 = 9.68

15.26 $pK_a + pK_b = 14.00$, $\therefore pK_a = 14.00 - 5.79 = 8.21$

Here the ionization of the parent base takes place according to the equation:

Mor + H_2O ⇌ H-Mor$^+$ + OH^-

and the ionization of the conjugate acid takes place according to the following equation:

H-Mor$^+$ ⇌ H^+ + Mor

and the value of K_a for the second of these equations is given by:

K_a = antilog($-pK_a$) = antilog(-8.21) = 6.2×10^{-9}

	H-Mor$^+$	⇌	H^+	+	Mor
initial conc. (M)	0.20		0.0		0.0
change in conc. (M)	-x		x		x
equilibrium conc. (M)	0.20 - x (≈ = 0.20)		x		x

Substituting the above values for equilibrium concentrations into the mass action expression for the ionization of H-Mor$^+$ gives:

$K_a = [H^+][Mor]/[H\text{-}Mor^+] = x^2/0.20 = 6.2 \times 10^{-9}$

$x = 3.5 \times 10^{-5}$ M = $[H^+]$

$pH = -\log[H^+] = -\log(3.5 \times 10^{-5}) = 4.46$

15.27 $pK_a + pK_b = 14.00$, $\therefore pK_a = 14.00 - 5.48 = 8.52$

Here the ionization of the parent base takes place according to the equation:

$Qu + H_2O \rightleftharpoons H\text{-}Qu^+ + OH^-$

and the ionization of the conjugate acid takes place according to the following equation:

$H\text{-}Qu^+ \rightleftharpoons H^+ + Qu$

and the value of K_a for the second of these equations is given by:

$K_a = \text{antilog}(-pK_a) = \text{antilog}(-8.52) = 3.0 \times 10^{-9}$

	$H\text{-}Qu^+$	$\rightleftharpoons$	H^+	+	Qu
initial conc. (M)	0.15		0.0		0.0
change in conc. (M)	-x		x		x
equilibrium conc. (M)	0.15 - x (≈ = 0.15)		x		x

Substituting the above values for equilibrium concentrations into the mass action expression for the ionization of $H\text{-}Qu^+$ gives:

$K_a = [H^+][Qu]/[H\text{-}Qu^+] = x^2/0.15 = 3.0 \times 10^{-9}$

$x = 2.1 \times 10^{-5}$ M = $[H^+]$

$pH = -\log[H^+] = -\log(2.1 \times 10^{-5}) = 4.68$

Buffers

15.28 (a) Added OH^- is consumed by the following equilibrium:

$H_2CO_3 + OH^- \rightleftharpoons H_2O + HCO_3^-$

Added H^+ is consumed by the following equilibrium:

$HCO_3^- + H^+ \rightleftharpoons H_2CO_3$

(b) Added OH^- is consumed by:

$H_2PO_4^- + OH^- \rightleftharpoons H_2O + HPO_4^{2-}$

Added H^+ is consumed by:

$HPO_4^{2-} + H^+ \rightleftharpoons H_2PO_4^-$

(c) Added OH^- is consumed by:

$NH_4^+ + OH^- \rightleftharpoons NH_3 + H_2O$

Added H^+ is consumed by:

$NH_3 + H^+ \rightleftharpoons NH_4^+$

15.29 We will proceed by demonstrating that one equation can be transformed to the other:

$pK_b = pK_w - pK_a$ such that the following equation:

$pOH = pK_b + \log([cation]/[base])$ becomes:

$pOH = pK_w - pK_a + \log([cation]/[base])$

Next, we replace pOH by its equivalent $pOH = pK_w - pH$:

$pK_w - pH = pK_w - pK_a + \log([cation]/[base])$ which simplifies to give:

$pH = pK_a - \log([cation]/[base])$

or

$pH = pK_a + \log([base]/[cation])$

15.30 First we consider the circumstance in which we have the general conjugate acid/base pair of the type BH^+ and B, where BH^+ is the acid and B is the base. Thus [base] means in general [conjugate base] and [cation] means in general [conjugate acid]. The equation of interest is:

$pH = pK_a + \log([base]/[cation])$

as derived in the answer to Review Exercise 15.29.

$pH = pK_a + \log([conjugate\ base]/[conjugate\ acid])$

where pK_a refers to the acid ionization equilibrium:

$BH^+ \rightleftharpoons H^+ + B$

Second, we consider the circumstance in which we have the general acid-base pair of the type HB and B^-, such that HB is the conjugate acid of B^-. The equation of interest is:

$pH = Pk_a + \log([anion]/[acid])$

Here [anion] means [conjugate base] and [acid] means [conjugate acid].

$pH = pK_a + \log([conjugate\ base]/[conjugate\ acid])$

Since this is the same result as that obtained above, the equations are both equivalent to the one given in the problem.

15.31 Buffer two has the greater concentration of acid neutralizing NH_3, and it is the buffer with the larger capacity.

15.32

$$pH = pK_a + \log\frac{[anion]}{[acid]} = pK_a + \log\frac{[A^-]}{[HA]}$$

$5.00 = 4.74 + \log([anion]/[acid])$

Taking the antilog of both sides of this equation gives:

$[anion]/[acid] = antilog(0.26) = 1.8$

Thus the ratio of the concentration of $NaC_2H_3O_2$ to the concentration of $HC_2H_3O_2$ should be 1.8.

$[C_2H_3O_2^-] = 1.8 \times [HC_2H_3O_2] = 1.8 \times (0.15\ M) = 0.27\ M$

Since this is to be 1 L of solution, we can use:

$0.27\ mol\ NaC_2H_3O_2 \times 82.0\ g/mol = 22\ g\ of\ NaC_2H_3O_2$

15.33

$$pH = pK_a + \log\frac{[anion]}{[acid]} = pK_a + \log\frac{[A^-]}{[HA]}$$

$3.80 = 3.75 + \log([NaCHO_2]/[HCHO_2])$

$[NaCHO_2]/[HCHO_2] = 1.1$

$[NaCHO_2] = 1.1 \times [HCHO_2] = 1.1 \times 0.12 = 0.13\ M$

Thus to the 1 L of formic acid solution we add:

$0.13 \text{ mol } NaCHO_2 \times 68.0 \text{ g/mol} = 8.8 \text{ g } NaCHO_2$

15.34

$$pOH = pK_b + \log \frac{[\text{cation}]}{[\text{base}]}$$

and we can determine the proper ratio of cation to base if we know the pOH:

$pOH = 14.00 - pH = 14.00 - 9.26 = 4.74$

$4.74 = 4.74 + \log([NH_4^+]/[NH_3])$

or, $\log([NH_4^+]/[NH_3]) = 0$

Taking the antilog of both sides of the equation gives:

$[NH_4^+]/[NH_3] = 1.0$

This is a general result. A ratio of base to the conjugate acid of 1.0 gives a solution whose pH is equal to pK_b for the conjugate base.

15.35

$$pOH = pK_b + \log \frac{[\text{cation}]}{[\text{base}]}$$

Now $pOH = 14.00 - pH = 14.00 - 10.00 = 4.00$

$4.00 = 4.74 + \log([NH_4^+]/[NH_3])$

$\log([NH_4^+]/[NH_3]) = -0.74$

Taking the antilog of both sides of this equation gives:

$[NH_4^+]/[NH_3] = \text{antilog}(-0.74) = 0.18$

Thus $[NH_4^+] = 0.18 \times [NH_3] = 0.18 \times 0.20 \text{ M} = 3.6 \times 10^{-2} \text{ M} = [NH_4Cl]$

Finally, 0.500 L of buffer solution would require:

$0.500 \text{ L} \times 3.6 \times 10^{-2} \text{ mol/L} \times 53.5 \text{ g/mol} = 0.96 \text{ g } NH_4Cl$

15.36 The approach to this sort of problem is outlined on pages 627 - 628.

(a) The equation that will be used is the normal Henderson-Hasselbach equation, namely:

$$pH = pK_a + \log \frac{[\text{anion}]}{[\text{acid}]} = pK_a + \log \frac{[A^-]}{[HA]}$$

where $A^- = C_2H_3O_2^-$ and $HA = HC_2H_3O_2$. We note further that the log term involves a ratio of concentrations, but that the volume remains constant in a process such as that to be analyzed here. Thus the log term may be replaced by a ratio of mole amounts, since volumes cancel:

$$pH = pK_a + \log (\text{moles } C_2H_3O_2^-)/(\text{moles } HC_2H_3O_2)$$

Thus we need only determine the number of moles of acid and conjugate base that remain in the buffer after the addition of a certain amount of H^+ or OH^-, in order to determine the pH of the buffer mixture after that addition.

The buffer is changed in the following way by the addition of OH^-:

$$HC_2H_3O_2 + OH^- \rightleftharpoons H_2O + C_2H_3O_2^-$$

In other words, if 0.0100 moles of OH^- are added to the buffer, the amount of $HC_2H_3O_2$ goes down by 0.0100 moles, whereas the amount of $C_2H_3O_2^-$ goes up by 0.0100 moles.

The buffer is changed in the following way by the addition of H^+:

$$C_2H_3O_2^- + H^+ \rightleftharpoons HC_2H_3O_2$$

If 0.0100 mol of H^+ are added, then the amount of $C_2H_3O_2^-$ goes down by 0.0100 moles and the amount of $HC_2H_3O_2$ goes up by 0.0100 moles.

For the general buffer mixture that contains x mol of $C_2H_3O_2^-$ and y mol of $HC_2H_3O_2$, we can apply the Henderson-Hasselbach equation, noting the maximum amount by which we want the pH to change (namely 0.10 units):

For the case of added base:

$$(5.12 + 0.10) = 4.74 + \log \frac{[x + 0.0100] \text{ mol } C_2H_3O_2^-}{[y - 0.0100] \text{ mol } HC_2H_3O_2}$$

Taking the antilog of both sides of the above equation gives:

$[x + 0.0100]/[y - 0.0100] = 3.0$, which is now designated eq. 1.

For the case of added acid:

$$(5.12 + 0.10) = 4.74 + \log \frac{[x - 0.0100] \text{ mol } C_2H_3O_2^-}{[y + 0.0100] \text{ mol } HC_2H_3O_2}$$

Taking the antilog of both sides of the equation gives:

[x - 0.0100]/[y + 0.0100] = 1.9, which is now designated eq. 2.

The equations 1 and 2 as designated above are solved simultaneously, since they are two equations containing two unknowns, giving:

x = initial mol $C_2H_3O_2^-$ = 0.15 mol

y = initial mol $HC_2H_3O_2$ = 6.3×10^{-2} mol

These values are converted to grams as follows:

6.3×10^{-2} mol × 60.1 g/mol = 3.8 g $HC_2H_3O_2$

0.15 mol $NaC_2H_3O_2$ × 118 g/mol = 18 g $NaC_2H_3O_2 \cdot 2H_2O$

These are the minimum amounts of acid and conjugate base that would be required in order to prepare a buffer that would change pH by only 0.10 units on addition of either 0.0100 mol of OH^- or 0.0100 mol of H^+.

(b) 6.3×10^{-2} mol/0.250 L = 0.25 M $HC_2H_3O_2$

0.15 mol $C_2H_3O_2^-$/0.250 L = 0.60 M $NaC_2H_3O_2$

(c) 0.010 mol OH^-/0.250 L = 4.0×10^{-2} M OH^-

pOH = $-\log[OH^-]$ = $-\log(4.0 \times 10^{-2})$ = 1.40

pH = 14.00 - pOH = 14.00 - 1.40 = 12.60

(d) 1.0×10^{-2} mol H^+/0.250 L = 4.0×10^{-2} M H^+

pH = $-\log[H^+]$ = $-\log(4.0 \times 10^{-2})$ = 1.40

15.37 (a) We pick the acid whose pK_a value is closest to the desired pH, here formic acid, pK_a = 3.74. The buffer would then contain also the sodium salt of the conjugate base, i.e. $NaHCO_2$.

(b) We follow the strategy developed in Review Exercise 15.36:

Let x = initial mol HCO_2^- and y = initial mol HCO_2H

For the case of added base we have:

$$(3.80 + 0.050) = 3.74 + \log \frac{[x + 0.0050]\ \text{mol } HCO_2^-}{[y - 0.0050]\ \text{mol } HCO_2H}$$

Taking the antilog of both sides of the above equation gives:

$[x + 0.0050]/[y - 0.0050] = 1.29$, which is now designated eq. 1.

For the case of added acid:

$$(3.80 + 0.050) = 3.74 + \log \frac{[x - 0.0050]\ \text{mol } HCO_2^-}{[y + 0.0050]\ \text{mol } HCO_2H}$$

Taking the antilog of both sides of the equation gives:

$[x - 0.0050]/[y + 0.0050] = 1.02$, which is now designated eq. 2.

The equations 1 and 2 as designated above are solved simultaneously, since they are two equations containing two unknowns, giving:

x = initial mol HCO_2^- = 9.2×10^{-2} mol

y = initial mol HCO_2H = 8.0×10^{-2} mol

These values are converted to grams as follows:

9.2×10^{-2} mol $NaHCO_2$ × 68.0 g/mol = 6.3 g $NaHCO_2$

8.0×10^{-2} mol HCO_2H × 46.0 g/mol = 3.7 g HCO_2H

(c) 9.2×10^{-2} mol/0.750 L = 0.12 M $NaHCO_2$

8.0×10^{-2} M HCO_2H/0.750 L = 0.11 M HCO_2H

(d) The $[H^+]$ in the original solution is:

$[H^+]$ = antilog(-pH) = antilog(-3.80) = 1.6×10^{-4} M

The number of moles of H^+ in the original mixture is thus:

1.6×10^{-4} M × 0.750 L = 1.2×10^{-4} mol H^+

The total number of moles after the addition is thus:

1.2×10^{-4} mol + 5.0×10^{-3} mol = 5.1×10^{-3} mol

and the concentration is:

$[H^+] = 5.1 \times 10^{-3}$ mol/0.750 L = 6.8×10^{-3} M

$pH = -\log[H^+] = -\log(6.8 \times 10^{-3}) = 2.17$

(e) The $[H^+]$ in the original solution is:

$[H^+] = \text{antilog}(-pH) = \text{antilog}(-3.80) = 1.6 \times 10^{-4}$ M

The number of moles of H^+ in the original mixture is thus:

1.6×10^{-4} M $\times$ 0.750 L = 1.2×10^{-4} mol H^+

After the addition of 5.0×10^{-3} mol of OH^-, the H^+ will be consumed completely, leaving $5.0 \times 10^{-3} - 1.2 \times 10^{-4} = 4.9 \times 10^{-3}$ mol of OH^-.

The concentration of OH^- is thus:

4.9×10^{-3} mol/0.750 L = 6.5×10^{-3} M OH^-

$pOH = -\log[OH^-] = -\log(6.5 \times 10^{-3}) = 2.19$

$pH = 14.00 - pOH = 14.00 - 2.19 = 11.81$

Polyprotic Acids

15.38 (a) The value of K_{a_1} (1.2×10^{-2}) is neither very large nor very small, and H_2SO_3 should accordingly be classified among the acids of moderate strength.

(b) $K_{a_1} = [H^+][HSO_3^-]/[H_2SO_3]$

$K_{a_2} = [H^+][SO_3^{2-}]/[HSO_3^-]$

(c) $H_2CO_3 \rightleftharpoons H^+ + HCO_3^-$

$K_{a_1} = [H^+][HCO_3^-]/[H_2CO_3]$

$HCO_3^- \rightleftharpoons H^+ + CO_3^{2-}$

$K_{a_2} = [H^+][CO_3^{2-}]/[HCO_3^-]$

(d) K_{a_1} for H_2CO_3 is 4.5×10^{-7}

K_{a_1} for H_2SO_3 is 1.2×10^{-2}

This means that H_2SO_3 is a stronger acid than is H_2CO_3.

K_{a_2} for H_2CO_3 is 4.7×10^{-11}

K_{a_2} for H_2SO_3 is 6.6×10^{-8}

This means that HCO_3^- is a weaker acid than HSO_3^-.

15.39 $H_2O \rightleftharpoons H^+ + OH^-$

$Ka_1 = [H^+][OH^-]/[H_2O]$

In using the constant K_w, we treat the molar concentration of H_2O as a constant that is incorporated into the numerical value of K_w. That is,

$K_w = K_{a_1} \times [H_2O]$

$OH^- \rightleftharpoons H^+ + O^{2-}$

$K_{a_2} = [H^+][O^{2-}]/[OH^-]$

15.40 Let x be the change in concentration due to the first ionization equilibrium and y be the change in concentration due to the second ionization equilibrium:

	H_3PO_3	$\rightleftharpoons$	$H_2PO_3^-$	+	H^+
initial conc. (M)	1.0		0.0		0.0
change in conc. (M)	-x		+ x - y		+ x + y
equilibrium conc. (M)	1.0 - x		x - y ($\approx$ x)		x + y ($\approx$ x)

Substituting the above values for equilibrium concentrations into the mass action expression that applies for K_{a_1} of phosphoric acid gives:

$$K_{a_1} = \frac{[H^+][H_2PO_3^-]}{[H_3PO_3]} = \frac{(x)(x)}{1.0 - x} = 1.0 \times 10^{-2}$$

$x^2 = 0.010 - (0.010)x$ or, in proper quadratic form,

$x^2 + (0.010)x - (0.010) = 0$ On solving for x we get:

$x = 0.095\ M = [H^+] = [H_2PO_3^-]$

$pH = -\log[H^+] = -\log(0.095) = 1.02$

The value of $[HPO_3^{2-}]$ is obtained from the mass action expression for the equilibrium:

$H_2PO_3^- \rightleftharpoons HPO_3^{2-} + H^+$

$K_{a_2} = [HPO_3^{2-}][H^+]/[H_2PO_3^-] = (x)[HPO_3^{2-}]/(x) = [HPO_3^{2-}]$

$[HPO_3^{2-}] = K_{a_2} = 2.6 \times 10^{-7}$

15.41 We proceed as in Review Exercise 15.40.

	H_6TeO_6	$\rightleftharpoons$ H^+	+ $H_5TeO_6^-$
initial conc. (M)	0.25	0.0	0.0
change in conc. (M)	-x	+ x + y	+ x - y
equilibrium conc. (M)	0.25 - x (≈ 0.25)	x + y (≈ x)	x - y (≈ x)

Substituting the above values for equilibrium concentrations into the mass action expression gives:

$$K_{a_1} = \frac{[H^+][H_5TeO_6^-]}{[H_6TeO_6]} = \frac{(x)(x)}{0.25} = 2.1 \times 10^{-8}$$

$x^2 = 5.3 \times 10^{-9}$ $\therefore$ $x = 7.2 \times 10^{-5} = [H^+] = [H_5TeO_6^-]$

$pH = -\log[H^+] = -\log(7.2 \times 10^{-5}) = 4.14$

$[H_4TeO_6^{2-}] = K_{a_2} = 6.5 \times 10^{-12}$ M

Hydrolysis of Ions

15.42 Since aspirin is a weak acid, it follows that its anion is a weak base which hydrolyzes in water solution according to the general reaction of any weak base A^-:

$A^- + H_2O \rightleftharpoons HA + OH^-$

15.43 The oxalate ion hydrolyzes in water according to the equation:

$C_2O_4^{2-} + H_2O \rightleftharpoons HC_2O_4^- + OH^-$

15.44 (a) Acidic solutions are formed by $(NH_4)_2SO_4$ and $HC_2H_3O_2$.

(b) Basic solutions are formed by KCN and KF.

(c) Neutral solutions are formed by NaI, $CsNO_3$, and KBr.

15.45

	H_2O +	CN^-	⇌	HCN	+	OH^-
initial conc. (M)		0.20		0.0		0.0
change in conc. (M)		-x		x		x
equilibrium conc. (M)		0.20 - x (≈ 0.20)		x		x

Substituting the above values for equilibrium concentrations into the mass action expression gives:

$K_b = 1.6 \times 10^{-5} = [HCN][OH^-]/[CN^-] = x^2/(0.20)$

$x^2 = 3.2 \times 10^{-6}$ $\therefore x = 1.8 \times 10^{-3}$ M $= [OH^-]$

$pOH = -\log[OH^-] = -\log(1.8 \times 10^{-3}) = 2.74$

$pH = 14.00 - pOH = 14.00 - 2.74 = 11.26$

15.46

	H_2O +	NO_2^-	⇌	HNO_2	+	OH^-
initial conc. (M)		0.10		0.0		0.0
change in conc. (M)		-x		x		x
equilibrium conc. (M)		0.10 - x (≈ 0.10)		x		x

Substituting the above values for equilibrium concentrations into the mass action expression gives:

$K_b = x^2/0.10 = 1.4 \times 10^{-11}$

$x^2 = 1.4 \times 10^{-12}$ $\therefore x = 1.2 \times 10^{-6} = [OH^-]$

$pOH = -\log[OH^-] = -\log(1.2 \times 10^{-6}) = 5.92$

$pH = 14.00 - pOH = 14.00 - 5.92 = 8.08$

15.47

	H_2O +	S^{2-}	⇌	HS^-	+	OH^-
initial conc. (M)		.15		0.0		0.0
change in conc. (M)		-x		x		x
equilibrium conc. (M)		0.15 - x		x		x

Substituting the above values for equilibrium concentrations into the mass action expression gives:

$K_b = [HS^-][OH^-]/[S^{2-}] = 7.7 \times 10^{-1} = x^2/(0.15 - x)$

$x^2 = 0.116 - 0.77(x)$ or $x^2 + 0.77x - 0.116 = 0$

Solving for x using the quadratic formula we get:

$x = [OH^-] = 0.13$ M

$pOH = -\log[OH^-] = -\log(0.13) = 0.89$

$pH = 14.00 - pOH = 14.00 - 0.89 = 13.11$

15.48

	H_2O +	CO_3^{2-} ⇌	HCO_3^- +	OH^-
initial conc. (M)		0.10	0.0	0.0
change in conc. (M)		-x	x	x
equilibrium conc. (M)		0.10 - x (≈ 0.10)	x	x

Substituting the above values for equilibrium concentrations into the mass action expression gives:

$K_b = [HCO_3^-][OH^-]/[CO_3^{2-}] = 2.1 \times 10^{-4} = x^2/(0.10)$

$x^2 = 2.1 \times 10^{-5}$ $\therefore x = 4.6 \times 10^{-3}$ M $= [OH^-]$

$pOH = -\log[OH^-] = -\log(4.6 \times 10^{-3}) = 2.34$

$pH = 14.00 - pOH = 14.00 - 2.34 = 11.66$

15.49 Only the first hydrolysis equilibrium needs to be considered:

$CO_3^{2-} + H_2O \rightleftharpoons HCO_3^- + OH^-$

$K_b = [HCO_3^-][OH^-]/[CO_3^{2-}] = 2.1 \times 10^{-4}$

$pK_w = pH + pOH = 14.0 = 14.00 - 11.70 = 2.30$

$[OH^-] = \text{antilog}(-pOH) = \text{antilog}(-2.30) = 5.0 \times 10^{-3}$ M

Since OH^- and HCO_3^- are formed on an equimolar basis, then we have also found that:

$[OH^-] = [HCO_3^-] = 5.0 \times 10^{-3}$ M

Substituting these values into the above K_a expression gives:

$(5.0 \times 10^{-3})(5.0 \times 10^{-3})/[CO_3^{2-}] = 2.1 \times 10^{-4}$

and $[CO_3^{2-}] = 0.12$ M

0.12 mol/L × 286 g/mol = 34 g/L of $Na_2CO_3 \cdot 10H_2O$ will be required.

15.50 The only hydrolysis that we need to consider is:

$NH_4^+ \rightleftharpoons NH_3 + H^+$

if pH = 5.15, then $[H^+]$ = antilog(-pH) - antilog(-5.15) = 7.1×10^{-6} M

This is also the required value for $[NH_3]$, since they are formed in a one to one mole ratio.

$K_a = [H^+][NH_3]/[NH_4^+] = 5.7 \times 10^{-10}$ and we arrive at:

$5.7 \times 10^{-10} = (7.1 \times 10^{-6})(7.1 \times 10^{-6})/[NH_4^+]$

$[NH_4^+] = 8.8 \times 10^{-2}$ M

8.8×10^{-2} mol/L × 98.0 g/mol = 8.6 g NH_4Br are needed per liter.

15.51 Only the first stage of hydrolysis needs to be considered:

$SO_3^{2-} + H_2O \rightleftharpoons HSO_3^- + OH^-$

pOH = 14.00 - pH = 14.00 - 10.00 = 4.00

$[OH^-]$ = antilog(-pOH) = antilog(-4.00) = 1.0×10^{-4} M

1.0×10^{-4} M is also equal to $[HSO_3^-]$

Thus $K_b = [HSO_3^-][OH^-]/[SO_3^{2-}] = 1.5 \times 10^{-7} = (1.0 \times 10^{-4})^2/[SO_3^{2-}]$

$[SO_3^{2-}] = 6.7 \times 10^{-2}$ M

6.7×10^{-2} mol/L × 252 g/mol = 17 g $Na_2SO_3 \cdot 7H_2O$ are needed.

Acid-Base Titrations

15.52 A pH other than 7 at the equivalence point can result from the hydrolysis of the salt that forms as a result of the titration. That is, the titration may form an ion that is normally hydrolyzed by water. This requires the selection of an indicator that changes color at the pH of the equivalence point.

15.53 The salt that forms as a result of the titration is NH_4Cl, which normally hydrolyzes to give a slightly acidic solution. Ethyl red is a better choice for an indicator, since its color change occurs at a pH less than 7.

15.54 The salt that forms as a result of this titration (KBr) is not hydrolyzed in water, and we choose the indicator that changes color closer to pH of 7.

15.55 The number of equivalents of acid is equal to the number of equivalents of base.

15.56 Since HCO_2H and NaOH both deliver 1 equivalent per mole:

$HCO_2H + NaOH \rightarrow NaHCO_2 + H_2O$

we can use the equation $V_a \times M_a = V_b \times M_b$ to determine the volume of NaOH

that is required to reach the equivalence point, i.e. the point at which the number of moles of NaOH is equal to the number of moles of HCO_2H:

$V_{NaOH} = 50 \text{ mL} \times 0.10/0.10 = 50 \text{ mL}$

Thus the final volume at the equivalence point will be 50 + 50 = 100 mL.

The concentration of $NaHCO_2$ would then be:

$0.10 \text{ mol/L} \times 0.050 \text{ L} = 5.0 \times 10^{-3} \text{ mol } HCO_2H = 5.0 \times 10^{-3} \text{ mol } NaHCO_2$

$5.0 \times 10^{-3} \text{ mol}/0.100 \text{ L} = 5.0 \times 10^{-2} \text{ M } NaHCO_2$

The hydrolysis of this salt at the equivalence point proceeds according to the following equilibrium:

	H_2O +	HCO_2^-	⇌	HCO_2H	+	OH^-
initial conc. (M)		0.050		0.0		0.0
change in conc. (M)		-x		x		x
equilibrium conc. (M)		0.050 - x (≈ 0.050)		x		x

Substituting the above values for equilibrium concentrations into the mass action expression gives:

$K_b = [HCO_2H][OH^-]/[HCO_2^-] = 5.6 \times 10^{-11} = x^2/(0.050)$

$x^2 = 2.8 \times 10^{-12} \quad \therefore x = 1.7 \times 10^{-6} \text{ M} = [OH^-] = [HCO_2H]$

$pOH = -\log[OH^-] = -\log(1.7 \times 10^{-6}) = 5.77$

pH = 14.00 - pOH = 14.00 - 5.77 = 8.23

Cresol red would be a good indicator, since it has a color change near the pH at the equivalence point.

15.57 $NH_3 + HBr \rightarrow NH_4Br$

Both NH_3 and HBr have 1 eq/mol and we use the relationship

$V_a \times M_a = V_b \times M_b$

to find the volume of HBr that is needed to reach the equivalence point:

$V_{HBr} = 25 \text{ mL} \times 0.10/0.10 = 25 \text{ mL}$

and the final volume will be 25 + 25 = 50 mL

The concentration of NH_4Br that is formed is:

$0.035 \text{ L} \times 0.10 \text{ mol/L} = 2.5 \times 10^{-3} \text{ mol } NH_4Br$

$2.5 \times 10^{-3} \text{ mol} \div 0.050 \text{ L} = 5.0 \times 10^{-2} \text{ M } NH_4Br$

The pH at the equivalence is, thus, that of a 0.050 M NH_4^+ solution:

	NH_4^+	⇌	NH_3	+	H^+
initial conc. (M)	0.050		0.0		0.0
change in conc. (M)	-x		x		x
equilibrium conc. (M)	0.050 - x (≈ 0.050)		x		x

Substituting the above values for equilibrium concentrations into the mass action expression gives:

$K_a = [H^+][NH_3]/[NH_4^+] = 5.7 \times 10^{-10} = x^2/(0.050)$

$x^2 = 2.9 \times 10^{-11} \quad \therefore \quad x = 5.3 \times 10^{-6} \text{ M} = [H^+] = [NH_3]$

$pH = -\log[H^+] = -\log(5.3 \times 10^{-6}) = 5.28$

A good indicator would therefore have a color change as close to pH = 5.28 as possible; ethyl red would be a good choice.

15.58 This is the titration of a strong acid by a strong base, each of which has 1 eq/mol. The salt that is formed is not hydrolyzed, and we expect that the equivalence point will have a pH of 7.0. The initial number of moles of acid is:

$0.1000\ mol/L \times 0.02500\ L = 2.500 \times 10^{-3}\ mol\ H^+ = mol\ HCl$

At each point up to, but not past, the equivalence point, we calculate the number of moles of OH^- that have been added (M × V), and subtract this number from the original number of moles of HCl to determine the remaining or excess number of moles of H^+. This number of moles is divided by the total volume at that point in the titration to give the $[H^+]$, which may then be converted to pH in the normal fashion.
At each point in the titration beyond the equivalence point, the number of moles of OH^- that have been added is greater than the original number of moles of H^+, and the solution is made basic by the amount of excess OH^-. The pH is determined by the amount of excess OH^-, and the $[OH^-]$ is given by the number of moles in excess of H^+, divided by the total volume that has accumulated.

(a) After adding 0 mL of 0.1000 N NaOH:

$[H^+] = 0.1000\ M$ and $pH = -\log[H^+] = -\log(0.1000) = 1.0000$

(b) After adding 10.00 mL of 0.1000 N NaOH:

added mol of $OH^- = 0.01000\ L \times 0.1000\ mol/L = 1.000 \times 10^{-3}\ mol$

remaining mol of $H^+ = 2.500 \times 10^{-3} - 1.000 \times 10^{-3} = 1.500 \times 10^{-3}\ mol$

$[H^+] = 1.500 \times 10^{-3}\ mol/0.03500\ L = 4.286 \times 10^{-2}\ M$

$pH = -\log[H^+] = -\log(4.286 \times 10^{-2}) = 1.3679$

(c) After adding 24.90 mL of 0.1000 N NaOH:

added mol of $OH^- = 0.02490\ L \times 0.1000\ mol/L = 2.490 \times 10^{-3}\ mol$

remaining mol of $H^+ = 2.500 \times 10^{-3} - 2.490 \times 10^{-3} = 1.0 \times 10^{-5}\ mol$

$[H^+] = 1.0 \times 10^{-5}\ mol/0.04990\ L = 2.0 \times 10^{-4}\ M$

$pH = -\log[H^+] = -\log(2.0 \times 10^{-4}) = 3.70$

(d) After adding 24.99 mL of 0.1000 N NaOH:

added mol of $OH^- = 0.02499\ L \times 0.1000\ mol/L = 2.499 \times 10^{-3}\ mol$

remaining mol of $H^+ = 2.500 \times 10^{-3} - 2.499 \times 10^{-3} = 1.0 \times 10^{-6}\ mol$

$[H^+] = 1.0 \times 10^{-6}\ mol/0.04999\ L = 2.0 \times 10^{-5}\ M$

$pH = -\log[H^+] = -\log(2.0 \times 10^{-5}) = 4.7$

(The $[H^+]$ that arises from the self-ionization of water is assumed to be negligibly small.)

(e) This is the equivalence point. Since the salt of a strong acid and a strong base will not hydrolyze, the pH is 7.0.

(f) After adding 25.01 mL of 0.1000 N NaOH:

added mol of $OH^- = 0.02501 \text{ L} \times 0.1000 \text{ mol/L} = 2.501 \times 10^{-3}$ mol

excess mol of $OH^- = 2.501 \times 10^{-3} - 2.500 \times 10^{-3} = 1 \times 10^{-6}$ mol

$[OH^-] = 1 \times 10^{-6} \text{ mol}/0.05001 \text{ L} = 2 \times 10^{-5}$ M

$pOH = -\log[OH^-] = -\log(2 \times 10^{-5}) = 4.7$

$pH = 14.00 - pOH = 14.00 - 4.7 = 9.3$

(g) After adding 25.10 mL of 0.1000 N NaOH:

added mol of $OH^- = 0.02510 \text{ L} \times 0.1000 \text{ mol/L} = 2.510 \times 10^{-3}$ mol

excess mol of $OH^- = 2.510 \times 10^{-3} - 2.500 \times 10^{-3} = 1.0 \times 10^{-5}$ mol

$[OH^-] = 1.0 \times 10^{-5} \text{ mol}/0.05010 \text{ L} = 2.0 \times 10^{-4}$ M

$pOH = -\log[OH^-] = -\log(2.0 \times 10^{-4}) = 3.70$

$pH = 14.00 - pOH = 14.00 - 3.70 = 10.30$

(h) After adding 26.00 mL of 0.1000 N NaOH:

added mol of $OH^- = 0.02510 \text{ L} \times 0.1000 \text{ mol/L} = 2.600 \times 10^{-3}$ mol

excess mol of $OH^- = 2.600 \times 10^{-3} - 2.500 \times 10^{-3} = 1.00 \times 10^{-4}$ mol

$[OH^-] = 1.00 \times 10^{-4} \text{ mol}/0.05100 \text{ L} = 1.96 \times 10^{-3}$ M

$pOH = -\log[OH^-] = -\log(1.96 \times 10^{-3}) = 2.708$

$pH = 14.00 - pOH = 14.00 - 2.708 = 11.292$

(i) After adding 50.00 mL of 0.1000 N NaOH:

added mol of $OH^- = 0.05000 \text{ L} \times 0.1000 \text{ mol/L} = 5.000 \times 10^{-3}$ mol

excess mol of $OH^- = 5.000 \times 10^{-3} - 2.500 \times 10^{-3} = 2.500 \times 10^{-3}$ mol

$[OH^-] = 2.500 \times 10^{-3}$ mol/0.07500 L = 3.333×10^{-2} M

$pOH = -\log[OH^-] = -\log(3.333 \times 10^{-2}) = 1.4772$

$pH = 14.00 - pOH = 14.00 - 1.4772 = 12.5228$

15.59 This is different from Review Exercise 15.58 in that we have here the titration of a weak acid with a strong base, and the salt that is formed ($NaC_2H_3O_2$) is hydrolyzed to give a slightly basic solution at the equivalence point. We have the same number of moles of OH^- added at each stage in the titration as in Review Exercise 15.58, as well as the same total volumes:

(a) Mol of NaOH added = 0 — total volume = 25.00 mL
(b) Mol of NaOH added = 1.000×10^{-3} — total volume = 35.00 mL
(c) Mol of NaOH added = 2.490×10^{-3} — total volume = 49.90 mL
(d) Mol of NaOH added = 2.499×10^{-3} — total volume = 49.99 mL
(e) Mol of NaOH added = 2.500×10^{-3} — total volume = 50.00 mL
(f) Mol of NaOH added = 2.501×10^{-3} — total volume = 50.01 mL
(g) Mol of NaOH added = 2.510×10^{-3} — total volume = 50.10 mL
(h) Mol of NaOH added = 2.600×10^{-3} — total volume = 51.00 mL
(i) Mol of NaOH added = 5.000×10^{-3} — total volume = 75.00 mL

The calculation of pH at each stage in the titration is then accomplished as follows:

(a) Before the addition of OH^-, we simply have the solution of a weak acid in water:

	$HC_2H_3O_2$	$\rightleftharpoons$	H^+	+	$C_2H_3O_2^-$
initial conc. (M)	0.1000		0.0		0.0
change in conc. (M)	-x		x		x
equilibrium conc. (M)	0.1000 - x (≈ 0.1000)		x		x

Substituting the above values for equilibrium concentrations into the mass action expression gives:

$K_a = 1.8 \times 10^{-5} = x^2/(0.1000)$

$x^2 = 1.8 \times 10^{-6} \quad \therefore x = 1.3 \times 10^{-3} = [H^+]$

$pH = -\log[H^+] = -\log(1.3 \times 10^{-3}) = 2.89$

(b) The amount of OH^- that is added (1.000×10^{-3} mol) is equal to the amount of $C_2H_3O_2^-$ that is formed. The amount of $HC_2H_3O_2$ that remains is:

$2.500 \times 10^{-3} - 1.000 \times 10^{-3} = 1.500 \times 10^{-3}$ mol $HC_2H_3O_2$

$$pH = pK_a + \log \frac{[\text{anion}]}{[\text{acid}]} = pK_a + \log \frac{[C_2H_3O_2^-]}{[HC_2H_3O_2]}$$

Since the log term in the above equation is a ratio of two molar quantities, and since the volume amounts are the same and therefore cancel, we can simply use mole amounts in the log term:

$pH = 4.74 + \log(1.000 \times 10^{-3})/(1.500 \times 10^{-3}) = 4.56$

(c) The amount of OH^- that is added (2.490×10^{-3} mol) is equal to the amount of $C_2H_3O_2^-$ that is formed. The amount of $HC_2H_3O_2$ that remains is:

$2.500 \times 10^{-3} - 2.490 \times 10^{-3} = 1.0 \times 10^{-5}$ mol $HC_2H_3O_2$

$$pH = pK_a + \log \frac{[\text{anion}]}{[\text{acid}]} = pK_a + \log \frac{[C_2H_3O_2^-]}{[HC_2H_3O_2]}$$

Since the log term in the above equation is a ratio of two molar quantities, and since the volume amounts are the same and therefore cancel, we can simply use mole amounts in the log term:

$pH = 4.74 + \log(2.490 \times 10^{-3})/(1.0 \times 10^{-5}) = 7.14$

(d) The amount of OH^- that is added (2.499×10^{-3} mol) is equal to the amount of $C_2H_3O_2^-$ that is formed. The amount of $HC_2H_3O_2$ that remains is:

$2.500 \times 10^{-3} - 2.499 \times 10^{-3} = 1 \times 10^{-6}$ mol $HC_2H_3O_2$

$$pH = pK_a + \log \frac{[\text{anion}]}{[\text{acid}]} = pK_a + \log \frac{[C_2H_3O_2^-]}{[HC_2H_3O_2]}$$

Since the log term in the above equation is a ratio of two molar quantities, and since the volume amounts are the same and therefore cancel, we can simply use mole amounts in the log term:

$pH = 4.74 + \log(2.499 \times 10^{-3})/(1 \times 10^{-6}) = 8.1$

(e) At the equivalence point, we have a solution that contains:

2.50×10^{-3} mol/0.0500 L = 0.05000 M $C_2H_3O_2^-$

	H_2O + $C_2H_3O_2^-$	⇌ $HC_2H_3O_2$	+ OH^-
initial conc. (M)	0.05000	0.0	0.0
change in conc. (M)	-x	x	x
equilibrium conc. (M)	0.05000 - x (≈ 0.05000)	x	x

Substituting the above values for equilibrium concentrations into the mass action expression derived from K_b gives:

$K_b = 5.7 \times 10^{-10} = x^2/0.05000$

$x = 5.3 \times 10^{-6}$ M = $[OH^-]$

$pOH = -\log[OH^-] = -\log(5.3 \times 10^{-6}) = 5.28$

$pH = 14.00 - pOH = 14.00 - 5.28 = 8.72$

(f) - (i) We can assume that the pH of the solutions following the equivalence point is governed predominantly by the amount of excess $[OH^-]$. Thus the pH values here are close to those already determined in Review Exercise 15.58:

(f) 9 (g) 10.30 (h) 11.292 (i) 12.5228

15.60 This is the titration of a weak base with a strong acid, i.e. the converse of Review Exercise 15.59. At first, we have a simple solution of a weak base in water:

$NH_3 + H_2O \rightleftharpoons NH_4^+ + OH^-$

At each point up to but not beyond the equivalence point we calculate the number of moles of H^+ that have been added, and subtract this from the total original number of moles of NH_3 to get the number of moles of NH_3 that remain. Also, the number of moles of NH_4^+ that are formed at any point in the titration (up to but not beyond the equivalence point) is equal to the number of moles of H^+ that have been added. Beyond the equivalence point, we assume that the pH is governed principally by the amount of excess acid that has been added, and pH can be calculated directly from the value for $[H^+]$.

The initial number of moles of NH_3 is:

$0.1000 \text{ mol/L} \times 0.02500 \text{ L} = 2.5 \times 10^{-3}$ mol

(a)

	H_2O + NH_3	⇌	NH_4^+	+	OH^-
initial conc. (M)	0.1000		0.0		0.0
change in conc. (M)	-x		x		x
equilibrium conc. (M)	0.1000 - x (≈ 0.1000)		x		x

Substituting the above values for equilibrium concentrations into the mass action expression gives:

$K_b = 1.8 \times 10^{-5} = x^2/(0.1000)$ ∴ $x = 1.3 \times 10^{-3}$ M = $[OH^-]$

$pOH = -\log[OH^-] = -\log(1.3 \times 10^{-3}) = 2.89$

$pH = 14.00 - pOH = 14.00 - 2.89 = 11.11$

(b) The number of moles of H^+ that have been added is:

$0.1000 \text{ mol/L} \times 0.01000 \text{ L} = 1.000 \times 10^{-3}$ mol H^+

which is also the number of moles of NH_4^+ that have been formed.

The number of moles of NH_3 that remain are thus:

$2.500 \times 10^{-3} - 1.000 \times 10^{-3} = 1.500 \times 10^{-3}$ mol NH_3

$$pOH = pK_b + \log \frac{[\text{cation}]}{[\text{base}]}$$

In the above log term, the ratio of molar concentrations can be replaced by a ratio of mole amounts, since the volumes are equal and cancel:

$pOH = 4.74 + \log(1.000 \times 10^{-3})/(1.500 \times 10^{-3}) = 4.56$

$pH = 14.00 - pOH = 14.00 - 4.56 = 9.44$

(c) The number of moles of H^+ that have been added is:

$0.1000 \text{ mol/L} \times 0.02490 \text{ L} = 2.490 \times 10^{-3}$ mol H^+

which is also the number of moles of NH_4^+ that have been formed.

The number of moles of NH_3 that remain are thus:

$2.500 \times 10^{-3} - 2.490 \times 10^{-3} = 1.0 \times 10^{-5}$ mol NH_3

$$pOH = pK_b + \log \frac{[cation]}{[base]}$$

In the above log term, the ratio of molar concentrations can be replaced by a ratio of mole amounts, since the volumes are equal and cancel:

$$pOH = 4.74 + \log(2.490 \times 10^{-3})/(1.0 \times 10^{-5}) = 7.14$$

$$pH = 14.00 - pOH = 14.00 - 7.14 = 6.86$$

(d) The number of moles of H^+ that have been added is:

$$0.1000 \text{ mol/L} \times 0.02499 \text{ L} = 2.499 \times 10^{-3} \text{ mol } H^+$$

which is also the number of moles of NH_4^+ that have been formed.

The number of moles of NH_3 that remain are thus:

$$2.500 \times 10^{-3} - 2.499 \times 10^{-3} = 1 \times 10^{-6} \text{ mol } NH_3$$

$$pOH = pK_b + \log \frac{[cation]}{[base]}$$

In the above log term, the ratio of molar concentrations can be replaced by a ratio of mole amounts, since the volumes are equal and cancel:

$$pOH = 4.74 + \log(2.499 \times 10^{-3})/(1 \times 10^{-6}) = 8.1$$

$$pH = 14.00 - pOH = 14.00 - 8.1 = 5.9$$

(e) This is the equivalence point, and we have 2.500×10^{-3} mol of NH_4^+ that have formed. The total volume is 50.00 mL, making the concentration of NH_4^+ become:

$$2.500 \times 10^{-3} \text{ mol}/0.0500 \text{ L} = 5.000 \times 10^{-2} \text{ M } NH_4^+$$

	NH_4^+	$\rightleftharpoons$ NH_3	+ H^+
initial conc. (M)	0.05000	0.0	0.0
change in conc. (M)	-x	x	x
equilibrium conc. (M)	0.05000 - x (≈ 0.05000)	x	x

Substituting the above values for equilibrium concentrations into the mass action expression gives:

$K_a = 5.7 \times 10^{-10} = x^2/(0.05000)$

$x = 5.3 \times 10^{-6}$ M = $[H^+]$

pH = $-\log[H^+] = -\log(5.3 \times 10^{-6}) = 5.28$

(f) The number of moles of H^+ that have been added is:

0.1000 mol/L × 0.02501 L = 2.501×10^{-3} mol

The number of moles of H^+ that are in excess is thus:

$2.501 \times 10^{-3} - 2.500 \times 10^{-3} = 1 \times 10^{-6}$ mol H^+

$[H^+] = 1 \times 10^{-6}$ mol/0.05001 L = 2×10^{-5} M

pH = $-\log[H^+] = -\log(2 \times 10^{-5}) = 4.7$

(g) The number of moles of H^+ that have been added is:

0.1000 mol/L × 0.02510 L = 2.510×10^{-3} mol

The number of moles of H^+ that are in excess is thus:

$2.510 \times 10^{-3} - 2.500 \times 10^{-3} = 1.0 \times 10^{-5}$ mol H^+

$[H^+] = 1.0 \times 10^{-5}$ mol/0.05010 L = 2.0×10^{-4} M

pH = $-\log[H^+] = -\log(2.0 \times 10^{-4}) = 3.70$

(h) The number of moles of H^+ that have been added is:

0.1000 mol/L × 0.02600 L = 2.600×10^{-3} mol

The number of moles of H^+ that are in excess is thus:

$2.600 \times 10^{-3} - 2.500 \times 10^{-3} = 1.0 \times 10^{-4}$ mol H^+

$[H^+] = 1.0 \times 10^{-4}$ mol/0.05100 L = 1.96×10^{-3} M

pH = $-\log[H^+] = -\log(1.96 \times 10^{-3}) = 2.708$

(i) The number of moles of H^+ that have been added is:

0.1000 mol/L × 0.05000 L = 5.000×10^{-3} mol

The number of moles of H^+ that are in excess is thus:

$5.000 \times 10^{-3} - 2.500 \times 10^{-3} = 2.500 \times 10^{-3}$ mol H^+

$[H^+] = 2.500 \times 10^{-3}$ mol/0.07500 L $= 3.333 \times 10^{-2}$ M

$pH = -\log[H^+] = -\log(3.333 \times 10^{-2}) = 1.4772$

CHAPTER SIXTEEN

PRACTICE EXERCISES

1. (a) ΔS is negative since the products have a lower entropy, i.e. a lower freedom of movement.

   (b) ΔS is positive since the products have a higher entropy, i.e. a higher freedom of movement.

2. (a) There are more reactant molecules of gas than there are product molecules of gas, and we conclude that the entropy decreases, i.e. that ΔS is negative.

   (b) Entropy increases as this reaction proceeds because two moles of gaseous material appear along with the solid product, where there was only solid reactant. ΔS is positive, and entropy increases.

3. $\Delta S° = (\text{sum } S°[\text{products}]) - (\text{sum } S°[\text{reactants}])$

   (a) $\Delta S° = \{S°[H_2O(\ell)] + S°[CaCl_2(s)]\} - \{S°[CaO(s)] + 2S°[HCl(g)]\}$

   $\Delta S° = \{1 \text{ mol} \times (69.96 \text{ J K}^{-1} \text{ mol}^{-1}) + 1 \text{ mol} \times (114 \text{ J K}^{-1} \text{ mol}^{-1})\}$

   $- \{1 \text{ mol} \times (40 \text{ J K}^{-1} \text{ mol}^{-1}) + 2 \text{ mol} \times (186.7 \text{ J K}^{-1} \text{ mol}^{-1})\}$

   $\Delta S° = -229 \text{ J/K}$

   (b) $\Delta S° = \{S°[C_2H_6(g)]\} - \{S°[H_2(g)] + S°[C_2H_4(g)]\}$

   $\Delta S° = \{1 \text{ mol} \times (229.5 \text{ J K}^{-1} \text{ mol}^{-1})\}$

   $- \{1 \text{ mol} \times (130.6 \text{ J K}^{-1} \text{ mol}^{-1}) + 1 \text{ mol} \times (219.8 \text{ J K}^{-1} \text{ mol}^{-1})\}$

   $\Delta S° = -120.9 \text{ J/K}$

4. First, we calculate $\Delta S°$, using the data of Table 16.1:

   $\Delta S° = \{2S°[Fe_2O_3(s)]\} - \{3S°[O_2(g)] + 4S°[Fe(s)]\}$

   $\Delta S° = \{2 \text{ mol} \times (90.0 \text{ J K}^{-1} \text{ mol}^{-1})\}$

   $- \{3 \text{ mol} \times (205.0 \text{ J K}^{-1} \text{ mol}^{-1}) + 4 \text{ mol} \times (27 \text{ J K}^{-1} \text{ mol}^{-1})\}$

   $\Delta S° = -543 \text{ J/K} = -0.543 \text{ kJ/mol}$

   Next, we calculate $\Delta H°$ using the data of Table 5.2:

$\Delta H° = (\text{sum } \Delta H_f°[\text{products}]) - (\text{sum } \Delta H_f°[\text{reactants}])$

$\Delta H° = \{2\Delta H_f°[Fe_2O_3(s)]\} - \{3\Delta H_f°[O_2(g)] + 4\Delta H_f°[Fe(s)]\}$

$\Delta H° = \{2 \text{ mol} \times (-822.2 \text{ kJ/mol})\} - \{3 \text{ mol} \times (0.0 \text{ kJ/mol}) + 4 \times (0.0 \text{ kJ/mol})\}$
$\Delta H° = -1644 \text{ kJ}$

Now the temperature is 25.0 + 273.15 = 298.15 K, and the calculation of $\Delta G°$ is as follows:

$\Delta G° = \Delta H° - T\Delta S° = -1644 \text{ kJ} - (298.15 \text{ K})(-0.543 \text{ kJ/K}) = -1482 \text{ kJ}$

5. $\Delta G° = (\text{sum } \Delta G_f°[\text{products}]) - (\text{sum } \Delta G_f°[\text{reactants}])$

(a) $\Delta G° = \{2\Delta G_f°[NO_2(g)]\} - \{\Delta G_f°[O_2(g)] + 2\Delta G_f°[NO(g)]\}$

$\Delta G° = \{2 \text{ mol} \times (51.84 \text{ kJ/mol})\}$
$- \{1 \text{ mol} \times (0.0 \text{ kJ/mol}) + 2 \text{ mol} \times (86.69 \text{ kJ/mol})\}$
$\Delta G° = -69.7 \text{ kJ}$

(b) $\Delta G° = \{2\Delta G_f°[H_2O(g)] + \Delta G_f°[CaCl_2(s)]\}$
$- \{2\Delta G_f°[HCl(g)] + \Delta G_f°[Ca(OH)_2(s)]\}$

$\Delta G° = \{2 \text{ mol} \times (-228.6 \text{ kJ/mol}) + 1 \text{ mol} \times (-750.2 \text{ kJ/mol})\}$
$- \{2 \text{ mol} \times (-95.27 \text{ kJ/mol}) + 1 \text{ mol} \times (-896.76 \text{ kJ/mol})\}$

$\Delta G° = -120.1 \text{ kJ}$

6. The maximum amount of work that is available is the free energy change for the process, in this case, the standard free energy change, $\Delta G°$, since the process occurs at 25 °C.

$4Al(s) + 3O_2(g) \rightarrow 2Al_2O_3(s)$

$\Delta G° = (\text{sum } \Delta G_f°[\text{products}]) - (\text{sum } \Delta G_f°[\text{reactants}])$

$\Delta G° = 2\Delta G_f°[Al_2O_3(s)] - \{3\Delta G_f°[O_2(g)] + 4\Delta G_f°[Al(s)]\}$

$\Delta G° = 2 \text{ mol} \times (-1576.4 \text{ kJ/mol})$
$- \{3 \text{ mol} \times (0.0 \text{ kJ/mol}) + 4 \text{ mol} \times (-3152.8 \text{ kJ/mol})\}$

$\Delta G° = -3152.8 \text{ kJ}$, for the reaction as written.

This calculation conforms to the reaction as written. This means that the above value of $\Delta G°$ applies to the equation involving 4 mol of Al. The conversion to give energy per mole of aluminum is then:

$-3152.8 \text{ kJ/4 mol Al} = -788 \text{ kJ/mol}$

The maximum amount of energy that may be obtained is thus 788 kJ.

7. For the vaporization process in particular, and for any process in general, we have:

$\Delta G = \Delta H - T\Delta S$

If the temperature is taken to be that at which equilibrium is obtained, that is the temperature of the boiling point (where liquid and vapor are in equilibrium with one another), then we also have the result that ΔG is equal to zero:

$\Delta G = 0 = \Delta H - T\Delta S$, or $T_{eq} = \Delta H/\Delta S$

We know ΔH to be 14.5 kcal/mol, and we need the value for ΔS in units kcal K^{-1} mol^{-1}:

$\Delta S° = (\text{sum } S°[\text{products}]) - (\text{sum } S°[\text{reactants}])$

$\Delta S° = S°[Hg(g)] - S°[Hg(\ell)]$

$\Delta S° = (41.8 \times 10^{-3} \text{ kcal K}^{-1} \text{ mol}^{-1}) - (18.2 \times 10^{-3} \text{ kcal K}^{-1} \text{ mol}^{-1})$

$\Delta S° = 23.6 \times 10^{-3} \text{ kcal K}^{-1} \text{ mol}^{-1}$

$T_{eq} = 14.5 \text{ kcal/mol} \div 23.6 \times 10^{-3} \text{ kcal/K mol} = 614 \text{ K}$

8. $\Delta G° = (\text{sum } \Delta G_f°[\text{products}]) - (\text{sum } \Delta G_f°[\text{reactants}])$

$\Delta G° = \Delta G_f°[SO_3(g)] - \{\Delta G_f°[SO_2(g)] + \frac{1}{2}\Delta G_f°[O_2(g)]\}$

$\Delta G° = 1 \text{ mol} \times (-370.4 \text{ kJ/mol})$
$- \{1 \text{ mol} \times (-300.4 \text{ kJ/mol}) + \frac{1}{2} \text{ mol } (0.0 \text{ kJ/mol})\}$

$\Delta G° = -70.0 \text{ kJ/mol}$

Since the sign of $\Delta G°$ is negative, the reaction should be spontaneous.

9. $\Delta G° = \Delta H° - T\Delta S°$

$\Delta G° = \{2\Delta G_f°[HCl(g)] + \Delta G_f°[CaCO_3(s)]\}$
$- \{\Delta G_f°[CaCl_2(s)] + \Delta G_f°[H_2O(g)] + \Delta G_f°[CO_2(g)]\}$

$\Delta G° = \{2 \text{ mol} \times (-95.27 \text{ kJ/mol}) + 1 \text{ mol} \times (-1128.8 \text{ kJ/mol})\}$
$- \{1 \text{ mol} \times (-750.2 \text{ kJ/mol}) + 1 \text{ mol} \times (-228.6 \text{ kJ/mol}) +$
$1 \text{ mol} \times (-394.4 \text{ kJ/mol})\}$

$\Delta G° = +53.9 \text{ kJ}$

Since $\Delta G°$ is positive, the reaction is not spontaneous, and we do not expect to see products formed from reactants.

10. First, we compute the standard free energy change for the reaction, based on the data of Table 16.2:

$\Delta G° = (\text{sum } \Delta G_f°[\text{products}]) - (\text{sum } \Delta G_f°[\text{reactants}])$

$\Delta G° = \{\Delta G_f°[H_2O(g)] + \Delta G_f°[CO_2(g)] + \Delta G_f°[Na_2CO_3(s)]\} - \{2\Delta G_f°[NaHCO_3(s)]\}$

$\Delta G° = \{1 \text{ mol} \times (-228.6 \text{ kJ/mol}) + 1 \text{ mol} \times (-394.4 \text{ kJ/mol})$
$+ 1 \text{ mol} \times (-1048 \text{ kJ/mol})\} - \{2 \text{ mol} \times (-851.9 \text{ kJ/mol})\}$

$\Delta G° = +33 \text{ kJ}$

Next, we determine values for $\Delta H°$ and $\Delta S°$:

$\Delta H° = (\text{sum } \Delta H_f°[\text{products}]) - (\text{sum } \Delta H_f°[\text{reactants}])$

$\Delta H° = \{\Delta H_f°[H_2O(g)] + \Delta H_f°[CO_2(g)] + \Delta H_f°[Na_2CO_3(s)]\} - \{2\Delta H_f°[NaHCO_3(s)]\}$

$\Delta H° = \{1 \text{ mol} \times (-241.8 \text{ kJ/mol}) + 1 \text{ mol} \times (-393.5 \text{ kJ/mol})$
$+ 1 \text{ mol} \times (-1131 \text{ kJ/mol})\} - \{2 \text{ mol} \times (-947.7 \text{ kJ/mol})\}$

$\Delta H° = +129 \text{ kJ}$

$\Delta S° = (\text{sum } S°[\text{products}]) - (\text{sum } S°[\text{reactants}])$

$\Delta S° = \{S°[H_2O(g)] + S°[CO_2(g)] + S°[Na_2CO_3(s)]\} - \{2S°[NaHCO_3(s)]\}$

$\Delta S° = \{1 \text{ mol} \times (188.7 \text{ J K}^{-1} \text{ mol}^{-1}) + 1 \text{ mol} \times (213.6 \text{ J K}^{-1} \text{ mol}^{-1})$

$+ 1 \text{ mol} \times (136 \text{ J K}^{-1} \text{ mol}^{-1})\} - \{2 \text{ mol} \times (102 \text{ J K}^{-1} \text{ mol}^{-1})\}$

$\Delta S° = 334 \text{ J/K} = 0.334 \text{ kJ/K}$

Next, we assume that both $\Delta H°$ and $\Delta S°$ are independent of temperature, and use these values to determine ΔG at a temperature of 200 + 273 = 473 K:

$\Delta G' = \Delta H° - T\Delta S° = 129 \text{ kJ} - (473 \text{ K})(0.334 \text{ kJ/K}) = -29 \text{ kJ}$

At the lower of these two temperatures (25 °C), the reaction has a positve value of ΔG. At the higher of these two temperatures (200 °C), the reaction has a negative value of ΔG. Thus ΔG becomes more negative as the temperature is raised, so the reaction becomes increasingly more favorable as the temperature is increased. In other words, the position of the equilibrium will be shifted more towards products at the higher temperature.

11. $\Delta G° = -RT \ln K$

$\Delta G° = -(8.314 \text{ J K}^{-1} \text{ mol}^{-1})(25 + 273 \text{ K}) \times \ln(6.9 \times 10^5) = -33.3 \times 10^3 \text{ J}$

$\Delta G° = -33.3 \text{ kJ}$

Alternatively, we can perform the calculation in calories, by using the different value for R:

$\Delta G° = -(1.987 \text{ cal K}^{-1} \text{ mol}^{-1})(298 \text{ K}) \times \ln(6.9 \times 10^5) = -7.96 \times 10^3 \text{ cal}$

$\Delta G° = -7.96 \text{ kcal}$

12. $\Delta G° = -RT \ln K_p$

$3.3 \times 10^3 \text{ J} = -(8.314 \text{ J K}^{-1} \text{ mol}^{-1})(298 \text{ K}) \times \ln(k_p)$

$\ln K_p = -3.3 \times 10^3 \text{ J}/(8.314 \text{ J K}^{-1} \text{ mol}^{-1})(298 \text{ K}) = -1.332$

Taking the antilog of both sides of the above equation gives:

$K_p = 0.26$

13. $\Delta G' = \Delta H° - T\Delta S° = -92.4 \times 10^3 \text{ J} - (323 \text{ K})(-198.3 \text{ J/K}) = -2.83 \times 10^4 \text{ J}$

$\Delta G = -RT \ln K$

Thus $-2.83 \times 10^4 \text{ J} = -(8.314 \text{ J K}^{-1} \text{ mol}^{-1})(323 \text{ K}) \times \ln(K_p)$

$\ln K_p = -2.38 \times 10^4 \text{ J}/(-8.314 \text{ J K}^{-1} \text{ mol}^{-1})(323 \text{ K}) = 10.54$

Taking the antilog of both sides of the above equation gives:

$K_p = 3.8 \times 10^4$

REVIEW EXERCISES

First Law of Thermodynamics

16.1 The term *thermodynamics* is meant to convey the two ideas of thermo (heat) and dynamics (motion), namely movement or transfer of heat.

16.2 The internal energy of a system consists of kinetic and potential energy. We cannot determine the amount of potential energy that a system has, nor can we know the precise speeds of the various particles in a system. We are therefore unable to determine the absolute total internal energy of a system. We can, however, determine changes in internal energy.

16.3 A state function is a thermodynamic quantity whose value is determined only by the state of the system currently, and is not determined by a system's prior condition or history. A change in a state function is the same regardless of the path that is used to arrive at the final state from the inital state. That is, changes in state function quantities are path-independent.

16.4 $\Delta E = E_{final} - E_{initial}$

16.5 If we refer to the equation in the answer to Review Exercise 16.4, we can see that if E_{final} is larger than $E_{initial}$, then ΔE must be positive, by definition. Thus ΔE for an endothermic process is positive.

16.6 $\Delta E = q - w$

ΔE is the change in internal energy, which is positive for an endothermic process, and negative for an exothermic process.

q is the heat absorbed by the system during the process, and it has a positive sign if the system absorbs heat during the process.

w is the work done by the system on the surroundings during the process, and it has a positive sign if the system does work on the surroundings, whereas it has a negative sign if the surroundings do work on the system during the process.

16.7 $\Delta E = q - w = 500\ J - (-200\ J) = +700\ J$

The overall process is endothermic, meaning that the internal energy of the system increases. Notice that both terms, q and w, contribute to the increase in internal energy of the system; the system gains heat (+q) and has work done on it (-w).

16.8 $\Delta E = q - w$

$-1250\ J = 603\ J - w$

$w = +1853\ J$

Since w is defined to be the work done by the system on the surroundings, then in this case, a positive amount of work is done by the system on the surrounding, namely + 1853 J of work.

16.9 ΔE is the heat of reaction at constant volume, and applies, for instance, to reactions performed in closed vessels such as bomb calorimeters.
ΔH is the heat of reaction at constant pressure, and applies, for instance, to reactions performed in open containers, such as "coffee cup

calorimeters."

16.10 The slower that the energy extraction is performed, the greater is the total amount of energy that can be obtained. This is the same as saying that the most energy is available from a process that occurs reversibly.

16.11 A pressure in Pascals, by definition, has the units Newtons/m^2, or N/m^2.

Thus P in pascals times ΔV in m^3 is:

$N/m^2 \times m^3 = N \cdot m$

The last result is the Newton • meter, or a force times a distance. By definition, a force of 1 N operating over a distance of 1 meter is the quantity 1 joule. Thus the quantity $P \times \Delta V$ has the units of energy, namely joules.

16.12 $P(atm) \times \Delta V(L) = 1$ L atm of work

Substituting the given information:

$P \times \Delta V = 101{,}325 \text{ Pa} \times 1 \text{ dm}^3$

Now it is true that 1 dm = 0.1 m, so:

$1 \text{ dm}^3 = (0.1 \text{ m})^3 = 1 \times 10^{-3} \text{ m}^3$

Also, $1 \text{ Pa} = 1 \text{ N/m}^2$

Thus we have:

$P(Pa) \times \Delta V(dm^3) = (101{,}325 \text{ N/m}^2)(1 \times 10^{-3} \text{ m}^3) = 101.325 \text{ N}\cdot\text{m}$

$101.325 \text{ N}\cdot\text{m} \times 1 \text{ J/1 N}\cdot\text{m} = 101.325 \text{ J}$

16.13 work = $P \times \Delta V$

The total pressure is atmospheric pressure plus that caused by the hand pump:

$P = (30.0 + 14.7) \text{ lb/in.}^2 = 44.7 \text{ lb/in.}^2$

Converting to atmospheres we get:

$P = 44.7 \text{ lb/in.}^2 \times 1 \text{ atm/14.7 lb/in.}^2 = 3.04 \text{ atm}$

Next we convert the volume change in units in.^{3} to units L:

$24.0 \text{ in.}^3 \times (2.54 \text{ cm/in.})^3 \times 1 \text{ L/1000 cm}^3 = 0.393 \text{ L}$

Hence $P \times \Delta V = (3.04 \text{ atm})(0.393 \text{ L}) = 1.19 \text{ L}\cdot\text{atm}$

1.19 L•atm × 101.3 J/L•atm = 121 J

16.14 $\Delta H = \Delta E + P \times \Delta V$

16.15 We can see from the equation that constitutes the answer to Review Exercise 16.14 that, since the quantity $P \times \Delta V$ is positive, ΔE must be the larger negative quantity. ΔE is a larger negative quantity than ΔH because, if the change is carried out at constant pressure, some of the total energy is lost as expansion work ($P \times \Delta V$), as the gases that are released by the reaction push back the atmosphere.

16.16 No work ($P \times \Delta V$) is accomplished in a bomb calorimeter because there is no change in volume.

16.17 This is a reaction that produces 1 mol of a single gaseous product, CO_2. Furthermore, this mole of gaseous product forms from

nongaseous materials. The volume that will be occupied by this gas, once it is formed, can be found by application of Charles' Law:

22.4 L × 298 K/273 K = 24.5 L

The work of gas expansion on forming the products is thus:

$P \times \Delta V$ = (1.00 atm)(24.5 L) = 24.5 L atm

24.5 l atm × 101 J/L atm = 2.47×10^3 J of work

Spontaneous Change and Enthalpy

16.18 A spontaneous change, in thermodynamic terms, is one for which the sign of ΔG is negative. It is a process that occurs by itself, without continued outside assistance.

16.19 Student answer

16.20 Student answer

16.21 A change that is characterized by a negative value for ΔH tends to be spontaneous. The only situation where this is not true is that in which the value of the product $T\Delta S$ is sufficiently negative to make the quantity $\Delta G = \Delta H - T\Delta S$ become positive.

16.22 In general, we have the equation:

$\Delta H° = (\text{sum } \Delta H_f°[\text{products}]) - (\text{sum } \Delta H_f°[\text{reactants}])$

and we take the answer to Review Exercise 16.21 as a guide to deciding on which reaction might favorably be expected to be spontaneous, that is have a chance for $\Delta G = \Delta G - T\Delta S$ to be negative.

(a) $\Delta H° = \{\Delta H_f^\circ[CaCO_3(s)]\} - \{\Delta H_f^\circ[CO_2(g)] + \Delta H_f^\circ[CaO(s)]\}$

$\Delta H° = \{1\text{ mol} \times (-1207\text{ kJ/mol})\} - \{1\text{ mol} \times (-393.5\text{ kJ/mol}) + 1\text{ mol} \times (-635.5\text{ kJ/mol})\}$

$\Delta H° = -178\text{ kJ}$ ∴ favored.

(b) $\Delta H° = \{\Delta H_f^\circ[C_2H_6(g)]\} - \{\Delta H_f^\circ[C_2H_2(g)] + 2\Delta H_f^\circ[H_2(g)]\}$

$\Delta H° = \{1\text{ mol} \times (-84.667\text{ kJ/mol})\} - \{1\text{ mol} \times (226.75\text{ kJ/mol}) + 2\text{ mol} \times (0.0\text{ kJ/mol})\}$

$\Delta H° = -311.42\text{ kJ}$ ∴ favored.

(c) $\Delta H° = \{\Delta H_f^\circ[Fe_2O_3(s)] + 3\Delta H_f^\circ[Ca(s)]\} - \{2\Delta H_f^\circ[Fe(s)] + 3\Delta H_f^\circ[CaO(s)]\}$

$\Delta H° = \{1\text{ mol} \times (-822.2\text{ kJ/mol}) + 3\text{ mol} \times (0.0\text{ kJ/mol})\} - \{2\text{ mol} \times (0.0\text{ kJ/mol}) + 3\text{ mol} \times (-635.5\text{ kJ/mol})\}$

$\Delta H° = +1084\text{ kJ}$ ∴ not favorable from the standpoint of enthalpy alone.

(d) $\Delta H° = \{\Delta H_f^\circ[H_2O(\ell)] + \Delta H_f^\circ[CaO(s)]\} - \{\Delta H_f^\circ[Ca(OH)_2(s)]\}$

$\Delta H° = \{1\text{ mol} \times (-285.9\text{ kJ/mol}) + 1\text{ mol} \times (-635.5\text{ kJ/mol})\} - \{1\text{ mol} \times (-986.59\text{ kJ/mol})\}$

$\Delta H° = +65.2\text{ kJ}$ ∴ not favored from the standpoint of enthalpy alone.

(e) $\Delta H° = \{2\Delta H_f^\circ[HCl(g)] + \Delta H_f^\circ[Na_2SO_4(s)]\} - \{2\Delta H_f^\circ[NaCl(s)] + \Delta H_f^\circ[H_2SO_4(\ell)]\}$

$\Delta H° = \{2\text{ mol} \times (-92.30\text{ kJ/mol}) + 1\text{ mol} \times (-1384.5\text{ kJ/mol})\} - \{2\text{ mol} \times (-411.0\text{ kJ/mol}) + 1\text{ mol} \times (-811.32\text{ kJ/mol})\}$

$\Delta H° = +64.2\text{ kJ}$ ∴ not favored from the standpoint of enthalpy alone.

16.23 We proceed as in the answer to Review Exercise 16.22:

(a) $\Delta H° = \{2\Delta H_f°[H_2O(g)] + 4\Delta H_f°[CO_2(g)]\}$

$- \{5\Delta H_f°[O_2(g)] + 2\Delta H_f°[C_2H_2(g)]\}$

$\Delta H° = \{2 \text{ mol} \times (-241.8 \text{ kJ/mol}) + 4 \text{ mol} \times (-393.5 \text{ kJ/mol})\}$
$- \{5 \text{ mol} \times (0.0 \text{ kJ/mol}) + 2 \text{ mol} \times (226.75 \text{ kJ/mol})\}$

$\Delta H° = -2511$ kJ ∴ favored from the standpoint of enthalpy alone.

(b) $\Delta H° = \{5\Delta H_f°[N_2(g)] + \Delta H_f°[H_2O(g)] + 2\Delta H_f°[CO_2(g)]\}$

$- \{5\Delta H_f°[N_2O(g)] + \Delta H_f°[C_2H_2(g)]\}$

$\Delta H° = \{5 \text{ mol} \times (0.0 \text{ kJ/mol}) + 1 \text{ mol} \times (241.8 \text{ kJ/mol})$

$+ 2 \text{ mol} \times (-393.5\text{kJ/mol})\} - \{5 \text{ mol} \times (81.57 \text{ kJ/mol})$

$+ 1 \text{ mol} \times (226.75 \text{ kJ/mol})\}$

$\Delta H° = -1179.8$ kJ ∴ favorable from the standpoint of enthalpy.

(c) $\Delta H° = \{2\Delta H_f°[Fe(s)] + \Delta H_f°[Al_2O_3(s)]\}$

$- \{2\Delta H_f°[Al(s)] + \Delta H_f°[Fe_2O_3(s)]\}$

$\Delta H° = \{2 \text{ mol} \times (0.0 \text{ kJ/mol}) + 1 \text{ mol} \times (-1669.8 \text{ kJ/mol})\}$
$- \{2 \text{ mol} \times (0.0 \text{ kJ/mol}) + 1 \text{ mol} \times (-822.2 \text{ kJ/mol})\}$

$\Delta H° = -847.6$ kJ ∴ favorable from the standpoint of enthalpy.

(d)

$\Delta H° = \{5 \text{ mol} \times (0.0 \text{ kJ/mol}) + 1 \text{ mol} \times (-241.8 \text{ kJ/mol}) +$
$2 \text{ mol} \times (-393.5 \text{ kJ/mol})\} - \{5 \text{ mol} \times (81.57 \text{ kJ/mol}) +$
$1 \text{ mol} \times (226.75 \text{ kJ/mol})\}$

$\Delta H° = -1663.4$ kJ ∴ favored from the standpoint of enthalpy alone.

(e) $\Delta H° = \{\Delta H_f°[HCl(g)] + \Delta H_f°[NH_3(g)]\} - \{\Delta H_f°[NH_4Cl(s)]\}$

$\Delta H° = \{1 \text{ mol} \times (-92.30 \text{ kJ/mol}) + 1 \text{ mol} \times (-46.19 \text{ kJ/mol})\}$
$- \{1 \text{ mol} \times (-315.4 \text{ kJ/mol})\}$

$\Delta H° = +176.91$ kJ ∴ not favorable from the standpoint of enthalpy.

Chapter 16

Entropy

16.24 The potassium and iodine atoms are in a highly ordered geometry in the crystalline KI sample. When KI dissolves, the K^+ and I^- ions become randomly dispersed throughout the solvent. The increase in randomness that attends the dissolving of the solid is responsible for the process's being favorable, or spontaneous, in spite of the fact that the enthalpy makes the process to be endothermic.

16.25 Entropy is a measure of the randomness of a system. An equivalent statement is that entropy is a measure of the statistical probability of a system.

16.26 (a) negative (b) negative (c) positive (d) negative (e) negative (f) positive

16.27 The statistical probability of a system in a given state, relative to all the other states, is the same regardless of how the system happened to have been formed.

16.28 The entropy of the universe increases when a spontaneous event occurs.

16.29 Two examples might be (1) a disorganized pile of bricks and boards jumping up to produce a house, or (2) a pile of bricks falling off a cliff, and landing in the form of a perfect cube. Both of these examples are accompanied by enormous decreases in entropy (randomness), and hence are not realistic spontaneous events.

16.30 The scattering of pollutants into the environment is characterized by a positive value for ΔS. It is a natural consequence of this value for ΔS that pollution is relentless, and that efforts to prevent it must employ other than spontaneous processes.

16.31 The probability is given by the number of possibilities that lead to the desired arrangement, divided by the total number of possible arrangements.

We list each of the possible results, heads (H) or tails (T) for each of the coins, and systematically write down all of the distinct arrangements:

There is only one arrangement that gives four heads: HHHH.

There is only one arrangement that gives four tails: TTTT

Four distinct arrangements can lead to three heads and one tail:

HHHT, THHH, HTHH, HHTH

There are similarly four distinct arrangements that lead to three tails and one head:

TTTH, HTTT, THTT, TTHT

There are six distinct arrangements that can lead to two heads and two tails:

HHTT, TTHH, THHT, THTH, HTTH, HTHT

Hence the probability of all heads (HHHH) is 1 in 16 or 1/16 = 0.0625.

The probability of two heads and two tails is 6 in 16 or 6/16 = 0.375.

16.32 This situation is completely analogous to that in Review Exercise 16.31, since there is an equal probability of finding a given molecule in either container. This arises because the containers have equal volumes. One container is thus analogous to a "head" and the other container is analogous to a "tail." Thus there are to be 16 possible distinct arrangements of the four molecules in the two containers. Only one of the 16 possible arrangements is one having all four molecules in the same container. Hence the probability of finding all 4 molecules in one container is 1 part in 16, and since there are two such containers, the probability of finding four molecules in one container is 2/16 = 1/8.

There are six arrangements in which each container will have two molecules, just as there are six possible ways of obtaining two heads and two tails when tossing four coins. Thus there is a 6/16 = 3/8 probability of finding two molecules in each container.

Since 3/8 (the probability of finding two molecules in each of two containers) is larger than 1/8 (the probability of finding all four molecules in one container), it is evident that the more likely distribution is the first. This suggests that molecules that are introduced into a container will undergo a spontaneous expansion to distribute themselves uniformly throughout the volume that is available to them.

16.33 (a) negative - since the number of gaseous materials decreases.
(b) negative - since the number of moles of gaseous material decreases.
(c) negative - since the number of moles of gas decreases.
(d) positive - since a gas appears where there formerly was none.

16.34 (a) ΔS is positive since randomness in a gas is higher than that in a solid.
(b) ΔS is positive. Although there is not a change in the number of moles of gaseous material, the product molecules are more complex and less random than the reactant molecules.
(c) ΔS is negative since gaseous material (which is highly random) is replaced by a solid (which is highly ordered).
(d) ΔS is negative since the relatively random liquid reactant disappears in a process that makes only a solid.

Chapter 16

Third Law of Thermodynamics

16.35 This is the statement that the entropy of a perfect crystalline solid at 0 K is equal to zero: S = 0 at 0 K.

16.36 The entropy of a mixture must be higher than that of two separate pure materials, because the mixture is guaranteed to have a higher degree of disorder. Said another way, a mixture is more disordered than either of its two separate components. Only a pure substance can have an entropy of zero, and then only at 0 K.

16.37 Entropy increases with increasing temperature because vibrations and movements within a solid lead to greater disorder at the higher temperatures. Melting especially produces more disorder, and vaporization even more.

16.38 $\Delta S° = (\text{sum } S°[\text{products}]) - (\text{sum } S°[\text{reactants}])$

(a) $\Delta S° = \{2S°[NH_3(g)]\} - \{3S°[H_2(g)] + S°[N_2(g)]\}$

$\Delta S° = \{2 \text{ mol} \times (192.5 \text{ J K}^{-1} \text{ mol}^{-1})\} - \{3 \text{ mol} \times (130.6 \text{ J K}^{-1} \text{ mol}^{-1})$

$+ 1 \text{ mol} \times (191.5 \text{ J K}^{-1} \text{ mol}^{-1})\}$

$\Delta S° = -198.3$ J/K $\therefore$ spontaneous from the standpoint of entropy alone.

(b) $\Delta S° = \{S°[CH_3OH(\ell)]\} - \{2S°[H_2(g)] + S°[CO(g)]\}$

$\Delta S° = \{1 \text{ mol} \times (126.8 \text{ J K}^{-1} \text{ mol}^{-1})\}$

$- \{2 \text{ mol} \times (130.6 \text{ J K}^{-1} \text{ mol}^{-1}) + 1 \text{ mol} \times (197.9 \text{ J K}^{-1} \text{ mol}^{-1})\}$

$\Delta S° = -332.3$ J/K $\therefore$ not favored from the standpoint of entropy alone.

(c) $\Delta S° = \{6S°[H_2O(g)] + 4S°[CO_2(g)]\} - \{7S°[O_2(g)] + 2S°[C_2H_6(g)]\}$

$\Delta S° = \{6 \text{ mol} \times (188.7 \text{ J K}^{-1} \text{ mol}^{-1}) + 4 \text{ mol} \times (213.6 \text{ J K}^{-1} \text{ mol}^{-1})\}$

$- \{7 \text{ mol} \times (205.0 \text{ J K}^{-1} \text{ mol}^{-1}) + 2 \text{ mol} \times (229.5 \text{ J K}^{-1} \text{ mol}^{-1})\}$

$\Delta S° = +92.6$ J/K $\therefore$ favorable from the standpoint of entropy alone.

(d) $\Delta S° = \{2S°[H_2O(\ell)] + S°[CaSO_4(s)]\} - \{S°[H_2SO_4(\ell)] + S°[Ca(OH)_2(s)]\}$

$\Delta S° = \{2\text{ mol} \times (69.96\text{ J K}^{-1}\text{ mol}^{-1}) + 1\text{ mol} \times (107\text{ J K}^{-1}\text{ mol}^{-1})\}$

$- \{1\text{ mol} \times (157\text{ J K}^{-1}\text{ mol}^{-1}) + 1\text{ mol} \times (76.1\text{ J K}^{-1}\text{ mol}^{-1})\}$

$\Delta S° = +14$ J/K ∴ favorable from the standpoint of entropy alone.

(e) $\Delta S° = \{2S°[N_2(g)] + S°[SO_2(g)]\} - \{2S°[N_2O(g)] + S°[S(s)]\}$

$\Delta S° = \{2\text{ mol} \times (191.5\text{ J K}^{-1}\text{ mol}^{-1}) + 1\text{ mol} \times (248.5\text{ J K}^{-1}\text{ mol}^{-1})\}$

$- \{2\text{ mol} \times (220.0\text{ J K}^{-1}\text{ mol}^{-1}) + 1\text{ mol} \times (31.9\text{ J K}^{-1}\text{ mol}^{-1})\}$

$\Delta S° = +159.6$ J/K ∴ favorable from the standpoint of entropy alone.

16.39 $\Delta S°$ = (sum S°[products]) - (sum S°[reactants])

(a) $\Delta S° = \{S°[AgCl(s)]\} - \{\frac{1}{2}S°[Cl_2(g)] + S°[Ag(s)]\}$

$\Delta S° = \{1\text{ mol} \times (96.2\text{ J K}^{-1}\text{ mol}^{-1})\}$

$- \{{}^1/_2\text{ mol} \times (223.0\text{ J K}^{-1}\text{ mol}^{-1}) + 1\text{ mol} \times (42.55\text{ J K}^{-1}\text{ mol}^{-1})\}$

$\Delta S° = -61.9$ J/K

(b) $\Delta S° = \{S°[H_2O(g)]\} - \{\frac{1}{2}S°[O_2(g)] + S°[H_2(g)]\}$

$\Delta S° = \{1\text{ mol} \times (188.7\text{ J K}^{-1}\text{ mol}^{-1})\}$

$- \{{}^1/_2\text{ mol} \times (205.0\text{ J K}^{-1}\text{ mol}^{-1}) + 1\text{ mol} \times (130.6\text{ J K}^{-1}\text{ mol}^{-1})\}$

$\Delta S° = -44.4$ J/K

(c) $\Delta S° = \{S°[H_2O(\ell)]\} - \{\frac{1}{2}S°[O_2(g)] + S°[H_2(g)]\}$

$\Delta S° = \{1\text{ mol} \times (69.96\text{ J K}^{-1}\text{ mol}^{-1})\}$

$- \{{}^1/_2\text{ mol} \times (205.0\text{ J K}^{-1}\text{ mol}^{-1}) + 1\text{ mol} \times (130.6\text{ J K}^{-1}\text{ mol}^{-1})\}$

$\Delta S° = -163.1$ J/K

(d) $\Delta S° = \{S°[CO_2(g)] + S°[H_2O(g)] + S°[CaSO_4(s)]\}$

$- \{S°[CaCO_3(s)] + S°[H_2SO_4(\ell)]\}$

$\Delta S° = \{1\text{ mol} \times (213.6\text{ J K}^{-1}\text{ mol}^{-1}) + 1\text{ mol} \times (188.7\text{ J K}^{-1}\text{ mol}^{-1})$

$+ 1\text{ mol} \times (107\text{ J K}^{-1}\text{ mol}^{-1})\} - \{1\text{ mol} \times (92.9\text{ J K}^{-1}\text{ mol}^{-1})$

$+ 1\text{ mol} \times (157\text{ J K}^{-1}\text{ mol}^{-1})\}$

$\Delta S° = +259\ J/K$

(e) $\Delta S° = \{S°[NH_4Cl(s)]\} - \{S°[HCl(g)] + S°[NH_3(g)]\}$

$\Delta S° = \{1\ mol \times (94.6\ J\ K^{-1}\ mol^{-1})\}$

$- \{1\ mol \times (186.7\ J\ K^{-1}\ mol^{-1}) + 1\ mol \times (192.5\ J\ K^{-1}\ mol^{-1})\}$

$\Delta S° = -284.6\ J/K$

16.40 The entropy change that is designated $\Delta S°_f$ is that which corresponds to the reaction in which one mole of a substance is formed from elements in their standard states. Since the value is understood to correspond to the reaction forming one mole of a single pure substance, the units may be written either $J\ K^{-1}$ or $J\ K^{-1}\ mol^{-1}$.

(a) $2C(s) + 2H_2(g) \rightarrow C_2H_4(g)$

$\Delta S° = \{S°[C_2H_4(g)]\} - \{2S°[C(s)] + 2S°[H_2(g)]\}$

$\Delta S° = \{1\ mol \times (219.8\ J\ K^{-1}\ mol^{-1})\}$

$- \{2\ mol \times (5.69\ J\ K^{-1}\ mol^{-1}) + 2\ mol \times (130.6\ J\ K^{-1}\ mol^{-1})\}$

$\Delta S° = -52.8\ J/K$ or $-52.8\ J\ K^{-1}\ mol^{-1}$

(b) $N_2(g) + \frac{1}{2}O_2(g) \rightarrow N_2O(g)$

$\Delta S° = \{S°[N_2O(g)]\} - \{S°[N_2(g)] + \frac{1}{2}S°[O_2(g)]\}$

$\Delta S° = \{1\ mol \times (220.0\ J\ K^{-1}\ mol^{-1})\} - \{1\ mol \times (191.5\ J\ K^{-1}\ mol^{-1}) +$

$^1/_2\ mol \times (205.0\ J\ K^{-1}\ mol^{-1})\}$

$\Delta S° = -74.0\ J/K$ or $-74.0\ J\ K^{-1}\ mol^{-1}$

(c) $Na(s) + \frac{1}{2}Cl_2(g) \rightarrow NaCl(s)$

$\Delta S° = \{S°[NaCl(s)]\} - \{\frac{1}{2}S°[Cl_2(g)] + S°[Na(s)]\}$

$\Delta S° = \{1\ mol \times (72.38\ J\ K^{-1}\ mol^{-1})\} - \{^1/_2 S°[223.0\ J\ K^{-1}\ mol^{-1})$

$+ 1\ mol \times (51.0\ J\ K^{-1}\ mol^{-1})\}$

$\Delta S° = -90.1\ J/K$ or $-90.1\ J\ K^{-1}\ mol^{-1}$

(d) $Ca(s) + S(s) + 3O_2(g) + 2H_2(g) \rightarrow CaSO_4 \cdot 2H_2O(s)$

$\Delta S° = \{S°[CaSO_4 \cdot 2H_2O(s)]\} - \{2S°[H_2(g)] + 3S°[O_2(g)] + S°[S(s)] + S°[Ca(s)]\}$

$\Delta S° = \{1 \text{ mol} \times (194.0 \text{ J K}^{-1} \text{ mol}^{-1})\} - \{2 \text{ mol} \times (130.6 \text{ J K}^{-1} \text{ mol}^{-1})$

$+ 3 \text{ mol} \times (205.0 \text{ J K}^{-1} \text{ mol}^{-1}) + 1 \text{ mol} \times (31.9 \text{ J K}^{-1} \text{ mol}^{-1})$

$+ 1 \text{ mol} \times (154.8 \text{ J K}^{-1} \text{ mol}^{-1})\}$

$\Delta S° = -868.9 \text{ J/K}$ or $-868.9 \text{ J K}^{-1} \text{ mol}^{-1}$

(e) $2H_2(g) + 2C(s) + O_2(g) \rightarrow HC_2H_3O_2(\ell)$

$\Delta S° = \{S°[HC_2H_3O_2(\ell)]\} - \{2S°[H_2(g)] + 2S°[C(s)] + S°[O_2(g)]\}$

$\Delta S° = \{1 \text{ mol} \times (160 \text{ J K}^{-1} \text{ mol}^{-1})\} - \{2 \text{ mol} \times (130.6 \text{ J K}^{-1} \text{ mol}^{-1})$

$+ 2 \text{ mol} \times (5.69 \text{ J K}^{-1} \text{ mol}^{-1}) + 1 \text{ mol} \times (205.0 \text{ J K}^{-1} \text{ mol}^{-1})\}$

$\Delta S° = -318 \text{ J/K}$ or $-318 \text{ J K}^{-1} \text{ mol}^{-1}$

16.41 The entropy change that is designated $\Delta S°_f$ is that which corresponds to the reaction in which one mole of a substance is formed from elements in their standard states. Since the value is understood to correspond to the reaction forming one mole of a single pure substance, the units may be written either J K^{-1} or $\text{J K}^{-1} \text{ mol}^{-1}$.

(a) $2Al(s) + \frac{3}{2}O_2(g) \rightarrow Al_2O_3(s)$

$\Delta S° = \{S°[Al_2O_3(s)]\} - \{2S°[Al(s)] + \frac{3}{2}S°[O_2(g)]\}$

$\Delta S° = \{1 \text{ mol} \times (51.00 \text{ J K}^{-1} \text{ mol}^{-1})\}$

$- \{2 \text{ mol} \times (28.3 \text{ J K}^{-1} \text{ mol}^{-1}) + {}^3/_2 \text{ mol} \times (205.0 \text{ J K}^{-1} \text{ mol}^{-1})\}$

$\Delta S° = -313 \text{ J/K}$ or $-313 \text{ J K}^{-1} \text{ mol}^{-1}$

(b) $Ca(s) + C(s) + \frac{3}{2}O_2(g) \rightarrow CaCO_3(s)$

$\Delta S° = \{S°[CaCO_3(s)]\} - \{S°[Ca(s)] + S°[C(s)] + \frac{3}{2}S°[O_2(g)]\}$

$\Delta S° = \{1 \text{ mol} \times (92.9 \text{ J K}^{-1} \text{ mol}^{-1})\} - \{1 \text{ mol} \times (154.8 \text{ J K}^{-1} \text{ mol}^{-1})$

$+ 1 \text{ mol} \times (5.69 \text{ J K}^{-1} \text{ mol}^{-1}) + {}^3/_2S° \text{ mol} \times (205.0 \text{ J K}^{-1} \text{ mol}^{-1})\}$

$\Delta S° = -375.1 \text{ J/K}$ or $-375.1 \text{ J K}^{-1} \text{ mol}^{-1}$

(c) $N_2(g) + 2O_2(g) \rightarrow N_2O_4(g)$

$\Delta S° = \{S°[N_2O_4(g)]\} - \{S°[N_2(g)] + 2S°[O_2(g)]\}$

$\Delta S° = \{1\ mol \times (304\ J\ K^{-1}\ mol^{-1})\} - \{1\ mol \times (191.5\ J\ K^{-1}\ mol^{-1})$
$+ 2\ mol \times (205.0\ J\ K^{-1}\ mol^{-1})\}$

$\Delta S° = -297.5\ J/K$ or $-297.5\ J\ K^{-1}\ mol^{-1}$

(d) $\frac{1}{2}N_2(g) + 2H_2(g) + \frac{1}{2}Cl_2(g) \rightarrow NH_4Cl(s)$

$\Delta S° = \{S°[NH_4Cl(s)]\} - \{\frac{1}{2}S°[N_2(g)] + 2S°[H_2(g)] + \frac{1}{2}S°[Cl_2(g)]\}$

$\Delta S° = \{1\ mol \times (94.6\ J\ K^{-1}\ mol^{-1})\} - \{^1/_2\ mol \times (191.5\ J\ K^{-1}\ mol^{-1})$

$+ 2\ mol \times (130.6\ J\ K^{-1}\ mol^{-1}) + {^1/_2}\ mol \times (223.0\ J\ K^{-1}\ mol^{-1})\}$

$\Delta S° = -373.9\ J/K$ or $-373.9\ J\ K^{-1}\ mol^{-1}$

(e) $Ca(s) + S(s) + \frac{9}{4}O_2(g) + \frac{1}{2}H_2(g) \rightarrow CaSO_4 \cdot \frac{1}{2}H_2O(s)$

$\Delta S° = \{S°[CaSO_4 \cdot \frac{1}{2}H_2O(s)]\} - \{S°[Ca(s)] + S°[S(s)] + \frac{9}{4}S°[O_2(g)]$

$+ {^1/_2}S°[H_2(g)]\}$

$\Delta S° = \{1\ mol \times (131\ J\ K^{-1}\ mol^{-1})\} - \{1\ mol \times (154.8\ J\ K^{-1}\ mol^{-1})$

$+ 1\ mol \times (31.9\ J\ K^{-1}\ mol^{-1}) + {^9/_4}\ mol \times (205.0\ J\ K^{-1}\ mol^{-1})$

$+ \frac{1}{2}\ mol \times (130.6\ J\ K^{J}\ mol^{m})\}$

$\Delta S° = -582.3\ J/K$ or $-582.3\ J\ K^{-1}\ mol^{-1}$

16.42 $\Delta S°$ = (sum S°[products]) - (sum S°[reactants])

$\Delta S° = \{2S°[HNO_3(\ell)] + S°[NO(g)]\} - \{3S°[NO_2(g)] + S°[H_2O(\ell)]\}$

$\Delta S° = \{2\ mol \times (155.6\ J\ K^{-1}\ mol^{-1}) + 1\ mol \times (210.6\ J\ K^{-1}\ mol^{-1})\}$

$- \{3\ mol \times (240.5\ J\ K^{-1}\ mol^{-1}) + 1\ mol \times (69.96\ J\ K^{-1}\ mol^{-1})\}$

$\Delta S° = -269.7\ J/K$

16.43 $\Delta S°$ = (sum S°[products]) - (sum S°[reactants])

$\Delta S° = \{S°[HC_2H_3O_2(\ell)] + S°[H_2O(\ell)]\} - \{S°[C_2H_5OH(\ell)] + S°[O_2(g)]\}$

$\Delta S° = \{1\ mol \times (160\ J\ K^{-1}\ mol^{-1}) + 1\ mol \times (69.96\ J\ K^{-1}\ mol^{-1})\}$

$- \{1\ mol \times (161\ J\ K^{-1}\ mol^{-1}) + 1\ mol \times (205.0\ J\ K^{-1}\ mol^{-1})\}$

$\Delta S° = -136$ J/K

Gibbs Free Energy

16.44 $\Delta G = \Delta H - T\Delta S$

16.45 (a) A change is spontaneous at all temperatures only if ΔH is negative and ΔS is positive.

(b) A change is spontaneous at low temperatures but not at high temperatures only if ΔH is negative and ΔS is negative.

(c) A change is spontaneous at high temperatures but not at low temperatures only if ΔH is positive and ΔS is positive.

16.46 A change is nonspontaneous regardless of the temperature, if ΔH is positive and ΔS is negative.

16.47 The quantity ΔG_f° applies to the equation in which one mole of pure phosgene is produced from the naturally occurring forms of the elements:

$C(s) + \frac{1}{2}O_2(g) + Cl_2(g) \rightarrow COCl_2(g)$, $\Delta G_f^\circ = ?$

We can determine ΔG_f° if we can find values for ΔH_f° and ΔS_f°, because:

$\Delta G° = \Delta H° - T\Delta S°$

The value of ΔS_f° is determined using S° for phosgene in the following way:

$\Delta S_f° = \{S°[COCl_2(g)]\} - \{S°[C(s)] + \frac{1}{2}S°[O_2(g)] + S°[Cl_2(g)]\}$

$\Delta S_f° = \{1 \text{ mol} \times (284 \text{ J K}^{-1} \text{ mol}^{-1})\} - \{1 \text{ mol} \times (5.69 \text{ J K}^{-1} \text{ mol}^{-1})$

$+ {}^1/_2 \text{ mol} \times (205.0 \text{ J K}^{-1} \text{ mol}^{-1}) + 1 \text{ mol} \times (223.0 \text{ J K}^{-1} \text{ mol}^{-1})\}$

$\Delta S_f° = -47 \text{ J K}^{-1} \text{ mol}^{-1}$ or -47 J/K

$\Delta G_f° = \Delta H_f^\circ - T\Delta S_f^\circ = -223$ kJ/mol $-$ (298 K)(-0.047 kJ/K mol)

$= -209$ kJ/mol

16.48 As in the answer to Review Exercise 16.47, the value of ΔS_f° is determined as follows:

$\Delta S° = $ (sum S°[products]) $-$ (sum S°[reactants])

$2Al(s) + {}^3/_2O_2(g) \rightarrow Al_2O_3(s)$

$\Delta S_f° = \{S°[Al_2O_3(s)]\} - \{2S°[Al(s)] + {}^3/_2S°[O_2(g)]\}$

$\Delta S_f° = \{1 \text{ mol} \times (51.00 \text{ J K}^{-1} \text{ mol}^{-1})\} - \{2 \text{ mol} \times (28.3 \text{ J K}^{-1} \text{ mol}^{-1})$
$+ {}^3/_2 \text{ mol} \times (205.0 \text{ J K}^{-1} \text{ mol}^{-1})\}$

$\Delta S_f° = -313 \text{ J/K or } -313 \text{ J K}^{-1} \text{ mol}^{-1} = -0.313 \text{ kJ K}^{-1} \text{ mol}^{-1}$

$\Delta G_f° = \Delta H_f° - T\Delta S_f° = -1669.8 \text{ kJ/mol} - (298 \text{ K})(-0.313 \text{ kJ K}^{-1} \text{ mol}^{-1})$

$\Delta G_f° = -1576.5 \text{ kJ/mol}$

This value agrees nicely with that listed in Table 16.2.

16.49 $\Delta G° = (\text{sum } \Delta G_f°[\text{products}]) - (\text{sum } \Delta G_f°[\text{reactants}])$

(a) $\Delta G° = \{\Delta G_f°[H_2SO_4(\ell)]\} - \{\Delta G_f°[H_2O(\ell)] + \Delta G_f°[SO_3(g)]\}$

$\Delta G° = \{1 \text{ mol} \times (-689.9 \text{ kJ/mol})\} - \{1 \text{ mol} \times (-237.2 \text{ kJ/mol})$
$+ 1 \text{ mol} \times (-370.4 \text{ kJ/mol})\}$

$\Delta G° = -82.3 \text{ kJ}$

(b) $\Delta G° = \{2\Delta G_f°[NH_3(g)] + \Delta G_f°[H_2O(\ell)] + \Delta G_f°[CaCl_2(s)]\}$
$- \{\Delta G_f°[CaO(s)] + 2\Delta G_f°[NH_4Cl(s)]\}$

$\Delta G° = \{2 \text{ mol} \times (-16.7 \text{ kJ/mol}) + 1 \text{ mol} \times (-237.2 \text{ kJ/mol})$
$+ 1 \text{ mol} \times (-750.2 \text{ kJ/mol})\} - \{1 \text{ mol} \times (-604.2 \text{ kJ/mol})$
$+ 2 \text{ mol} \times (-203.9 \text{ kJ/mol})\}$

$\Delta G° = -8.8 \text{ kJ}$

(c) $\Delta G° = \{\Delta G_f°[H_2SO_4(\ell)] + \Delta G_f°[CaCl_2(s)]\} - \{\Delta G_f°[CaSO_4(s)]$
$+ 2\Delta G_f°[HCl(g)]\}$

$\Delta G° = \{1 \text{ mol} \times (-689.9 \text{ kJ/mol}) + 1 \text{ mol} \times (-750.2 \text{ kJ/mol})\}$
$- \{1 \text{ mol} \times (-1320.3 \text{ kJ/mol}) + 2 \text{ mol} \times (-95.27 \text{ kJ/mol})\}$

$\Delta G° = +70.7 \text{ kJ}$

(d) $\Delta G° = \{\Delta G_f°[C_2H_5OH(\ell)]\} - \{\Delta G_f°[H_2O(g)] + \Delta G_f°[C_2H_4(g)]\}$

$\Delta G° = \{1 \text{ mol} \times (-174.8 \text{ kJ/mol})\} - \{1 \text{ mol} \times (-228.6 \text{ kJ/mol})$
$+ 1 \text{ mol} \times (68.12 \text{ kJ/mol})\}$

$\Delta G° = -14.3 \text{ kJ}$

(e) $\Delta G° = \{2\Delta G_f^\circ[H_2O(\ell)] + \Delta G_f^\circ[SO_2(g)] + \Delta G_f^\circ[CaSO_4(s)]\}$
$- \{2\Delta G_f^\circ[H_2SO_4(\ell)] + \Delta G_f^\circ[Ca(s)]\}$

$\Delta G° = \{2 \text{ mol} \times (-237.2 \text{ kJ/mol}) + 1 \text{ mol} \times (-300.4 \text{ kJ/mol}) + 1 \text{ mol} \times (-1320.3 \text{ kJ/mol})\} - \{2 \text{ mol} \times (-689.9 \text{ kJ/mol}) + 1 \text{ mol} \times (0.0 \text{ kJ/mol})\}$

$\Delta G° = -715.3 \text{ kJ}$

16.50 $\Delta G° = (\text{sum } \Delta G_f^\circ[\text{products}]) - (\text{sum } \Delta G_f^\circ[\text{reactants}])$

(a) $\Delta G° = \{\Delta G_f^\circ[H_2O(g)] + \Delta G_f^\circ[CaCl_2(s)]\}$
$- \{2\Delta G_f^\circ[HCl(g)] + \Delta G_f^\circ[CaO(s)]\}$

$\Delta G° = \{1 \text{ mol} \times (-228.6 \text{ kJ/mol}) + 1 \text{ mol} \times (-750.2 \text{ kJ/mol})\} - \{2 \text{ mol} \times (-95.27 \text{ kJ/mol}) + 1 \text{ mol} \times (-604.2 \text{ kJ/mol})\}$

$\Delta G° = -184.1 \text{ kJ}$

(b) $\Delta G° = \{\Delta G_f^\circ[Na_2SO_4(s)] + 2\Delta G_f^\circ[HCl(g)]\}$
$- \{\Delta G_f^\circ[H_2SO_4(\ell)] + 2\Delta G_f^\circ[NaCl(s)]\}$

$\Delta G° = \{1 \text{ mol} \times (-1266.8 \text{ kJ/mol}) + 2 \text{ mol} \times (-95.27 \text{ kJ/mol})\} - \{1 \text{ mol} \times (-689.9 \text{ kJ/mol}) + 2 \text{ mol} \times (-384.0 \text{ kJ/mol})\}$

$\Delta G° = +0.6 \text{ kJ}$

(c) $\Delta G° = \{\Delta G_f^\circ[NO(g)] + 2\Delta G_f^\circ[HNO_3(\ell)]\}$
$- \{3\Delta G_f^\circ[NO_2(g)] + \Delta G_f^\circ[H_2O(\ell)]\}$

$\Delta G° = \{1 \text{ mol} \times (86.69 \text{ kJ/mol}) + 2 \text{ mol} \times (-79.91 \text{ kJ/mol})\} - \{3 \text{ mol} \times (51.84 \text{ kJ/mol}) + 1 \text{ mol} \times (-237.2 \text{ kJ/mol})\}$

$\Delta G° = +8.6 \text{ kJ}$

(d) $\Delta G° = \{2\Delta G_f^\circ[Ag(s)] + \Delta G_f^\circ[CaCl_2(s)]\}$
$- \{2\Delta G_f^\circ[AgCl(s)] + \Delta G_f^\circ[Ca(s)]\}$

$\Delta G° = \{2 \text{ mol} \times (0.0 \text{ kJ/mol}) + 1 \text{ mol} \times (-750.2 \text{ kJ/mol})\} - \{2 \text{ mol} \times (-109.7 \text{ kJ/mol}) + 1 \text{ mol} \times (0.0 \text{ kJ/mol})\}$

$\Delta G° = -530.8 \text{ kJ}$

(e) $\Delta G° = \{\Delta G_f^\circ[NH_4Cl(s)]\} - \{\Delta G_f^\circ[HCl(g)] + \Delta G_f^\circ[NH_3(g)]\}$

$\Delta G° = \{1 \text{ mol} \times (-203.9 \text{ kJ/mol})\} - \{1 \text{ mol} \times (-95.27 \text{ kJ/mol}) + 1 \text{ mol} \times (-16.7 \text{ kJ/mol})\}$

$\Delta G° = -91.9$ kJ

16.51 $2CaSO_4 \cdot \frac{1}{2}H_2O(s) + 3H_2O(\ell) \rightarrow 2CaSO_4 \cdot 2H_2O(s)$

$\Delta G° = (\text{sum } \Delta G°_f[\text{products}]) - (\text{sum } \Delta G°_f[\text{reactants}])$

$\Delta G° = \{2\Delta G°_f[CaSO_4 \cdot 2H_2O(s)]\} - \{2\Delta G°_f[CaSO_4 \cdot \frac{1}{2}H_2O(s)] + 3\Delta G°_f[H_2O(\ell)]\}$

$$\Delta G° = \{2 \text{ mol} \times (-1795.7 \text{ kJ/mol})\} \\ - \{2 \text{ mol} \times (-1435.2 \text{ kJ/mol}) + 3 \text{ mol} \times (-237.2 \text{ kJ/mol})\}$$

$\Delta G° = -9.4$ kJ

16.52 $\Delta G° = (\text{sum } \Delta G°_f[\text{products}]) - (\text{sum } \Delta G°_f[\text{reactants}])$

$COCl_2(g) + H_2O(g) \rightarrow CO_2(g) + 2HCl(g)$

$\Delta G° = \{2\Delta G°_f[HCl(g)] + \Delta G°_f[CO_2(g)]\} - \{\Delta G°_f[H_2O(g)] + \Delta G°_f[COCl_2(g)]\}$

$$\Delta G° = \{2 \text{ mol} \times (-95.27 \text{ kJ/mol}) + 1 \text{ mol} \times (-394.4 \text{ kJ/mol})\} \\ - \{1 \text{ mol} \times (-228.6 \text{ kJ/mol}) + 1 \text{ mol} \times (-210 \text{ kJ/mol})\}$$

$\Delta G° = -146$ kJ

16.53 Multiply the reverse of the second equation by 2 (remembering to multiply the associated free energy change by -2), and add the result to the first equation:

$4NO(g) \rightarrow 2N_2O(g) + O_2(g)$, $\Delta G° = -139.56$ kJ

$4NO_2(g) \rightarrow 4NO(g) + 2O_2(g)$, $\Delta G° = +139.4$ kJ

$4NO_2 \rightarrow 3O_2(g) + 2N_2O(g)$, $\Delta G° = -0.16$ kJ

This result is the reverse of the desired reaction, which must then have

$\Delta G° = +0.16$ kJ

16.54 Add the reverse of the first equation to the second equation plus twice the third equation:

$CO(NH_2)_2(s) + 2NH_4Cl(s) \rightarrow COCl_2(g) + 4NH_3(g)$, $\Delta G° = +79.36$ kcal

$COCl_2(g) + H_2O(\ell) \rightarrow CO_2(g) + 2HCl(g)$, $\Delta G° = -33.89$ kcal

$2NH_3(g) + 2HCl(g) \rightarrow 2NH_4Cl(s)$, $\Delta G° = -43.96$ kcal

$CO(NH_2)_2(s) + H_2O(\ell) \rightarrow 2NH_3(g) + CO_2(g)$, $\Delta G° = +1.51$ kcal

16.55 The desired reaction is obtained simply by adding the two equations, and the resulting $\Delta G°$ is the simple sum of the two values given in this problem:

$$\Delta G° = 13.13 \text{ kJ} + (-32.22 \text{ kJ}) = -19.09 \text{ kJ}$$

Free Energy and Work

16.56 The value of ΔG is equal to the maximum amount of work that may be obtained from any process.

16.57 A reversible process is one in which the driving force of the process is nearly completely balanced by an opposing force. The situation is a bit esoteric, since no process can be run in a completely reversible manner. Nevertheless, the closer one obtains to reversibility, the more efficient the system becomes as a source of the maximum amount of useful work that can be achieved with the process. See also the answer to Review Exercise 16.59.

16.58 As with other processes that are not carried out in a reversible fashion, the energy is lost to the environment and becomes unavailable for use.

16.59 Part of the answer lies in that given for Review Exercise 16.57. A truly reversible process would take forever to occur. Thus if we can observe an event happening, it cannot be a reversible one.

16.60 The maximum work obtainable from a reaction is equal in magnitude to the value of ΔG for the reaction. Thus we need only determine $\Delta G°$ for the process:

$$\Delta G° = (\text{sum } \Delta G_f°[\text{products}]) - (\text{sum } \Delta G_f°[\text{reactants}])$$

$$\Delta G° = \{3\Delta G_f°[H_2O(g)] + 2\Delta G_f°[CO_2(g)]\} - \{3\Delta G_f°[O_2(g)] + \Delta G_f°[C_2H_5OH(\ell)]\}$$

$$\Delta G° = \{3 \text{ mol} \times (-228.6 \text{ kJ/mol}) + 2 \text{ mol} \times (-394.4 \text{ kJ/mol})\} - \{3 \text{ mol} \times (0.0 \text{ kJ/mol})\ 1 \text{ mol} \times (-174.8 \text{ kJ/mol})\}$$

$$\Delta G° = -1299.8 \text{ kJ}$$

16.61 We must first determine $\Delta G°$ for the reaction:

$$CH_4(g) + 2O_2(g) \rightarrow CO_2(g) + 2H_2O(g)$$

$$\Delta G° = (\text{sum } \Delta G_f°[\text{products}]) - (\text{sum } \Delta G_f°[\text{reactants}])$$

$\Delta G° = \{\Delta G_f°[CO_2(g)] + 2\Delta G_f°[H_2O(g)]\} - \{\Delta G_f°[CH_4(g)] + 2\Delta G_f°[O_2(g)]\}$

$\Delta G° = \{1 \text{ mol} \times (-394.4 \text{ kJ/mol}) + 2 \text{ mol} \times (-228.6 \text{ kJ/mol})\}$
$\qquad - \{1 \text{ mol} \times (-50.79 \text{ kJ/mol}) + 2 \text{ mol} \times (0.0 \text{ kJ/mol})\}$

$\Delta G° = -800.8 \text{ kJ/mol}$

Next, we determine the amount of work available from the combustion of 27.0 g of CH_4:

$$27.0 \text{ g } CH_4 \times \frac{1 \text{ mol } CH_4}{16.0 \text{ g } CH_4} \times \frac{800.8 \text{ kJ}}{1 \text{ mol } CH_4} = 1.35 \times 10^3 \text{ kJ}$$

16.62 $\Delta G°$ for the combustion of C_2H_5OH is -1299.8 kJ/mol, as determined in Example 16.5. $\Delta G°$ for the combustion of octane is -5307 kJ/mol as determined in Example 16.6.

Next, determine the number of moles in 3.78 L of each material:

For C_2H_5OH:

$$3.78 \text{ L} \times \frac{1000 \text{ mL}}{1 \text{ L}} \times \frac{0.7893 \text{ g}}{1.000 \text{ mL}} \times \frac{1 \text{ mol}}{46.07 \text{ g}} = 64.76 \text{ mol } C_2H_5OH$$

For octane:

$$3.78 \text{ L} \times \frac{1000 \text{ mL}}{1 \text{ L}} \times \frac{0.7025 \text{ g}}{1.000 \text{ mL}} \times \frac{1 \text{ mol}}{114.2 \text{ g}} = 23.25 \text{ mol octane}$$

The maximum amount of work for each material is calculated as follows:

For C_2H_5OH:

$$1299.8 \text{ kJ/mol} \times 64.76 \text{ mol} = 8.42 \times 10^4 \text{ kJ}$$

For octane:

$$5307 \text{ kJ/mol} \times 23.25 \text{ mol} = 1.23 \times 10^5 \text{ kJ}$$

Thus octane has the capacity to supply more work upon combustion per gallon than does ethanol.

Free Energy and Equilibrium

16.63 At equilibrium, the value of ΔG is zero.

16.64 The process of bond breaking always has an associated value for ΔH that is positive. Also, since bond breaking increases disorder, it always has an associated value for ΔS that is positive. A process such as this, with a positive value for ΔH and a positive value for ΔS, becomes spontaneous at high temperatures.

16.65 Before the heat transfer, the molecules in the hot object vibrate and move more violently than do those of the cooler object. When in contact with one another, the objects transfer heat through collisions, and eventually, some of the kinetic energy of the molecules in the hot object is transferred to the molecules of the cool object. This process of energy transfer continues until the objects have the same temperature. The heat transfer is spontaneous because the scattering of kinetic energy among the molecules of both objects is a process with a positive value for its associated ΔS.

16.66

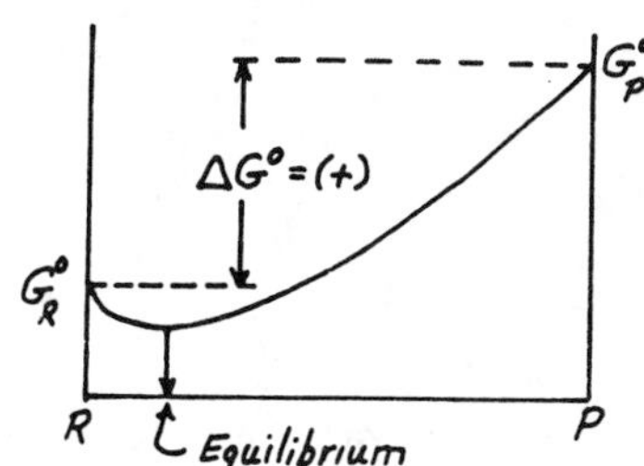

16.67

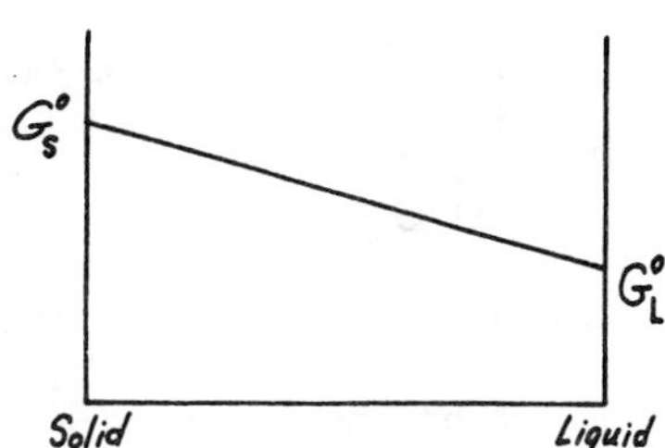

16.68 At equilibrium, $\Delta G = 0 = \Delta H - T\Delta S$

$T_{eq} = \Delta H/\Delta S$, and, assuming that ΔS is independent of temperature, we have:

$T_{eq} = (31.4 \times 10^3\ J\ mol^{-1}) \div (94.2\ J\ K^{-1}\ mol^{-1}) = 333\ K$

16.69 At equilibrium, $\Delta G = 0 = \Delta H - T\Delta S$

$T_{eq} = \Delta H/\Delta S$, and, assuming that ΔS and ΔH are independent of temperature, we have:

$T_{eq} = (10.0 \times 10^3\ J\ mol^{-1}) \div (9.50\ J\ K^{-1}\ mol^{-1}) = 1.05 \times 10^3\ K$

Apparently ΔH and ΔS are not independent of temperature!

16.70 At equilibrium, $\Delta G = 0 = \Delta H - T\Delta S$

Thus $\Delta H = T\Delta S$, and if we assume that both ΔH and ΔS are independent of temperature, we have:

$\Delta S = \Delta H/T_{eq} = (9.01 \times 10^3\ cal/mol) \div (99.3 + 273.15\ K)$

$\Delta S = 24.2 \text{ cal K}^{-1} \text{ mol}^{-1}$

16.71 We proceed as in the answer to Review Exercise 16.70:

$$\Delta S = \Delta H/T_{eq} = (31.9 \times 10^3 \text{ J/mol}) \div (56.2 + 273.15 \text{ K})$$
$$= 96.9 \text{ J K}^{-1} \text{ mol}^{-1}$$

16.72 The reaction is spontaneous if its associated value for $\Delta G°$ is negative.

$$\Delta G° = (\text{sum } \Delta G°_f[\text{products}]) - (\text{sum } \Delta G°_f[\text{reactants}])$$

$$\Delta G° = \{\Delta G°_f[HC_2H_3O_2(\ell)] + \Delta G°_f[H_2O(\ell)] + \Delta G°_f[NO(g)] + \Delta G°_f[NO_2(g)]\} - \{\Delta G°_f[C_2H_4(g)] + 2\Delta G°_f[HNO_3(\ell)]\}$$

$$\Delta G° = \{1 \text{ mol} \times (-392.5 \text{ kJ/mol}) + 1 \text{ mol} \times (-237.2 \text{ kJ/mol}) + 1 \text{ mol} \times (86.69 \text{ kJ/mol}) + 1 \text{ mol} \times (51.84 \text{ kJ/mol})\} - \{1 \text{ mol} \times (68.12 \text{ kJ/mol}) + 2 \text{ mol} \times (-79.91 \text{ kJ/mol})\}$$

$$\Delta G° = -399.5 \text{ kJ}$$

Yes, the reaction is spontaneous.

16.73 We first balance each equation, and then calculate a value of $\Delta G°$. If $\Delta G°$ is a negative number, then the reaction is spontaneous.

(a) $3PbO(s) + 2NH_3(g) \rightarrow 3Pb(s) + N_2(g) + 3H_2O(g)$

$$\Delta G° = \{3\Delta G°_f[Pb(s)] + \Delta G°_f[N_2(g)] + 3\Delta G°_f[H_2O(g)]\} - \{3\Delta G°_f[PbO(s)] + 2\Delta G°_f NH_3(g)]\}$$

$$\Delta G° = \{3 \text{ mol} \times (0.0 \text{ kJ/mol}) + 1 \text{ mol} \times 0.0 \text{ kJ/mol} + 3 \text{ mol} \times (-228.6 \text{ kJ/mol})\} - \{3 \text{ mol} \times (-189.3 \text{ kJ/mol}) + 2 \text{ mol} \times (-16.7 \text{ kJ/mol})\}$$

$\Delta G° = -84.5 \text{ kJ}$ $\therefore$ the reaction is spontaneous.

(b) $NaOH(s) + HCl(g) \rightarrow NaCl(s) + H_2O(\ell)$

$$\Delta G° = \{\Delta G°_f[NaCl(s)] + \Delta G°_f[H_2O(\ell)]\} - \{\Delta G°_f[NaOH(s)] + \Delta G°_f[HCl(g)]\}$$

$$\Delta G° = \{1 \text{ mol} \times (-384.0 \text{ kJ/mol}) + 1 \text{ mol} \times (-237.2 \text{ kJ/mol})\} - \{1 \text{ mol} \times (-382 \text{ kJ/mol}) + 1 \text{ mol} \times (-95.27)\}$$

$\Delta G° = -144 \text{ kJ}$ $\therefore$ the reaction is spontaneous.

(c) $Al_2O_3(s) + 2Fe(s) \rightarrow Fe_2O_3(s) + 2Al(s)$

$$\Delta G° = \{\Delta G_f°[Fe_2O_3(s)] + 2\Delta G_f°[Al(s)]\} - \{\Delta G_f°[Al_2O_3(s)] + 2\Delta G_f°[Fe(s)]\}$$

$$\Delta G° = \{1 \text{ mol} \times (-741.0 \text{ kJ/mol}) + 2 \text{ mol} \times (0.0 \text{ kJ/mol})\} - \{1 \text{ mol} \times (-1576.4 \text{ kJ/mol}) + 2 \text{ mol} \times (0.0 \text{ kJ/mol})\}$$

$\Delta G° = +835.4$ kJ $\therefore$ the reaction is not spontaneous.

(d) $2CH_4(g) \rightarrow C_2H_6(g) + H_2(g)$

$$\Delta G° = \{\Delta G_f°[C_2H_6(g)] + \Delta G_f°[H_2(g)]\} - \{2\Delta G_f°[CH_4(g)]\}$$

$$\Delta G° = \{1 \text{ mol} \times (-32.9 \text{ kJ/mol}) + 1 \text{ mol} \times (0.0 \text{ kJ/mol})\} - \{2 \text{ mol} \times (-50.79 \text{ kJ/mol})\}$$

$\Delta G° = +68.7$ kJ $\therefore$ the reaction is not spontaneous.

16.74 Although a reaction may have a favorable ΔG, and therefore be a spontaneous reaction in the thermodynamic sense of the word, it may develop that the reaction is too slow at normal temperatures to be observed.

Effect of Temperature on the Free Energy Change

16.75 ΔG depends explicitly on temperature: $\Delta G = \Delta H - T\Delta S$

16.76 $\Delta G' = \Delta H° - T\Delta S°$

16.77 As the temperature is raised, $\Delta G'$ will become less negative, if $\Delta H°$ is negative and $\Delta S°$ is negative. Accordingly, less product will be present at equilibrium.

16.78 $\Delta G' = \Delta H° - T\Delta S°$, where $T = 373$ K in all cases, and where the values of $\Delta H°$ and $\Delta S°$ are obtained in the usual manner, i.e.:

$$\Delta H° = (\text{sum } \Delta H_f°[\text{products}]) - (\text{sum } \Delta H_f°[\text{reactants}])$$

$$\Delta S° = (\text{sum } S°[\text{products}]) - (\text{sum } S°[\text{reactants}])$$

(a) $\Delta H° = \{\Delta H_f°[C_2H_6(g)]\} - \{\Delta H_f°[H_2(g)] + \Delta H_f°[C_2H_4(g)]\}$

$$\Delta H° = \{1 \text{ mol} \times (-84.667 \text{ kJ/mol})\} - \{1 \text{ mol} \times (0.0 \text{ kJ/mol}) + 1 \text{ mol} \times (52.284 \text{ kJ/mol})\}$$

$\Delta H° = -136.951\ kJ$

$\Delta S° = \{S°[C_2H_6(g)]\} - \{S°[H_2(g)] + S°[C_2H_4(g)]\}$

$\Delta S° = \{1\ mol \times (229.5\ J\ K^{-1}\ mol^{-1})\} - \{1\ mol \times (130.6\ J\ K^{-1}\ mol^{-1})$

$+ 1\ mol \times (219.8\ J\ K^{-1}\ mol^{-1})\}$

$\Delta S° = -120.9\ J/K = -0.1209\ kJ/K$

$\Delta G' = \Delta H° - T\Delta S° = -136.951\ kJ - (373\ K)(-0.1209\ kJ/K) = -91.9\ kJ$

(b) $\Delta H° = \{3\Delta H_f°[H_2O(g)] + 5\Delta H_f°[SO_2(g)] + 2\Delta H_f°[NO(g)]\}$

$- \{5\Delta H_f°[SO_3(g)] + 2\Delta H_f°[NH_3(g)]\}$

$\Delta H° = \{3\ mol \times (-241.8\ kJ/mol) + 5\ mol \times (-296.9\ kJ/mol)$
$+ 2\ mol \times (90.37\ kJ/mol)\} - \{5\ mol \times (-395.2\ kJ/mol)$
$+ 2\ mol \times (-46.19\ kJ/mol)\}$

$\Delta H° = +39.2\ kJ$

$\Delta S° = \{3S°[H_2O(g)] + 5S°[SO_2(g)] + 2S°[NO(g)]\}$

$- \{5S°[SO_3(g)] + 2S°[NH_3(g)]\}$

$\Delta S° = \{3\ mol \times (188.7\ J\ K^{-1}\ mol^{-1}) + 5\ mol \times (248.5\ J\ K^{-1}\ mol^{-1})$

$+ 2\ mol \times (210.6\ J\ K^{-1}\ mol^{-1})\} - \{5\ mol \times (256.2\ J\ K^{-1}\ mol^{-1})$

$+ 2\ mol \times (192.5\ J\ K^{-1}\ mol^{-1})\}$

$\Delta S° = +563.8\ J/K = 0.5638\ kJ/K$

$\Delta G° = \Delta H° - T\Delta S°$

$\Delta G' = 39.2\ kJ - (373\ K)(0.5638\ kJ/K) = -171\ kJ$

(c) $\Delta H° = \{3\Delta H_f°[NO(g)]\} - \{\Delta H_f°[NO_2(g)] + \Delta H_f°[N_2O(g)]\}$

$\Delta H° = \{3\ mol \times (90.37\ kJ/mol)\}$
$- \{1\ mol \times (33.8\ kJ/mol) + 1\ mol \times (81.57\ kJ/mol)\}$

$\Delta H° = +155.7\ kJ$

$\Delta S° = \{3S°[NO(g)]\} - \{S°[NO_2(g)] + S°[N_2O(g)]\}$

$\Delta S° = \{3\ mol \times (210.6\ J\ K^{-1}\ mol^{-1})\}$

$- \{1\ mol \times (240.5\ J\ K^{-1}\ mol^{-1}) + 1\ mol \times (220.0\ J\ K^{-1}\ mol^{-1})\}$

$\Delta S° = +171.3\ J/K = 0.1713\ kJ/K$

$\Delta G' = \Delta H° - T\Delta S° = 155.7 \text{ kJ} - (373 \text{ K})(0.1713 \text{ kJ/K}) = 91.8 \text{ kJ}$

<u>Thermodynamic Equilibrium Constants</u>

16.79 $\Delta G = -RT \ln K$

16.80 This is an equilibrium constant whose numerical value is calculated using thermodynamic data.

16.81 When K = 1, $\Delta G° = 0$.

16.82 $\Delta G° = -RT \ln K_p$

$-50.79 \times 10^3 \text{ J} = -(8.314 \text{ J K}^{-1} \text{ mol}^{-1})(298 \text{ K}) \times \ln K_p$

$\ln K_p = 20.50$

Taking the antiln of both sides of this equation gives:

$K_p = 8.000 \times 10^8$

This is a favorable reaction, since the equilibrium lies far to the side favoring products.

16.83 T = 37 + 273 = 310 K

$\Delta G' = -RT \ln K_c$

$-8 \times 10^3 \text{ cal} = -(1.987 \text{ cal K}^{-1} \text{ mol}^{-1})(310 \text{ K}) \times \ln K_c$

$\ln K_c = -13.0$

$K_c = 4 \times 10^5$

16.84 For such a circumstance, the value of ln K must be equal to zero, because no other term in the following equation can be equal to zero:

$\Delta G = -RT \ln K$

If ln K = 0, then K = 1.

16.85 $\Delta G° = -RT \ln K_p$

$\Delta G° = -(8.314 \text{ J K}^{-1} \text{ mol}^{-1})(298 \text{ K}) \times \ln(10.0)$

$\Delta G° = 5.70 \times 10^3 \text{ J}$

16.86 $\Delta G' = -RT \ln K_p$

$\Delta G' = -(8.314\ J\ K^{-1}\ mol^{-1})(500\ K)\ \ln (6.25 \times 10^{-3})$

$\Delta G' = 2.11 \times 10^4\ J = 21.1\ kJ$

16.87 $\Delta G° = (\text{sum}\ \Delta G_f°[\text{products}]) - (\text{sum}\ \Delta G_f°[\text{reactants}])$

$\Delta G° = \{\Delta G_f°[N_2(g)] + 2\Delta G_f°[CO_2(g)]\} - \{2\Delta G_f°[NO(g)] + 2\Delta G_f°[CO(g)]\}$

$\Delta G° = \{1\ mol \times (0.0\ kJ/mol) + 2\ mol \times (-394.4\ kJ/mol)\}$
$\qquad - \{2\ mol \times (86.69\ kJ/mol) + 2\ mol \times (-137.3\ kJ/mol)\}$

$\Delta G° = -687.6\ kJ$

It is understood that this is the $\Delta G°$ value that corresponds to the equation in which one mole of N_2 is formed.

$\Delta G° = -RT \ln K_p$

$-687.6 \times 10^3\ J = -(8.314\ J\ K^{-1}\ mol^{-1})(298\ K)\ \ln K_p$

$\ln K_p = 278$

Taking the antiln of both sides of this last equation gives:

$K_p = 1 \times 10^{121}$

16.88 First we determine $\Delta H°$ by using the normal approach:

$\Delta H°$ = (sum ΔH°_f[products]) - (sum ΔH°_f[reactants])

$\Delta H°$ = $\{\Delta H^\circ_f[N_2(g)] + 2\Delta H^\circ_f[CO_2(g)]\} - \{2\Delta H^\circ_f[NO(g)] + 2\Delta H^\circ_f[CO(g)]\}$

$\Delta H°$ = {1 mol × (0.0 kJ/mol) + 2 mol × (-393.5 kJ/mol)}
- {2 mol × (90.37 kJ/mol) + 2 mol × (-110.5 kJ/mol)}

$\Delta H°$ = -746.7 kJ for the formation of one mole of N_2.

$\Delta S°$ = $\{S°[N_2(g)] + 2S°[CO_2(g)]\} - \{2S°[NO(g)] + 2S°[CO(g)]\}$

$\Delta S°$ = {1 mol × (191.5 J K^{-1} mol^{-1}) + 2 mol × (213.6 J K^{-1} mol^{-1})}

- {2 mol × (210.6 J K^{-1} mol^{-1}) + 2 mol × (197.9 J K^{-1} mol^{-1})}

$\Delta S°$ = -198.3 J/K for the formation of one mole of N_2.

$\Delta G' = \Delta H° - T\Delta S°$

$\Delta G'$ = -746.7 kJ - (773 K)(-0.1983 kJ/K) = -593 kJ/mol

$\Delta G = -RT \ln K_p$

-593 kJ/mol = -(8.314 J K^{-1} mol^{-1})(773 K) $\ln K_p$

$\ln K_p = 92.3$

Taking the antiln of both sides of this equation gives:

$K_p = 1 \times 10^{40}$

CHAPTER SEVENTEEN

PRACTICE EXERCISES

1. $2H_2O(\ell) + 2e^- \rightarrow H_2(g) + 2OH^-(aq)$ reduction

   $2Br^-(aq) \rightarrow Br_2(aq) + 2e^-$ oxidation

   $2H_2O(\ell) + 2Br^-(aq) \rightarrow Br_2(aq) + H_2(g) + 2OH^-(aq)$ net cell reaction

2. The number of Coulombs is:

   $4.00\ A \times 200\ s = 800\ C$

   The number of moles is:

   $$800\ C \times \frac{1\ F}{96{,}500\ C} \times \frac{1\ mol\ OH^-}{1\ F} = 8.29 \times 10^{-3}\ mol\ OH^-$$

3. The number of moles of Au to be deposited is:

   $3.00\ g\ Au \div 197\ g/mol = 0.0152\ mol\ Au$

   The number of Coulombs (A•s) is:

   $$0.0152\ mol\ Au \times \frac{3\ F}{1\ mol\ Au} \times \frac{96{,}500\ C}{1\ F} = 4.40 \times 10^3\ C$$

   The number of minutes is:

   $$\frac{4.40 \times 10^3\ A \cdot s}{10.0\ A} \times \frac{1\ min}{60\ s} = 7.33\ min$$

4. As in Practice Exercise 3 above, the number of Coulombs is 4.40×10^3 C. This corresponds to a current of:

   $$\frac{4.40 \times 10^3\ A \cdot s}{20.0\ min} \times \frac{1\ min}{60\ s} = 3.67\ A$$

5.

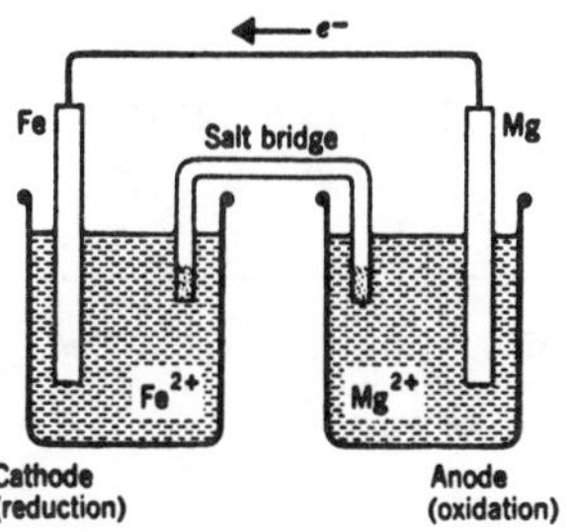

6. $E°_{cell} = E°_{substance\ reduced} - E°_{substance\ oxidized}$

$E°_{cell} = E°_{Fe^{2+}} - E°_{Mg^{2+}}$

$1.96\ V = (-0.41\ V) - E°_{Mg^{2+}}$

$E°_{Mg^{2+}} = -0.41 - 1.96 = -2.37\ V$

7. The half-reaction having the more positive value for E° will occur as a reduction. The other half-reaction should be reversed, so as to appear as an oxidation.

$NiO_2(s) + 2H_2O(\ell) + 2e^- \rightarrow Ni(OH)_2(s) + 2OH^-(aq)$ reduction

$2OH^-(aq) + Fe(s) \rightarrow 2e^- + Fe(OH)_2(s)$ oxidation

$NiO_2(s) + Fe(s) + 2H_2O(\ell) \rightarrow Ni(OH)_2(s) + Fe(OH)_2(s)$ net reaction

$E°_{cell} = E°_{substance\ reduced} - E°_{substance\ oxidized}$

$E°_{cell} = E°_{NiO_2} - E°_{Fe(OH)_2}$

$E°_{cell} = 0.49 - (-0.88) = 1.37\ V$

8. The half-reaction having the more positive value for E° will occur as a reduction. The other half-reaction should be reversed, so as to appear as an oxidation.

$3 \times (MnO_4^-(aq) + 8H^+(aq) + 5e^- \rightarrow Mn^{2+}(aq) + 4H_2O(\ell))$ reduction

$5 \times (Cr(s) \rightarrow Cr^{3+}(aq) + 3e^-)$ oxidation

$3MnO_4^-(aq) + 24H^+(aq) + 5Cr(s) \rightarrow 5Cr^{3+}(aq) + 3Mn^{2+}(aq) + 12H_2O(\ell)$ net reaction

$E^\circ_{cell} = E^\circ_{substance\ reduced} - E^\circ_{substance\ oxidized}$

$E^\circ_{cell} = E^\circ_{MnO_4^-} - E^\circ_{Cr^{3+}}$

$E^\circ_{cell} = 1.49\ V - (-0.74\ V) = 2.23\ V$

9. The half-reaction with the more positive value of E° (listed higher in Table 17.1) will occur as a reduction. The half-reaction having the less positive (more negative) value of E° (listed lower in Table 17.1) will be reversed and occur as an oxidation.

$Br_2(aq) + 2e^- \rightarrow 2Br^-(aq)$ reduction

$H_2SO_3(aq) + H_2O(\ell) \rightarrow SO_4^{2-}(aq) + 4H^+(aq) + 2e^-$ oxidation

$Br_2(aq) + H_2SO_3(aq) + H_2O(\ell) \rightarrow 2Br^-(aq) + SO_4^{2-}(aq) + 4H^+(aq)$ net reaction

10. A reaction will occur spontaneously in the forward direction if the value of E° is positive. We therefore evaluate E° for each reaction using:

$E^\circ_{cell} = E^\circ_{substance\ reduced} - E^\circ_{substance\ oxidized}$

(a) $Br_2(aq) + 2e^- \rightarrow 2Br^-(aq)$ reduction

$Cl_2(aq) + 2H_2O(\ell) \rightarrow 2HOCl(aq) + 2H^+(aq) + 2e^-$ oxidation

$E^\circ_{cell} = E^\circ_{Br_2} - E^\circ_{HOCl}$

$E^\circ_{cell} = 1.07\ V - (1.63) = -0.56$ ∴ nonspontaneous

(b) $2Cr^{3+}(aq) + 6e^- \rightarrow 2Cr(s)$ reduction

$3Zn(s) \rightarrow 3Zn^{2+}(aq) + 6e^-$ oxidation

$$E°_{cell} = E°_{Cr^{3+}} - E°_{Zn^{2+}}$$

$$E°_{cell} = -0.74 - (-0.76) = +0.02\ V \quad \therefore \text{ spontaneous}$$

11. It was determined in Practice Exercise 8 that the reaction in question had a standard cell potential of 2.23 V, or 2.23 J/C. Since 15 mole of e^- are involved (i.e. n = 15), we have:

$$\Delta G° = -nFE°_{cell} = -(15)(96{,}500)(2.23) = -3.23 \times 10^6\ J = -3.23 \times 10^3\ kJ$$

12.

$$E°_{cell} = \frac{0.0592}{n} \log K_c$$

$$-0.46 = (0.0592)/2 \times \log K_c$$

$$\log K_c = -15.5$$

Taking the antilog of both sides of the above equation gives:

$$K_c = 3.2 \times 10^{-16}$$

This very small value for the equilibrium constant means that the products of the reaction are not formed spontaneously. The equilibrium lies far to the left, favoring reactants, and we do not expect much product to form.

13. The expression for Q is:

$$Q = \frac{[Zn^{2+}]}{[Cu^{2+}]} = \frac{1.00}{0.0100} = 100$$

$$E_{cell} = E°_{cell} - \frac{0.0592}{n} \log Q$$

$$E_{cell} = 1.10 - (0.0592)/2 \times \log(100) = 1.04\ V$$

14. The Nernst equation for this cell is:

$$E_{cell} = E°_{cell} - \frac{0.0592}{n} \times \log \frac{[Cu^{2+}]}{[Ag^+]^2}$$

In each case n = 2, $E°_{cell} = 0.46$ V, and $[Ag^+] = 1.00$, so the above equation becomes:

$$E_{cell} = 0.46 - (0.0296) \times \log [Cu^{2+}]$$

For the first solution, we have:

$$E_{cell} = 0.57\ V = 0.46 - (0.0296) \times \log [Cu^{2+}]$$

$\log [Cu^{2+}] = -3.72$ or $[Cu^{2+}] = \text{antilog}(-3.72) = 1.9 \times 10^{-4}$ M

For the second solution, we have:

$$E_{cell} = 0.82\ V = 0.46 - (0.0296) \times \log [Cu^{2+}]$$

$\log [Cu^{2+}] = -12.2$ or $[Cu^{2+}] = \text{antilog}(-12.2) = 6.3 \times 10^{-13}$ M

15. We are told that, in this galvanic cell, the chromium electrode is the anode, meaning that oxidation occurs at the chromium electrode.

Now in general, we have the equation:

$$E^\circ_{cell} = E^\circ_{reduction} - E^\circ_{oxidation}$$

which becomes, in particular for this case:

$$E^\circ_{cell} = E^\circ_{Ni^{2+}} - E^\circ_{Cr^{3+}}$$

or $E^\circ_{cell} = -0.25\ V - (-0.74\ V) = +0.49\ V$

The net cell reaction is given by the sum of the reduction and the oxidation half-reactions, multiplied in each case so as to eliminate electrons from the result:

$3 \times (2e^- + Ni^{2+}(aq) \rightarrow Ni(s))$	reduction
$2 \times (Cr(s) \rightarrow Cr^{3+}(aq) + 3e^-)$	oxidation
$3Ni^{2+}(aq) + 2Cr(s) \rightarrow 2Cr^{3+}(aq) + 3NI(s)$	net reaction

In this reaction, n = 6, and the value of Q is given by the expresssion:

$$Q = \frac{[Cr^{3+}]^2}{[Ni^{2+}]^3}$$

and the Nernst equation becomes:

$$0.55\ V = 0.49\ V - \frac{0.0592}{6} \log \frac{[Cr^{3+}]^2}{[Ni^{2+}]^3}$$

Solving for the log of Q, and substituting the fact that $[Ni^{2+}] = 1.20$ M, we get:

$$\log \frac{[Cr^{3+}]^2}{(1.20)^3} = -6$$

or $[Cr^{3+}] = 1 \times 10^{-3}$ M

REVIEW EXERCISES

Electrolysis

17.1 Oxidation is a loss of electrons, whereas reduction is a gain of electrons.

17.2 An electrochemical change is one that is caused by or that produces electricity.

17.3 The anode is defined to be the electrode at which oxidation occurs, whereas the cathode is the electrode at which reduction occurs.

17.4 In an electrolysis cell, the cathode is negative, and the anode is positive. The opposite is true of a galvanic cell.

17.5 In electrolytic conduction, charge is carried by the movement of positive and negative ions. In metallic conduction, charge is transported by the flow of electrons.

17.6 The flow of electrons in the external circuit must be accompanied by the electrolysis reaction. Otherwise the electrodes would accumulate charge, and the system would cease to function.

17.7 In direct current, electrons flow in only one direction. In alternating current, electron flow periodically reverses, flowing alternately in one direction and then in the other.

17.8 The cell reaction is the chemical reaction that occurs in an electrolysis or a galvanic cell.

17.9 This arrow indicates that an otherwise nonspontaneous reaction (unfavorable from the standpoint of thermodynamics) is being driven by the electrolytic apparatus.

17.10 In solid NaCl, the ions are held in place and cannot move about. In molten NaCl, the crystal lattice of the solid has been destroyed, and the ions are free to move, and consequently to conduct current by migrating either to the anode or to the cathode.

17.11 An inert electrode is one that does not react either with the electrolyte or with the products of the electrochemical reaction.

17.12 anode: $2Cl^-(\ell) = Cl_2(g) + 2e^-$

cathode: $2Na^+(\ell) + 2e^- = 2Na(\ell)$

overall: $2Na^+(\ell) + 2Cl^-(\ell) = Cl_2(g) + 2Na(\ell)$

17.13 oxidation: $2H_2O(\ell) = 4H^+(aq) + 4e^- + O_2(g)$

reduction: $2H_2O(\ell) + 2e^- = H_2(g) + 2OH^-(aq)$

17.14 It is reduction that occurs at the cathode, and near it, the pH increases due to the formation of $OH^-(aq)$. At the anode, where the oxidation of water occurs, the pH decreases due to the production of $H^+(aq)$. See the equations given in the answer to Review Exercise 17.13.

17.15 The separate ions migrate so as to maintain electrical neutrality at the two electrodes.

17.16 (a) Water is more readily oxidized than NO_3^-, so we have:

$H_2O(\ell) \rightarrow 4H^+(aq) + 4e^- + O_2(g)$

(b) Br^- is more readily oxidized than water, so we have:

$2Br^-(aq) \rightarrow Br_2(aq) + 2e^-$

(c) $Br^-(aq)$ is more readily oxidized than NO_3^-, so we have:

$2Br^-(aq) \rightarrow Br_2(aq) + 2e^-$

17.17 (a) Water is more easily reduced than K^+, so we write:

$2H_2O(\ell) + 2e^- \rightarrow H_2(g) + 2OH^-(aq)$

(b) $Cu^{2+}(aq)$ is more readily reduced than water, so we write:

$Cu^{2+} + 2e^- \rightarrow Cu(s)$

(c) $Cu^{2+}(aq)$ is more readily reduced than $K^{+}(aq)$, so we write:

$Cu^{2+}(aq) + 2e^{-} \rightarrow Cu(s)$

17.18 The answers to Review Exercises 17.16 and 17.17 guide us here:

At the cathode, where reduction occurs, we expect Cu(s).

At the anode, where oxidation occurs, we expect Br_2.

Stoichiometric Relationships in Electrolysis

17.19 One Faraday (F) is equivalent to one mole of electrons. Also, one Faraday is equal to 96,500 Coulombs, and a Coulomb is equivalent to an Ampere•second:

1 F = 96,500 C and 1 C = 1 A•s

17.20 1 C = 1 A•s

(a) $4.00 \text{ A} \times 600 \text{ s} = 2.40 \times 10^{3} \text{ C}$

(b) $10.0 \text{ A} \times 20 \text{ min} \times 60 \text{ s/min} = 1.2 \times 10^{4} \text{ C}$

(c) $1.50 \text{ A} \times 6.00 \text{ hr} \times 3600 \text{ s/hr} = 3.24 \times 10^{4} \text{ C}$

17.21 Multiply each value times the factor 1 F/96,500 C, i.e. divide by 96,500.

(a) 2.49 F (b) 0.12 F (c) 0.336 F

17.22 (a) $Fe^{2+} + 2e^{-} \rightarrow Fe$; Thus there are two Faradays per mole of Fe.

$0.20 \text{ mol } Fe^{2+} \times 2 \text{ F/mol} = 0.40 \text{ F}$

(b) $Cl^{-} \rightarrow {}^{1}/_{2}Cl_2 + e^{-}$; Thus there is one Faraday per mole of Cl^{-}.

$0.70 \text{ mol } Cl^{-} \times 1 \text{ F/mol} = 0.70 \text{ F}$

(c) $Cr^{3+} + 3e^{-} \rightarrow Cr$; Thus there are three Faradays per mole of Cr.

$1.50 \text{ mol} \times 3 \text{ F/mol} = 4.50 \text{ F}$

(d) $MnO_4^{-}(aq) + 8H^{+}(aq) + 5e^{-} \rightarrow Mn^{2+}(aq) + 4H_2O(\ell)$; Thus there are 5 Faradays per equivalent of MnO_4^{-}.

$$1.0 \times 10^{-2} \text{ mol} \times 5 \text{ F/mol} = 5.0 \times 10^{-2} \text{ F}$$

17.23 (a) $Mg^{2+} + 2e^- \rightarrow Mg$; Thus there are two Faradays per mole of Mg.

$$5.00 \text{ g Mg} \times \frac{1 \text{ mol Mg}}{24.3 \text{ g Mg}} \times \frac{2 \text{ F}}{1 \text{ mol Mg}} = 0.412 \text{ F}$$

(b) $Cu^{2+} + 2e^- \rightarrow Cu$; Thus there are 2 Faradays per mole of Cu.

$$41.0 \text{ g Cu} \times \frac{1 \text{ mol Cu}}{63.5 \text{ g Cu}} \times \frac{2 \text{ F}}{1 \text{ mol Cu}} = 1.29 \text{ F}$$

17.24 $Ag^+ + e^- \rightarrow Ag$, and $Cr^{3+} + 3e^- \rightarrow Cr$

This shows that there are three Faradays per mole of Cr but only one Faraday per mole of Ag.

The number of Faradays involved in the silver reaction are:

$$10.0 \text{ g Ag} \times \frac{1 \text{ mol Ag}}{108 \text{ g Ag}} \times \frac{1 \text{ F}}{1 \text{ mol Ag}} = 0.0926 \text{ F}$$

The amount of Cr is then:

$$0.0926 \text{ F} \times \frac{1 \text{ mol } Cr^{3+}}{3 \text{ F}} = 0.0309 \text{ mol } Cr^{3+}$$

17.25 $Fe(s) + 2OH^-(aq) \rightarrow Fe(OH)_2(s) + 2e^-$; Thus there are two Faradays per mole of Fe.

The number of Coulombs is:

$$10.0 \text{ min} \times 60 \text{ s/min} \times 12.0 \text{ A} = 7.20 \times 10^3 \text{ C}$$

The number of grams of $Fe(OH)_2$ is:

$$7.20 \times 10^3 \text{ C} \times \frac{1 \text{ F}}{96,500 \text{ C}} \times \frac{1 \text{ mol}}{2 \text{ F}} \times \frac{90.0 \text{ g}}{1 \text{ mol}} = 3.36 \text{ g Fe}$$

17.26 $2H^+(aq) + 2e^- \rightarrow H_2(g)$; Thus there are two Faradays per mole of H_2.

The number of Coulombs is:

$$5.00 \text{ min} \times 60 \text{ s/min} \times 1.50 \text{ A} = 450 \text{ C}$$

The number of moles of H_2 is:

$$450 \text{ C} \times \frac{1 \text{ F}}{96{,}500 \text{ C}} \times \frac{1 \text{ mol}}{2 \text{ F}} = 2.33 \times 10^{-3} \text{ mol } H_2$$

Finally, we calculate the volume of H_2 gas:

$$2.33 \times 10^{-3} \text{ mol} \times 22.4 \text{ L/mol} \times 1000 \text{ ml/L} = 52.2 \text{ mL}$$

17.27 The electrolysis of NaCl solution results in the reduction of water, together with the formation of hydroxide ion:

$$2H_2O(\ell) + 2e^- \rightarrow H_2(g) + 2OH^-(aq)$$

The number of Coulombs is:

$$3.00 \text{ A} \times 10.0 \text{ min} \times 60 \text{ s/min} = 1.80 \times 10^3 \text{ C}$$

The number of moles of OH^- is:

$$1.8 \times 10^3 \text{ C} \times \frac{1 \text{ F}}{96{,}500 \text{ C}} \times \frac{1 \text{ mol}}{2 \text{ F}} = 0.0187 \text{ mol } OH^-$$

The volume of acid solution that will neutralize this much OH^- is:

$$0.0187 \text{ mol } OH^- \times \frac{1 \text{ mol HCl}}{1 \text{ mol } OH^-} \times \frac{1 \text{ L HCl}}{0.100 \text{ mol}} = 0.187 \text{ L} = 187 \text{ mL HCl}$$

17.28 $Cr^{3+}(aq) + 3e^- \rightarrow Cr(s)$; Thus there are 3 Faradays per mole of Cr.

The number of Coulombs that will be required is:

$$25.0 \text{ g Cr} \times \frac{1 \text{ mol}}{52.0 \text{ g}} \times \frac{3 \text{ F}}{1 \text{ mol}} \times \frac{96{,}500 \text{ C}}{1 \text{ F}} = 1.39 \times 10^5 \text{ C}$$

The time that will be required is:

$$\frac{1.39 \times 10^5 \text{ A}\cdot\text{s}}{1.25 \text{ A}} \times \frac{1 \text{ hr}}{3600 \text{ s}} = 30.9 \text{ hr}$$

17.29 From the half-reaction, we see that there are 2 Faradays per mole of Pb.

The number of Coulombs that will be required is:

$$25.0 \text{ g Pb} \times \frac{1 \text{ mol}}{207 \text{ g}} \times \frac{2 \text{ F}}{1 \text{ mol}} \times \frac{96{,}500 \text{ C}}{1 \text{ F}} = 2.33 \times 10^4 \text{ C}$$

The time that will be required is:

$$\frac{2.33 \times 10^4\ \text{A}\cdot\text{s}}{0.50\ \text{A}} \times \frac{1\ \text{hr}}{3600\ \text{s}} = 13\ \text{hr}$$

17.30 $Mg^{2+} + 2e^- \rightarrow Mg(\ell)$; There are two Faradays per mole of Mg.

The number of Coulombs that will be required is:

$$48.0\ \text{g} \times \frac{1\ \text{mol}}{24.3\ \text{g}} \times \frac{2\ \text{F}}{1\ \text{mol}} \times \frac{96{,}500\ \text{C}}{1\ \text{F}} = 3.81 \times 10^5\ \text{C}$$

The number of amperes is:

$$3.81 \times 10^5\ \text{A}\cdot\text{s} \div 3600\ \text{s} = 106\ \text{A}$$

17.31 $Al^{3+} + 3e^- \rightarrow Al(s)$ Thus there are 3 Faradays per mole of Al. The number of Coulombs that are required is:

$$409 \times 10^3\ \text{g} \times \frac{1\ \text{mol}}{27.0\ \text{g}} \times \frac{3\ \text{F}}{1\ \text{mol}} \times \frac{96{,}500\ \text{C}}{1\ \text{F}} = 4.39 \times 10^9\ \text{C}$$

The number of amperes is:

$$4.39 \times 10^9\ \text{A}\cdot\text{s} \div 8.64 \times 10^4\ \text{s} = 5.1 \times 10^4\ \text{A}$$

17.32 $2Cl^-(\ell) \rightarrow Cl_2 + 2e^-$ There are thus 2 Faradays per mole of Cl_2.

The number of Coulombs is:

$$2.50\ \text{A} \times 40.0\ \text{min} \times 60\ \text{s/min} = 6.00 \times 10^3\ \text{C}$$

The number of grams of Cl_2 that will be produced is:

$$6.00 \times 10^3\ \text{C} \times \frac{1\ \text{F}}{96{,}500\ \text{C}} \times \frac{1\ \text{mol}}{2\ \text{F}} \times \frac{70.9\ \text{g}}{1\ \text{mol}} = 2.20\ \text{g}\ Cl_2$$

17.33 (a) By definition, one Faraday is equal to one equivalent, and one Eq is equal to 96,500 Coulombs:

$$1.50\ \text{A} \times 30.0\ \text{min} \times 60\ \text{s/min} = 2.70 \times 10^3\ \text{A}\cdot\text{s} = 2.70 \times 10^3\ \text{C}$$

$$2.70 \times 10^3\ \text{C} \times \frac{1\ \text{Eq}}{96{,}500\ \text{C}} = 0.0280\ \text{Eq}$$

(b) $0.475\ \text{g} \div 50.9\ \text{g/mol} = 9.33 \times 10^{-3}\ \text{mol V}$

(c) $2.80 \times 10^{-2}\ \text{Eq}/9.33 \times 10^{-3}\ \text{mol} = 3.00\ \text{Eq/mol}$

Applications of Electrolysis

17.34 Electroplating is a procedure by which a metal is deposited on another metallic surface. See Figure 17.4 of the text, page 703.

17.35 $Al_2O_3(s)$ is dissolved in molten cryolite, Na_3AlF_6. The liquid mixture is electrolyzed to drive the following reaction:

$$2Al_2O_3(\ell) \rightarrow 4Al(\ell) + 3O_2(g)$$

See also Figure 17.5 on page 703 of the text.

17.36 Sea water is made basic in order to cause the precipitation of magnesium hydroxide:

$$Mg^{2+}(aq) + 2OH^-(aq) \rightarrow Mg(OH)_2(s)$$

This is purified by redissolving the hydroxide and precipitating the chloride by evaporation of water:

$$Mg(OH)_2 + 2\ HCl \rightarrow Mg^{2+} + 2Cl^- + 2H_2O$$

The solid $MgCl_2$ that is so obtained is electrolyzed as a melt in order to recover the metal.

$$MgCl_2(\ell) \rightarrow Mg + Cl_2(g)$$

17.37 Sodium is obtained from electrolysis of molten NaCl using the Downs cell that is diagramed in Figure 17.6 on page 704 of the text. Some of the uses of sodium are to make tetraethyl lead, for sodium vapor lamps, and as a coolant in nuclear reactors.

$Na^+(\ell) + e^- \rightarrow Na(\ell)$ cathode

$Cl^-(\ell) \rightarrow {}^1/_2Cl_2(g) + e^-$ anode

$NaCl(\ell) \rightarrow Na(\ell) + {}^1/_2Cl_2(g)$ net reaction

17.38 This is shown in the photo and the diagram (Figure 17.7) on page 705 of the text. Impure copper is the anode, which dissolves during the process. Pure copper is deposited at the cathode. Anode sludge contains precious metals, whose value makes the process cost effective.

Typical reactions occurring at the anode are:

$$Cu(s) \rightarrow Cu^{2+}(aq) + 2e^-$$

$Zn(s) \rightarrow Zn^{2+}(aq) + 2e^-$

$Fe(s) \rightarrow Fe^{2+}(aq) + 2e^-$

The reaction that occurs at the cathode is:

$Cu^{2+}(aq) + 2e^- \rightarrow Cu(s)$

17.39 The various methods are diagramed in Figures 17.8, 17.9 and 17.10 of the text. The physical apparatus influences the products that are obtained:

The unstirred reaction is the one that applies to Figure 17.9:

$NaCl(aq) + H_2O(\ell) \rightarrow NaOH(aq) + Cl_2 + H_2$

The stirred reaction is different in that NaOCl is formed instead of NaOH:

$NaCl(aq) + H_2O \rightarrow NaOCl + H_2$

17.40 The diaphragm cell offers the advantage that H_2 and Cl_2 may be captured separately, the reaction of Cl_2 and OH^- to form OCL^- is prevented, and the sodium hydroxide solution is conveniently obtained at the bottom of the apparatus. The disadvantages include the fact that the NaOH solution that is obtained is contaminated with some unreacted NaCl, and that the cell itself is complicated and difficlut to maintain.

The mercury cell diagramed in Figure 17.10 offers the advantage that pure NaOH can be obtained. The disadvantage is the environmental hazard that mercury imposes.

Galvanic Cells

17.41 A galvanic cell is one in which a spontaneous redox reaction occurs, producing electricity. A half-cell is one of either the cathode or the anode, together with the accompanying electrolyte.

17.42 The salt bridge connects two half cells, and allows for electrical neutrality to be maintained by a flow of appropriate ions.

17.43 These must be kept separate, because otherwise Ag^+ ions would be reduced directly by Cu metal, and no current would be produced.

17.44 As in the electrolytic cell, the anode is the electrode at which oxidation takes place, and the cathode is the electrode at which reduction takes place. The charges of the electrodes in a galvanic cell are opposite to those in the electrolysis cell; the cathode is positive and the anode is negative.

17.45 In both the galvanic and the electrolysis cells, the anions move away from the cathode toward the anode, and the cations move away from the anode toward the cathode.

17.46 Magnesium metal is oxidized at the anode and copper ions are reduced at the cathode to give copper ions. The anode half cell is a magnesium wire dipping into a solution of Mg^{2+} ions, and the cathode half cell is a copper wire dipping into a solution of Cu^{2+} ions. Additionally, a salt bridge must connect the two half cell compartments.

17.47 Aluminum metal constitutes the anode:

$$Al \rightarrow Al^{3+} + 3e^-$$

The cathode is tin metal:

$$Sn^{2+} + 2e^- \rightarrow Sn$$

17.48 The charge is not balanced.

Cell Potentials and Reduction Potentials

17.49 A cell potential is a standard potential only if the temperature is 25 °C, the pressure is 1 atm, and all ions have a concentration of 1 M.

17.50 The cell potential for the anode half reaction is subtracted from the cell potential for the cathode half cell:

$$E^\circ_{cell} = E^\circ_{substance\ reduced} - E^\circ_{substance\ oxidized}$$

$$E^\circ_{cell} = E^\circ_{reduction} - E^\circ_{oxidation}$$

17.51 This is electromotive force, which has the units Volts.

17.52 1 v = 1 joule per Coulomb: V = J/C

17.53 An amp (A) is a Coulomb per second (C/s). A volt (V) is a joule per Coulomb (J/C). The product of amp × V × sec is:

C/s × J/C × s = J (joules)

17.54 No. Emf measurements require a completed circuit for current flow, and there must always be two half cells, connected with a salt bridge.

17.55 The standard hydrogen electrode is diagramed in Figure 17.14 on page 713 of the text. It consists of a platinum wire in contact with a solution having $[H^+]$ equal to 1 M, and hydrogen gas at a pressure of 1 atm is placed over the system. The half cell potential is 0 V.

17.56 A positive reduction potential indicates that the substance is more easily reduced than the hydrogen ion. Conversely, a negative reduction potential indicates that the substance comprising the half cell is less easily reduced than the hydrogen ion.

17.57 The difference between the reduction potentials for hydrogen and copper is a constant that is independent of the choice for the reference potential. In other words, the reduction half cell potential for copper is to be 0.34 units higher for copper than for hydrogen, regardless of the chosen point of reference. If E° for copper is taken to be 0 V, then E° for hydrogen must be -0.34 V.

17.58 The negative terminal of the voltmeter must be connected to the anode in order to obtain correct readings of the voltage that is generated by the cell.

Using Standard Reduction Potentials

17.59 The reactions are spontaneous if the overall cell potential is positive.

$$E^\circ_{cell} = E^\circ_{substance\ reduced} - E^\circ_{substance\ oxidized}$$

(a) $E^\circ_{cell} = 1.42 - (0.54) = 0.88$ V $\therefore$ spontaneous

(b) $E^\circ_{cell} = -0.44 - (0.96) = 1.40$ V $\therefore$ not spontaneous

(c) $E^\circ_{cell} = -0.74 - (-2.76) = 2.02$ V $\therefore$ spontaneous

17.60 A reaction is spontaneous if its net cell potential is positive:

$$E^\circ_{cell} = E^\circ_{substance\ reduced} - E^\circ_{substance\ oxidized}$$

(a) $E^\circ_{cell} = 1.07 - (1.36) = -0.29$ V $\therefore$ not spontaneous

(b) $E^\circ_{cell} = -0.25 - (-0.44) = +0.19$ V $\therefore$ spontaneous

(c) $E^\circ_{cell} = 1.07 - (0.17) = +0.90$ V $\therefore$ spontaneous

17.61 The metals are placed into the activity series based on their values of standard reduction potentials.

17.62 The given equation is separated into its two half reactions:

$NO_3^-(aq) + 4H^+(aq) + 3e^- \rightarrow NO(g) + 2H_2O(\ell)$ reduction

$3Fe^{2+}(aq) \rightarrow 3Fe^{3+}(aq) + 3e^-$ oxidation

$E°_{cell} = E°_{reduction} - E°_{oxidation} = 0.96 - 0.77 = 0.19$ V

17.63 As in the answer to Review Exercise 17.62:

$Cd^{2+}(aq) + 2e^- \rightarrow Cd(s)$ reduction

$Fe(s) \rightarrow Fe^{2+}(aq) + 2e^-$ oxidation

$E°_{cell} = E°_{reduction} - E°_{oxidation} = -0.40 - (-0.44) = 0.04$ V

17.64 The half cell with the more positive $E°_{cell}$ will appear as a reduction, and the other half reaction is reversed, to appear as an oxidation:

$BrO_3^- + 6H^+ + 6e^- \rightarrow Br^- + 3H_2O$ reduction

$3 \times (2I^- \rightarrow I_2 + 2e^-)$ oxidation

$BrO_3^- + 6I^- + 6H^+ \rightarrow 3I_2 + Br^- + 3H_2O$ net reaction

$E°_{cell} = E°_{substance\ reduced} - E°_{substance\ oxidized}$

or

$E°_{cell} = E°_{reduction} - E°_{oxidation} = 1.44 - (0.54) = 0.90$ V

17.65 The half reaction having the more positive reduction potential is the reduction half reaction, and the other is reversed to become the oxidation half reaction:

$MnO_2 + 4H^+ + 2e^- \rightarrow Mn^{2+} + 2H_2O$ reduction

$Pb + 2Cl^- \rightarrow PbCl_2 + 2e^-$ oxidation

$MnO_2 + 4H^+ + Pb + 2Cl^- \rightarrow Mn^{2+} + 2H_2O + PbCl_2$ net reaction

$E°_{cell} = E°_{reduction} - E°_{oxidation} = 1.23 - (-0.27) = 1.50$ V

17.66 The half reaction having the more positive standard reduction potential is the one that occurs as a reduction, and the other one is written as an oxidation:

$2 \times (2HOCl + 2H^+ + 2e^- \rightarrow Cl_2 + 2H_2O)$ reduction

$3H_2O + S_2O_3^{2-} \rightarrow 2H_2SO_3 + 2H^+ + 4e^-$ oxidation

$$4HOCl + 4H^+ + 3H_2O + S_2O_3^{2-} \rightarrow 2Cl_2 + 4H_2O + 2H_2SO_3 + 2H^+ + 4e^-$$

which simplifies to give the following net reaction:

$$4HOCl + 2H^+ + S_2O_3^{2-} \rightarrow 2Cl_2 + H_2O + 2H_2SO_3$$

17.67 $Br_2 + 2e^- \rightarrow 2Br^-$ $E°_{red} = 1.07\ V$

$I_2 + 2e^- \rightarrow 2I^-$ $E°_{red} = 0.54\ V$

Since the first of these has the larger reduction half cell potential, it occurs as a reduction, and the second is reversed to become an oxidation:

$$Br_2 + 2I^- \rightarrow I_2 + 2Br^-$$

17.68 The two half reactions are:

$SO_4^{2-}(aq) + 2e^- + 4H^+(aq) \rightarrow H_2SO_3(aq) + H_2O(\ell)$ reduction

$2I^- \rightarrow I_2 + 2e^-$ oxidation

$$E°_{cell} = E°_{reduction} - E°_{oxidation} = 0.17 - (0.54) = -0.37\ V$$

Since the overall cell potential is negative, we conclude that the reaction is not spontaneous in the direction written.

17.69 The two half reactions are:

$S_2O_8^{2-} + 2e^- \rightarrow 2SO_4^{2-}$ reduction

$Ni(OH)_2 + 2OH^- \rightarrow NiO_2 + 2H_2O + 2e^-$ oxidation

$$E°_{cell} = E°_{reduction} - E°_{oxidation} = 2.01 - (0.49) = 1.52\ V$$

Since the overall cell potential is positive, we conclude that the reaction is spontaneous in the direction written.

Cell Potentials and Thermodynamics

17.70 $\Delta G° = -nFE°_{cell}$

17.71

$$E°_{cell} = \frac{0.0592}{n} \log K_c$$

17.72 The maximum amount of work that can be obtained from the cell, per mole of Ag_2O reacting, is given by the absolute value of $\Delta G°$ for the reaction,

where it must be remembered that the unit volt (V) is equivalent to a joule per Coulomb (J/C):

$$\Delta G° = -nFE°_{cell} = -(2)(96{,}500\ C)(1.50\ J/C) = -2.90 \times 10^5\ J$$

$$1.00\ g \times \frac{1\ mol}{232\ g} \times \frac{2.90 \times 10^5\ J}{1\ mol} = 1.25 \times 10^3\ J$$

17.73 $\Delta G° = -nFE°_{cell}$, $E°_{cell} = 1.34\ V = 1.34\ J/C$, and $n = 2$

$$\Delta G° = -(2)(96{,}500\ C)(1.34\ J/C) = -2.59 \times 10^5\ J \text{ per mol of HgO}$$

The maximum amount of work that can be derived from this cell, on using 1.00 g of HgO, is thus:

$$1.00\ g\ HgO \times \frac{1\ mol\ HgO}{217\ g\ HgO} \times \frac{2.59 \times 10^5\ J}{1\ mol\ HgO} = 1.19 \times 10^3\ J \text{ of work}$$

Now, since 1 watt = 1 J/s, then 5×10^{-4} watt = 5×10^{-4} J/s, and the time required for this process is:

$$1.19 \times 10^3\ J \times \frac{1\ s}{5 \times 10^{-4}\ J} \times \frac{1\ min}{60\ s} \times \frac{1\ hr}{60\ min} = 7 \times 10^2\ hr$$

17.74 (a) $E°_{cell} = E°_{reduction} - E°_{oxidation} = 2.01 - (1.47) = 0.54\ V$

(b) Since $n = 10$,

$$\Delta G° = -nFE°_{cell} = -(10)(96{,}500\ C)(0.54\ V) = -5.2 \times 10^5\ J$$

$$\Delta G° = -5.2 \times 10^2\ kJ$$

(c)

$$E°_{cell} = \frac{0.0592}{n} \log K_c$$

$0.54 = (0.0592)/10 \times \log K_c$

$\log K_c = 91.22$ Taking the antilog of both sides of this equation:

$K_c = 1.7 \times 10^{91}$

17.75 On separating this reaction into two half reactions, we discover that n = 10.

$\Delta G° = -nFE°_{cell} = -(10)(96{,}500\ C)(1.69) = -1.63 \times 10^6\ J = -1.63 \times 10^3\ kJ$

17.76 First, separate the overall reaction into its two half reactions:

$2Br^- \rightarrow Br_2 + 2e^-$ oxidation

$I_2 + 2e^- \rightarrow 2I^-$ reduction

$E°_{cell} = E°_{reduction} - E°_{oxidation} = 0.54 - (1.07) = -0.53\ V$

The value of n is 2:

$\Delta G° = -nFE°_{cell} = -(2)(96{,}500\ C)(-0.53\ J) = 1.0 \times 10^5\ J = 1.0 \times 10^2\ kJ$

17.77 Ni^{2+} is reduced by two electrons and Co is oxidized by two electrons.

$E°_{cell} = E°_{substance\ reduced} - E°_{substance\ oxidized} = -0.25 - (-0.28)$

$= +0.03\ V$

$$E°_{cell} = \frac{0.0592}{n} \log K_c$$

$+0.03 = 0.0592/2 \times \log K_c$

$\log K_c = 1$ and $K_c = \text{antilog}(1) = 10$

17.78 Sn is oxidized by two electrons and Ag is reduced by two electrons:

$$E°_{cell} = \frac{0.0592}{n} \log K_c$$

$-0.015 = 0.0592/2 \times \log K_c$

$\log K_c = -0.51$

$K_c = \text{antilog}(-0.51) = 0.31$

17.79 First, separate the overall reaction into two half reactions:

$2H_2O \rightarrow 4H^+ + 4e^- + O_2$ oxidation

$2 \times (Cl_2 + 2e^- \rightarrow 2Cl^-)$ reduction

$E°_{cell} = E°_{reduction} - E°_{oxidation} = 1.36 - (1.23) = 0.13$ V

Since n = 4, we write:

$$E°_{cell} = \frac{0.0592}{n} \log K_c$$

$0.13 = 0.0592/4 \times \log K_c$

$\log K_c = 8.78$ and $K_c = \text{antilog}(8.78) = 6.0 \times 10^8$

The Effect of Concentration on Cell Potentials

17.80 $\Delta G = \Delta G° + RT \ln Q$

But it is also true that: $\Delta G = -nFE_{cell}$ and $\Delta G° = -nFE°_{cell}$, which allows the following substitution:

$-nFE_{cell} = -nFE°_{cell} + RT \ln Q$

Dividing each side of the above equation by the quantity -nF gives:

$E_{cell} = E°_{cell} - RT/nF \ln Q$

The various values to be used are: R = 8.3144 J
T = 298.15 K
F = 96494 C/mol

Thus:

$$E_{cell} = E°_{cell} - \frac{(8.3144)(298.15)(2.303)}{n(96494)} \log Q$$

$E_{cell} = E°_{cell} - 0.0592/n \log Q$ Q.E.D.

17.81 We begin by separating the reaction into its two half reactions, in order to obtain the value of n.

$Pb(s) + SO_4^{2-} \rightarrow PbSO_4 + 2e^-$

$$PbO_2 + 4H^+ + SO_4^{2-} + 2e^- \rightarrow PbSO_4 + 2H_2O$$

Thus n is equal to 2, and the equation that we are to use is:

$$E_{cell} = E^\circ_{cell} - \frac{0.0592}{n} \log Q$$

$$E_{cell} = 2.05 - \frac{0.0592}{2} \log \frac{1}{[H^+]^4[SO_4^{2-}]^2}$$

17.82 This reaction involves the oxidation of Ag by two electrons and the reduction of Ni by two electrons. The concentration of the hydrogen ion is derived from the pH of the solution:

$$[H^+] = \text{antilog}\ (-\text{pH}) = \text{antilog}\ (-6) = 1 \times 10^{-6}\ M$$

$$E_{cell} = 2.48 - \frac{0.0592}{2} \log \frac{[Ag^+]^2[Ni^{2+}]}{[H^+]^4}$$

$$E_{cell} = 2.48 - \frac{0.0592}{2} \log \frac{(1.0 \times 10^{-1})^2(1.0 \times 10^{-1})}{(1 \times 10^{-6})^4}$$

$$E_{cell} = 2.48 - 0.62 = 1.86\ V$$

17.83 The following half reactions indicate that the value of n is 30:

$$5Cr_2O_7^{2-} + 70H^+ + 30e^- \rightarrow 10Cr^{3+} + 35H_2O$$

$$3I_2 + 18H_2O \rightarrow 6IO_3^- + 36H^+ + 30e^-$$

$$E_{cell} = 0.135 - \frac{0.0592}{30} \log \frac{[Cr^{3+}]^{10}[IO_3^-]^6}{[H^+]^{34}[Cr_2O_7^{2-}]^5}$$

$$E_{cell} = 0.135 - \frac{0.0592}{30} \log \frac{(1.0 \times 10^{-4})^{10}(1.0 \times 10^{-3})^6}{(1.0 \times 10^{-2})^{34}(1.0 \times 10^{-1})^5}$$

$$E_{cell} = 0.135 - 0.0592/30 \times \log(1.0 \times 10^{15}) = 0.135 - 0.030 = 0.105\ V$$

17.84 The initial numbers of moles of Ag^+ and Zn^{2+} are:

$$1.00\ mol/L \times 0.100\ L = 0.100\ mol$$

The number of Coulombs (A•s) that have been employed is:

$0.10\ A \times 10.00\ hr \times 3600\ s/hr = 3.600 \times 10^3\ C$

The number of Faradays is:

$3.600 \times 10^3\ C \div 96{,}500\ C/F = 3.7 \times 10^{-2}\ F$

For Ag^+, there is 1 mol per Faraday, and for Zn^{2+}, there are two Faradays per mol. This means that the number of moles of the two ions that have been consumed or formed is given by:

For Ag^+, $3.7 \times 10^{-2}\ F \times 1\ mol/F = 3.7 \times 10^{-2}$ mol Ag^+ have reacted.

For Zn^{2+}, $3.7 \times 10^{-2} \times 1\ mol/2\ F = 1.9 \times 10^{-2}$ mol Zn^{2+} have reacted.

The number of moles of Ag^+ that remain is:

$0.100 - 0.037 = 0.063$ mol of Ag^+, and the final concentration of silver ion is:

$[Ag^+] = 0.063\ mol/0.100\ L = 0.63\ M$

The number of moles of Zn^{2+} that are present is:

$0.100 + 0.019 = 0.119$ mol Zn^{2+}, and the final concentration of zinc ion is:

$[Zn^{2+}] = 0.119\ mol/0.100\ L = 1.19\ M$

The standard cell potential should be:

$E°_{cell} = E°_{reduction} - E°_{oxidation} = 0.80 - (-0.76) = 1.56\ V$

We now apply the Nernst equation:

$$E_{cell} = 1.56 - \frac{0.0592}{2} \log \frac{(1.19)}{(0.63)^2}$$

$$E_{cell} = 1.56 - 0.014 = 1.55\ V$$

17.85

$$E_{cell} = E°_{cell} - \frac{0.0592}{2} \log \frac{[Mg^{2+}]}{[Cd^{2+}]}$$

$$1.67 = 1.97 - \frac{0.0592}{2} \log \frac{1}{[Cd^{2+}]}$$

$\log [Cd^{2+}] = -10.14$, and $[Cd^{2+}] = \text{antilog}(-10.24) = 7.2 \times 10^{-11}\ M$

17.86 Since the copper half cell is the cathode, this is the half cell in which reduction takes place. The silver half cell is therefore the anode, where oxidation of silver ion occurs. The standard cell potential is:

$$E°_{cell} = E°_{reduction} - E°_{oxidation} = 0.34 - 0.22 = 0.12\ V$$

The overall cell reaction is:

$Cu^{2+} + 2Ag + 2Cl^- \rightarrow Cu + 2\ AgCl$,

and the Nernst equation becomes:

$$E_{cell} = 0.12 - \frac{0.0592}{2} \log \frac{1}{[Cu^{2+}][Cl^-]^2}$$

If we use the various values given in the exercise, we arrive at:

$0.09 = 0.12 - 0.0296 \times \log(1/[Cl^-]^2)$, which rearranges to give:

$0.0296 \times \log([Cl^-]^2) = -0.03$

$\log\ [Cl^-] = -0.5$

$[Cl^-] = \text{antilog}(-0.5) = 0.3\ M$

17.87 In the iron half cell, we are initially given:

$50.0\ mL \times 0.10\ mol/L = 5.00 \times 10^{-3}\ mol\ Fe^{2+}(aq)$

The precipitation of $Fe(OH)_2$ consumes some of the added hydroxide ion, as well as some of the iron ion:

$Fe^{2+} + 2OH^- \rightarrow Fe(OH)_2(s)$

The number of moles of OH^- that have been added to the iron half cell is:

$0.50\ m/L \times 0.0500\ L = 2.50 \times 10^{-2}\ mol\ OH^-$

The stoichiometry of the precipitation reaction requires that the following number of moles of OH^- be consumed on precipitation of 5.00×10^{-3} mol of $Fe(OH)_2(s)$:

$5.00 \times 10^{-3}\ mol\ Fe(OH)_2 \times 2\ mol\ OH^-/mol\ Fe(OH)_2 = 1.00 \times 10^{-2}\ mol\ OH^-$

The number of moles of OH^- that are unprecipitated in the iron half cell is:

$2.50 \times 10^{-2}\ mol - 1.00 \times 10^{-2}\ mol = 1.50 \times 10^{-2}\ mol\ OH^-$

Since the resulting volume is 50.0 mL + 50.0 mL, the concentration of hydroxide ion in the iron half cell becomes, upon precipitation of the $Fe(OH)_2$:

$[OH^-] = 1.50 \times 10^{-2}$ mol/0.100 L = 0.150 M OH^-

We have assumed that the iron hydroxide that forms in the above precipitation reaction is completely insoluble. This is not accurate, though, because some small amount does dissolve in water according to the following equilibrium:

$Fe(OH)_2(s) \rightleftharpoons Fe^{2+}(aq) + 2OH^-(aq)$

This means that the true $[OH^-]$ is slightly higher than 0.150 M as calculated above. Thus we must set up the usual equilibrium table, in order to analyze the extent to which $Fe(OH)_2(s)$ dissolves in 0.150 M OH^- solution:

	$Fe(OH)_2$	$\rightleftharpoons$	Fe^{2+}	+	$2OH^-$
initial conc., M			0.0		0.150
change in conc., M			+x		+2x
equilibrium conc., M			x		0.150 + 2x

The quantity x in the above table is the molar solubility of $Fe(OH)_2$ in the solution that is formed in the iron half cell.

$K_{sp} = [Fe^{2+}][OH^-]^2 = (x)(0.150 + 2x)^2$

The standard cell potential is:

$E°_{cell} = E°_{reduction} - E°_{oxidation} = 0.337 - (-0.440) = 0.777$

The Nernst equation is:

$$E_{cell} = E°_{cell} - \frac{0.0592}{2} \log \frac{[Fe^{2+}]}{[Cu^{2+}]}$$

which becomes:

$0.398 = 0.777 - 0.0296 \times \log [Fe^{2+}]$

$[Fe^{2+}] = 3.58 \times 10^{-14}$ M

This is the concentration of Fe^{2+} in the saturated solution, and it is the value to be used for x in the above expression for K_{sp}.

$$K_{sp} = (x)(0.150 + 2x)^2 = (3.58 \times 10^{-14})[0.150 + (2)(3.58 \times 10^{-14})]^2$$

$$K_{sp} = 8.06 \times 10^{-16}$$

17.88 The half cell reactions and the overall cell reaction are:

$Cu^{2+} + 2e^- \rightarrow Cu$ $\qquad E°_{red} = +0.337$ V

$H_2 \rightarrow 2H^+ + 2e^-$ $\qquad E°_{ox} = 0.000$ V

$Cu^{2+} + H_2 \rightarrow Cu + 2H^+$

(a) The standard cell potential is:

$$E°_{cell} = E°_{reduction} - E°_{oxidation} = 0.377 - 0 = +0.377 \text{ V}$$

The Nernst equation for this system is:

$$E_{cell} = E°_{cell} - \frac{0.0592}{n} \log \frac{[H^+]^2}{[Cu^{2+}]}$$

which becomes, under the circumstances defined in the problem:

$$E_{cell} = E°_{cell} - \frac{0.0592}{2} \log [H^+]^2$$

Rearranging the last equation gives:

$$\frac{2 \times (E_{cell} - E°_{cell})}{0.0592} = -\log [H^+]^2$$

which becomes the desired relationship:

$$\frac{(E_{cell} - E°_{cell})}{0.0592} = -\log [H^+] = pH$$

(b) The equation derived in the answer to part (a) of this question is conveniently rearranged to give:

$$E_{cell} = (0.0592)(pH) + E°_{cell} = (0.0592)(4.25) + 0.337 = 0.589 \text{ V}$$

(c) The equation that was derived in the answer to part (a) of this question may be used directly:

$$pH = (E_{cell} - E°_{cell}) \div 0.0592 = (0.660 - 0.337) \div 0.0592 = 5.46$$

Practical Galvanic Cells

17.89 $Pb(s) + SO_4^{2-}(aq) \rightarrow PbSO_4(s) + 2e^-$ anode

$PbO_2(s) + 4H^+(aq) + SO_4^{2-}(aq) + 2e^- \rightarrow PbSO_4(s) + 2H_2O(\ell)$ cathode

Connecting six cells in series produces 12 volts.

17.90 $PbSO_4(s) + 2e^- \rightarrow Pb(s) + SO_4^{2-}(aq)$ cathode

$PbSO_4(s) + 2H_2O(\ell) \rightarrow 2e^- + PbO_2(s) + 4H^+(aq) + SO_4^{2-}(aq)$ anode

17.91 The anodic reaction is: $Zn(s) \rightarrow Zn^{2+}(aq) + 2e^-$

Several reactions take place at the cathode; one of the important ones is:

$2MnO_2(s) + 2NH_4^+(aq) + 2e^- \rightarrow Mn_2O_3(s) + 2NH_3(aq) + H_2O(\ell)$

17.92 $Zn(s) + 2OH^-(aq) \rightarrow ZnO(s) + H_2O(\ell) + e^-$ anode

$2MnO_2(s) + H_2O(\ell) + 2e^- \rightarrow Mn_2O_3(s) + 2OH^-(aq)$ cathode

17.93 $Cd(s) + 2OH^-(aq) \rightarrow Cd(OH)_2(s) + 2e^-$ anode

$2e^- + NiO_2(s) + 2H_2O(\ell) \rightarrow Ni(OH)_2(s) + 2OH^-(aq)$ cathode

The overall cell reaction on discharge of the battery is:

$Cd(s) + NiO_2(s) + 2H_2O(\ell) \rightarrow Cd(OH)_2(s) + Ni(OH)_2(s)$

17.94 (a) $Zn(s) + 2OH^-(aq) \rightarrow Zn(OH)_2(s) + 2e^-$ anode

$Ag_2O(s) + H_2O(\ell) + 2e^- \rightarrow 2Ag(s) + 2OH^-(aq)$ cathode

$Zn(s) + Ag_2O(s) + H_2O(\ell) \rightarrow Zn(OH)_2(s) + 2Ag(s)$ net cell reaction

(b) $Zn(s) + 2OH^-(aq) \rightarrow ZnO(s) + H_2O(\ell) + 2e^-$ anode

$HgO(s) + H_2O(\ell) + 2e^- \rightarrow Hg(\ell) + 2OH^-(aq)$ cathode

$Zn(s) + HgO(s) \rightarrow ZnO(s) + Hg(\ell)$ net cell reaction

17.95 Fuel cells are more efficient thermodynamically, and more of the energy of the reaction can be made available for useful work.

CHAPTER EIGHTEEN

PRACTICE EXERCISES

1. The rate of the reaction after 250 seconds have elapsed is equal to the slope of the tangent to the curve at 250 seconds. First draw the tangent, and then estimate its slope as follows, where A is taken to represent one point on the tangent, and B is taken to represent another point on the tangent:

$$\text{rate} = \frac{A(\text{mol/L}) - B(\text{mol/L})}{A(s) - B(s)} = \frac{\text{change in concentration}}{\text{change in time}}$$

   A value near 9.4×10^{-5} mol L^{-1} s^{-1} is correct.

2. On a mole basis, 1 mol of N_2O_4 produces two moles of NO_2, and we conclude that NO_2 forms twice as fast as N_2O_4 disappears. Thus the rate of formation of NO_2 is 2 × the rate of disappearance of N_2O_4, or the rate of disappearance of N_2O_4 is half the rate of formation of NO_2:

   rate of decomposition of N_2O_4 = $^1/_2$(rate of formation of NO_2)

   = $^1/_2$ × (0.010 mol L^{-1} s^{-1})

   = 5.0×10^{-3} mol L^{-1} s^{-1}

3. (a) First use the given data in the rate law:

   rate = $k[HI]^2$

   2.5×10^{-4} mol L^{-1} s^{-1} = $k[5.58 \times 10^{-2}$ mol/L$]^2$

   k = 8.0×10^{-2} L mol^{-1} s^{-1}

   (b) L mol^{-1} s^{-1}

4. The order of the reaction with respect to a given substance is the exponent to which that substance is raised in the rate law:

   order of the reaction with respect to $[BrO_3^-]$ = 1

   order of the reaction with respect to $[SO_3^{2-}]$ = 1

overall order of the reaction = 1 + 1 = 2

This is a second order reaction overall, being first order in each of two reactants.

5. In each case, $k = \text{rate}/[A][B]^2$, and the units of k are $L^2\ mol^{-2}\ s^{-1}$.

Each calculation is performed as follows, using the second data set as the example:

$$k = \frac{0.40\ mol\ L^{-1}\ s^{-1}}{(0.20\ mol/L)(0.10\ mol/L)^2} = 2.0 \times 10^2\ L^2\ mol^{-2}\ s^{-1}$$

Each of the other data sets also gives the value:

$$k = 2.0 \times 10^2\ L^2\ mol^{-2}\ s^{-1}.$$

6. The rate will likely take the form rate = $k[C_{12}H_{22}O_{11}]^n$, where n is the order of the reaction with respect to sucrose.

A comparison of the first two lines of data shows that increasing the sucrose concentration by a factor of 2 (from 0.10 to 0.20) causes rate to increase also by nearly a factor of two (from 6.17×10^{-5} to 12.3×10^{-5}). This corresponds to the case in Table 18.3 for which a concentration increase by a factor of 2 causes a rate increase by a factor of $2 = 2^1$, and we conclude that the order of the reaction with respect to sucrose is one.

The same conclusion is reached on examining the data of the second and third rows. Increasing the concentration by a factor of 5 causes an increase in the rate by a factor of $5 = 5^1$, and we conclude that n = 1.

7. The rate law will likely take the form rate = $k[A]^n[B]^{n'}$, where n and n′ are the order of the reaction with respect to A and B, respectively. On comparing the first two lines of data, in which the concentration of B is held constant, we note that increasing the concentration of A by a factor of 2 (from 0.40 to 0.80) causes an increase in the rate by a factor of 4 (from 1.0×10^{-4} to 4.0×10^{-4}). Thus we have a rate increase by 2^2, caused by a concentration increase by a factor of 2. This corresponds to the case in Table 18.3 for which n = 2, and we conclude that the reaction is second order with respect to A.
On comparing the second and third lines of data (wherein the concentration of A is held constant), we note that increasing the concentration of B by a factor of 2 (from 0.30 to 0.60) causes an increase in the rate by a factor of 4 (from 4.0×10^{-4} to 16.0×10^{-4}). This is an increase in rate by a

factor of 2^2, and it forces us to conclude, using the information of Table 18.3, that the value of n' is 2. Thus the reaction is also second order with respect to B. The rate law is then written:

$$\text{rate} = k[A]^2[B]^2$$

The value of k may be computed using any of the lines of data that are given in the exercise. Using the data of the third line, we get:

$$k = \frac{\text{rate}}{[A]^2[B]^2} = \frac{1.6 \times 10^{-3}\ \text{mol L}^{-1}\ \text{s}^{-1}}{[0.80\ \text{mol/L}]^2[0.60\ \text{mol/L}]^2} = 6.9 \times 10^{-3}\ \text{mol}^{-3}\ \text{L}^3\ \text{s}^{-1}$$

8. (a) We substitute into equation 18.7, first converting time into the units seconds:

$$t = 2\ \text{hr} \times 3600\ \text{s/hr} = 7.20 \times 10^3\ \text{s}$$

$$[\text{sucrose}]_t = [\text{sucrose}]_0\ e^{-kt}$$

$$[\text{sucrose}]_t = [0.40\ \text{M}]\ e^{-(6.2 \times 10^{-5}\ \text{s}^{-1})(7.2 \times 10^3\ \text{s})}$$

$$[\text{sucrose}]_{2\ \text{hr}} = 0.26\ \text{M}$$

(b) Use equation 18.5:

$$\ln \frac{[\text{sucrose}]_0}{[\text{sucrose}]_t} = kt$$

$$\ln \frac{[0.40\ \text{M}]}{[0.30\ \text{M}]} = 0.29 = (6.2 \times 10^{-5}\ \text{s}^{-1}) \times t$$

Solving for t we get: $t = 4.7 \times 10^3$ s

$4.7 \times 10^3\ \text{s} \times 1\ \text{min}/60\ \text{s} = 78\ \text{min}$

9. This is a second order reaction, and we use equation 18.8:

$$\frac{1}{[\text{NOCl}]_t} - \frac{1}{[\text{NOCl}]_0} = kt$$

$$\frac{1}{[0.010\ \text{M}]} - \frac{1}{[0.040\ \text{M}]} = (2.0 \times 10^{-2}\ \text{L mol}^{-1}\ \text{s}^{-1}) \times t$$

$$t = 3.8 \times 10^3\ \text{s}$$

$$t = 3.8 \times 10^3\ \text{s} \times 1\ \text{min}/60\ \text{s} = 63\ \text{min}$$

10. For a first order reaction:

$$t_{1/2} = \frac{0.693}{k} = \frac{0.693}{(6.17 \times 10^{-4}\ s^{-1})} = 1.12 \times 10^{3}\ s$$

$$t_{1/2} = 1.12 \times 10^{3}\ s \times \frac{1\ min}{60\ s} = 18.7\ min$$

If we refer to the chart given in the text in example 18.9, we see that two half lives will have passed if there is to be only one quarter of the original amount of material remaining. This corresponds to:

18.7 min per half life × 2 half lives = 37.4 min

11. The reaction is first order. A second order reaction should have a half life that depends on the initial concentration according to equation 18.11.

12. (a)

$$\ln\left[\frac{k_2}{k_1}\right] = \frac{-E_a}{R}\left[\frac{1}{T_2} - \frac{1}{T_1}\right]$$

$$\ln\left[\frac{23\ L\ mol^{-1}\ s^{-1}}{3.2\ L\ mol^{-1}\ s^{-1}}\right] = \frac{-E_a}{8.314\ J\ K^{-1}\ mol^{-1}}\left[\frac{1}{673\ K} - \frac{1}{623\ K}\right]$$

Solving for E_a gives 1.4×10^{5} J/mol = 1.4×10^{2} kJ/mol

(b) We again use equation 18.14, substituting the values:

$k_1 = 3.2\ L\ mol^{-1}\ s^{-1}$ at $T_1 = 623\ K$

$k_2 = ?$ at $T_2 = 573\ K$

$$\ln\left[\frac{k_2}{3.2\ L\ mol^{-1}\ s^{-1}}\right] = \frac{-140 \times 10^{3}\ J\ mol^{-1}}{8.314\ J\ K^{-1}\ mol^{-1}}\left[\frac{1}{573\ K} - \frac{1}{623\ K}\right]$$

$\ln (k_2 / 3.2) = -2.36$

Taking the antiln of both sides of this equation gives:

$k_2 / 3.2\ L\ mol^{-1}\ s^{-1} = 0.0944$ or $k_2 = 0.30\ L\ mol^{-1}\ s^{-1}$

13. If it is known that the entire process occurs by a single, elementary mechanistic step, then we have the special circumstance that the order of

the reaction with respect to each chemical participating in the elementary step is given directly by the coefficient of that chemical in the balanced equation for the elementary step:

$$\text{rate} = k[NO]^1[O_3]^1$$

REVIEW EXERCISES

Factors That Affect Reaction Rate

18.1 Reaction rate constitutes the speed with which a reactant or product changes concentration during a chemical reaction.

18.2 (a) combustion of gasoline
(b) cooking an egg in boiling water
(c) curing of cement

18.3 An explosion is an extremely rapid reaction in which the energy that is released cannot be dissipated, causing the products to expand, pushing everything outward.

18.4 Industrialists need to be able to adjust the conditions so as to make products as rapidly as possible.

18.5 A collision between only two molecules is much more probable than the simultaneous collision of three molecules. We therefore conclude that a reaction involving a two-body collision is faster (i.e. will occur more frequently) than one requiring a three-body collision.

18.6 Five factors that affect reaction rate are:
- the nature of the reactants
- the ability of the reactants to collide effectively
- the concentration of the reactants
- the temperature
- the presence of a catalyst

18.7 A homogeneous reaction is one in which all reactants and products are in the same phase. An example would be:

$$2H_2(g) + O_2(g) \rightarrow H_2O(g)$$

18.8 A heterogeneous reaction is one in which all reactants and products are not in the same phase. An example would be:

$$25O_2(g) + 2C_2H_{18}(\ell) \rightarrow 16CO_2(g) + 18H_2O(g)$$

18.9 Chemical reactions that are carried out in solution take place smoothly because the reactants can mingle effectively at the molecular level. Also, studies of concentration and temperature effects on the rate are made easy.

18.10 Heterogeneous reactions are most affected by the extent of surface contact between the reactant phases.

18.11 In a heterogeneous reaction, the smaller the particle size, the faster the rate. This is because decreasing the particle size increases the surface area of the material, thereby increasing contact with another reactant phase.

18.12 This illustrates the effect of concentration.

18.13 Reaction rate generally increases with increasing temperature.

18.14 A catalyst is a substance that affects the rate of a reaction without being used up by the reaction.

18.15 In cool weather, the rates of metabolic reactions of cold-blooded insects decrease, because of the affect of temperature on rate.

18.16 Water boils at a higher temperature under pressure, and foods cook faster at the higher temperature.

18.17 The low temperature causes the rate of metabolism to be very low.

<u>Measuring Rates of Reaction</u>

18.18 Reaction rate has the units mol L^{-1} s^{-1}, or molar per second ($M\ s^{-1}$).

18.19 This is determined by the coefficients of the balanced chemical equation. For every mole of N_2 that reacts, 3 mol of H_2 will react. Thus the rate of disappearance of hydrogen is three times the rate of disappearance of nitrogen. Similarly, the rate of disappearance of N_2 is half the rate of appearance of NH_3, or NH_3 appears twice as fast as N_2 disappears.

18.20

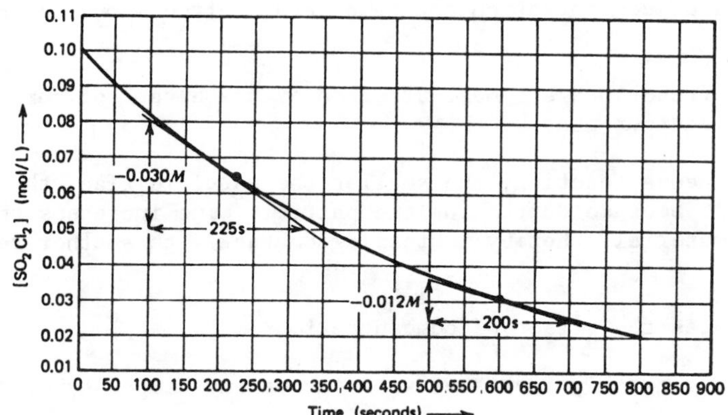

The rate at time t is the slope of the tangent at time t, which is the change in concentration divided by the change in time:

At 200 s, rate = $-0.030 \text{ mol L}^{-1}/225 \text{ s}$

$= -1.3 \times 10^{-4} \text{ mol L}^{-1} \text{ s}^{-1}$

At 600 s, rate = $-0.012 \text{ mol L}^{-1}/200 \text{ s}$

$= -6.0 \times 10^{-5} \text{ mol L}^{-1} \text{ s}^{-1}$

18.21

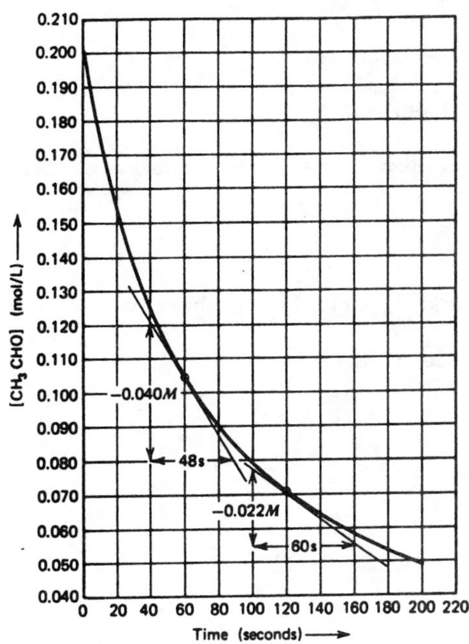

The rate at time = t is the slope of the tangent at time = t, or the rate is given by the change in concentration divided by the change in time.

At t = 60 s, rate = -0.040 mol L^{-1}/48 s = -8.3×10^{-4} mol L^{-1} s^{-1}

At t = 120 s, rate = -0.022 mol L^{-1}/60 s = -3.7×10^{-4} mol L^{-1} s^{-1}

18.22 From the coefficients in the balanced equation we see that, for every mole of B that reacts, 2 mol of A are consumed, and three mol of C are produced. This means that A will be consumed twice as fast as is B, and C will be produced three times faster than B is consumed.

rate of disappearance of A = 2(0.30) = 0.60 mol L^{-1} s^{-1}

rate of appearance of C = 3(0.30) = 0.90 mol L^{-1} s^{-1}

18.23 We rewrite the balanced chemical equation to make the problem easier to answer:

$N_2O_5 \rightarrow 2NO_2 + \frac{1}{2}O_2$

Thus the rates of formation of NO_2 and O_2 will be, respectively, twice and one half the rate of disappearance of N_2O_5.

rate formation NO_2 = $2(2.5 \times 10^{-6}) = 5.0 \times 10^{-6}$ mol L^{-1} s^{-1}

rate formation O_2 = $\frac{1}{2}(2.5 \times 10^{-6}) = 1.3 \times 10^{-6}$ mol L^{-1} s^{-1}

Concentration and Rate: Rate Laws

18.24 A rate law is an equation that states the experimentally determined relationship between the rate of a reaction (for which the units are mol L^{-1} s^{-1}, or M/s) to the concentrations of reactants and products that are found to influence reaction rate. The proportionality constant is k, the rate constant.

18.25 The overall order of a reaction is the sum of the exponents for the concentration terms of the rate law.

18.26 The units are, in each case, whatever is required to give the units of rate (mol L^{-1} s^{-1}) to the overall rate law:

(a) s^{-1} (b) L mol^{-1} s^{-1} (c) L^2 mol^{-2} s^{-1}

18.27 The exponents in a rate law (i.e. the order of reaction with respect to each of the reactants) must be determined experimentally.

18.28 No.

18.29 This is the fourth case in Table 18.3 of the text, viz. a first order reaction.

18.30 This is the seventh case in Table 18.3, and the rate increases by a factor of $2^2 = 4$.

18.31 The order of reaction for any given reactant is determined by measuring the rate change that is obtained upon making changes in reactant concentrations.

18.32 rate = $(7.1 \times 10^9\ L^2\ mol^{-2}\ s^{-1})(1.0 \times 10^{-3}\ mol/L)^2(3.4 \times 10^{-2}\ mol/L)$

rate = $2.4 \times 10^2\ mol\ L^{-1}\ s^{-1}$

18.33 rate = $(1.0 \times 10^{-5}\ s^{-1})(1.0 \times 10^{-3}\ mol/L) = 1.0 \times 10^{-8}\ mol\ L^{-1}\ s^{-1}$

18.34 In each case, the order with respect to a reactant is the exponent to which that reactant's concentration is raised in the rate law.

(a) For $HCrO_4^-$, the order is 1.

For HSO_3^-, the order is 2.

For H^+, the order is 1.

(b) The overall order is 1 + 2 + 1 = 4.

18.35 In each case, the order with respect to a reactant is the exponent to which that reactant's concentration is raised in the rate law. The order of reaction with respect to each reactant is 1. The overall order is 1 + 1 = 2.

18.36 Since the substrate concentration does not influence rate, its concentration does not appear in the rate law, and the order of the reaction with respect to substrate is zero.

18.37 On comparing the data of the first and second experiments, we find that, whereas the concentration of N is unchanged, the concentration of M has been doubled, causing a doubling of the rate. This corresponds to the fourth case in Table 18.3, and we conclude that the order of the reaction with respect to M is 1. In the second and third experiments, we have the converse, namely that whereas the concentration of N is held constant, the concentration of M is tripled, causing an increase in the rate by a factor of nine. This constitutes the eighth case in Table 18.3, and we conclude that the order of the reaction with respect to M is 2. This means that the overall rate expression is:

rate = $k[M][N]^2$ and we can solve for the value of k by substituting the appropriate data:

$5.0 \times 10^{-3}\ mol\ L^{-1}\ s^{-1} = k \times [0.020\ mol/L][0.010\ mol/L]^2$

$k = 2.5 \times 10^3\ mol^{-2}\ L^2\ s^{-1}$

18.38 First compare the first and second experiments, in which there has been an increase in concentration by a factor of 2. This has caused an increase in rate of:

$$\frac{5.90 \times 10^{-5}}{2.95 \times 10^{-5}} = 2.00 = 2^1$$

This is the fourth case in Table 18.3, and the order of the reaction is found to be 1. Similarly, on comparing the first and third experiments, an increase in concentration by a factor of 3 has caused an increase in rate by a factor of:

$$\frac{8.85 \times 10^{-5}}{2.95 \times 10^{-5}} = 3.00 = 3^1$$

This is the fifth case in Table 18.3, and we again conclude that the order is 1.

rate $= k[C_3H_6]$

18.39 The reaction is first order in OCl^-, because an increase in concentration by a factor of two, while holding the concentration of I^- constant (compare the first and second experiments of the table), has caused an increase in rate by a factor of $2^1 = 2$. The order of reaction with respect to I^- is also 1, as is demonstrated by a comparison of the second and third experiments.

rate $= k[OCl^-][I^-]$

Using the last data set:

$3.5 \times 10^4\ mol\ L^{-1}\ s^{-1} = k[1.7 \times 10^{-3}\ mol/L][3.4 \times 10^{-3}\ mol/L]$

$k = 6.1 \times 10^9\ L\ mol^{-1}\ s^{-1}$

18.40 Compare the data of the first and second experiments, in which the concentration of NO is held constant and the concentration of O_2 is increased by a factor of 4. Since this caused a rate increase by a factor of $28.4/7.10 = 4^1$, we conclude that the order of the reaction with respect to O_2 is one (case number six in Table 18.3). In the second and third experiments, an increase in the concentration of NO by a factor of 3 (while holding the concentration of O_2 constant) caused a rate increase by a factor of 255.6/28.4 = 9. This is the eighth case in Table 18.3, and the order is seen to be two.

rate = $k[O_2][NO]^2$

We can use any of the three sets of data to solve for k. Using the first data set gives:

$7.10\ mol\ L^{-1}\ s^{-1} = k[1.0 \times 10^{-3}\ mol/L][1.0 \times 10^{-3}\ mol/L]^2$

$k = 7.10 \times 10^9\ L^2\ mol^{-2}\ s^{-1}$

18.41 Compare the first and second experiments. On doubling the ICl concentration, the rate is found to increase by a factor of 2 = 2^1, and the order of the reaction with respect to ICl is 1 (case number four in Table 18.3). In the first and third experiments, the concentration of ICl is constant, whereas the concentration of H_2 in the first experiment is twice that in the third. This causes a change in the rate by a factor of 2 also, and the rate law is found to be:

rate = $k[ICl][H_2]$

Using the data of the first experiment:

$1.5 \times 10^{-3}\ mol\ L^{-1}\ s^{-1} = k[0.10][0.10]$

$k = 1.5 \times 10^{-1}\ L\ mol^{-1}\ s^{-1}$

18.42 In the first, fourth, and fifth experiments, the concentration of OH^- has been made to increase, causing no change in rate. This means that the rate is independent of $[OH^-]$, and the order of reaction with respect to OH^- is zero. The concentration of $(CH_3)_3CBr$ doubles from the first to the second experiment, and triples from the first to the third experiment, as the OH^- concentration is held constant. There is a corresponding 2-fold (i.e. 2^1) increase in rate from the first to the second experiment, and there is a 3-fold (i.e. 3^1) increase in rate from the first to the third experiment. In both cases, we conclude that the order with respect to $(CH_3)_3CBr$ is one.

rate = $k[(CH_3)_3CBr]$

Using the third set of data gives:

3.0×10^{-3} mol L^{-1} s^{-1} = k[3.0×10^{-1} mol/L]

k = 1.0×10^{-2} s^{-1}

Concentration and Time

18.43 First order:

$$\ln \frac{[A]_0}{[A]_t} = kt$$

Second order:

$$\frac{1}{[B]_t} - \frac{1}{[B]_0} = kt$$

18.44

$[A]_t = [A]_0 \; e^{-kt}$

Taking the ln of both sides of this equation gives:

$\ln [A]_t = \ln [A]_0 - kt$

Rearranging gives:

$\ln [A]_t - \ln [A]_0 = -kt$

Multiplying each term by -1 gives:

$\ln [A]_0 - \ln [A]_t = kt$

which is equivalent to:

$\ln \frac{[A]_0}{[A]_t} = kt$

Q.E.D.

18.45 (a) The time involved must be converted to a value in seconds:

$1\ \text{hr} \times 3600\ \text{s/hr} = 3.6 \times 10^3\ \text{s}$,

and then we make use of equation 18.7, where A is taken to represent SO_2Cl_2:

$$[A]_t = [A]_0\, e^{-kt} = [0.0040\ M] \times e^{-(2.2 \times 10^{-5}\ s^{-1})(3.6 \times 10^3\ s)}$$

$$[SO_2Cl_2]_t = [0.0040\ M][0.924] = 3.7 \times 10^{-3}\ M$$

(b) The time is converted to a value having the units seconds

$24\ \text{hr} \times 3600\ \text{s/hr} = 8.64 \times 10^4\ \text{s}$,

and then we use equation 18.7, where A is taken to represent SO_2Cl_2:

$$[A]_t = [A]_0\, e^{-kt} = [0.0040\ M] \times e^{-(2.2 \times 10^{-5}\ s^{-1})(8.64 \times 10^4\ s)}$$

$$[SO_2Cl_2]_t = [0.0040\ M][0.149] = 6.0 \times 10^{-4}\ M$$

18.46 We use equation 18.8, letting B represent HI:

$$\frac{1}{[B]_t} - \frac{1}{[B]_0} = kt$$

$$\frac{1}{[5.0 \times 10^{-4}\ M]} - \frac{1}{[1.1 \times 10^{-2}\ M]} = (1.6 \times 10^{-3}\ L\ mol^{-1}\ s^{-1}) \times t$$

Solving for t gives:

$t = 1.2 \times 10^6\ s$ or $t = 1.2 \times 10^6\ s \times 1\ min/60\ s = 2.0 \times 10^4\ min$

18.47 Use equation 18.8, where the time is:

$2.0 \times 10^3\ min \times 60\ s/min = 1.2 \times 10^5\ s$

$$\frac{1}{[B]_t} - \frac{1}{[B]_0} = kt$$

$$\frac{1}{[3.0 \times 10^{-4}\ M]} - \frac{1}{[HI]_0} = (1.6 \times 10^{-3}\ L\ mol^{-1}\ s^{-1})(1.2 \times 10^5\ s)$$

$$1/[HI]_0 = 3.14 \times 10^3\ L\ mol^{-1}$$

$$[HI]_0 = 3.2 \times 10^{-4}\ M$$

18.48 Any consistent set of units for expressing concentration may be used in equation 18.5, where we let A represent the drug that is involved:

$$\ln \frac{[A]_0}{[A]_t} = kt$$

$$\ln \frac{[30.0]}{[10.0]} = k(120 \text{ min})$$

$$1.10 = k(120 \text{ min}) \quad \therefore k = 9.17 \times 10^{-3} \text{ min}^{-1}$$

18.49 Again, let A represent the drug, and make use of equation 18.7, and use time in the units minutes:

$$3.00 \text{ hr} \times 60 \text{ min/hr} = 180 \text{ min}$$

$$[A]_t = [A]_0 \, e^{-kt}$$

$$[10.0] = [A]_0 \times e^{-(9.17 \times 10^{-3} \text{ min}^{-1})(180 \text{ min})}$$

$$[A]_0 = [10.0]/0.192 = 52.1 \text{ mg/kg of body weight}$$

18.50 Use equation 18.5, taking time in minutes, as given:

$$\ln \frac{[A]_0}{[A]_t} = kt$$

$$\ln (100/10) = k(45.0 \text{ min})$$

$$k = 5.1 \times 10^{-2} \text{ min}^{-1}$$

Half Lives

18.51 The half life of a reaction is the time required for the concentration of a reactant to be reduced to half of its initial value.

18.52 The half life of a first order reaction is unaffected by the initial concentration.

18.53 The half life of a second order reaction is inversely proportional to the initial concentration, as expressed in equation 18.11.

18.54 We use equation 18.5, substituting exactly one half the value of $[A]_0$ for the value of $[A]_t$:

$\ln \frac{[A]_0}{[A]_t} = kt$ or $\ln \frac{1.0}{0.5} = kt$

This simplifies to give: $kt = \ln(2) = 0.693$. Thus $t = 0.693/k$

For the second order reaction, we also substitute half the value of $[B]_0$ for the value of $[B]_t$ in equation 18.8:

$$\frac{1}{[B]_t} - \frac{1}{[B]_0} = kt$$

$$\frac{1}{0.5} - \frac{1}{1.0} = kt$$

$kt = 1/[B]_0 \quad \therefore \quad t = 1/k[B]_0$

18.55 $t = 0.693/k = 0.693/1.6 \times 10^{-3}\ s^{-1} = 4.3 \times 10^{2}$ seconds

18.56

$$t = \frac{1}{(6.7 \times 10^{-4} \text{mol L}^{-1}\ s^{-1})(0.20 \text{ mol L}^{-1})} = 7.5 \times 10^{3} \text{ seconds}$$

18.57 It requires approximately 350 s (as determined from the graph) for the concentration of SO_2CL_2 to decrease from 0.100 M to 0.050 M, i.e. to

decrease to half its initial concentration. Likewise, in another 350 seconds, the concentration decreases by half again, i.e. from 0.050 M to 0.025 M. This means that the half life of the reaction is independent of the initial concentration, and we conclude that the reaction is first order in SO_2Cl_2.

18.58 From the graph, we see that the first half life is achieved in about 66 seconds, because this is the amount of time it requires for the concentration to decrease from its initial value (0.200 M) to half its initial value (0.100 M). The second half life period (i.e. the time required for the concentration to decrease from 0.100 M to 0.050 M) is longer, namely about 128 seconds. Since the half life does depend on concentration, we conclude that the reaction is not first order. In fact, the data are consistent with second order kinetics, because the value of the half life decreases in proportion to the inverse of the initial concentration. That is, as the initial concentration for any half life period becomes smaller, the half life becomes larger.

18.59

$$2.0 \text{ hrs} \times \frac{60 \text{ min}}{1 \text{ hr}} \times \frac{1.0 \text{ half life}}{15 \text{ min}} = 8.0 \text{ half lives}$$

Eight half lives correspond to the following fraction of original material remaining:

Number of half lives	Fraction Remaining
1	1/2
2	1/4
3	1/8
4	1/16
5	1/32
6	1/64
7	1/128
8	1/256

18.60 As listed in the answer to Review Exercise 18.59, a remaining fraction of 1/32 of the original concentration represents 5 half lives:

5.0 half lives × 28 yr/half life = 140 years

Effect of Temperature on Rate

18.61 Rates are typically about doubled when a 10 °C temperature increase is involved.

18.62 According to collision theory, the rate is proportional to the number of collisions per second among the reactants.

18.63 The effectiveness of collisions is influenced by the orientation of the reactants and by the activation energy.

18.64 This happens because a larger fraction of the reactant molecules possess the minimum energy necessary to surpass E_a.

18.65

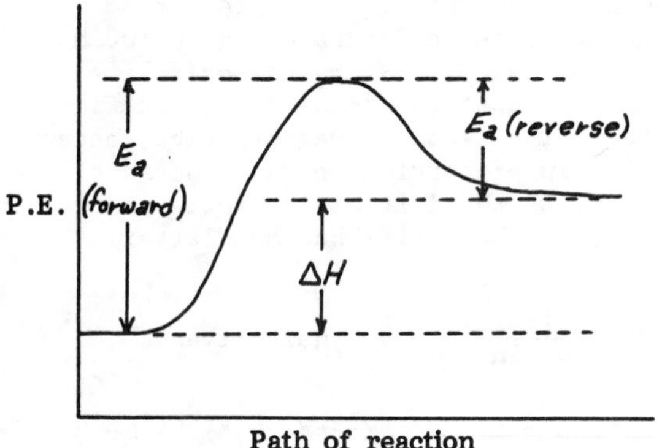

18.66 The energy that is necessary for the reaction is obtained at the expense of the total kinetic energy of the system. Thus kinetic energy is lost on converting it into potential energy, and the temperature goes down.

18.67 Transition state - the high point on the potential energy diagram of the reaction.

Activated complex - the chemical arrangement that exists at the transition state.

18.68

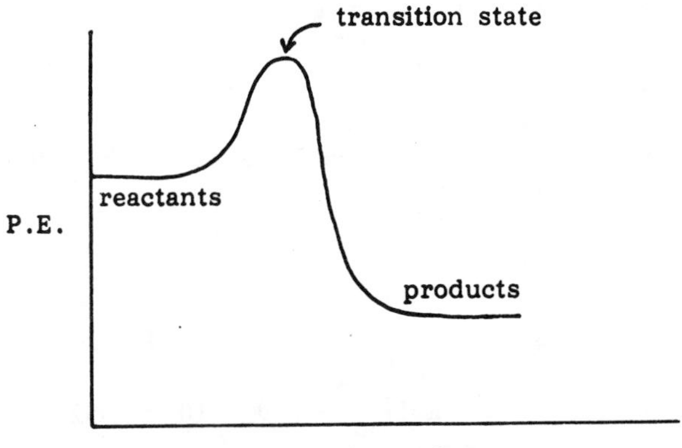

18.69 Breaking a strong bond requires a large input of energy, hence a large E_a.

Calculations Involving the Activation Energy

18.70

$$k = Ae^{-E_a/RT}$$

where k = the rate constant
A = a frequency factor
e = the base for natural logarithms
E_a = the activation energy

R = the gas constant in energy units
T = absolute temperature (Kelvins)

18.71 The graph is prepared exactly as in Example 18.11 on page 760 of the text. The slope is found to be, roughly:

$$\text{slope} = \frac{\Delta(\ln k)}{\Delta(1/T)} = -9.5 \times 10^3\ K^{-1}$$

Thus $-9.5 \times 10^3\ K^{-1} = -E_a/R$

$$E_a = (9.5 \times 10^3\ K^{-1})(8.314\ J\ K^{-1}\ mol^{-1}) = 7.9 \times 10^4\ J/mol = 79\ kJ/mol$$

Using equation 18.14, we proceed as follows:

$$\ln\left[\frac{k_2}{k_1}\right] = \frac{-E_a}{R}\left[\frac{1}{T_2} - \frac{1}{T_1}\right]$$

$$\ln\left[\frac{1.94 \times 10^{-3}\ s^{-1}}{2.88 \times 10^{-4}\ s^{-1}}\right] = \frac{-E_a}{8.314\ J\ K^{-1}\ mol^{-1}}\left[\frac{1}{673\ K} - \frac{1}{593\ K}\right]$$

$$1.907 = \frac{(2.00 \times 10^{-4}\ K^{-1})}{(8.314\ J\ K^{-1}\ mol^{-1})} \times E_a$$

$$E_a = 7.93 \times 10^4\ J/mol = 79.3\ kJ/mol$$

18.72 The graph is prepared exactly as in Example 18.11 on page 760 of the text. The slope is found to be, roughly:

$$\text{slope} = \frac{\Delta(\ln k)}{\Delta(1/T)} = -1.66 \times 10^{4}\ K^{-1}$$

Thus $-1.66 \times 10^{4}\ K^{-1} = -E_a/R$

$$E_a = (1.66 \times 10^{4}\ s^{-1})(8.314\ J\ K^{-1}\ mol^{-1}) = 1.38 \times 10^{5}\ J/mol$$

Using equation 18.14 we have:

$$\ln\left[\frac{k_2}{k_1}\right] = \frac{-E_a}{R}\left[\frac{1}{T_2} - \frac{1}{T_1}\right]$$

$$\ln\left[\frac{0.108\ L\ mol^{-1}\ s^{-1}}{0.0191\ L\ mol^{-1}\ s^{-1}}\right] = \frac{-E_a}{8.314\ J\ K^{-1}\ mol^{-1}}\left[\frac{1}{503\ K} - \frac{1}{478\ K}\right]$$

$$1.732 = \frac{(1.00 \times 10^{-4}\ K^{-1})}{(8.314\ J\ K^{-1}\ mol^{-1})} \times E_a$$

$$E_a = 1.44 \times 10^{5}\ J/mol = 144\ kJ/mol$$

18.73

$$\ln\left[\frac{k_2}{k_1}\right] = \frac{-E_a}{R}\left[\frac{1}{T_2} - \frac{1}{T_1}\right]$$

$$\ln\left[\frac{1.0 \times 10^{-3}\ L\ mol^{-1}\ s^{-1}}{9.3 \times 10^{-5}\ L\ mol^{-1}\ s^{-1}}\right] = \frac{-E_a}{8.314\ J\ K^{-1}\ mol^{-1}}\left[\frac{1}{403\ K} - \frac{1}{373\ K}\right]$$

$(2.40 \times 10^{-5}) \times E_a = 2.38$ J/mol or $E_a = 9.9 \times 10^{4}$ J/mol or 99 kJ/mol

If we take the ln of both sides of the Arrhenius equation we obtain equation 18.13:

$$\ln k = \ln A - E_a/RT$$

$$\ln (9.3 \times 10^{-5}) = \ln A - \frac{9.9 \times 10^{4}\ J/mol}{(8.314\ J/K\ mol)(373\ K)}$$

$$\ln A = -9.28 + 31.92 = 22.64$$

$$A = \text{antiln}\ (22.64) = 6.8 \times 10^{9}\ L\ mol^{-1}\ s^{-1}$$

18.74 (a) Substituting into equation 18.14:

$$\ln\left[\frac{1.1 \times 10^{-5}\ s^{-1}}{1.3 \times 10^{-6}\ s^{-1}}\right] = \frac{-E_a}{8.314\ J\ K^{-1}\ mol^{-1}}\left[\frac{1}{703\ K} - \frac{1}{673\ K}\right]$$

$(7.63 \times 10^{-6})(E_a) = 2.14$ J/mol

$E_a = 2.8 \times 10^5$ J/mol $= 2.8 \times 10^2$ kJ/mol

(b) Equation 18.12 can be conveniently rearranged to give equation 18.13:

$\ln k = \ln A - E_a/RT$, or $\ln A = \ln k + E_a/RT$

$$\ln A = \ln (1.1 \times 10^{-5}) + \frac{2.8 \times 10^5\ J\ mol^{-1}}{(8.314\ J\ K^{-1}\ mol^{-1})(703\ K)}$$

$\ln A = -11.42 + 47.91 = 36.49$

$A = \text{antiln}\ (36.49) = 7.0 \times 10^{15}\ s^{-1}$

(c) Equation 18.12 can be used directly now that the value of A is known:

$\ln k = \ln A - E_a/RT$

$$\ln k = 36.49 - \frac{2.8 \times 10^5\ J\ mol^{-1}}{(8.314\ J\ K^{-1}\ mol^{-1})(623\ K)} = -17.57$$

Taking the antiln of both sides of this equation gives:

$k = \text{antiln}\ (-17.57) = 2.3 \times 10^{-8}\ s^{-1}$

18.75 Substituting into equation 18.14:

$$\ln\left[\frac{3.75 \times 10^{-2}\ s^{-1}}{2.1 \times 10^{-3}\ s^{-1}}\right] = \frac{-E_a}{8.314\ J\ K^{-1}\ mol^{-1}}\left[\frac{1}{298\ K} - \frac{1}{273\ K}\right]$$

$(3.70 \times 10^{-5})(E_a) = 2.88$ J/mol

$E_a = 7.8 \times 10^4$ J/mol $= 78$ kJ/mol

18.76 Since we do not know the value of A, we cannot use equation 18.13. Substituting into equation 18.14 gives:

$$E_a = \frac{25.0\ \text{kcal}}{\text{mol}} \times \frac{4.184\ \text{kJ}}{1\ \text{kcal}} \times \frac{1000\ \text{J}}{1\ \text{kJ}} = 1.05 \times 10^5\ \text{J/mol}$$

$$\ln\left[\frac{k_2}{3.0 \times 10^{-4}\ \text{s}^{-1}}\right] = \frac{-1.05 \times 10^5\ \text{J/mol}}{8.314\ \text{J K}^{-1}\ \text{mol}^{-1}}\left[\frac{1}{323\ \text{K}} - \frac{1}{298\ \text{K}}\right]$$

$$\ln (k_2)/(3.0 \times 10^{-4}\ \text{s}^{-1}) = 3.28$$

Taking the antiln of both sides of this equation gives:

$$k_2/3.0 \times 10^{-4}\ \text{s}^{-1} = \text{antiln}(3.28) = 26.6$$

$$k_2 = 8.0 \times 10^{-3}\ \text{s}^{-1}$$

18.77

$$\ln\left[\frac{k_2}{k_1}\right] = \frac{-E_a}{R}\left[\frac{1}{T_2} - \frac{1}{T_1}\right]$$

$$\ln\left[\frac{2}{1}\right] = \frac{-E_a}{8.314\ \text{J K}^{-1}\ \text{mol}^{-1}}\left[\frac{1}{308\ \text{K}} - \frac{1}{298\ \text{K}}\right]$$

$$(1.31 \times 10^{-5})(E_a) = 0.693\ \text{J/mol}$$

$$E_a = 5.29 \times 10^4\ \text{J/mol} = 52.9\ \text{kJ/mol}$$

18.78 We can use equation 18.13:

(a) $\ln k = \ln A - E_a/RT$

$$\ln k = \ln(4.3 \times 10^{13}\ \text{s}^{-1}) - \frac{103 \times 10^3\ \text{J/mol}}{(8.314\ \text{J K}^{-1}\ \text{mol}^{-1})(293\ \text{K})} = -10.89$$

$$k = \text{antiln}(-10.89) = 1.9 \times 10^{-5}\ \text{s}^{-1}$$

(b)

$$\ln k = \ln(4.3 \times 10^{13}\ \text{s}^{-1}) - \frac{103 \times 10^3\ \text{J/mol}}{(8.314\ \text{J K}^{-1}\ \text{mol}^{-1})(373\ \text{K})} = -1.82$$

$k = \text{antiln}(-1.82) = 1.6 \times 10^{-1}\ s^{-1}$

18.79 Substitute into equation 18.14:

$$\ln\left[\frac{k_2}{k_1}\right] = \frac{-E_a}{R}\left[\frac{1}{T_2} - \frac{1}{T_1}\right]$$

$$\ln\left[\frac{k_2}{6.2 \times 10^{-5}\ s^{-1}}\right] = \frac{-108 \times 10^3\ J/mol}{8.314\ J\ K^{-1}\ mol^{-1}}\left[\frac{1}{318\ K} - \frac{1}{308\ K}\right]$$

$\ln (k_2/6.2 \times 10^{-5}\ s^{-1}) = 1.33$

$k_2 = 6.2 \times 10^{-5}\ s^{-1} \times \text{antiln}(1.33) = 2.3 \times 10^{-4}\ s^{-1}$

Reaction Mechanism

18.80 An elementary process is an actual collision event that occurs during the reaction. It is one of the key events that moves the reaction along in the stepwise process that leads to the overall reaction that is observed. It is thus one step in a potentially multi-step mechanism.

18.81 The rate-determining step in a mechanism is the slowest step.

18.82 The rate law for a reaction is based on the rate-determining step.

18.83 Adding all of the steps gives:

$2NO + 2H_2 \rightarrow N_2 + 2H_2O$

18.84 For such a mechanism, the rate law should be:

rate = $k[NO_2][CO]$

Since this is not the same as the observed rate law, this is not a reasonable mechanism to propose.

18.85 Add all of the steps together:

$CO + O_2 + NO \rightarrow CO_2 + NO_2$

18.86 On adding together all of the steps in each separate mechanism, we get, in each case:

$OCl^- + I^- \rightarrow Cl^- + OI^-$

18.87 The predicted rate law is based on the rate-determining step:

$\text{rate} = k[NO_2]^2$

Free Radical Reactions

18.88 A free radical is a substance having unpaired electrons. They can be generated at high temperatures or by the absorption of a photon having sufficient energy to ionize (eject) an electron from a stable substance.

18.89 A chain reaction is one in which the reaction of one free radical produces a product, plus another free radical, thereby causing further reaction to give more product, plus another free radical, and so on, in consecutive fashion.

18.90 (a) The initiating step is the first step, because it is the one that produces the first radical.

(b) The propagating steps are the second and third ones, because they are responsible not only for the formation of products, but also for the generation of additional radicals.

(c) The terminating step is the one that causes the radical to be quenched, namely the third step.

The other reactions slow down the rate of overall reaction, by providing alternative paths for loss of a radical.

18.91 The Cl that is produced in the second step is required as a reactant in the first step.

Catalysts

18.92 A catalyst changes the mechanism of a reaction, and provides, thereby, a reaction path having a smaller activation energy.

19.93 A homogeneous catalyst is one that is present in the same phase as the reactants. It is used in one step of a cycle, but regenerated in a subsequent step, so that in a net sense, it is not consumed.

18.94 Heterogeneous catalysts are present in a phase that is different from that of the reactants. They function by an adsorption of reactants, where the catalyst causes the activation energy to become smaller than would otherwise be true.

18.95 Adsorption is a clinging to a surface. Absorption involves a penetration below the surface, as in the action of a sponge. Heterogeneous catalysis involves adsorption.

18.96 The catalytic converter promotes oxidation of unburned hydrocarbons, as well as the decomposition of nitrogen oxide pollutants. Lead poisons the catalyst and renders it ineffective.

CHAPTER NINETEEN

REVIEW EXERCISES

Occurrence of Nonmetals

19.1 The initial body, according to the big bang theory, consisted of protons, neutrons, and electrons.

19.2 The two most abundant elements are hydrogen and helium.

19.3 Helium formed by fusion of hydrogen atoms.

19.4 Oxygen has the highest atom percent, as shown in the data of Table 19.3.

19.5 The two most abundant elements in the earth's atmosphere are N_2 (78%) and O_2 (21%). See the data of Table 19.5.

19.6 The general processes that consume oxygen are the various oxidations, i.e. respiration, decay, combustion, etc.

19.7 Oxygen is replenished by photosynthesis.

19.8 The four most prevalent nonmetallic elements in the human body are hydrogen, carbon, nitrogen and oxygen.

Noble Gases

19.9 These are given in Table 19.7: Helium, He; Neon, Ne; Argon, Ar; Krypton, Kr; and Xenon, Xe.

19.10 Among the noble gas elements, it is mainly xenon and krypton that form compounds with other elements. Some examples are the following fluorides and oxides: XeF_2, XeF_4, XeF_6, XeO_3, XeO_4, $XeOF_4$, XeO_2F_2.

19.11 Xenon is more reactive because it has a larger atomic radius and a lower ionization potential.

19.12 The noble gas elements are useful as air-excluding gases in arc welding, as interior gases for light bulbs, as protectors that are used in the atmospheres above air sensitive chemicals and reactions, in electric discharge tubes, in cryogenics, and in lighter-than-air balloons.

19.13 Helium and argon are more widely used than the other noble gases because the former are more abundant.

19.14 The noble gas remains inert to the filament, whereas oxygen or nitrogen would soon react with and destroy the filament at high temperatures.

19.15 Liquid helium is useful in reaching the sorts of low temperatures that are required for superconduction. It is the only substance that cannot be solidified by cooling at one atmosphere of pressure.

Hydrogen

19.16 Hydrogen is obtained chiefly from reactions of steam over catalysts with hydrocarbons and other oil-based materials, as described on page 786 of the text.

19.17 These are listed in Table 19.8 of the text: protium, H, mass number 1; deuterium, D, mass number 2; tritium, T, mass number 3.

19.18 The various isotopes of hydrogen have differing chemical properties that are used to advantage in, for instance, studies employing the deuterium isotope effect.

19.19 The deuterium isotope effect is the fact that reactions that require the breaking of a C-D bond are measurably slower than those that require the breaking of a C-H bond.

19.20 The tightly confined positive charge of a proton has a strong tendency to polarize electron density in neighboring substances so as to cause electrons from the neighbor to migrate towards the proton. This means that a proton will form a bond to anything that is near it.

19.21 The three types of hydrides are ionic hydrides (e.g. NaH, sodium hydride), covalent hydrides (e.g. CH_4, methane or H_2O, water), and metallic hydrides (e.g. NiH_2, nickel hydride or UH_3, uranium hydride).

19.22 Ionic hydrides have much higher melting points than do molecular hydrides. Ionic hydrides tend to be sources of H^-, whereas molecular hydrides are more likely to be sources of H^+. Remember that ionic hydrides are, by definition, not molecular substances.

19.23 (a) $NaH + H_2O \rightarrow NaOH + H_2$

(b) $CaH_2 + 2H_2O \rightarrow Ca(OH)_2 + 2H_2$

(c) $HCl(g) + H_2O \rightarrow H_3O^+(aq) + Cl^-(aq)$

(d) $2Na(s) + H_2(g) \rightarrow 2NaH(s)$

(e) $Mg(s) + H_2(g) \rightarrow MgH_2(s)$

19.24 Such hydrides are powerful bases as well as powerful reducing agents. They react with water to give metal hydroxide solutions and hydrogen gas.

19.25 The largest use of hydrogen in industry is for the synthesis of ammonia.

19.26 $N_2 + 3H_2 \rightarrow 2NH_3$

19.27 On forming products from reactants, there is a net decrease in the number of moles of gas. By the Principle of Le Chatelier, increasing the pressure should favor a shift from left to right, decreasing the volume required by the materials.

19.28 The two hydrogen atoms are added across the double bond, and the process requires a high temperature, a high pressure, and a catalyst:

$H_2C{=}CH_2 + H_2 \rightarrow H_3C{-}CH_3$

19.29 The OXO reaction converts a mixture of CO and H_2 to larger organic

substances containing a C=O group, as described on page 791 of the text. It is a reaction that is employed in the synthesis of detergents, plastics, etc.

Oxygen

19.30 These are given in Table 19.9: Oxygen, O; sulfur, S; selenium, Se; tellurium, Te; and polonium, Po.

19.31 Priestly made oxygen by igniting mercury(II) oxide:

$2HgO(s) \rightarrow 2Hg(\ell) + O_2(g)$

19.32 On the industrial scale, oxygen is obtained from the distillation of liquid air.

19.33 Oxygen is used primarily in the manufacture of steel, rocket propulsion, and in the health care industry.

19.34 The allotrope of oxygen other than O_2 is O_3, ozone.

19.35 An isotope is a form of an atom that differs from other forms of the same element in the number of neutrons in the nucleus. An example is H and D, two of the isotopes of hydrogen. An allotrope is a different structure or physical form of an element. In addition to the two allotropes of oxygen (O_2 and O_3) we also have allotropes of elements such as sulfur,

phosphorus, etc.

19.36 Ozone is made by the action of an electric discharge or ultraviolet light on dioxygen.

19.37 Neutral particles (M in the following equations) serve to quench some of the excess collisional energy that ozone molecules possess soon after they form by reaction with an O atom:

$$O_2 \rightarrow 2O$$

$$O + O_2 + M \rightarrow O_3 + M$$

19.38 Ozone reacts with many things in a harmful way. Examples are plants, fabrics, plastics, and humans, all of which are readily oxidized by ozone.

19.39 Ozone is effective as an antibacterial and an antiviral agent, but it is difficult to detect when present in undesirable amounts.

Oxides

19.40 (a) The most ionic oxides are those of the Group IA and IIA metals.
(b) The most basic oxides are those of the Group IA and IIA metals.
(c) The Group IIA metal oxides (beryllium excepted) are insoluble in water, although their basic nature is indicated by the fact that they do dissolve in aqueous acid solutions.
(d) The most acidic oxides are the covalent oxides, typically of a nonmetal such as sulfur.

19.41 This change in color indicates that MgO is basic. Although MgO is insoluble in water, it does react as a base:

$$MgO(s) + 2H^+(aq) \rightarrow Mg^{2+}(aq) + H_2O(\ell)$$

19.42 The material with the higher boiling point is the ionic one, i.e. TiO_2.

The higher oxidation state of osmium in OsO_4 causes the bonds to osmium to be more covalent than the bonds to titanium in TiO_2.

19.43 The high oxidation state of osmium in OsO_4 causes the bonds in this substance to be more covalent than ionic. This is why osmium tetroxide is a covalent, acidic oxide.

19.44 (a) $[Al(H_2O)_6]^{3+}(aq) + 4OH^-(aq) \rightarrow [Al(H_2O)_2(OH)_4]^-(aq) + 4H_2O(\ell)$

(b) The high charge and small size of the Al^{3+} ion is more polarizing than the sodium ion, and, consequently, the aluminum ion more strongly attracts the electrons of an H_2O molecule to itself. This enhances ionization of a water group attached to aluminum, giving an OH group in place of the water group, and a proton.

19.45 This is a reaction of a metal oxide with OH^-, and we conclude that the metal oxide acts as an acid. This means that the oxide is an acidic anhydride, or a covalent oxide. The high oxidation state of molybdenum causes the bonds to oxygen to be covalent.

19.46 An acidic oxide is one that reacts with water to give an acid. It is also termed a covalent oxide, or an acidic anhydride.

19.47 An amphoteric material is capable of reacting either as an acid or as a base. A reaction of chromium hydroxide with an acid might be:

$Cr(OH)_3 + 3HCl \rightarrow CrCl_3 + 3H_2O$

A reaction of chromium hydroxide with a base might be:

$Cr(OH)_3 + 3NaOH \rightarrow Na_3CrO_3 + 3H_2O$

Peroxides and Superoxides

19.48 Peroxides contain an O_2^{2-} group that may be regarded as derived from the parent peroxide, hydrogen peroxide.

19.49 Dilute aqueous hydrogen peroxide solution is useful as a mild bleaching agent, and as a disinfectant. More concentrated solutions are dangerous to handle, because of its ready decomposition to give water and oxygen. Some industrial uses are as a bleach for paper, textiles, and leather products, and in the manufacture of organic chemicals for the polymer, plastic and pharmaceutical industries.

19.50 If each hydrogen in H_2O_2 is assigned the oxidation number +1, then oxygen is seen to have oxidation state -1.

19.51 Pure hydrogen peroxide is dangerous both as an oxidizing agent and because of its ready decomposition to give water and molecular oxygen, often accompanied by an explosion or fire that is sustained by the oxygen that is produced.

19.52 (a)

$$K_a = \frac{[HO_2^-][H^+]}{[H_2O_2]}$$

(b) $pK_a = -\log K_a \quad \therefore K_a = \text{antilog}(-pK_a)$

$K_a = \text{antilog}(-11.75) = 1.8 \times 10^{-12}$

(c)

	H_2O_2	$\rightleftharpoons$	H^+	+	HO_2^-
initial conc., M	1.0		0		0
change in conc., M	-x		+x		+x
equilibrium conc., M	1.0 - x		x		x

$K_a = 1.8 \times 10^{-12} = x^2/(1.0 - x) \approx x^2 \quad \therefore x = 1.3 \times 10^{-6} = [H^+]$

$pH = -\log [H^+] = -\log(1.3 \times 10^{-6}) = 5.87$

19.53 (a) Since this is an acidic solution, we use equation 19.1 for the hydrogen peroxide half reaction:

reduction $\quad H_2O_2 + 2H^+ + 2e^- \rightarrow 2H_2O$

We find the half reaction for H_2SO_3 in Table 17.1:

oxidation $\quad SO_4^{2-} + 4H^+ + 2e^- \rightarrow H_2SO_3 + H_2O$

The net cell reaction is the sum of the above two half reactions:

net cell reaction $\quad H_2O_2 + H_2SO_3 \rightarrow SO_4^{2-} + 2H^+ + H_2O$

(b) $E°_{cell} = E°_{reduction} - E°_{oxidation} = 1.77 - (0.17) = 1.60$ V

(c) $\frac{34.0 \text{ g}}{1 \text{ mol}} \times \frac{1 \text{ mol}}{2 \text{ eq}} = 17.0$ g/eq

(d)

$$5.25\ g\ Na_2SO_3 \times \frac{1\ mol\ Na_2SO_3}{126\ g\ Na_2SO_3} \times \frac{1\ mol\ H_2O_2}{1\ mol\ Na_2SO_3} \times \frac{34.0\ g\ H_2O_2}{1\ mol\ H_2O_2} = 1.42\ g$$

19.54 This change in color of the litmus paper indicates that the substance is a bleach rather than a base. The compound must therefore be Na_2O_2, which reacts in water to give hydrogen peroxide:

$Na_2O_2 + 2H_2O \rightarrow 2NaOH + H_2O_2$

Had the compound been the oxide Na_2O, it would have turned the red litmus paper blue because it is a base only, not also a bleach:

$Na_2O + H_2O \rightarrow 2NaOH$

19.55 (a) Sodium reacts with oxygen to give a peroxide.
(b) Lithium reacts with oxygen to give an oxide.
(c) Potassium, rubidium, and cesium react with oxygen to give superoxides.

19.56 (a) $4KO_2 + 2CO_2 \rightarrow 2K_2CO_3 + 3O_2$

(b) $2KO_2 + H_2O + CO_2 \rightarrow O_2 + KHO_2 + KHCO_3$

Reactions Involving Oxygen and Its Compounds

19.57 (a) $2NaH(s) + O_2(g) \rightarrow Na_2O(s) + H_2O$

(b) $H^- + H_2O \rightarrow OH^-(aq) + H_2(g)$

(c) $2HgO(s) \rightarrow 2Hg(\ell) + O_2(g)$

(d) $2KClO_3(s) \rightarrow 2KCl(s) + 3O_2(g)$

(e) $Na_2O_2(s) + 2H_2O \rightarrow 2NaOH(aq) + H_2O_2(aq)$

(f) $4Li(s) + O_2(g) \rightarrow 2Li_2O(s)$

(g) $2H_2O_2(\ell) \rightarrow O_2(g) + H_2O(\ell)$

19.58 (a) $Na_2O(s) + H_2O(\ell) \rightarrow 2NaOH(aq)$

(b) $Na_2O(s) + 2HCl(aq) \rightarrow 2NaCl(aq) + H_2O(\ell)$

(c) $Na_2O(s) + 2HNO_3(aq) \rightarrow 2NaNO_3(aq) + H_2O(\ell)$

(d) $Al_2O_3(s) + 6HCl(aq) \rightarrow 2AlCl_3(aq) + 3H_2O(\ell)$

(e) $Al_2O_3(s) + 2NaOH(aq) + 7H_2O(\ell) \rightarrow 2Na[Al(H_2O)_2(OH)_4](aq)$

(f) $K_2O(s) + 2HCl(aq) \rightarrow 2KCl(aq) + H_2O$

(g) $K_2O(s) + H_2O(\ell) \rightarrow 2KOH(aq)$

(h) $Na_2O_2(s) + 2H_2O \rightarrow 2NaOH(aq) + H_2O_2(aq)$

(i) $Al_2O_3(s) + 6HBr(aq) \rightarrow 2AlBr_3(aq) + 3H_2O(\ell)$

19.59 Either solid would form a basic solution in water, but only the peroxide would dissolve to generate a gas, which is O_2.

$2Na_2O_2 + 2H_2O \rightarrow 4NaOH + O_2$

This is the net reaction, obtained on adding:

$2Na_2O_2 + 4H_2O \rightarrow 4NaOH + 2H_2O_2$ and $2H_2O_2 \rightarrow 2H_2O + O_2$

Nitrogen

19.60 Large quantities of elemental nitrogen are obtained by the fractional distillation of liquified air.

19.61 (a) Enhanced oil recovery is a process in which a gas such as nitrogen, which does not react with petroleum, is used under pressure in order to force additional amounts of crude from a geological formation.

(b) Air, which contains oxygen, would cause some degradation of the petroleum, because of ready oxidation at high pressures.

19.62 (a) An oxidation state of +6 would require the loss of more electrons than are present in the valence shell of a nitrogen atom. The charge is simply too high.

(b) An oxidation state of -4 would require a greater number of electrons on nitrogen than is present in the next noble gas. The charge is simply too high.

19.63 Nitrogen fixation is the conversion of elemental nitrogen into useful, reduced forms of nitrogen. The process is accomplished by bacteria in the soil.

Ammonia

19.64 The smaller value for the heat of vaporization would make ammonia easier to vaporize, and, hence, harder to handle as a liquid.

19.65 (a) A low temperature favors the formation of ammonia because the reaction is exothermic.
(b) A low temperature causes the reaction to be uselessly slow.
(c) According to the principle of Le Chatelier, a high pressure favors the direction of reaction that reduces the total number of moles of gaseous materaial.

19.66 This is shown on page 803 of the text.

19.67 (a) $NH_3 + H_2O \rightarrow NH_4^+ + OH^-$

(b) $NH_3(aq) + HCl(aq) \rightarrow NH_4^+(aq) + Cl^-(aq)$

(c) $4NH_3(g) + 3O_2(g) \rightarrow 2N_2(g) + 6H_2O(g)$

19.68 (a) $NH_3(aq) + HCl(aq) \rightarrow NH_4Cl(aq)$

(b) $NH_3(aq) + HBr(aq) \rightarrow NH_4Br(aq)$

(c) $NH_3(aq) + HI(aq) \rightarrow NH_4I(aq)$

(d) $2NH_3(aq) + H_2SO_4(aq) \rightarrow (NH_4)_2SO_4(aq)$

(e) $NH_3(aq) + HNO_3(aq) \rightarrow NH_4NO_3(aq)$

19.69 An ammoniated substance is one that is surrounded by molecules of ammonia, which is taken to be the solvent.

19.70 The proton donor is NH_4^+, and the proton acceptor is NH_2^-.

19.71 Sodium amine is prepared by dissolving sodium metal in liquid ammonia:

$2Na + 2NH_3 \rightarrow 2NaNH_2 + H_2$

19.72 $NH_4^+ + NH_2^- \rightarrow 2NH_3$

19.73 The amide anion is a stronger base than hydroxide, because the amide ion reacts with water to form the hydroxide ion:

$NH_2^- + H_2O \rightarrow NH_3 + OH^-$

19.74 molecular: $KNH_2(s) + H_2O(\ell) \rightarrow NH_3(aq) + KOH(aq)$

net ionic: $NH_2^- + H_2O \rightarrow NH_3 + OH^-$

19.75 $NH_4(s) \rightarrow NH_3(g) + HCl(g)$

19.76 $NH_4^+ + H_2O \rightarrow NH_3 + H_3O^+$

Nitrides

19.77 A nitride is a compound of nitrogen (in the 3- oxidation state) with an element other than hydrogen. A nitride may be an ionic substance, in which case it contains the nitride ion, N^{3-}. Covalent nitrides are also known, and they adopt a wide variety of structures and properties. Some of the latter are polymers, whereas others are molecular.

19.78 (a) The nitride formed by a Group IIA element has the general formula M_3N_2.

(b) The oxidation number of nitrogen in a Group IIA nitride, M_3N_2, is -3.

(c) As do all ionic nitrides, a nitride of a Group IIA element should react with water to give a metal hydroxide solution:

$M_3N_2(s) + 6H_2O(\ell) \rightarrow 3M(OH)_2(aq) + 2NH_3(aq)$

19.79 (a) $Mg_3N_2(s) + 6H_2O(\ell) \rightarrow 3Mg(OH)_2(aq) + 2NH_3(aq)$

(b) $2Mg(s) + O_2(g) \rightarrow 2MgO(s)$

Hydrazine

19.80 $2NH_3(aq) + NaOCl(aq) \rightarrow N_2H_4(aq) + NaCl(aq) + H_2O(\ell)$

19.81 The oxidation number of nitrogen in hydrazine is 2-.

19.82 These two materials react to give hydrazine, which is toxic.

19.83 The thrust in hydrazine-powered rocket engines is the expansion of hot gases, formed from the combustion of hydrazine:

$N_2H_4(\ell) + O_2(g) \rightarrow N_2(g) + 2H_2O(g)$

19.84 (a) $N_2(g) + 2H_2(g) \rightarrow N_2H_4(\ell)$, $\Delta H° = +50.6$ kJ

(b) It is unstable, since it is formed endothermically.

(c) The activation energy for the reaction is so high that the rate of reaction is low.

(d) $\Delta G° = \Delta H° - T\Delta S°$

$$149.2 \text{ kJ} = 50.6 \text{ kJ} - (298 \text{ K})(\Delta S°)$$

$$\Delta S° = -0.331 \text{ kJ K}^{-1} \text{ mol}^{-1} = -331 \text{ J K}^{-1} \text{ mol}^{-1}$$

Since the value of $\Delta S°$ is negative, the reaction is not spontaneous from the standpoint of entropy. This is because, on going from reactants to products, there is a decrease in the number of molecules of gas, and, hence, a decrease in disorder.

19.85 (a) $N_2H_4(aq) + O_2(aq) \rightarrow N_2(g) + 2H_2O(\ell)$

(b) The products are not reactive with metals, and they are not harmful.

19.86 These ions have structures based on the addition of one and two protons to the structure of hydrazine itself, as diagramed on page 807 of the text.

```
[      H     H      ] +          [      H     H        ] 2+
       |     |                          |     |
[ H —  N  —  N :    ]            [ H —  N  —  N  — H   ]
       |     |                          |     |
[      H     H      ]            [      H     H        ]
```

The geometry at each nitrogen atom in $N_2H_6^{2+}$ is likely tetrahedral.

19.87 (a) $N_2(g) + 4H_2O(g) + 4e^- \rightleftharpoons N_2H_4(aq) + 4OH^-(aq)$, $E° = -1.16$ V

(b) The reduction half reaction that is given in the answer to part (a) must be reversed, to become the oxidation half reaction. We obtain the silver half reaction from the data in Table 17.1:

$$Ag^+ + e^- \rightarrow Ag(s), \quad E° = +0.80 \text{ V}$$

$$E°_{cell} = E°_{reduction} - E°_{oxidation} = +0.80 - (-1.16) = +1.96$$

The following reaction therefore has a positive cell potential, and we conclude that hydrazine is able to reduce silver ion to silver metal:

$$4Ag^+(aq) + N_2H_4(aq) + 4OH^-(aq) \rightarrow 4Ag(s) + N_2(g) + 4H_2O(\ell)$$

Hydroxylamine and Hydrazoic Acid

19.88 $NH_2OH(aq) + H_2O(\ell) \rightarrow NH_3OH^+(aq) + OH^-(aq)$

$$K_a = \frac{[OH^-][NH_3OH^+]}{[NH_2OH]}$$

19.89 Ammonia is the stronger base since its value for K_b is larger than that for hydroxylamine.

19.90 Ammonia is more basic than water, and we conclude that the site of protonation of hydroxylamine is the nitrogen atom:

```
[      H           ] +
       |     ..
  H -- N --  O -- H
       |     ..
       H
]
```

19.91 $HN_3(aq) + H_2O(\ell) \rightarrow H_3O^+(aq) + N_3^-(aq)$

$$K_a = \frac{[H_3O^+][N_3^-]}{[HN_3]}$$

19.92 From the value of pK_a, we can calculate pK_b for the azide ion:

$pK_a + pK_b = 14.00$

pK_b for azide anion is: $14.00 - 4.75 = 9.25$

We conclude that a solution of lithium azide should be basic.

19.93 These are contact explosives.

19.94 These are shown at the bottom of page 807 of the text.

Nitric Acid and Nitrates

19.95 $4NH_3(g) + 5O_2(g) \rightarrow 4NO(g) + 6H_2O(g)$

$2NO(g) + O_2(g) \rightarrow 2NO_2(g)$

$H_2O(g) + 3NO_2(g) \rightarrow 2HNO_3(aq) + NO(g)$

19.96 Nitric acid is used mostly in the synthesis of ammonium nitrate:

$NH_3 + HNO_3 \rightarrow NH_4NO_3$

19.97 The red-brown color is due to the formation of NO_2:

$4HNO_3(aq) \rightarrow 4NO_2(g) + O_2(g) + 2H_2O(\ell)$

19.98 (a) $3Cu(s) + 8HNO_3(aq) \rightarrow 3Cu(NO_3)_2(aq) + 2NO(g) + 4H_2O(\ell)$

(b) $Cu(s) + 4HNO_3(aq) \rightarrow Cu(NO_3)_2(aq) + 2NO_2(g) + 2H_2O(\ell)$

19.99 (a) oxidation:

$NO + 2H_2O \rightarrow NO_3^- + 4H^+ + 3e^-$, $E° = 0.96$ V

reduction:

$NO_3^- + 2H^+ + e^- \rightarrow NO_2 + H_2O$, $E° = +0.80$ V

(b) Add the two equations given in the answer to part (a):

$2NO_3^- + NO + 2H^+ \rightarrow 3NO_2 + H_2O$

(c) $E°_{cell} = E°_{reduction} - E°_{oxidation} = 0.80 - (0.96) = -0.04$ V

The negative value of E° indicates that the reaction is not spontaneous under standard conditions.

(d) The value of $E°_{cell}$ is only slightly negative, and the second term in the Nernst equation can be made to be sufficiently positive so as to offset the negative value of $E°_{cell}$.

19.100 (a) A small number of moles of the explosive must form, on reaction, a very large number of moles of hot gaseous products.

(b) The explosive must not detonate prematurely or accidentally. Thus it must be stable on storage, in transportation, and in contact with devices and materials that are used in handling it.

(c) Nobel invented the relatively stable mixture known as dynamite: nitroglycerin, ammonium nitrate, sodium nitrate, wood pulp, and calcium carbonate.

19.101 Sodium nitrite is thermally stable, even at temperatures as high as its melting point, 284 °C. Sodium nitrite is the product when sodium nitrate is decomposed thermally:

$2NaNO_3(s) \rightarrow 2NaNO_2(s) + O_2(g)$

On the other hand, potassium nitrate decomposes thermally at temperatures above 500 °C to give, not the nitrite, but the oxide:

$4KNO_3(s) \rightarrow 2K_2O(s) + 2N_2(g) + 5O_2(g)$

This indicates that potassium nitrite is not as stable as sodium nitrite.

19.102 These nitrates are very stable above their melting points, although they do decompose at extremely high (> 500 °C) temperatures.

19.103 Bismuth nitrate decomposes in water to give a subnitrate:

$Bi(NO_3)_3(s) + H_2O(\ell) \rightarrow BiO(NO_3)(s) + 2HNO_3(aq)$

The pentahydrated form of bismuth nitrate, $Bi(NO_3)_3 \cdot 5H_2O$ also forms the subnitrate in water.

19.104 (a) $NH_4NO_3(s) \rightarrow N_2O(g) + 2H_2O(g)$

(b) $2NH_4NO_3(s) \rightarrow 2N_2(g) + O_2(g) + 4H_2O(g)$

(c) N_2O is unstable towards the formation of N_2 and $\frac{1}{2}O_2$.

Oxides of Nitrogen and Nitrous Acid

19.105 The bond order in NO is 2.5, whereas it is 3.0 in N_2, because an extra antibonding, or π^* electron in NO cancels some of the bond order from three lower lying, bonding molecular orbitals.

19.106 The decomposition of N_2O produces more moles of gaseous products than there are moles of gas from the reactant:

$2N_2O \rightarrow 2N_2 + O_2$

19.107 As shown at the top of page 814, the dimerization of NO_2 to give N_2O_4 causes the pairing of two odd electrons.

19.108 $3HNO_2(aq) \rightarrow HNO_3(aq) + 2NO(g) + H_2O(\ell)$

19.109 (a) $k_a = \text{antilog}(-pK_a) = \text{antilog}(-3.35) = 4.5 \times 10^{-4}$

(b) $HNO_2(aq) = H^+(aq) + NO_2^-(aq)$

19.110 $NaNO_2$ is a weak base, which causes its aqueous solutions to have a pH greater than 7:

$NO_2^-(aq) + H_2O(\ell) = HNO_2(aq) + OH^-(aq)$

19.111 This is prepared by the dehydration of nitric acid using P_4O_{10}:

$4HNO_3(aq) + P_4O_{10}(s) \rightarrow 2N_2O_5(s) + 4HPO_3(\ell)$

19.112

:Ö:
|
N
H — O — N = O

19.113 There are sixteen valence electrons to be distributed in this linear ion:

$[\ :\!O = N = O\!:\ ]^+$

19.114 (a) dinitrogen pentoxide, N_2O_5

(b) nitrogen dioxide, NO_2

(c) dinitrogen monoxide, N_2O

(d) nitrogen monoxide, NO

(e) nitrogen monoxide, NO

(f) dinitrogen trioxide, N_2O_3

(g) nitrogen dioxide, NO_2

(h) dinitrogen trioxide, N_2O_3

(i) dinitrogen monoxide, N_2O

(j) nitrogen dioxide, NO_2 and nitrogen monoxide, NO

(k) dinitrogen pentoxide, N_2O_5

(l) nitrogen monoxide, NO

(m) nitrogen monoxide, NO

(n) nitrogen dioxide, NO_2

(o) nitrogen dioxide, NO_2

Formulas, Names and Reactions of Substances

19.115 (a) H^- (b) KNH_2 (c) $(NH_4)_2SO_4$

(d) O_3 (e) NH_2^- (f) Na_2O

(g) Na_2O_2 (h) NaH (i) NH_4NO_3

(j) $LiNH_2$ (k) Na_3N (l) HNO_3

(m) $NaNH_2$ (n) HNO_2 (o) H_2O_2

(p) N_2H_4 (q) NO (r) $NaNO_2$

(s) N_2O (t) N_2O_5 (u) NH_2OH

(v) HN_3

19.116 (a) $Zn(s) + 2HCl(aq) \rightarrow ZnCl_2(aq) + H_2(g)$

(b) $Ca(s) + H_2(g) \rightarrow CaH_2(s)$

(c) $NaH(s) + H_2O(\ell) \rightarrow NaOH(aq) + H_2(g)$

(d) $Mg(s) + 2HCl(aq) \rightarrow MgCl_2(aq) + H_2(g)$

(e) $KH(s) + H_2O(\ell) \rightarrow KOH(aq) + H_2(g)$

(f) $CH_3{-}CH{=}CH_2(g) + H_2(g) \rightarrow CH_3{-}CH_2{-}CH_3(g)$

(g) $CH_2{=}CH_2(g) + CO(g) + H_2(g) \rightarrow CH_3{-}CH_2{-}CH{=}O(\ell)$

(h) $2HgO(s) \rightarrow 2Hg(\ell) + O_2(g)$

(i) $Na_2O(s) + H_2O(\ell) \rightarrow 2NaOH(aq)$

(j) $MgO(s) + 2HBr(aq) \rightarrow MgBr_2(aq) + H_2O(\ell)$

(k) $2KClO_3(s) \rightarrow 2KCl(s) + 3O_2(g)$

(l) $Al_2O_3(s) + 6HBr(aq) \rightarrow 2AlBr_3(aq) + 3H_2O(\ell)$

(m) $Al_2O_3(s) + 2KOH(aq) + 7H_2O(\ell) \rightarrow 2K[Al(H_2O)_2(OH)_4](aq)$

(n) $2CrO_3(s) + H_2O(\ell) \rightarrow H_2Cr_2O_7(aq)$

(o) $2H_2O_2(aq) \rightarrow 2H_2O(\ell) + O_2(g)$

(p) $K_2O_2(s) + 2H_2O(\ell) \rightarrow 2KOH(aq) + H_2O_2(aq)$

19.117 (a) $NH_3(aq) + HCl(aq) \rightarrow NH_4Cl(aq)$

(b) $NH_4Br(s) \rightarrow NH_3(g) + HBr(g)$

(c) $NH_4Cl(aq) + NaOH(aq) \rightarrow NH_3(g) + NaCl(aq) + H_2O(\ell)$

(d) $NH_4Cl(am) + NaNH_2(am) \rightarrow 2NH_3(\ell) + NaCl(am)$

(e) $(NH_4)_2Cr_2O_7(s) \rightarrow N_2(g) + Cr_2O_3(s) + 4H_2O(g)$

(f) $Li_3N(s) + 3H_2O(\ell) \rightarrow 3LiOH(aq) + NH_3(aq)$

(g) $2NH_3(aq) + NaOCl(aq) \rightarrow N_2H_4(aq) + NaCl(aq) + H_2O(\ell)$

(h) $N_2H_4(\ell) + O_2(g) \rightarrow N_2(g) + 2H_2O(g)$

(i) $NH_2OH(aq) + HCl(aq) \rightarrow NH_3OHCl(aq)$

(j) $HN_3(\ell) + LiOH(aq) \rightarrow LiN_3(aq) + H_2O(\ell)$

19.118 (a) $2NO(g) + O_2(g) \rightarrow 2NO_2(g)$

(b) $KOH(aq) + HNO_3(aq) \rightarrow KNO_3(aq) + H_2O(\ell)$

(c) $2N_2O(g) \rightarrow 2N_2(g) + O_2(g)$

(d) $2NO_2(g) + H_2O(\ell) \rightarrow HNO_2(aq) + HNO_3(aq)$

(e) $N_2O_3(g) \rightarrow NO(g) + NO_2(g)$

(f) $N_2O_5(s) + H_2O(\ell) \rightarrow 2HNO_3(aq)$

(g) $N_2O_4(g) \rightarrow 2NO_2(g)$

(h) $4HNO_3(\ell) + P_4O_{10}(s) \rightarrow 2N_2O_5(s) + 4HPO_3(\ell)$

(i) $2NaNO_3(s) \rightarrow 2NaNO_2(s) + O_2(g)$

(j) $Bi(NO_3)_3(aq) + H_2O(\ell) \rightarrow BiO(NO_3)(s) + 2HNO_3(aq)$

(k) $N_2O_3(g) + 2NaOH(aq) \rightarrow 2NaNO_2(aq) + H_2O(\ell)$

(l) $NaNO_2(aq) + HCl(aq) \rightarrow HNO_2(aq) + H_2O(\ell)$

(m) $3NaNO_2(aq) + 3HCl(aq) \rightarrow HNO_3(aq) + 2NO(g) + 3NaCl(aq) + H_2O(\ell)$

CHAPTER TWENTY

REVIEW EXERCISES

Sulfur

20.1 Sulfur occurs in nature in a variety of forms: as the free element, in sulfide and sulfate minerals, as H_2S and other inorganic forms in petroleum, as various organic forms in coal and petroleum, and as -SH and —S—S— groups in proteins.

20.2 $2H_2S(g) + SO_2(g) \rightarrow 3S(s) + 2H_2O(g)$

20.3 This is sulfur dioxide, SO_2.

$2CuS(s) + 3O_2(g) \rightarrow 2CuO(s) + 2SO_2(g)$

The SO_2 is used as in Review Exercise 20.2 (i.e. the formation of sulfur from H_2S) and in the production of sulfuric acid.

20.4 This is orthorhombic sulfur, of S_8 molecules in the S_α form.

20.5 The S_8 molecule in the S_α form is an eight-membered ring having the crown conformation.

20.6 The various S_8 molecules become arranged in the crystal in a different pattern.

20.7 This is also the S_8 molecule, but the arrangement of the various molecules is different, giving crystals having a different morphology, as depicted in Figure 20.4 of the text.

20.8 It reverts to the more stable α form.

20.9 The S_8 molecules open up and join with others to create chains that are entangled.

20.10 Above 200 °C, the large polymer molecules that cause the low temperature melt to be viscous, become broken into smaller molecules which are more free to move about.

Chapter 20

Sulfur Dioxide, Sulfurous Acid, Acid Rain, and the Sulfites

20.11 This is mostly due to trace amounts of sulfur found in coal, which when burned, produces SO_2.

20.12 This is termed sulfurous acid, $H_2SO_3(aq)$, although H_2SO_3 molecules have never been found. Such solutions contain hydrated sulfur dioxide, $SO_2 \cdot nH_2O$, $H^+(aq)$, and $HSO_3^-(aq)$.

20.13 Acid rain is rain with a lower than normal pH. It is mostly composed of rain having added amounts of sulfurous and sulfuric acids, or some nitrous or nitric acids, or both.

20.14 The nonmetal oxides that contribute to acid deposition (both acid rain and solid particle deposition) are the acidic anhydrides of sulfur and nitrogen, namely NO_2, SO_2, and SO_3.

20.15 The smokestack gases are scrubbed with wet calcium hydroxide pellets:

$$SO_2(g) + Ca(OH)_2(s) \rightarrow CaSO_3(s) + H_2O(\ell)$$

20.16 (a) $NaOH(aq) + SO_2(g) \rightarrow NaHSO_3(aq)$

(b) $NaHCO_3(aq) + SO_2(g) \rightarrow NaHSO_3(aq) + CO_2(g)$

20.17 (a) $S_2O_5^{2-}$ The structure is shown on page 831 of the text.

(b) SO_3^{2-} and $S_2O_5^{2-}$

(c) HSO_3^-

(d) SO_3^{-2} and $S_2O_5^{2-}$

(e) SO_3^{2-}

(f) SO_3^{2-} and HSO_3^-

20.18 $2HSO_3^-(aq) \rightleftharpoons S_2O_5^{2-}(aq) + H_2O(\ell)$

20.19 $I_2(s) + H_2SO_3(aq) + H_2O(\ell) \rightarrow 2I^-(aq) + SO_4^{2-}(aq) + 4H^+(aq)$

$E°_{cell} = E°_{red} - E°_{ox}$, and the standard half cell potentials are obtained from Table 17.1 on page 715 of the text.

$E°_{cell} = 0.54 - (0.17) = 0.37$ V

The reaction is spontaneous in the direction written, because the overall cell potential is greater than zero.

20.20 The gas was sulfur dioxide, SO_2. The solid was $NaHSO_3(s)$.

$NaHSO_3(s) + HCl(aq) \rightarrow NaCl(aq) + H_2O(\ell) + SO_2(g)$

$HSO_3^-(aq) + H^+(aq) \rightarrow H_2O(\ell) + SO_2(g)$

20.21 (a) $NaHSO_3(aq) + HCl(aq) \rightarrow H_2O(\ell) + SO_2(g) + NaCl(aq)$

$HSO_3^-(aq) + H^+(aq) \rightarrow H_2O(\ell) + SO_2(g)$

(b) $Na_2SO_3(aq) + 2HCl(aq) \rightarrow H_2O(\ell) + SO_2(g) + 2NaCl(aq)$

$SO_3^{2-}(aq) + 2H^+(aq) \rightarrow SO_2(g) + H_2O(\ell)$

20.22 (a) $NaHSO_3(aq) + HBr(aq) \rightarrow H_2O(\ell) + SO_2(g) + NaBr(aq)$

$HSO_3^-(aq) + H^+(aq) \rightarrow H_2O(\ell) + SO_2(g)$

(b) $Na_2SO_3(aq) + 2HBr(aq) \rightarrow H_2O(\ell) + SO_2(g) + 2NaBr(aq)$

$SO_3^{2-}(aq) + 2H^+(aq) \rightarrow H_2O(\ell) + SO_2(g)$

(c) $K_2SO_3(aq) + 2HBr(aq) \rightarrow H_2O(\ell) + SO_2(g) + 2KBr(aq)$

$SO_3^{2-}(aq) + 2H^+(aq) \rightarrow H_2O(\ell) + SO_2(g)$

(d) $KHSO_3(aq) + HBr(aq) \rightarrow H_2O(\ell) + SO_2(g) + KBr(aq)$

$HSO_3^-(aq) + H^+(aq) \rightarrow H_2O(\ell) + SO_2(g)$

Sulfur Trioxide, Sulfuric Acid, and the Sulfates

20.23 (a) $SO_3(g) + H_2O(\ell) \rightarrow H_2SO_4(aq)$

(b) $SO_3(g) + 2NaHCO_3(aq) \rightarrow Na_2SO_4(aq) + H_2O(\ell) + 2CO_2(g)$

(c) $SO_3(g) + Na_2CO_3(aq) \rightarrow Na_2SO_4(aq) + CO_2(g)$

(d) $SO_3(g) + 2NaOH(aq) \rightarrow Na_2SO_4(aq) + H_2O(\ell)$

20.24 $S(s) + O_2(g) \rightarrow SO_2(g)$

$2SO_2(g) + O_2(g) \rightarrow 2SO_3(g)$

$SO_3(g) + H_2SO_4(\ell) \rightarrow H_2S_2O_7(\ell)$

20.25 Most forms of phosphate rock contain the insoluble material $Ca_5(PO_4)_3F$, which renders the material insoluble in water. Sulfuric acid is used to transform this into superphosphate, which is more soluble:

$2Ca_5(PO_4)_3F + 7H_2SO_4 + 17H_2O \rightarrow 7CaSO_4{\cdot}2H_2O + 3Ca(H_2PO_4)_2{\cdot}H_2O + 2HF$

20.26 Pure sulfuric acid readily solidifies, because its freezing point is only 10.4 °C. This makes the pure material harder to handle and transport.

20.27 Sulfuric acid is (a) a strong acid, (b) inexpensive, (c) easy to make and ship in relatively pure form, (d) stable and not easily vaporized, (e) not normally an oxidizing agent, and (f) a useful dehydrating agent.

20.28 Since the acid is thick and syrupy, it is difficult to wash away. It also reacts exothermically, especially with water.

20.29 The solid was the acid $NaHSO_4$, reacting with sodium carbonate to release $CO_2(g)$.

$2NaHSO_4(aq) + Na_2CO_3(s) \rightarrow 2Na_2SO_4(aq) + H_2O(\ell) + CO_2(g)$

20.30 Such a solution would be acidic:

$HSO_4^-(aq) \rightarrow H^+(aq) + SO_4^{2-}(aq)$

20.31 Such a solution would be neutral.

20.32 Drierite is solid $MgSO_4$. It serves as a drying agent by absorbing water

to form hydrates, and it can be regenerated because it does not decompose or melt when heated at a high enough temperature to drive off the water.

20.33 It combines with water to form hydrates.

Other Compounds of Sulfur

20.34 $S(s) + Na_2SO_3(aq) \rightarrow Na_2S_2O_3(aq)$

20.35 The parent acid (thiosulfuric acid, or "$H_2S_2O_3$") is not stable.

20.36 (a) $S_2O_3^{2-}(aq) + 4Cl_2(g) + 5H_2O(\ell) \rightarrow 2HSO_4^-(aq) + 8H^+(aq) + 8Cl^-(aq)$

(b) $2S_2O_3^{2-}(aq) + I_2(aq) \rightarrow S_4O_6^{2-}(aq) + 2I^-(aq)$

20.37 H_2S, hydrogen sulfide

20.38 On reaction with HCl, the sulfide mineral would generate H_2S:

$FeS(s) + 2HCl(aq) \rightarrow FeCl_2(aq) + H_2S(g)$

On reaction with HCl, the carbonate mineral would generate $CO_2(g)$:

$FeCO_3(s) + 2HCl(aq) \rightarrow FeCl_2(aq) + CO_2(g) + H_2O(\ell)$

20.39 This would be O^{2-}, because H_2S is a stronger acid than H_2O.

20.40 The polysulfanes, H_2S_x, are molecules containing S—S linkages, made by melting hydrated sodium sulfide and sulfur, followed by pouring the melt into dilute HCl at -10 °C.

20.41 SF_6

20.42 disulfur dichloride, S_2Cl_2 and sulfur dichloride, SCl_2

20.43 (a) H_2SeO_3 (b) H_2TeO_3 (c) H_2SeO_4 (d) $Te(OH)_6$ or H_6TeO_6

20.44 Selenium forms many allotropes, e.g. Se_8 in α and β forms. There is also amorphous Se.

20.45 H_2Se and H_2Te

20.46 $H_2SeO_4(aq) + 2NaOH(aq) \rightarrow Na_2SeO_4(aq) + 2H_2O(\ell)$

Phosphorus and Some of its Binary Compounds

20.47 Phosphorus readily forms bonds to itself more than once, whereas nitrogen does not, because the formation of π bonds from one phosphorus atom to another (as in N≡N) is not favored, the phosphorus atoms being too large. Rather, the phosphorus to phosphorus bond is typically a σ bond.

20.48 This is white phosphorus, P_4, whose tetrahedral structure is shown on page 840 of the text.

20.49 The small angle in the P_4 tetrahedron causes the P—P bond to be strained and therefore weak. Thus it is susceptible to the sort of chemical attack that opens the P_4 molecule.

20.50 Red phosphorus does not contain the strained and unstable P_4 unit.

20.51 This is phosphine, PH_3.

(a) It is less soluble than ammonia.
(b) It is less basic in water than ammonia.
(c) It is more reactive with oxygen than ammonia.
(d) It has a different odor entirely, smelling like garlic.

20.52 PCl_3, phosphorus trichloride and PCl_5, phosphorus pentachloride

$2P(s) + 3Cl_2(g) \rightarrow 2PCl_3(\ell)$

$2P(s) + 5Cl_2(g) \rightarrow 2PCl_5(s)$, which is better written: $(PCl_4)(PCl_6)(s)$.

20.53 (a) In the first case, we have the formation of phosphorous acid:

$PCl_3(\ell) + 3H_2O(\ell) \rightarrow H_3PO_3(aq) + 3HCl(aq)$

(b) In the second case, we have the formation of phosphoric acid:

$PCl_5(\ell) + 4H_2O(\ell) \rightarrow H_3PO_4(aq) + 5HCl(aq)$

20.54 $2PBr_3(\ell) + 6H_2O(\ell) \rightarrow 2H_3PO_3(aq) + 6HBr(aq)$

$PBr_5(s) + 4H_2O(\ell) \rightarrow H_3PO_4(aq) + 5HBr(aq)$

20.55 $2PCl_3(\ell) + O_2(g) \rightarrow 2OPCl_3(\ell)$

20.56 (a) $POCl_3(\ell) + 3H_2O(\ell) \rightarrow (HO)_3PO(aq) + 3HCl(aq)$

(b) $POCl_3(\ell) + 3CH_3OH(\ell) \rightarrow (CH_3O)_3PO(\ell) + 3HCl(g)$

(c) $POCl_3(\ell) + 3CH_3CH_2OH \rightarrow (CH_3CH_2O)_3PO(\ell) + 3HCl(g)$

Oxides of Phosphorus and Oxoacids

20.57 $4P(s) + 5O_2(g) \rightarrow P_4O_{10}(s)$

20.58 (a) $P_4O_{10}(s) + 6H_2O(\ell) \rightarrow 4H_3PO_4(aq)$

(b) This is done by increasing the proportion of H_2SO_4 that is employed

in the production of superphosphate, page 835 of the text. Thus the following further reaction is driven:

$$Ca(H_2PO_4)_2 \cdot H_2O + H_2SO_4 \rightarrow CaSO_4(s) + H_2O(\ell) + 2H_3PO_4(\ell)$$

20.59 $H_4P_2O_7$

$$2H_3PO_4 \rightarrow H_4P_2O_7 + H_2O$$

Thus loss of one water molecule from between two molecules of phosphoric acid gives diphosphoric acid.

20.60 (a) The production of triple superphosphate employs phosphoric acid rather than sulfuric acid.

(b) They differ in that the triple superphosphate contains no calcium sulfate, as does superphosphate.

20.61 Its aqueous solutions are acidic:

$$H_2PO_4^- \rightleftharpoons H^+ + HPO_4^{2-}$$

20.62 This is a substance that will commence acting as an acid when water is added.

$$H_2PO_4^{2-}(aq) + HCO_3^-(aq) \rightarrow CO_2(g) + H_2O(\ell) + HPO_4^{2-}(aq)$$

20.63 (a) KH_2PO_4

(b) Na_3PO_4

(c) Na_2HPO_4

(d) $Ca_3(PO_4)_2$

(e) $Na_4P_2O_7$

(f) $Na_5P_3O_{10}$

(g) $Na_3P_3O_9$

(h) KH_2PO_2

(i) H_3PO_3

(j) AsH_3

(k) SbH_3

20.64 (a) orthophosphoric acid or simply phosphoric acid
(b) phosphorous acid
(c) hypophosphorous acid
(d) potassium dihydrogen phosphate
(e) magnesium monohydrogen phosphate
(f) sodium monohydrogen phosphate
(g) sodium hypophosphite
(h) sodium dihydrogen phosphite

20.65 (a) arsenic acid
(b) sodium arsenate
(c) antimonic acid
(d) sodium dihydrogen arsenate
(e) arsenic trichloride
(f) antimony pentachloride

The Halogens and Their Compounds

20.66 The halogens are fluorine, F_2; chlorine, Cl_2; bromine, Br_2; iodine, I_2; and astatine, At_2.

20.67 (a) Fluorine is made by the electrolysis of KHF_2 in liquid HF:

$$2F^- + 2H^+ \rightarrow H_2 + F_2$$

(b) Chlorine is obtained by the electrolysis of aqueous sodium chloride solution:

$$2Na^+ + 2Cl^- + 2H_2O \rightarrow Cl_2 + 2Na^+ + 2OH^- + H_2$$

(c) Bromine is obtained from salt marsh or brine, by oxidation using Cl_2:

$$Cl_2(g) + 2Br^-(aq) \rightarrow Br_2(\ell) + 2Cl^-(aq)$$

(d) Iodine is also obtained by the oxidation of the anion:

$$Cl_2(g) + 2I^-(aq) \rightarrow I_2(s) + 2Cl^-(aq)$$

20.68 $Cl_2 + 2I^- \rightarrow I_2 + 2Cl^-$

20.69 $2F_2(g) + 2H_2O(\ell) \rightarrow 4HF(g) + O_2(g)$

$$Cl_2(aq) + H_2O(\ell) \rightarrow HCl(aq) + HOCl(aq)$$

$$Br_2(aq) + H_2O(\ell) \rightarrow HBr(aq) + HOBr(aq)$$

20.70 $4HF(g) + SiO_2(s) \rightarrow SiF_4(g) + 2H_2O(\ell)$

20.71 Fluorine is a small atom, with the largest electronegativity of any element. Thus the covalent H—F bond is very strongly polarized, causing one HF molecule to be strongly H-bonded to another.

20.72 There is some replacement by F^- ions of the OH^- ions that are normally present in the mineral (hydroxyapatite) of tooth enamel, giving a harder material that is also more resistant to acids.

20.73 $2Ca(OH)_2(s) + 2Cl_2(g) \rightarrow Ca(OCl)_2(s) + CaCl_2(s) + 2H_2O(\ell)$

20.74 dichlorine monoxide, Cl_2O; chlorine dioxide, ClO_2;

dichlorine heptoxide, Cl_2O_7

20.75 In a saturated aqueous solution of Cl_2, the equilibrium concentration of Cl_2 is 0.061 M. This is different from the maximum solubility of Cl_2 in water (0.91 M) because 37 % of the chlorine that dissolves reacts according to the following equilibrium:

$Cl_2 + H_2O \rightleftharpoons HCl + HOCl$

In contrast, although Br_2 is more soluble in water than is chlorine, less of the halogen reacts, although some does, according to the following equation:

$Br_2 + H_2O \rightleftharpoons HBr + HOBr$

Thus a saturated solution of Br_2 in water has a bromine concentration of 0.21 M, but the concentration of HOBr is only 0.0012 M.

20.76 (a) $3ClO^-(aq) \rightleftharpoons 2Cl^-(aq) + ClO_3^-(aq)$

(b) $2ClO_3^-(aq) + 2Cl^-(aq) + 4H^+(aq) \rightleftharpoons 2ClO_2(aq) + Cl_2(g) + 2H_2O(\ell)$

20.77 I_2 reacts with I^- to form I_3^-, which is soluble:

$I_2(aq) + I^-(aq) \rightleftharpoons I_3^-(aq)$

20.78 When pure, perchloric acid is a very powerful oxidizing agent that explodes on contact with virtually any organic material.

20.79 Sulfuric acid would oxidize HI to I_2, defeating the synthesis. In contrast, HBr is not oxidized by sulfuric acid. Phosphoric acid does not oxidize HI, so it is used as the acid source in the following reaction:

$NaI(s) + H_3PO_4 \rightarrow HI(g) + NaH_2PO_4(s)$

20.80 (a) Y^- is oxidized.
(b) X_2 is reduced.
(c) No. Cl_2 is known to oxidize Br^- instead.
(d) Yes. Cl_2 is a stronger oxidizing agent than Br_2.
(e) Yes. Br_2 is a stronger oxidizing agent than I_2.

20.81 (a) $I_2 < Br_2 < Cl_2 < F_2$

(b) $HF < HCl < HBr < HI$

(c) I < Br < Cl < F
(d) I < Br < Cl < F

(e) $I^- < Br^- < Cl^- < F^-$
(f) $I_2 < Br_2 < Cl_2 < F_2$

(g) $F^- < Cl^- < Br^- < I^-$

20.82 Tincture of iodine is a solution of I_2 in aqueous alcohol.

20.83 Iodine is required in order for the biosynthesis of thyroxin, a hormone, which serves to prevent growth of a goiter.

20.84 This is the oxidation of iodide ion by oxygen:

$$O_2(g) + 4HI(aq) \rightarrow 2I_2(s) + 2H_2O(\ell)$$

20.85 These are iodic acid, HIO_3 and periodic acid, H_5IO_6.

Carbon and Its Inorganic Compounds

20.86 carbon, C; silicon, Si; germanium, Ge; tin, Sn; and lead, Pb

20.87 (a) There are three possibilities:

```
    H   H   H   H   H
    |   |   |   |   |
H — C — C — C — C — C — H
    |   |   |   |   |
    H   H   H   H   H
```

which may also be written: $CH_3CH_2CH_2CH_2CH_3$, plus the following two:

```
      CH3                                CH3
      |                                  |
CH3 — CH — CH2 — CH3               CH3 — C — CH3
                                         |
                                         CH3
```

(b) $H_3Sn{-}SnH_3$ (c) $CH_3{-}CH{=}CH_2$

(d)

$$\begin{array}{ccccccc} & & H & & H & & H & & \\ & & | & & | & & | & & \\ H & — & Ge & — & Ge & — & Ge & — & H \\ & & | & & | & & | & & \\ & & H & & H & & H & & \end{array}$$

which may also be written $H_3Ge–GeH_2–GeH_3$.

(e) $CH_3–C≡C–CH_3$ and $CH_3–CH_2–C≡C–H$

20.88 Carbon dioxide is a gas, whereas SiO_2 is a network or macromolecular solid.

20.89 The atoms of graphite are arranged in sheets, with relatively weak forces existing between the adjoining sheets. This causes graphite to be slippery, because one sheet can slip by another with relative ease.

20.90 This process, which is accomplished at high temperature and pressure, is possible because the density of diamond is greater than that of graphite. Thus the less dense graphite can be forced to adopt the diamond structure, under pressure.

20.91 (a) $2C(s) + O_2(g) \rightarrow 2CO(g)$

(b) $C(s) + H_2O \rightarrow CO(g) + H_2(g)$

20.92 Concentrated sulfuric acid is used as a dehydrating agent in order to drive the following reaction:

$HCHO_2 \rightarrow CO(g) + H_2O$

20.93 Carbon monoxide is used as a reducing agent in recovering metals from their ores:

$FeO + CO \rightarrow Fe + CO_2$ $CuO + CO \rightarrow Cu + CO_2$

20.94 Carbon monoxide serves as a poison by binding tightly to hemoglobin, and thereby preventing oxygen from entering the respiration process.

20.95 Iron pentacarbonyl, $Fe(CO)_5$, spontaneously ignites in air:

$4Fe(CO)_5 + 13O_2 \rightarrow 2Fe_2O_3 + 20CO_2$

20.96 Methane is the raw material for the production of CO_2, which is used to make urea:

$CO_2 + 2NH_3 \rightarrow (NH_2)_2CO + H_2O$

20.97 $CaCO_3(s) \rightarrow CaO(s) + CO_2(g)$

20.98 The process for making Portland cement is described on page 863. Kilns are used to heat an intimate mixture of limestone, sand and clay. It contains calcium silicates and calcium aluminates, examples being Ca_2SiO_4, Ca_3SiO_5, and $Ca_3Al_2O_6$.

20.99 Concrete is formed by interlocking of —O—Si—O—Si—O—Si— units, which gives a macromolecular or network material that is very hard.

20.100 Liquid CO_2 can be made only at high pressures, typically 5.2 atm at -56°C.

20.101 Each of the materials in equilibrium with H_2CO_3 can neutralize base:

$$OH^- + H_2CO_3 \rightarrow HCO_3^- + H_2O$$

$$OH^- + CO_2 \rightarrow HCO_3^-$$

$$OH^- + HCO_3^- \rightarrow CO_3^{2-} + H_2O$$

20.102 $2H^+(aq) + CO_3^{2-}(aq) \rightarrow H_2O(\ell) + CO_2(g)$

$H^+(aq) + HCO_3^-(aq) \rightarrow H_2O(\ell) + CO_2(g)$

20.103 $CH_4 + 4S \rightarrow CS_2 + 2H_2S$

20.104 Carbon disulfide is treacherously explosive and toxic.

20.105 $CN^-(aq) + H_2O(\ell) \rightarrow HCN(aq) + OH^-(aq)$

20.106 $H^+(aq) + CN^-(aq) \rightarrow HCN(g)$

20.107 Cyanide irreversibly binds to hemoglobin, preventing respiration by blocking the sites normally used for O_2 transport from the lungs. It also blocks an essential copper-containing cell enzyme.

20.108 First we have the carbides of the type C_2^{2-}, such as CaC_2, calcium carbide. Second there are the carbides containing C^{4-}, such as Mg_2C, magnesium carbide. These first two examples are ionic or salt-like carbides. Then there are carbides of the type SiC, silicon carbide, which are best regarded as covalent carbides. Last, there are interstitial carbides, tungsten carbide being an example.

20.109 (a) no reaction

(b) $CaC_2(s) + 2H_2O(\ell) \rightarrow HC{\equiv}CH(g) + Ca(OH)_2(s)$

(c) no reaction

(d) $Mg_2C(s) + 4H_2O(\ell) \rightarrow CH_4(g) + 2Mg(OH)_2(s)$

20.110 Carborundum is the α form of silicon carbide, SiC(s).

Names, Formulas, and Reactions

20.111
(a) SO_3^{2-}
(b) HSO_4^-
(c) $HClO_4$
(d) CaC_2
(e) H_2SO_3
(f) $HClO_3$
(g) SO_4^{2-}
(h) HF
(i) HSO_3^-
(j) $S_2O_3^{2-}$
(k) HF_2^-
(l) $HClO_2$
(m) H_3PO_4
(n) $P_3O_{10}^{5-}$
(o) BrO_3^-
(p) ClO_4^-
(q) PH_3
(r) H_3PO_3
(s) H_3PO_2

20.112
(a) H_3AsO_4
(b) CaC_2
(c) SbH_3
(d) HPO_3
(e) $SnCl_4$
(f) Mg_2C
(g) $H_4P_2O_7$
(h) $Na_2S_2O_5$
(i) S_8
(j) $S_2O_5^{2-}$
(k) $CaSO_4 \cdot 2H_2O$
(l) H_3SeO_3
(m) P_4
(n) Na_2HPO_4
(o) H_3SeO_4
(p) CaF_2
(q) $Na_3(AlF_6)$
(r) NaCl
(s) NaOBr
(t) $NaClO_3$
(u) $KClO_4$
(v) $Ca(OCl)_2$
(w) $NaBrO_2$
(x) NaH_2PO_2

20.113 (a) phosphorus trichloride
(b) hypochlorite ion
(c) sulfuric acid
(d) diphosphoric acid
(e) sodium perchlorate
(f) phosphoric acid
(g) sulfurous acid
(h) phosphine
(i) metaphosphoric acid
(j) dichlorine heptoxide
(k) perchloric acid
(l) phosphorus pentabromide
(m) chlorate ion
(n) bisulfite ion
(o) bicarbonate ion
(p) magnesium carbide
(q) chloric acid
(r) dichlorine monoxide
(s) bromic acid
(t) hypophorous acid

20.114 (a) sodium disulfite
(b) sodium thiosulfate
(c) sodium sulfite
(d) sodium hydrogen sulfate
(e) sodium sulfate
(f) sodium phosphate
(g) sodium monohydrogen phosphate
(h) sodium dihydrogen phosphate
(i) sodium diphosphate
(j) sodium cyclo-triphosphate
(k) sodium tripolyphosphate
(l) sodium bicarbonate
(m) sodium carbonate
(n) sodium arsenate
(o) sodium selenite
(p) sodium selenate
(q) sodium carbide
(r) sodium hypochlorite
(s) sodium bromate
(t) sodium perchlorate
(u) sodium chlorite
(v) sodium hypophosphite

20.115 (a) $SO_2(g) + H_2O(\ell) \rightarrow H_2SO_3(aq) \rightleftharpoons H^+(aq) + HSO_3^-(aq)$

(b) $SO_3(g) + H_2O(\ell) \rightarrow H_2SO_4(aq)$

(c) $CO_2(g) + H_2O(\ell) \rightarrow H_2CO_3(aq) \rightleftharpoons H^+(aq) + HCO_3^-(aq)$

(d) $SO_2(g) + 2NaOH(aq) \rightarrow Na_2SO_3(aq) + H_2O(\ell)$

(e) $SO_3(g) + 2NaOH(aq) \rightarrow Na_2SO_4(aq) + H_2O(\ell)$

(f) $CO_2(g) + 2NaOH(aq) \rightarrow Na_2CO_3(aq) + H_2O(\ell)$

(g) $S(s) + O_2(g) \rightarrow SO_2(g)$

(h) $SO_2(g) + {}^1/_2O_2(g) \rightarrow SO_3(g)$

(i) $SO_2(g) + Na_2CO_3(aq) \rightarrow Na_2SO_3(aq) + CO_2(g)$

(j) $Na_2SO_3(aq) + 2HCl(aq) \rightarrow 2NaCl(aq) + SO_2(g) + H_2O(\ell)$

(k) $S(s) + Na_2SO_3(aq) \rightarrow Na_2S_2O_3(aq)$

(l) $K_2S(aq) + 2HCl(aq) \rightarrow H_2S(g) + 2KCl(aq)$

(m) $KHS(aq) + HCl(aq) \rightarrow H_2S(g) + KCl(aq)$

(n) $P_4O_{10}(s) + 6H_2O(\ell) \rightarrow 4H_3PO_4(aq)$

(o) $P_4O_6(s) + 6H_2O(\ell) \rightarrow 4H_3PO_3(aq)$

20.116 (a) $2KOH(aq) + SO_2(g) \rightarrow K_2SO_3(aq) + H_2O(\ell)$

$2OH^-(aq) + SO_2(g) \rightarrow SO_3^{2-}(aq) + H_2O(\ell)$

(b) $2KOH(aq) + SO_3(g) \rightarrow K_2SO_4(aq) + H_2O(\ell)$

$2OH^-(aq) + SO_3(g) \rightarrow SO_4^{2-}(aq) + H_2O(\ell)$

(c) $2KOH(aq) + CO_2(g) \rightarrow K_2CO_3(aq) + H_2O(\ell)$

$2OH^-(aq) + CO_2(g) \rightarrow CO_3^{2-}(aq) + H_2O(\ell)$

20.117 (a) $HCl(aq) + NaOH(aq) \rightarrow H_2O(\ell) + NaCl(aq)$

$H^+(aq) + OH^-(aq) \rightarrow H_2O(\ell)$

(b) $HCl(aq) + NaHCO_3(aq) \rightarrow H_2O(\ell) + CO_2(g) + NaCl(aq)$

$H^+(aq) + HCO_3^-(aq) \rightarrow H_2O(\ell) + CO_2(g)$

(c) $2HCl(aq) + Na_2CO_3(aq) \rightarrow H_2O(\ell) + CO_2(g) + 2NaCl(aq)$

$2H^+(aq) + CO_3^{2-}(aq) \rightarrow H_2O(\ell) + CO_2(g)$

(d) $HCl(aq) + KOH(aq) \rightarrow H_2O(\ell) + KCl(aq)$

$H^+(aq) + OH^-(aq) \rightarrow H_2O(\ell)$

(e) $2HCl(aq) + K_2CO_3(aq) \rightarrow H_2O(\ell) + CO_2(g) + 2KCl(aq)$

$2H^+(aq) + CO_3^{2-}(aq) \rightarrow H_2O(\ell) + CO_2(g)$

(f) $HCl(aq) + KHCO_3(aq) \rightarrow H_2O(\ell) + CO_2(g) + KCl(aq)$

$H^+(aq) + HCO_3^-(aq) \rightarrow H_2O(\ell) + CO_2(g)$

(g) $2HCl(aq) + CaCO_3(s) \rightarrow H_2O(\ell) + CO_2(g) + CaCl_2(aq)$

$2H^+(aq) + CaCO_3(s) \rightarrow H_2O(\ell) + CO_2(g) + Ca^{2+}(aq)$

(h) $2HCl(aq) + Ca(OH)_2(s) \rightarrow 2H_2O(\ell) + CaCl_2(aq)$

$2H^+(aq) + Ca(OH)_2(s) \rightarrow 2H_2O(\ell) + Ca^{2+}(aq)$

(i) $2HCl(aq) + Mg(OH)_2(s) \rightarrow 2H_2O(\ell) + MgCl_2(aq)$

$2H^+(aq) + Mg(OH)_2(s) \rightarrow 2H_2O(\ell) + Mg^{2+}(aq)$

(j) $2HCl(aq) + MgCO_3(s) \rightarrow H_2O(\ell) + CO_2(g) + MgCl_2(aq)$

$2H^+(aq) + MgCO_3(s) \rightarrow H_2O(\ell) + CO_2(g) + Mg^{2+}(aq)$

20.118 (a) $HBr(aq) + NaOH(aq) \rightarrow H_2O(\ell) + NaBr(aq)$

$H^+(aq) + OH^-(aq) \rightarrow H_2O(\ell)$

(b) $HBr(aq) + NaHCO_3(aq) \rightarrow H_2O(\ell) + CO_2(g) + NaBr(aq)$

$H^+(aq) + HCO_3^-(aq) \rightarrow H_2O(\ell) + CO_2(g)$

(c) $2HBr(aq) + Na_2CO_3(aq) \rightarrow H_2O(\ell) + CO_2(g) + 2NaBr(aq)$

$2H^+(aq) + CO_3^{2-}(aq) \rightarrow H_2O(\ell) + CO_2(g)$

(d) $HBr(aq) + KOH(aq) \rightarrow H_2O(\ell) + KBr(aq)$

$H^+(aq) + OH^-(aq) \rightarrow H_2O(\ell)$

(e) $2HBr(aq) + K_2CO_3(aq) \rightarrow H_2O(\ell) + CO_2(g) + 2KBr(aq)$

$2H^+(aq) + CO_3^{2-}(aq) \rightarrow H_2O(\ell) + CO_2(g)$

(f) $HBr(aq) + KHCO_3(aq) \rightarrow H_2O(\ell) + CO_2(g) + KBr(aq)$

$H^+(aq) + HCO_3^-(aq) \rightarrow H_2O(\ell) + CO_2(g)$

(g) $2HBr(aq) + CaCO_3(s) \rightarrow H_2O(\ell) + CO_2(g) + CaBr_2(aq)$

$2H^+(aq) + CaCO_3(s) \rightarrow H_2O(\ell) + CO_2(g) + Ca^{2+}(aq)$

(h) $2HBr(aq) + Ca(OH)_2(s) \rightarrow 2H_2O(\ell) + CaBr_2(aq)$

$2H^+(aq) + Ca(OH)_2(s) \rightarrow 2H_2O(\ell) + Ca^{2+}(aq)$

(i) $2HBr(aq) + Mg(OH)_2(s) \rightarrow 2H_2O(\ell) + MgBr_2(aq)$

$2H^+(aq) + Mg(OH)_2(s) \rightarrow 2H_2O(\ell) + Mg^{2+}(aq)$

(j) $2HBr(aq) + MgCO_3(s) \rightarrow H_2O(\ell) + CO_2(g) + MgBr_2(aq)$

$2H^+(aq) + MgCO_3(s) \rightarrow H_2O(\ell) + CO_2(g) + Mg^{2+}(aq)$

20.119 (a) $HNO_3(aq) + NaOH(aq) \rightarrow H_2O(\ell) + NaNO_3(aq)$

$H^+(aq) + OH^-(aq) \rightarrow H_2O(\ell)$

(b) $HNO_3(aq) + NaHCO_3(aq) \rightarrow H_2O(\ell) + CO_2(g) + NaNO_3(aq)$

$H^+(aq) + HCO_3^-(aq) \rightarrow H_2O(\ell) + CO_2(g)$

(c) $2HNO_3(aq) + Na_2CO_3(aq) \rightarrow H_2O(\ell) + CO_2(g) + 2NaNO_3(aq)$

$2H^+(aq) + CO_3^{2-}(aq) \rightarrow H_2O(\ell) + CO_2(g)$

(d) $HNO_3(aq) + KOH(aq) \rightarrow H_2O(\ell) + KNO_3(aq)$

$H^+(aq) + OH^-(aq) \rightarrow H_2O(\ell)$

(e) $2HNO_3(aq) + K_2CO_3(aq) \rightarrow H_2O(\ell) + CO_2(g) + 2KNO_3(aq)$

$2H^+(aq) + CO_3^{2-}(aq) \rightarrow H_2O(\ell) + CO_2(g)$

(f) $HNO_3(aq) + KHCO_3(aq) \rightarrow H_2O(\ell) + CO_2(g) + KNO_3(aq)$

$H^+(aq) + HCO_3^-(aq) \rightarrow H_2O(\ell) + CO_2(g)$

(g) $2HNO_3(aq) + CaCO_3(s) \rightarrow H_2O(\ell) + CO_2(g) + Ca(NO_3)_2(aq)$

$2H^+(aq) + CaCO_3(s) \rightarrow H_2O(\ell) + CO_2(g) + Ca^{2+}(aq)$

(h) $2HNO_3(aq) + Ca(OH)_2(s) \rightarrow 2H_2O(\ell) + Ca(NO_3)_2(aq)$

$2H^+(aq) + Ca(OH)_2(s) \rightarrow 2H_2O(\ell) + Ca^{2+}(aq)$

(i) $2HNO_3(aq) + Mg(OH)_2(s) \rightarrow 2H_2O(\ell) + Mg(NO_3)_2(aq)$

$2H^+(aq) + Mg(OH)_2(s) \rightarrow 2H_2O(\ell) + Mg^{2+}(aq)$

(j) $2HNO_3(aq) + MgCO_3(s) \rightarrow H_2O(\ell) + CO_2(g) + Mg(NO_3)_2(aq)$

$2H^+(aq) + MgCO_3(s) \rightarrow H_2O(\ell) + CO_2(g) + Mg^{2+}(aq)$

20.120 (a) $H_2SO_4(aq) + 2NaOH(aq) \rightarrow 2H_2O(\ell) + Na_2SO_4(aq)$

$H^+(aq) + OH^-(aq) \rightarrow H_2O(\ell)$

(b) $H_2SO_4(aq) + 2NaHCO_3(aq) \rightarrow 2H_2O(\ell) + 2CO_2(g) + Na_2SO_4(aq)$

$H^+(aq) + HCO_3^-(aq) \rightarrow H_2O(\ell) + CO_2(g)$

(c) $H_2SO_4(aq) + Na_2CO_3(s) \rightarrow H_2O(\ell) + CO_2(g) + Na_2SO_4(aq)$

$2H^+(aq) + CO_3^{2-}(aq) \rightarrow H_2O(\ell) + CO_2(g)$

(d) $H_2SO_4(aq) + 2KOH(aq) \rightarrow 2H_2O(\ell) + K_2SO_4(aq)$

$H^+(aq) + OH^-(aq) \rightarrow H_2O(\ell)$

(e) $H_2SO_4(aq) + K_2CO_3(aq) \rightarrow H_2O(\ell) + CO_2(g) + K_2SO_4(aq)$

$2H^+(aq) + CO_3^{2-}(aq) \rightarrow H_2O(\ell) + CO_2(g)$

(f) $H_2SO_4(aq) + 2KHCO_3(aq) \rightarrow 2H_2O(\ell) + 2CO_2(g) + K_2SO_4(aq)$

$H^+(aq) + HCO_3^-(aq) \rightarrow H_2O(\ell) + CO_2(g)$

20.121 (a) $H^+(aq) + OH^-(aq) \rightarrow H_2O(\ell)$

(b) $H^+(aq) + HCO_3^-(aq) \rightarrow H_2O(\ell) + CO_2(g)$

(c) $2H^+(aq) + CO_3^{2-}(aq) \rightarrow H_2O(\ell) + CO_2(g)$

(d) $H^+(aq) + HSO_3^-(aq) \rightarrow H_2O(\ell) + SO_2(g)$

(e) $2H^+(aq) + SO_3^{2-}(aq) \rightarrow H_2O(\ell) + SO_2(g)$

(f) $H^+(aq) + CN^-(aq) \rightarrow HCN(g)$

(g) $H^+(aq) + HS^-(aq) \rightarrow H_2S(g)$

(h) $2H^+(aq) + S^{2-}(aq) \rightarrow H_2S(g)$

20.122 (a) $Zn(s) + 2HCl(aq) \rightarrow H_2(g) + ZnCl_2(aq)$

$Zn(s) + 2H^+(aq) \rightarrow H_2(g) + Zn^{2+}(aq)$

(b) $ZnO(s) + 2HCl(aq) \rightarrow H_2O(\ell) + ZnCl_2(aq)$

$ZnO(s) + 2H^+(aq) \rightarrow H_2O(\ell) + Zn^{2+}(aq)$

(c) $Zn(OH)_2(s) + 2HCl(aq) \rightarrow 2H_2O(\ell) + ZnCl_2(aq)$

$Zn(OH)_2(s) + 2H^+(aq) \rightarrow 2H_2O(\ell) + Zn^{2+}(aq)$

(d) No neutralization results.

(e) $Zn(HCO_3)_2(s) + 2HCl(aq) \rightarrow 2H_2O(\ell) + 2CO_2(g) + ZnCl_2(aq)$

$Zn(HCO_3)_2(s) + 2H^+(aq) \rightarrow 2H_2O(\ell) + 2CO_2(g) + Zn^{2+}(aq)$

CHAPTER TWENTY-ONE

REVIEW EXERCISES

Metallic Character and the Periodic Table

21.1 A metallic lattice is the regular and repetitious arrangement of metal atoms in a metallic solid. The atoms are packed and arranged next to one another in a repeating, three-dimensional array. Such a lattice geometry is formed in order to allow each atom to interact and bond to the others by sharing valence electrons as much as possible throughout the extended space of the lattice. This happens because metal atoms in general have too few valence electrons to achieve an octet otherwise.

21.2 (a) Metallic character decreases from left to right across a period.
(b) Metallic character increases from top to bottom in a group.

21.3 (a) Tl (b) Ba (c) K

21.4 RbCl

21.5 MgO

Metallurgy

21.6 Metallurgy is the science and technology of metals.

21.7 An ore is a naturally occurring material from which a desired substance can be extracted in an economical way. The ability to realize a profit by using a given source of the desired substance is what distinguishes an ore from other types of material that may also contain the desired substance.

21.8 Sodium and magnesium are present in sea water in relatively large concentrations, and sea water is not expensive.

21.9 $Mg^{2+}(aq) + 2OH^-(aq) \rightarrow Mg(OH)_2(s)$

$Mg(OH)_2(s) + 2HCl(aq) \rightarrow MgCl_2(aq)$

$MgCl_2(\ell) \rightarrow Mg(\ell) + Cl_2(g)$

21.10 Lime is CaO, and it is obtained by heating calcium carbonate:

$CaCO_3(s) \rightarrow CaO(s) + CO_2(g)$

21.11 These are solid deposits (that can be recovered from the ocean floor) containing manganese and iron compounds. They are typically about the size and shape of an orange. This potential ore has not yet been exploited because there are still less expensive sources available in land deposits.

21.12 Gangue is the unwanted material that is discarded when an ore is purified and the desirable substances are extracted.

21.13 Panning is effective because gold is more dense than sand, etc.

21.14 The ore is mixed thoroughly with water, oil and a detergent. The oil coats the ore particles, but not the gangue. Air is blown through the mixture, and bubbles cling to the oil-coated particles, floating them to the surface, where they can be collected. The detergent aids in stabilizing the froth containing the ore particles, and the sand and gangue settle to the bottom of the tank.

21.15 $Cu_2S(s) + 2O_2(g) \rightarrow SO_2(g) + 2CuO(s)$

$2PbS(s) + 3O_2(g) \rightarrow 2PbO(s) + 2SO_2(g)$

$CaO(s) + SO_2(g) \rightarrow CaSO_3(s)$

The $SO_2(g)$ that is produced in the above reactions can be conveniently diverted to make sulfuric acid.

21.16 $Al_2O_3(s) + 2OH^-(aq) \rightarrow 2AlO_2^-(aq) + H_2O(\ell)$

$AlO_2^-(aq) + H^+(aq) + H_2O(\ell) \rightarrow Al(OH)_3(s)$

$2Al(OH)_3(s) \rightarrow Al_2O_3 + 3H_2O(g)$

21.17 The metals that are to be recovered from an ore generally exist in the ore as compounds of the metals in positive oxidation states. Thus the oxidation number of the metal must be reduced to zero in order to obtain the pure metal.

21.18 These metals are, themselves, powerful reducing agents, so that even stronger reducing agents would be required in order to reduce an ion of one of these metals to obtain the free metal.

21.19 Carbon is plentiful and inexpensive.

21.20 $2PbO(s) + C(s) \rightarrow 2Pb(s) + CO_2(g)$

$2CuO(s) + C(s) \rightarrow 2Cu(s) + CO_2(g)$

21.21 This can be accomplished by reaction with oxygen, followed by heating the resulting mixture of the sulfide and the oxide in the absence of additional oxygen:

$2Cu_2S(s) + 3O_2(g) \rightarrow 2Cu_2O(s) + 2SO_2(g)$

$Cu_2S(s) + 2Cu_2O(s) \rightarrow 6Cu(s) + SO_2(g)$

21.22 The name is taken from the blast of hot air that is forced into the bottom of the furnace.

21.23 The charge consists of a mixture of iron ore (Fe_2O_3), $CaCO_3$, and coke (C).

21.24 The active reducing agent is carbon monoxide:

$3Fe_2O_3 + CO \rightarrow 2Fe_3O_4 + CO_2$

$Fe_3O_4 + CO \rightarrow 3FeO + CO_2$

$FeO + CO \rightarrow Fe + CO_2$

21.25 Slag is calcium silicate, $CaSiO_3$. It is formed by the reaction of calcium oxide with sand:

$CaO + SiO_2 \rightarrow CaSiO_3$

It is useful in the manufacture of cement and insulating materials.

21.26 Refining means purifying and bringing the metal to the state in which it can be of commercial use.

21.27 Pig iron is unrefined, in that it has a higher carbon content and more undesirable impurities than steel.

21.28 The Bessemer converter is an old method for producing steel, in which oxygen is blown through molten pig iron. The molten iron during the process is contained within a pear-shaped vessel that is lined with CaO or MgO. The open hearth furnace gave a more reliable level of purity in the steel product, although it is an essentially slower process.

21.29 An open hearth furnace is a large basin containing molten iron, some Fe_2O_3 and $CaCO_3$. Hot gases keep the iron molten, while the excess Fe_2O_3 reacts with carbon and the $CaCO_3$ forms a slag with other impurities.

These are thus essentially the same reactions that are accomplished in a Bessemer furnace, although the outcome of the process is a more reliably pure steel.

21.30 In the basic oxygen process, scrap iron and molten pig iron are placed in a pear-shaped vessel that is lined with bricks. Oxygen and powdered limestone are blown through the mixture. The oxygen burns off excess carbon, and the $CaCO_3$ reacts with impurities to form slag.

The Alkali Metals

21.31 (a) $1s^22s^1$ (b) $[Ar]4s^1$ (c) $[Rn]7s^1$

21.32 These elements are found in the oceans, brine wells and salt deposits. They are not found free in nature because they are very reactive elements.

21.33 Saltpeter is KNO_3, and Chile saltpeter is $NaNO_3$.

21.34 All of its isotopes are radioactive and have very short half lives.

21.35 Na and potassium are of greatest biological importance. Potassium is most important in plants.

21.36 These metals are soft because the metallic lattice is weak, being made up of large atoms, with few valence electrons.

21.37 Sodium has been used because it has a relatively low melting point, high fluidity, and a large heat capacity.

21.38 Lithium ion has a much larger hydration energy than Na^+, because it has a higher charge density; the charge is the same, but the ion is smaller. Thus lithium cation is hydrated more strongly and stabilized more in water

than is the sodium ion. This makes Li^+ more difficult to reduce, and in the presence of water, lithium is more easily oxidized than sodium. In the absence of water, however, sodium should be the better reducing agent, being more readily oxidized than lithium.

21.39 Metallic sodium is prepared by the electrolysis of molten NaCl. Molten KCl can be reduced by sodium vapor (as a means of preparing metallic potassium) because the equilibrium:

$$KCl + Na \rightleftharpoons NaCl + K$$

is shifted to the right at high temperature. This happens because K is more volatile than Na.

21.40 (a) $2Li + 2H_2O \rightarrow 2LiOH + H_2$

(b) $2K + 2H_2O \rightarrow 2KOH + H_2$

(c) $2Rb + 2H_2O \rightarrow 2RbOH + H_2$

21.41 (a) $4Li + O_2 \rightarrow 2Li_2O$

(b) $2Na + O_2 \rightarrow Na_2O_2$

(c) $Rb + O_2 \rightarrow RbO_2$

21.42 (a) $Na_2O_2 + H_2O \rightarrow 2Na^+ + OH^- + HO_2^-$

(b) $Na_2O + H_2O \rightarrow 2Na^+ + 2OH^-$

(c) $2KO_2 + H_2O \rightarrow 2K^+ + OH^- + HO_2^- + O_2$

(d) $NaOH + CO_2 \rightarrow NaHCO_3$

21.43 KO_2 reacts with H_2 and CO_2 to generate O_2.

21.44 $6Li(s) + N_2(g) \rightarrow 2Li_3N(s)$

21.45 Both (a) and (b) are NaOH.

21.46 Anode: $2Cl^- \rightarrow Cl_2(g) + 2e^-$

Cathode: $Na^+(aq) + e^- \rightarrow Na$ (in mercury)

The NaOH is formed by reaction of Na (dissolved in mercury) with water, at the surface of the mercury:

$2Na + 2H_2O \rightarrow 2NaOH + H_2$

21.47 Sodium carbonate is used principally in the manufacture of glass. Its ore is trona. The Solvay process is:

$CaCO_3 \rightarrow CaO + CO_2$

$CO_2 + NaCl + NH_3 + H_2O \rightarrow NaHCO_3 + NH_4Cl$

$2NaHCO_3 \rightarrow Na_2CO_3 + H_2O + CO_2$

$2NH_4Cl + CaO \rightarrow 2NH_3 + CaCl_2 + H_2O$

21.48 $2NaHCO_3 \rightarrow Na_2CO_3 + H_2O + CO_2$

21.49 $HCO_3^- + H^+ \rightarrow H_2CO_3$

$HCO_3^- \rightarrow H^+ + CO_3^{2-}$

21.50 $CO_3^{2-} + H_2O \rightleftharpoons HCO_3^- + OH^-$, $K_b = 2.1 \times 10^{-4}$

$K_b = 2.1 \times 10^{-4} = x^2/1.0 \quad \therefore x = 1.5 \times 10^{-2}$

$pOH = -\log [OH^-] = -\log (1.5 \times 10^{-2}) = 1.84$

$pH = 14.00 - pOH = 14.00 - 1.84 = 12.16$

21.51 (a) $NaClO_4$ (b) $KC_2H_3O_2$ (c) $KC_2H_3O_2$

21.52 (a) yellow (b) violet (c) red

21.53 $Na_2CO_3 + 2HNO_3 \rightarrow 2NaNO_3 + H_2O + CO_2$

21.54 1.41×10^5 lb

Alkaline Earth Metals

21.55 (a) $[Ne]1s^2$ (b) $[Kr]5s^2$

21.56 The alkaline earth elements are never found in nature in an uncombined or elemental state because they are so very reactive.

21.57 (a) Limestone is $CaCO_3$.

(b) Dolomite is $CaCO_3 \cdot MgCO_3$.

(c) Gypsum is $CaSO_4 \cdot 2H_2O$.

(d) Beryl is $Be_3Al_2(SiO_3)_6$.

21.58 Radium is obtained from uranium ores.

21.59 Both emerald and aquamarine are composed of beryl.

21.60 The group IIA metals have "cations" with a +2 charge, and the electrons in the lattice are held more tightly. Also, the atoms are smaller, and there are two valence electrons available for bonding. The alkaline earth metals are generally less reactive than the alkali metals. Thus the alkaline earth metals are more difficult to oxidize.

21.61 The alkaline earth metals are recovered by electrolysis.

21.62 The alkaline earth metals that react with cold water are calcium, strontium and barium. Magnesium reacts with hot water, and beryllium does not react with water at all.

21.63 Only Be and Mg are sufficiently unreactive to make their commercial use practical. The other metals are, therefore, not made commercially, because they are too reactive, especially with water.

21.64 Beryllium metal is prepared by the electrolysis of $BeCl_2$ in molten NaCl.

21.65 Beryllium imparts hardness to the alloy. A typical use is in springs, electrical contacts, and spark-free tools.

21.66 $CaCO_3 \cdot MgCO_3 \rightarrow CaO \cdot MgO + 2CO_2$

$CaO(s) + MgO(s) + 2H_2O(\ell) \rightarrow Ca(OH)_2(aq) + Mg(OH)_2(s)$

$Mg(OH)_2(s) + 2HCl(aq) \rightarrow MgCl_2(s) + 2H_2O(\ell)$

$MgCl_2(\ell) \rightarrow Mg(\ell) + Cl_2(g)$

21.67 Magnesium is used in flashbulbs and flares, because it burns brightly in air:

$Mg(s) + {}^1/_2O_2(g) \rightarrow MgO(s)$

21.68 Lithium and magnesium ions have similar charge to radius ratios, causing them to be like one another in forming ionic compounds, in solubilities of their ionic compounds, and in forming organometallic compounds that are largely covalent. Magnesium is a harder, less reactive metal.

21.69 Since Be is smaller than Mg, the beryllium atom polarizes an attached chlorine atom more than the magnesium atom does. In other words, the beryllium-to-halogen bond is more polar. This makes the Be-Cl bond more covalent. The tendency to form covalent bonds decreases on descending Group IIA. Conversely, the tendency to form ionic compounds increases on descending Group IIA.

21.70 See Figure 21.10 on page 894 of the text.

21.71 $Mg(OH)_2$

$Mg(OH)_2 + 2HCl \rightarrow MgCl_2 + 2H_2O$

21.72 The solubilities of the sulfates decrease on descending Group IIA. The solubilities of the hydroxides increase from top to bottom in Group IIA.

21.73 Deliquescence is the ability of a material to absorb moisture from the air, resulting in the formation of an aqueous solution of the deliquescent material.

21.74 $2Mg + O_2 \rightarrow 2MgO$

$MgCO_3 \rightarrow MgO + CO_2$

$Mg(OH)_2 \rightarrow MgO + H_2O$

21.75 $CaO + H_2O \rightarrow Ca(OH)_2$

$Mg^{2+} + 2OH^- \rightarrow Mg(OH)_2$

21.76 (a) plaster board
(b) chalk, and in gardens to neutralize soil that is too acidic
(c) in X ray photographs of intestines, and as a whitener in paint
(d) fertilizers, fireproofing fabrics, and tanning of leather
(e) as a dehumidifying agent
(f) antacid
(g) as window material in X ray machines, and in alloys
(h) flashbulbs, fireworks, and in high strength alloys

21.77 Plaster of paris, $CaSO_4 \cdot {}^1/_2H_2O$, is made by heating gypsum:

$2CaSO_4 \cdot 2H_2O(s) \rightarrow (CaSO_4)_2 \cdot H_2O(s) + 3H_2O(g)$

Its reaction with water is simply the reverse of the above reaction.

21.78 (a) orange-red (b) crimson (c) yellow-green

Aluminum

21.79 (a) $[Ne]3s^2 3p^1$ (b) [Ne]

21.80 (a) Aluminum is the third most abundant element.
(b) Aluminum is the most abundant metal.

21.81 Aluminum is obtained from its ores by the Hall process (Chapter 17). In brief, the purified ore (page 880 of the text) is electrolyzed in molten cryolite, $Na_3AlF_6(\ell)$.

21.82 Cryolite is Na_3AlF_6. It contains the ions Na^+ and AlF_6^{3-}.

21.83 A protective oxide film forms on the surface of aluminum, preventing oxidation of the underlying metal.

21.84 Anodized aluminum is metal that has been purposefully coated with a thick oxide layer.

21.85 $2Al + Fe_2O_3 \rightarrow 2Fe + Al_2O_3$

21.86 The formation of Al_2O_3 from Al is an extremely exothermic process. The exhaust gases are therefore produced at a very high temperature. It is the thermal expansion of these exhaust gases that provides the thrust.

21.87 (a) $Al(s) + 3H^+(aq) \rightarrow Al^{3+}(aq) + {}^3/{}_2H_2(g)$

(b) $Al(s) + OH^-(aq) + H_2O(\ell) \rightarrow AlO_2^-(aq) + {}^3/{}_2H_2(g)$

21.88 The ion is $Al(H_2O)_6^{3+}(aq)$, and upon crystallization, the water groups are retained.

21.89 This ion is acidic because of the high degree of polarity in the OH bond, caused by the high charge density of the small Al^{3+} ion.

$Al(H_2O)_6^{3+}(aq) \rightarrow Al(H_2O)_5OH^{2+}(aq) + H^+(aq)$

In base, we have precipitation, caused by neutralization:

$Al(H_2O)_6^{3+}(aq) + 3OH^-(aq) \rightarrow Al(H_2O)_3(OH)_3(s) + 3H_2O(\ell)$

21.90 (a) $Al(H_2O)_3(OH)_3(s) + OH^-(aq) \rightarrow Al(H_2O)_2(OH)_4^-(aq) + H_2O(\ell)$

(b) In acid, there are three step-wise neutralization equilibria:

$Al(H_2O)_3(OH)_3(s) + H_3O^+(aq) \rightarrow Al(H_2O)_4(OH)_2^+(aq) + H_2O(\ell)$

$Al(H_2O)_4(OH)_2^+(aq) + H_3O^+(aq) \rightarrow Al(H_2O)_5OH^{2+}(aq) + H_2O(\ell)$

$Al(H_2O)_5OH^{2+}(aq) + H_3O^+(\ell) \rightarrow Al(H_2O)_6^{3+}(aq) + H_2O(\ell)$

21.91 The precipitation of gelatinous aluminum hydroxide removes solids and bacteria from the water.

21.92 $Al(H_2O)_2(OH)_4^-$

21.93 γ-Al_2O_3 is prepared by mild heating of the gelatinous hydroxide, and it is soluble in both acidic and basic solution. α-Al_2O_3 is extremely hard, and resistant to most forms of chemical attack. It is prepared by heating γ-Al_2O_3 at temperatures above 1000 °C. The sapphire is naturally occurring α-Al_2O_3 (corundum) with traces of iron and titanium impurities. Ruby is also composed mostly of corundum, but the trace impurity is chromium, rather than titanium and iron.

21.94 The alums are mixed-metal aluminum sulfates having the general formula $M^+Al^{3+}(SO_4)_2 \cdot 12H_2O$, where M is a univalent metal such as K or Na. They are termed double salts because they contain two different metals. Potassium alum is used to treat cotton materials so as to make them more absorbing towards dyes, and sodium alum is employed in baking powders.

21.95 The chlorine atoms are arranged in roughly a tetrahedral fashion around each aluminum atom, as diagramed on page 901 of the text.

21.96 This is best seen by comparing Figures 21.18 (page 901) and 21.10 (page 894). Beryllium and aluminum illustrate the Periodic Table diagonal effect, in that their chemistries are alike. Beryllium and aluminum are both amphoteric, and their oxides are too.

Other Metals in Groups IIA, IVA, and VA

21.97 The post transition metals are the metals of Groups IIA, IVA, and VA, which all directly follow one of the rows of transition metals. Their electron configurations are:

Al = $[Ne]3s^2 3p^1$

Ga = $[Ar]3d^{10} 4s^2 4p^1$

In = $[Kr]4d^{10} 5s^2 5p^1$

Tl = $[Xe]4f^{14} 5d^{10} 6s^2 6p^1$

Sn = $[Kr]4d^{10} 5s^2 5p^2$

Pb = $[Xe]4f^{14} 5d^{10} 6s^2 6p^2$

Bi = $[Xe]4f^{14} 5d^{10} 6s^2 6p^3$

21.98 (a) Tl^+, $[Xe]4f^{14} 5d^{10} 6s^2$

Tl^{3+}, $[Xe]4f^{14} 5d^{10}$

(b) Sn^{2+}, $[Kr]4d^{10} 5s^2$

Sn^{4+}, $[Kr]4d^{10}$

(c) Pb^{2+}, $[Xe]4f^{14} 5d^{10} 6s^2$

Pb^{4+}, $[Xe]4f^{14} 5d^{10}$

(d) Bi^{3+}, $[Xe]4f^{14}5d^{10}6s^2$

Bi^{5+}, $[Xe]4f^{14}5d^{10}$

21.99 On going down this periodic table group, the 1+ oxidation state becomes increasingly more stable, and the 3+ oxidation state becomes increasingly less stable, due to increasing size of the atoms. The latter effect causes the covalent bond to atoms to be weaker in general.

21.100 The electrons of the post transition metals are held to the atom more tightly, and these metals have higher ionization energies than the transition metals in general. This arises because of the increasingly imperfect shielding by the d-orbital electrons.

21.101 $SnO_2 + C \rightarrow Sn + CO_2$

$2PbO + C \rightarrow 2Pb + CO_2$

$2Bi_2O_3 + 3C \rightarrow 4Bi + 3CO_2$

21.102 At high temperatures, there exists a metallic allotrope, called white tin. At low temperatures, this changes slowly to a non-metallic allotrope called grey tin.

21.103 Alloys of bismuth expand on cooling. Wood's metal is an alloy of bismuth, lead, tin, and cadmium; it has a low melting point.

21.104 The small Sn^{4+} ion polarizes the chlorine atoms, making the Sn-Cl bond to be largely covalent. Divalent tin cannot do this so well, so its compounds with chlorine are more ionic in nature.

21.105 $Sn(s) + 4HNO_3(aq) \rightarrow SnO_2(s) + 4NO_2(g) + 2H_2O(\ell)$

$3Pb(s) + 8HNO_3(aq) \rightarrow 3Pb(NO_3)_2(aq) + 2NO(g) + 4H_2O(\ell)$

21.106 PbO_2 is a better oxidizing agent than SnO_2.

21.107 The stannite ion is $Sn(OH)_4^{2-}$, and the plumbite ion is $Pb(OH)_4^{2-}$. The stannite ion is the better reducing agent.

21.108 (a) $Sn(s) + 2OH^-(aq) + 2H_2O(\ell) \rightarrow Sn(OH)_4^{2-}(aq) + H_2(g)$

(b) $SnO(s) + 4OH^-(aq) + H_2O(\ell) \rightarrow [Sn(OH)_6]^{4-}(aq)$

(c) $Pb(s) + 2OH^-(aq) + 2H_2O(\ell) \rightarrow Pb(OH)_4^{2-}(aq) + H_2(g)$

(d) $PbO(s) + 2OH^-(aq) + H_2O(\ell) \rightarrow [Pb(OH)_4]^{2-}(aq)$

21.109 (a) $2Bi(s) + {}^{3}/{}_{2}O_2(g) \rightarrow Bi_2O_3(s)$

(b) $Bi + {}^{3}/{}_{2}Cl_2(g) \rightarrow BiCl_3$

21.110 In each case, $K_{sp} = 4x^3$, where x is the molar solubility.

(a) $1.7 \times 10^{-5} = 4x^3$, $\therefore x = 1.6 \times 10^{-2}$ M for $PbCl_2$

(b) $2.1 \times 10^{-6} = 4x^3$, $\therefore x = 8.1 \times 10^{-3}$ M for $PbBr_2$

(c) $7.9 \times 10^{-9} = 4x^3$, $\therefore x = 1.3 \times 10^{-3}$ M for PbI_2

21.111 The 5+ oxidation state of Bi is not stable. Rather we have $BiCl_3$.

21.112 Pb^{4+} is such a powerful oxidizing agent that it tends to oxidize the halide ion to the element:

$PbBr_4 \rightarrow PbBr_2 + Br_2$

21.113 White lead is $Pb_3(OH)_2(CO_3)_2$, which darkens in air owing to reaction with H_2S, to form lead sulfide, PbS(s), which is black.

21.114 This is lead chromate, $PbCrO_4$.

21.115 Litharge, PbO, is used in pottery glazes and lead crystal. Red lead, Pb_3O_4, is used in corrosion-resistant coatings for steel. The simple oxide of lead, PbO_2, finds use in lead storage batteries.

21.116 The hydrolysis reaction is:

$BiCl_3 + H_2O \rightleftharpoons BiOCl(s) + 2H^+ + 2Cl^-$

Addition of HCl(aq) causes BiOCl(s) to redissolve by forcing the above equilibrium to shift to the left.

21.117 The test for manganese causes the violet permanganate to form:

$14H^+ + 5BiO_3^- + 2Mn^{2+} \rightarrow 2MnO_4^- + 5Bi^{3+} + 7H_2O$

CHAPTER TWENTY-TWO

PRACTICE EXERCISES

1. (a) 6 (b) 6 (c) 6 (d) 6

REVIEW EXERCISES

General Properties of Transition Metals

22.1

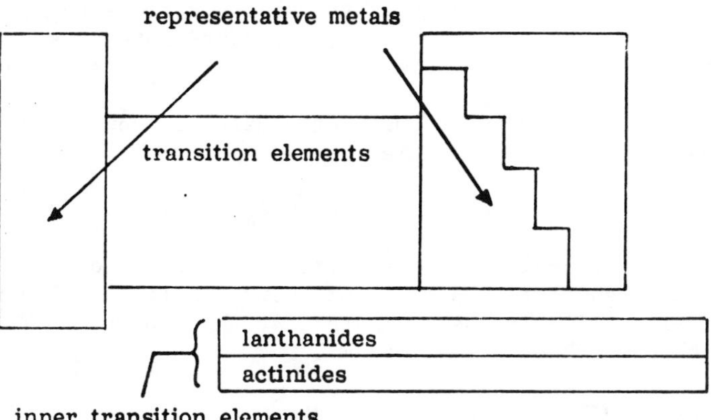

22.2 The transition metals characteristically display multiple oxidation states, colored compounds, and complex ion formation.

22.3 The 2+ oxidation state is common because these metals have, in addition to the d-electrons that typify each element, an ns^2 electron configuration as part of their valence shell. Loss of these two electrons leads to the 2+ ion.

22.4 A complex ion results from the bonding of one or more groups (typically either molecules of anions) to a central metal ion. The resulting coordination compound may have an overall charge that is positive or negative, or it may be neutral. The complex mentioned in the problem has an overall charge of 2-, and it is written: $[Cu(OH)_4]^{2-}$.

22.5 The transition metals typically form more than one oxidation state, because in addition to ionization of the valence ns^2 electrons (leading to oxidation state +2), the metals possess filled or partially filled d subshells that are close in energy to the ns^2 valence electrons, and which, additionally, may be ionized, resulting in oxidation states higher than 2+.

22.6 As shown on page 915, Figure 22.2, there is a gradual decrease in size, until Group VIII, followed by an increase in size.

22.7 The lanthanide contraction is the decrease in size that occurs for the elements from Z = 58 to 71, corresponding to the step-wise filling of the 4f subshell. This contraction causes the elements that follow the lanthanides to be unexpectedly small. In fact, the post lanthanide elements are only about as large as the elements above them, in row 5 of the periodic table. As a consequence, the post lanthanide elements have high densities, high melting points, and they are generally unreactive.

22.8 There is a similarity in the formula of a metal compound from the A and the B Groups of the periodic table:

(a) Na_2SO_4 and Na_2CrO_4 (b) $KMnO_4$ and $KClO_4$ (c) TiO_2 and CO_2

22.9 If the atoms have roughly the same size, then the ratio of their atomic masses should be equal to the ratio of their densities. The mass ratio is $180.95 \div 92.91 = 1.948$, so we estimate that the density of Ta should be nearly twice that of Nb, or 16.7 g/cm³.

22.10 There are three horizontal rows of three related elements, or triads, as they are called. These are Fe, Co, and Ni, as well as the corresponding elements in the second and third rows of transition elements.

22.11 The elements that are ferromagnetic are iron, cobalt and nickel.

22.12 Paramagnetism is about a million times less intense a magnetic effect because, although there are unpaired electrons in the atoms of both paramagnetic and ferromagnetic substances, it is only in the ferromagnetic substances that the atoms are so arranged as to allow the unpaired electrons of neighboring atoms to interact with one another.

22.13 Melting disrupts the arrangements of the domains that are responsible for the large magnetism of ferromagnetic materials, because the structure and arrangement of the domains is lost in the liquid state.

22.14 The magnetic domains that are responsible for the ferromagnetic effect are disrupted when the atoms are made to adopt more random arrangements. This allows the domains to become aligned with the earth's magnetic field. See Figure 22.3, on page 917 of the text.

Chemistry of Some Important Transition Elements:

Titanium, Vanadium, and Chromium

22.15 Titanium is very strong and resistant to corrosion, and it is therefore very useful in high temperature applications requiring strength, i.e. in aircraft engines, etc. It reacts with oxygen and carbon, as reduction of the oxide with carbon is not a practical method of recovering the pure metal from its ore. Rather, titanium metal is obtained from the tetrachloride, by reduction with magnesium:

$TiCl_4 + 2Mg \rightarrow Ti + 2MgCl_2$

22.16 $TiCl_4$

$TiCl_4 + 2H_2O \rightarrow TiO_2 + 4HCl$

22.17 The titanium ion, being small and highly charged, polarizes the electrons of the chloride ion so much that the Ti—Cl linkage is essentially covalent, rather than ionic.

22.18 This is titanium(IV) oxide, TiO_2. It is used as a white pigment in paint and paper.

22.19 Vanadium is used as an additive to steel in order to make the metal more ductile, and it is employed in the preparation of the catalyst V_2O_5.

22.20 V_2O_5 is a catalyst, most notably for the contact process for H_2SO_4.

22.21 A water molecule that is attached to the Bi^{3+} center finds its O—H bonds polarized by the bismuth ion, and loss of two protons from this water ligand gives the O atom of BiO^+:

$Bi{-}OH_2^{3+} \rightarrow Bi{-}O^+ + 2H^+$

22.22 This is the oxocation VO_2^+, diagramed on page 919 of the text.

22.23 Chromium is hard, lustrous and corrosion resistant.

22.24 Nichrome is an alloy of chromium and nickel, which is used in electrical heating wires.

22.25 The common oxidation states are +2, +3, and +6.

Cr^{2+} = $[Ar]3d^4$

Cr^{3+} = $[Ar]3d^3$

Cr^{6+} = [Ar]

22.26 Chrome alum is the violet product of co-precipitation of potassium and chromium sulfate solution, giving $KCr(SO_4)_2 \cdot 12H_2O$.

22.27 $4H^+(aq) + 4Cr^{2+}(aq) + O_2(g) \rightarrow 4Cr^{3+}(aq) + 2H_2O(\ell)$

22.28 Cr^{2+} is the better reducing agent, because this ion is readily oxidized to Cr^{3+}.

22.29 $Cr(H_2O)_6{}^{3+}(aq) + H_2O(\ell) \rightleftharpoons Cr(H_2O)_5(OH)^{2+}(aq) + H_3O^+(aq)$

22.30 First there is formation of a blue-violet gelatinous hydroxide:

$Cr(H_2O)_6{}^{3+}(aq) + 3OH^-(aq) \rightarrow Cr(H_2O)_3(OH)_3(s) + 3H_2O(\ell)$

followed by the formation of a green tetrahydroxo ion:

$Cr(H_2O)_3(OH)_3(s) + OH^-(aq) \rightarrow Cr(H_2O)_2(OH)_4{}^-(aq) + H_2O(\ell)$

The original hexaaquo ion is violet.

22.31 CrO_3 is the acidic anhydride of chromic acid, H_2CrO_4, and the following acid-base equilibria exist:

$CrO_3 + H_2O \rightleftharpoons H_2CrO_4$

$H_2CrO_4 \rightleftharpoons 2H^+ + CrO_4{}^{2-}$

$2CrO_4{}^{2-} + 2H^+ \rightleftharpoons Cr_2O_7{}^{2-} + H_2O$

22.32 $CrO_4{}^{2-}$ is yellow, and $Cr_2O_7{}^{2-}$ is orange-red.

22.33 $H_2CrO_4 \rightleftharpoons H^+ + HCrO_4{}^-$ and $HCrO_4{}^- \rightleftharpoons H^+ + CrO_4{}^{2-}$

22.34 The high oxidation state of chromium in CrO_3 causes this oxide to be covalent, and therefore it is purely an acidic anhydride. On the other hand, Cr_2O_3 is more ionic, and it is therefore amphoteric.

22.35 As diagramed on page 921, a water molecule is eliminated from between two $HCrO_4^-$ ions.

22.36 Cr_2O_3 is green, and it is used as a coloring agent in paints and the like.

22.37 (a) chromic acid
(b) dichromate ion
(c) chromate ion
(d) chromium(III) oxide

22.38 $ZnCrO_4$ and $PbCrO_4$

Chemistry of Some Important Transition Elements:

Manganese, Iron, Cobalt, and Nickel

22.39 Of the common oxidation states of manganese (+2, +3, +4, +6, and +7), Mn^{2+} is the most stable, and the most easily reduced is manganese(VII).

22.40 Manganese is more reactive than chromium.

22.41 (a) $Mn(s) + 2HCl(aq) \rightarrow MnCl_2(aq) + H_2(g)$

(b) $MnCl_2(aq) + 2NaOH(aq) \rightarrow Mn(OH)_2(s) + 2NaCl(aq)$

22.42 MnO(OH)

22.43 (a) pale pink (b) deep purple (c) green

22.44 (a) permanganate ion (b) manganate ion (c) manganese dioxide

22.45 The manganate ion is stable only in very basic solution.

22.46 The permanganate ion serves as its own indicator of the end point in its titration. This is becasuse permanganate has an intense purple color that is changed sharply, at the end point, to the light pink color of manganese(II).

22.47 A nonstoichiometric compound is one in which the various elements are not present in exact integer (whole number) ratios. An example is manganese dioxide, which is not always composed simply of one manganese for every two oxygen atoms.

22.48 $MnO_4^- + 8H^+ + 5e^- \rightarrow Mn^{2+} + 4H_2O$ acidic

$MnO_4^- + 2H_2O + 3e^- \rightarrow MnO_2 + 4OH^-$ basic

22.49 A disproportionation reaction is one in which the same substance (reactant) is both oxidized and reduced.

$3MnO_4^{2-} + 4H^+ \rightarrow 2MnO_4^- + MnO_2 + 2H_2O$

22.50 MnO_2 is used in the common dry cell, where it serves as the cathode.

22.51 The ores of iron include hematite, Fe_2O_3, magnetite, Fe_3O_4, and taconite, which is a silicate-containing magnetite ore.

22.52 Iron is dissolved by concentrated hydrochloric acid:

$Fe + 2HCl \rightarrow Fe^{2+} + 2Cl^- + H_2$

Iron is passivated by nitric acid, apparently due to the formation of a protective oxide coating on the surface of the metal.

22.53 The common ions of iron are Fe^{2+} and Fe^{3+}.

22.54 The oxides of iron are FeO, Fe_3O_4, and Fe_2O_3.

22.55 (a) $Fe(H_2O)_6^{2+}(aq) + 2OH^-(aq) \rightarrow Fe(OH)_2(s) + 6H_2O(\ell)$

(b) $Fe(H_2O)_6^{3+}(aq) + 3OH^-(aq) \rightarrow Fe(H_2O)_3(OH)_3(s) + 3H_2O(\ell)$

22.56 On exposure to air, $Fe(OH)_2$ is rapidly transformed to the hydrous oxide, $Fe_2O_3 \cdot xH_2O$.

22.57 (a) Fe_2O_3 (b) $Fe(H_2O)_6^{2+}$ (c) $Fe_4[Fe(CN)_6]_3 \cdot 16H_2O$

(d) $Fe(H_2O)_3(OH)_3$, or, more correctly, $Fe_2O_3 \cdot xH_2O$

22.58 Aqueous solutions of iron(III) are acidic due to hydrolysis:

$Fe(H_2O)_6^{3+}(aq) + H_2O(\ell) \rightleftharpoons Fe(H_2O)_5(OH)^{2+}(aq) + H_3O^+(aq)$

22.59 The ion produced upon hydrolysis (as in the answer to exercise 22.58) is yellow.

22.60 Both ions cause hydrolysis of one or more attached water groups, but the Fe^{3+} ion does this more strongly. This is because the Fe^{3+} ion has a higher charge and a smaller size, i.e. a greater charge density. It therefore causes a greater polarization of the O—H bond in the attached water ligands.

22.61 These are potassium ferricyanide and potassium ferrocyanide, respectively.

22.62 $Fe(s) \rightarrow Fe^{2+} + 2e^-$

$^1/_2O_2 + H_2O + 2e^- \rightarrow 2OH^-$

$Fe^{2+}(aq) + 2OH^-(aq) \rightarrow Fe(OH)_2(s)$

$Fe(OH)_2(s) \xrightarrow{H_2O,\ O_2} Fe_2O_3 \cdot xH_2O(s)$

22.63 The unexposed paper contains ammonium ferricyanide, $(NH_4)_3[Fe(CN)_6]$, and iron(III) citrate. Exposure to light causes reduction of iron(III) to iron(II), by the citrate ion. The iron(II) that is formed reacts with ferricyanide ion to give prussian blue, $Fe_4[Fe(CN)_6]_3 \cdot 16H_2O(s)$.

22.64 Cobalt is a hard metal that dissolves slowly in HCl(aq). It is ferromagnetic. Its common oxidation states are +2 and +3. It is used in catalysts, and in the preparation of a number of useful alloys, some examples of which are stellite (used in high-temperature machining applications), and alnico (used in the preparation of permanent magnets).

22.65 Stellite is an alloy of cobalt tungsten that is employed for high-temperature machining devices. Alnico is an alloy of cobalt, nickel, iron, aluminum, and copper; it is used in the manufacture of magnets.

22.66 The pink ion in aqueous solutions of cobalt(II) is $Co(H_2O)_6{}^{2+}(aq)$.

22.67 The color changes from pink to blue on heating:

$[Co(H_2O)_6]Cl_2(s) \rightleftharpoons [Co(H_2O)_4]Cl_2(s) + 2H_2O(g)$

(pink) (blue)

22.68 $4Co^{3+}(aq) + 2H_2O(\ell) \rightarrow 4Co^{2+}(aq) + O_2(g) + 4H^+(aq)$

22.69 Nickel is corrosion resistant, makes strong alloys with iron, and is employed in the manufacture of stainless steel. The "nickel" coin is an alloy of copper and nickel. Monel is another alloy of nickel (and copper) that is very resistant to corrosion, even on prolonged exposure to sea water.

22.70 $Ni + 4CO \rightarrow Ni(CO)_4$

$Ni(CO)_4 \rightarrow Ni + 4CO$

22.71 The oxidation state of nickel in $Ni(CO)_4$ is zero. The most important oxidation state of nickel is +2.

22.72 $Ni(s) + 2HCl(aq) \rightarrow Ni^{2+}(aq) + 2Cl^-(aq) + H_2(g)$

22.73 $Ni(H_2O)_6^{2+}$

22.74 Stainless steel is usually composed of iron, chromium, and nickel, plus small amounts of phosphorus, manganese, carbon, sulfur, and silicon.

Chemistry of Some Important Transition Elements:

The Coinage Metals, Zinc, Cadmium, and Mercury

22.75 Copper is used in coins (the U.S. penny), electrical wiring, and in plumbing applications.

22.76 Copper ores contain Cu_2S. The ore is recovered by floatation (Figure 21.3, page 879), and the sulfide is roasted to give the crude metal. The metal is further purified by electrolysis (Figure 17.7, page 705).

22.77 Silver has the highest electrical conductivity of any metal.

22.78 The coinage metals are copper, silver, and gold.

22.79 Copper is alloyed with silver in sterling silver jewelry.

22.80 Gold leaf is composed of very thin sheets of metallic gold.

22.81 Each forms the +1 oxidation state readily. In addition, copper forms the +2 ion, and gold forms the +3 ion.

22.82 (a) no reaction

(b) $Cu + 2H_2SO_4 \rightarrow CuSO_4 + SO_2 + 2H_2O$

(c) $3Cu + 8HNO_3 \rightarrow 3Cu(NO_3)_2 + 2NO + 4H_2O$

(d) $Cu + 4HNO_3 \rightarrow Cu(NO_3)_2 + 2NO_2 + 2H_2O$

22.83 $Cu_2(OH)_2CO_3$

22.84 $Cu(s) + Cu^{2+}(aq) + 2Cl^-(aq) \rightarrow 2CuCl(s)$

22.85 These crystals contain the ion $Cu(H_2O)_4^{2+}$, which is blue.

22.86 Concentrated solutions of $CuCl_2$ are green owing to the presence of two distinct ions: $CuCl_4^{2-}(aq)$, which is yellow, and $Cu(H_2O)_4^{2+}$, which is blue. When diluted, these solutions become blue because the

$Cu(H_2O)_4^{2+}(aq)$ becomes the predominant substance. This happens because the following equilibrium is shifted to the left when the proportion of water to chloride ion is increased:

$$Cu(H_2O)_4^{2+}(aq) + 4Cl^-(aq) \rightleftharpoons CuCl_4^{2-}(aq) + 4H_2O(\ell)$$

22.87 $Cu^{2+}(aq) + 2OH^-(aq) \rightarrow Cu(OH)_2(s)$

$Cu(OH)_2(s) + 2OH^-(aq) \rightarrow Cu(OH)_4^{2-}(aq)$

22.88 $Cu(CN)_4^{2-}$

22.89 The test for copper ion is the formation of the characteristically deep blue ammonia complex ion:

$$Cu^{2+}(aq) + 4NH_3(aq) \rightarrow Cu(NH_3)_4^{2+}(aq)$$

22.90 $Ag + 2H^+ + NO_3^- \rightarrow Ag^+ + NO_2 + H_2O$

$3Ag + 4H^+ + NO_3^- \rightarrow 3Ag^+ + NO + 2H_2O$

22.91 $Ag_2O(s)$

22.92 Silver fluoride is quite soluble in water. The remaining silver halides have solubilities that decrease in the following order: AgCl > AgBr > AgI.

22.93 This is a metathesis reaction that results from exchange of halide in the following equilibrium:

$$AgCl(s) + Br^-(aq) \rightleftharpoons AgBr(s) + Cl^-(aq)$$

The reaction is made possible by the greater insolubility of silver bromide.

22.94 Three equations must be written in order to account for all steps of the test:

$$Ag^+(aq) + Cl^-(aq) \rightarrow AgCl(s)$$

$$AgCl(s) + 2NH_3(aq) \rightarrow Ag(NH_3)_2^+(aq) + Cl^-(aq)$$

$$Ag(NH_3)_2^+(aq) + 2H^+(aq) + Cl^-(aq) \rightarrow AgCl(s) + 2NH_4^+(aq)$$

22.95 Gold(III) is readily reduced.

22.96 Gold follows the lanthanide elements, and is therefore subject to the lanthanide contraction, which makes gold to be an unusually small atom with a high ionization potential.

22.97 The ore of zinc is zinc blende, ZnS. Cadmium is present as an impurity in zinc ores. The ore of mercury is cinnabar, HgS.

22.98 $Zn_2(OH)_2CO_3$

22.99 (a) $Zn + 2H^+ \rightarrow Zn^{2+} + H_2$

(b) $Zn + 2OH^- + 2H_2O \rightarrow Zn(OH)_4^{2-} + H_2$

22.100 $Zn^{2+}(aq) + 2OH^-(aq) \rightarrow Zn(OH)_2(s)$

$Zn(OH)_2(s) + 2OH^-(aq) \rightarrow Zn(OH)_4^{2-}(aq)$

22.101 Zinc is more readily oxidized than iron, and even if the surface zinc is scratched away, there is operation of a galvanic cell, in which iron is the cathode (being reduced) and zinc is the anode (being oxidized). This means that all of the zinc will be sacrificially oxidized before any rusting of the iron can occur. Tin has a more favorable reduction potential than iron.

22.102 Cadmium does not react with base, and it can be used as a protective coating in alkaline systems which would otherwise cause degradation of a zinc coating.

22.103 $4Zn(s) + NO_3^-(aq) + 10H^+(aq) \rightarrow 4Zn^{2+}(aq) + NH_4^+(aq) + 3H_2O(\ell)$

22.104 (a) $Zn + 2H^+ \rightarrow Zn^{2+} + H_2$

(b) $Cd + 2H^+ \rightarrow Cd^{2+} + H_2$

(c) no reaction

22.105 $Hg + 2HNO_3 \rightarrow Hg^{2+} + 2NO_3^- + H_2$

22.106 An amalgam is a solution of a metal in mercury. Gold readily amalgamates in mercury, and thus can be removed from other materials that contaminate gold deposits. The dense amalgam of gold is recovered, the mercury is distilled from the elemental gold, and used again.

22.107 Hg_2Cl_2 is much less soluble than $HgCl_2$.

22.108 These are molecular substances whose structures are shown at the top of page 932 of the text.

22.109 The solid product of this reaction is an intimate mixture of liquid mercury (black) and $Hg(NH_2)Cl$ (white), and it normally is dark grey or black:

$$Hg_2Cl_2(s) + 2NH_3(aq) \rightarrow Hg(\ell) + Hg(NH_2)Cl(s) + NH_4^+(aq) + Cl^-(aq)$$

22.110 Vermilion is the deep red form of mercury sulfide that is obtained on heating HgS, which is normally black.

Complexes of the Transition Metals

22.111 These ligands (shown at the bottom of page 932 of the text) attach themselves to a metal atom by using more than one atom in the ligand, and a ring that includes the metal atom is formed.

22.112 A monodentate ligand that attaches to a metal by only one atom.

22.113 It is the porphyrin ligand that is found in chlorophyll, heme and vitamin B_{12}.

22.114 The coordination number of a metal complex is the number of ligand atoms that are attached to the metal. The common coordination geometries of complex ions having coordination number 4 are the square plane and the tetrahedron.

22.115 The coordination number is six, and the oxidation number of the iron atom is +2.

22.116 The chelate effect is the greater strength of a complex formed with a metal and a polydentate ligand than the complex formed between a metal and monodentate ligands. It arises because of the greater probability of a polydentate remaining attached to the metal if one of its donor atoms becomes momentarily detached from the metal. A monodentate ligand, once detached from the metal, does not have as great a probability of becoming reattached. A polydentate ligand, once detached from the metal at one atom only, is still bound to the metal through other donor atoms, and the likelihood of reattachment is high.

22.117 The process for doing this is outlined at the bottom of page 935, Figure 22.12.

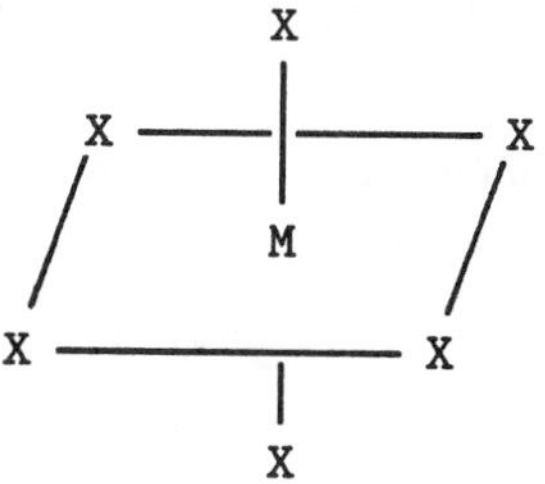

22.118

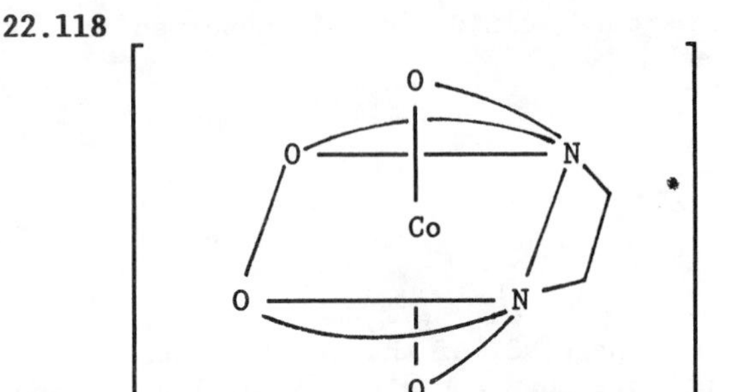

22.119

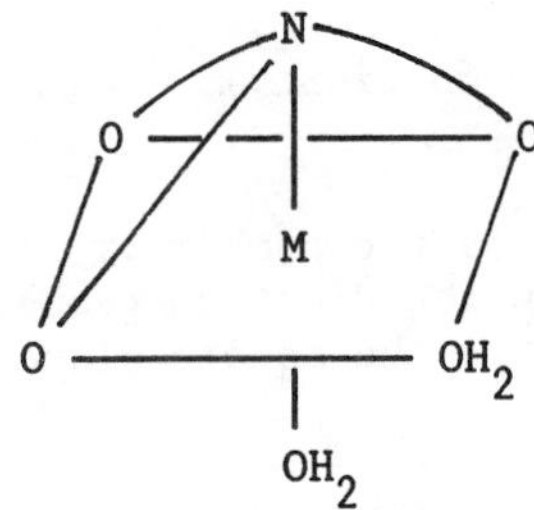

Isomers of Coordination Compounds

22.120 *Isomers* are compounds that have the same chemical formula, but that are also distinct and different substances, owing to differences in the arrangements of the atoms.

22.121 *Stereoisomerism* is the existence of isomers that differ only in the spatial orientation of their atoms.

Geometric isomerism is the existence of isomers that differ because their molecules have different geometries.

Chiral isomers are isomers that differ because one is the non-superimposable mirror image of the other.

Enantiomers are chiral isomers of one another, i.e. they are non-superimposable mirror images of one another.

22.122 This is illustrated on page 937 of the text. In a cis isomer, like groups are located adjacent to one another, whereas in a trans isomer, like groups are located opposite one another.

22.123 Cisplatin is the cis geometrical isomer of $Pt(NH_3)_2Cl_2$, whose structure is shown on page 937 of the text.

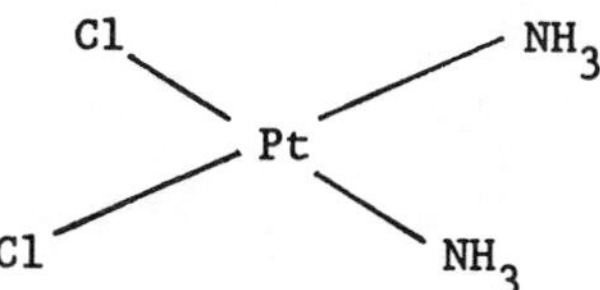

22.124

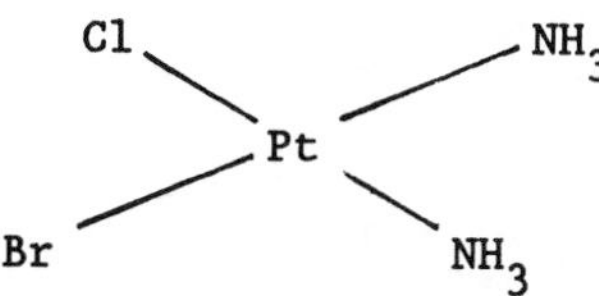

cis isomer

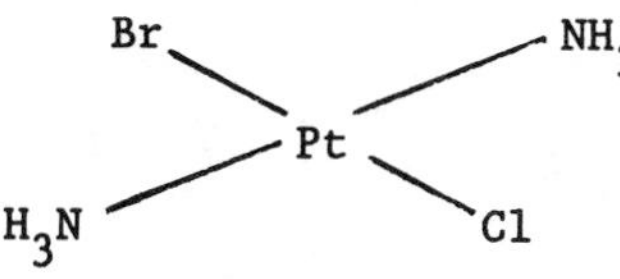

trans isomer

22.125 Student answer.

22.126

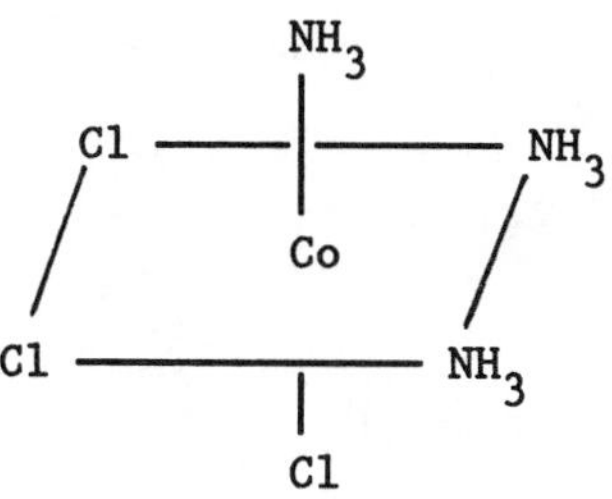

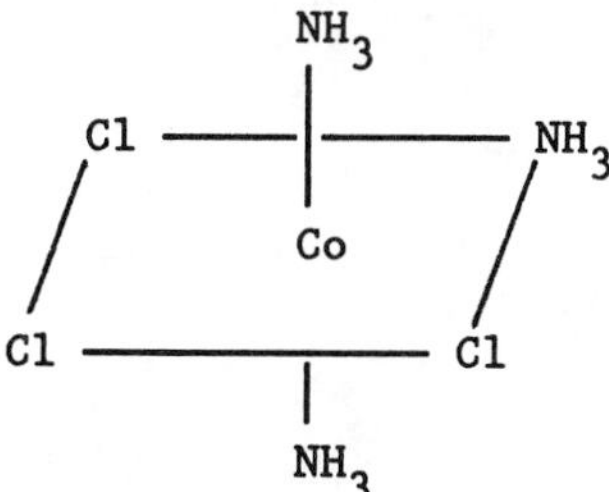

22.127 There are two chiral isomers of the cis geometrical isomer:

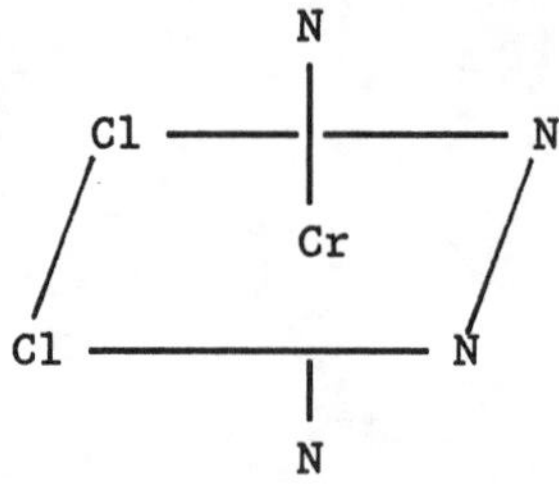

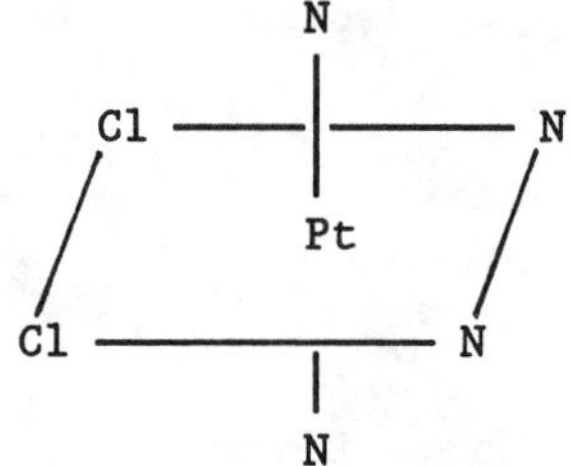

The trans geometrical isomer is not chiral:

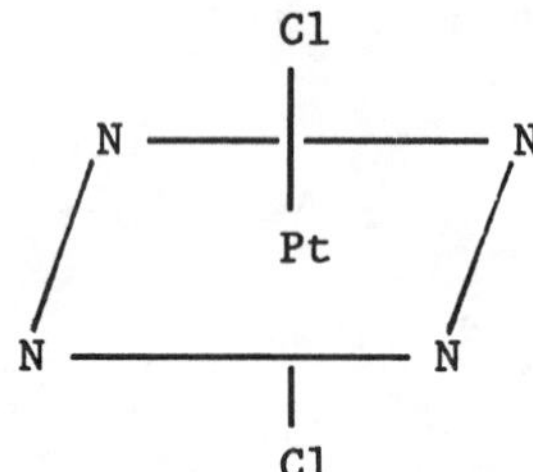

22.128 In the following diagram, let O⌒O stand for $C_2O_4^{2-}$:

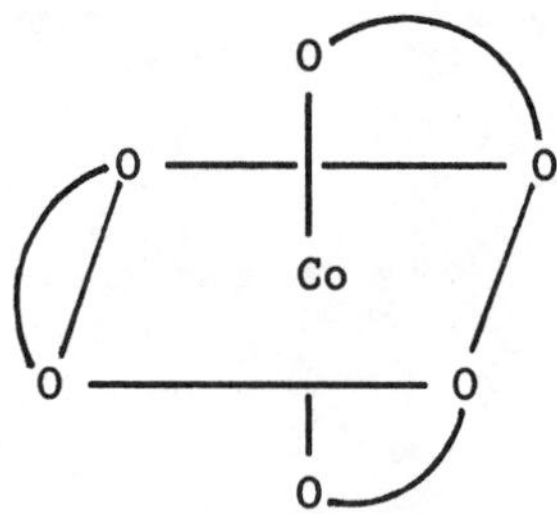

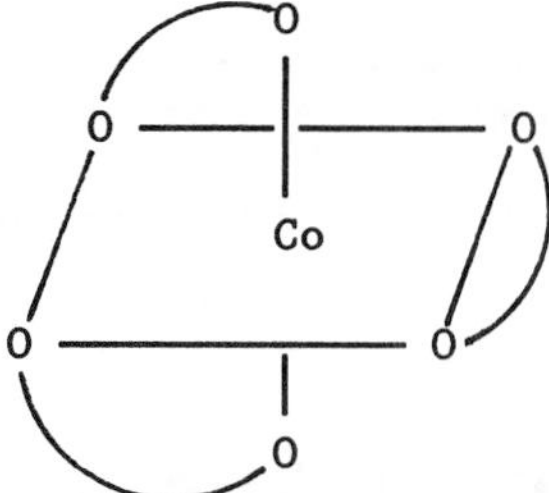

22.129 Optical isomers are chiral isomers, i.e. isomers that differ because they are non-superimposable mirror images on one another.

Bonding in Complexes

22.130 See Figure 22.19.

22.131 The $d_{x^2-y^2}$ and d_{z^2} orbitals are located near the ligands, whereas the d_{xy}, d_{xz}, and d_{yz} orbitals are located between (farther from) the ligands.

22.132 See Figure 22.20.

22.133 See Figure 22.21. The oxidation of Co^{2+} to give Co^{3+} removes a high energy electron, and also gives a larger value for Δ. Both of these factors favor the oxidation.

22.134 (a) $Cr(H_2O)_6{}^{3+}$ (b) $Cr(en)_3{}^{3+}$

22.135 Various ligands may be bound to the metal, giving different complexes, each having its own characteristic value of Δ. As diagramed in Figure 22.22, as the value of Δ changes, so too does the frequency of light that the complex absorbs.

22.136 The spectrochemical series is a list of ligands, arranged in the order of their increasing ability to produce a large value of Δ. It is determined by measuring the frequencies of light that are absorbed by complexes having various ligands.

22.137 (a) $Fe(H_2O)_6{}^{2+}$ (b) $Mn(CN)_6{}^{4-}$

22.138 $Cr(CN)_6{}^{3-}$

22.139 The high spin case is the one in which the maximum number of unpaired electrons is found. The converse is true of the low spin case; it is the one having the fewest possible number of unpaired electrons.

22.140 Both high and low spin configurations are possible only for the following configurations:

d^4, d^5, d^6, and d^7.

22.141 For $Fe(H_2O)_6^{3+}$, we expect a relatively small value for Δ, and we predict the high spin case having five unpaired electrons:

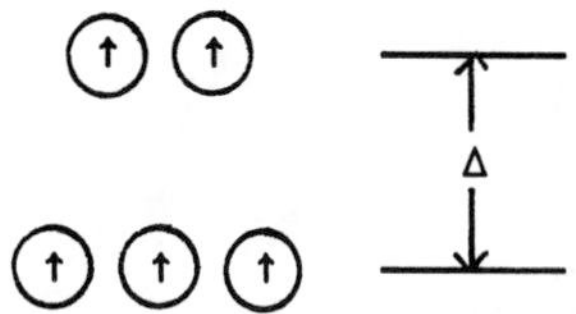

For $Fe(CN)_6^{3-}$, we expect a relatively large value for Δ, and we predict the low spin case having one unpaired electron:

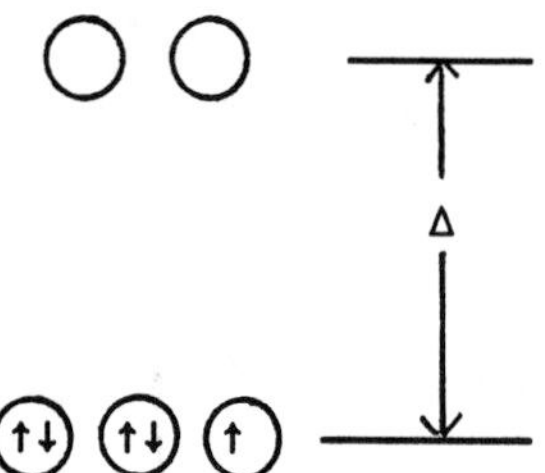

22.142 There is promotion of an electron into the upper energy level, by absorption of a photon having energy equal to $E = \Delta = h\nu$:

22.143 This is a complex of Co^{3+}. The complex is known to be diamagnetic, so no unpaired electrons are present. We conclude that it must be a low spin complex:

CHAPTER TWENTY-THREE

PRACTICE EXERCISES

1. $${}^{226}_{88}\text{Ra} \rightarrow {}^{0}_{0}\gamma + {}^{4}_{2}\text{He} + {}^{222}_{86}\text{Rn}$$

2. $${}^{90}_{38}\text{Sr} \rightarrow {}^{0}_{-1}\text{e} + {}^{90}_{39}\text{Y}$$

3. We make use of the Inverse Square Law:

$$\frac{I_1}{I_2} = \frac{d_2^{\,2}}{d_1^{\,2}}$$

$$\frac{1.4}{I_2} = \frac{(1.2)^2}{(10)^2}$$

$$I_2 = (1.4)(100)/(1.4) = 1.0 \times 10^2 \text{ units}$$

4. We shall employ equation 23.4, and we shall express the activities S_0 and S in units μCi/mL, i.e. in terms of the activities found per mL of solution.

$$S_0 = \frac{0.348 \text{ mg}}{495 \text{ mL}} \times \frac{4.15\ \mu\text{Ci}}{1 \text{ mg}} = 2.92 \times 10^{-3}\ \mu\text{Ci/mL}$$

$$S = \frac{9.00 \times 10^{-3}\ \mu\text{Ci}}{9.70 \text{ mL}} = 9.28 \times 10^{-4}\ \mu\text{Ci/mL}$$

$$n = 0.348 \text{ mg} \times \left[\frac{2.92 \times 10^{-3}\ \mu\text{Ci/mL}}{9.28 \times 10^{-4}\ \mu\text{Ci/mL}} - 1 \right]$$

$$n = 0.747 \text{ mg of non-radioactive vitamin } B_{12}$$

If we express this in terms of μg of non-radioactive vitamin per mL of broth, we have:

$$0.747 \text{ mg} \times 1000\ \mu\text{g/mg} = 7.47 \times 10^2\ \mu\text{g of vitamin, and:}$$

$$7.47 \times 10^2\ \mu\text{g} \div 495 \text{ mL} = 1.51\ \mu\text{g/mL of non-radioactive vitamin } B_{12}$$

5. We use equation 23.6, in which s is defined to be the specific activity in the sample after t years since death have elapsed:

$t = 8.26 \times 10^3 \text{ yr} \times \ln (918/s)$

$t = 8.26 \times 10^3 \text{ yr} \times \ln (918/265) = 1 \times 10^4 \text{ yr}$

REVIEW EXERCISES

Conservation of Mass-Energy

23.1 We do not normally, in the course of everyday life, encounter nuclear reactions. In typical chemical reactions, the mass of the various materials that are involved is unaffected by the relatively slow velocities that these materials possess.

23.2 We make a distinction between an object's mass and its rest mass when considering nuclear reactions. The absolute mass of an object is dependent upon its velocity, according to equation 23.1. When the velocity of the object is relatively low, however, the denominator on the right side of equation 23.1 becomes equal to 1.0, and the mass is equal to the rest mass, m_0. The distinction between mass and rest mass is necessary only for particles having extremely high velocities.

23.3 If the velocity of an object approaches the speed of light, then the denominator on the right side of equation 23.1 would take on a value increasingly close to zero. This would require the mass m of the object to increase, and in the limit, become infinity.

23.4 Equation 23.1 becomes:

$$m = \frac{1.00}{\sqrt{1 - (v/c)^2}}$$

On substituting the various velocities for the value of v in the above expression, and using the value $c = 3.00 \times 10^8 \text{ m s}^{-1}$, we get:

(a) m = 1.01 kg (b) 3.90 kg (c) 12.26 kg

23.5 The sum of all of the energy in the universe plus all of the mass equivalent of energy is a constant.

23.6 $\Delta E = \Delta m_0 \times c^2$

23.7 Solve the Einstein equation for Δm:

$\Delta m = \Delta E/c^2$

$1 \text{ kJ} = 10^3 \text{ J} = 10^3 \text{ kg-m}^2/\text{s}^2$

$\Delta m = 10^3 \text{ kg-m}^2/\text{s}^2 \div (3.00 \times 10^8 \text{ m/s})^2 = 1.11 \times 10^{14} \text{ kg} = 1.11 \times 10^{-17} \text{ g}$

23.8 The changes in mass that accompany normal (non-nuclear) chemical reactions are so small as to be insignificant.

23.9 As illustrated in the text at the top of page 959, the joule is equal to a $\text{kg-m}^2/\text{s}^2$, and this is employed directly in the Einstein equation:

$\Delta m = \Delta E/c^2$,

where ΔE is the enthalpy of formation of liquid water, which is available in Table 5.2:

$H_2(g) + {}^1/_2O_2(g) \rightarrow H_2O(g)$, $\Delta H° = -285.9 \text{ kJ/mol}$

$\Delta m = (-285.9 \times 10^3 \text{ kg-m}^2/\text{s}^2) \div (3.00 \times 10^8 \text{ m/s}^2) = -3.18 \times 10^{-12} \text{ kg}$

$-3.18 \times 10^{-12} \text{ kg} \times 1000 \text{ g/kg} \times 10^9 \text{ ng/g} = -3.18 \text{ ng}$

Nuclear Binding Energies

23.10 A small amount of mass is converted to energy, and lost from the system, when the nucleons assemble into a stable atomic nucleus. The amount of mass that is converted into energy corresponds to the gain in stability that the nucleus achieves by loss of this amount of energy.

23.11 The mass of the deuterium nucleus is the mass of the proton (1.007277 amu) plus that of a neutron (1.008665 amu), or 2.015942 amu. The difference between this calculated value and the observed value is equal to Δm:

$\Delta m = (2.015942 - 2.0135) = 2.4 \times 10^{-3} \text{ amu}$

$\Delta E = \Delta mc^2 = (2.4 \times 10^{-3} \text{ amu})(1.6606 \times 10^{-27} \text{ kg/amu})(3.00 \times 10^8 \text{ m/s})^2$

$\Delta E = 3.6 \times 10^{-13} \text{ kg-m}^2/\text{s}^2 = 3.6 \times 10^{-13} \text{ J}$

Since there are two neucleons per deuterium nucleus, we have:

$\Delta E = 3.6 \times 10^{-13}$ J/2 nucleons = 1.8×10^{-13} J per nucleon

23.12 The mass of the tritium nucleus is the mass of one proton plus that of two neutrons: 1.007277 + 2(1.008665) = 3.024607 amu. The difference between this calculated value and the observed value is equal to Δm:

$\Delta m = 3.024607 - 3.01550 = 9.11 \times 10^{-3}$ amu

$\Delta E = \Delta mc^2 = (9.11 \times 10^{-3}\text{ amu})(1.6606 \times 10^{-27}\text{ kg/amu})(3.00 \times 10^8\text{ m/s})^2$

$\Delta E = 1.36 \times 10^{-12}\text{ kg-m}^2/\text{s}^2 = 1.36 \times 10^{-12}$ J

Since there are three nucleons per tritium nucleus, the energy per nucleon is:

1.36×10^{-12} J/3 nucleons = 4.53×10^{-13} J per nucleon

23.13 The total mass of the various nucleons is:

protons:	26 × 1.007277 =	26.189202
neutrons:	30 × 1.008665 =	30.259950
		56.449152 amu

$\Delta m = 56.449152 - 55.9349 = 0.5142$ amu

$\Delta E = \Delta mc^2 = (0.5142\text{ amu})(1.6606 \times 10^{-27}\text{ kg/amu})(3.00 \times 10^8\text{ m/s})^2$

$\Delta E = 7.68 \times 10^{-11}\text{ kg-m}^2/\text{s}^2 = 7.68 \times 10^{-11}$ J

Converting to joules per nucleon gives:

7.68×10^{-11} J/56 nucleons = 1.37×10^{-12} J per nucleon

The data of Figure 23.1 on page 960 of the text indicate that iron has the largest binding energy per nucleon. This means that iron has the most stable nucleus.

23.14 The total mass of the ^{235}U nucleus is the observed mass of the atom, minus the mass due to 92 electrons, which is a significantly large number for this heavy atom:

observed mass = 235.0439 - 92(0.0005486) = 234.9934 amu

The calculated mass of the nucleus is the sum of the masses of 92 protons and 143 neutrons:

protons: $92 \times 1.007277 = 92.66948$ amu
neutrons: $143 \times 1.008665 = 144.2391$ amu

236.9086 amu

$\Delta m = 236.9086 - 234.9934 = 1.9152$ amu

$\Delta E = \Delta mc^2 = (1.9152 \text{ amu})(1.6606 \times 10^{-27} \text{ kg/amu})(3.00 \times 10^8 \text{ m/s})^2$

$\Delta E = 2.86 \times 10^{-10} \text{ kg-m}^2/\text{s}^2 = 2.86 \times 10^{-10}$ J

The energy per nucleon is:

2.86×10^{-10} J/235 nucleons $= 1.22 \times 10^{-12}$ J per nucleon

Radioactivity

23.15 The two forces in an atomic nucleus are the electrostatic repulsion among the protons of the nucleus, and the so-called strong force that operates to bind the nucleons together in spite of the repulsive forces. The strong force decreases more rapidly as a function of distance than does the electrostatic force, and the result is that the strong force of a nucleus does not extend to neighboring nuclei.

23.16 The nuclear strong force and the intervening effects of the neutrons cause the half lives of these nuclei to be very long.

23.17 Radionuclides are isotopes that are radioactive.

23.18 Naturally occurring radionuclides emit alpha particles (alpha radiation), beta particles, and gamma radiation.

23.19 (a) The alpha particle is composed of helium nuclei, and has a 2+ charge.
(b) Beta particles are electrons.
(c) Positrons are particles with a +1 charge, and have the same mass as the electron.
(d) A deuteron consists of one neutron and one proton, i.e. the nucleus of a deuterium atom.

23.20 The alpha particle is more massive than the other particles, and this, coupled with its high positive charge, makes it likely that an alpha particle will collide with something soon after it is ejected. This collision normally transforms the alpha particle into a helium atom, by gain of two electrons.

23.21 (a) ${}^{211}_{83}Bi$ (b) ${}^{177}_{72}Hf$ (c) ${}^{216}_{84}Po$ (d) ${}^{19}_{9}F$

23.22 (a) ${}^{241}_{94}Pu$ (b) ${}^{140}_{57}La$ (c) ${}^{58}_{28}Ni$ (d) ${}^{68}_{31}Ga$

23.23 (a) ${}^{242}_{94}Pu \rightarrow {}^{4}_{2}He + {}^{238}_{92}U$

(b) ${}^{28}_{12}Mg \rightarrow {}^{0}_{-1}e + {}^{28}_{13}Al$

(c) ${}^{26}_{14}Si \rightarrow {}^{0}_{1}e + {}^{26}_{13}Al$

(d) ${}^{37}_{18}Ar + {}^{0}_{-1}e \rightarrow {}^{37}_{17}Cl$

23.24 (a) ${}^{55}_{26}Fe + {}^{0}_{-1}e \rightarrow {}^{55}_{25}Mn$

(b) ${}^{42}_{19}K \rightarrow {}^{0}_{-1}e + {}^{42}_{20}Ca$

(c) ${}^{93}_{44}Ru \rightarrow {}^{0}_{+1}e + {}^{93}_{43}Tc$

(d) ${}^{251}_{98}Cf \rightarrow {}^{4}_{2}He + {}^{247}_{96}Cm$

23.25 (a) ${}^{30}_{13}Al \rightarrow {}^{0}_{-1}e + {}^{30}_{14}Si$

(b) ${}^{252}_{99}Es \rightarrow {}^{4}_{2}He + {}^{248}_{97}Bk$

(c) ${}^{93}_{42}Mo + {}^{0}_{-1}e \rightarrow {}^{93}_{41}Nb$

(d) ${}^{28}_{15}P \rightarrow {}^{0}_{+1}e + {}^{28}_{14}Si$

23.26 (a) ${}^{10}_{6}C \rightarrow {}^{0}_{+1}e + {}^{10}_{5}B$

(b) ${}^{243}_{96}Cm \rightarrow {}^{4}_{2}He + {}^{239}_{94}Pu$

(c) ${}^{49}_{23}V + {}^{0}_{-1}e \rightarrow {}^{49}_{22}Ti$

(d) ${}^{20}_{8}O \rightarrow {}^{0}_{-1}e + {}^{20}_{9}F$

23.27 (a) ${}^{261}_{102}No$ (b) ${}^{211}_{82}Pb$ (c) ${}^{141}_{61}Pm$ (d) ${}^{179}_{74}W$

23.28 (a) $^{80}_{38}Sr$ (b) $^{121}_{50}Sn$ (c) $^{50}_{25}Mn$ (d) $^{257}_{100}Fm$

23.29 (a) Because iodine-123 has a shorter half life, the body is subjected to radioactivity for a shorter time than if iodine-131 were used. Furthermore, since the mode of radioactive decay of iodine-123 is by electron capture, the radiation that is produced (X rays) is completely diagnostically useful. Iodine-131, on the other hand, decays by a route (beta emission) that is not useful for diagnostic purposes.

(b) $^{123}_{53}I + ^{0}_{-1}e \rightarrow ^{123}_{52}Te$

23.30 $^{58}_{26}Fe$

23.31 Gamma rays are emitted from the nucleus, whereas X rays arise from electronic transitions.

23.32 Electron capture of a low-level electron by the nucleus creates a low-level hole in the electron configuration that is filled by descent of an electron from an upper electronic level. The atom emits radiation corresponding to the difference in energies between the two electronic levels, and this radiation lies in the X ray region of the electromagnetic spectrum.

Nuclear Stability

23.33 On a plot of the number of neutrons versus the number of protons for all known nuclides, the band of stability constitutes the region in which all stabile nuclides are found to reside.

23.34 The band of stability is used to judge whether or not a particular nuclide might be stable, and, therefore, whether or not attempts to make the nuclide can be expected to be successful. Also, whether the nuclide lies above or below the band of stability can be used to predict the likely mode of decay of the nuclide.

23.35 Barium-140 should have the longer half life because both its mass number and atomic number are even.

23.36 Both the mass number and the atomic number of tin-112 (atomic number 50) are even. Only one of these is even for indium-112 (atomic number 49).

23.37 Lanthanum-139 is a stable nuclide because it has 82 neutrons, a magic number.

23.38 There must be a sufficient number of neutrons to shield the protons from one another, especially so as the number of protons becomes increasingly large.

23.39 The loss of an alpha particle is the most effective way to move the nuclide toward the band of stability, if the unstable nuclide lies above the band of stability, because the alpha particle has a relatively large mass.

23.40 The neutron-to-proton ratio of lead-164 is too low.

23.41 The loss of a beta particle is most likely, because the net nuclear effect is the conversion of a neutron into a proton, and this reduces the neutron-to-proton ratio.

23.42 The loss of a positron is most likely, because the net nuclear effect is the conversion of a proton into a neutron, and this increases the neutron-to-proton ratio.

23.43 The net effect of neutron capture is the conversion of a proton into a neutron, and this increases the neutron-to-proton ratio. Radionuclides lying below the band of stability are more likely to undergo electron capture than those lying above the band of stability.

23.44 The more likely process is beta emission, because this produces a product having a lower neutron-to-proton ratio.

$$^{38}_{19}K \rightarrow \ ^{0}_{-1}e + \ ^{38}_{20}Ca$$

23.45 Electron capture is the more likely event:

$$^{37}_{18}Ar + \ ^{0}_{-1}e \rightarrow \ ^{37}_{17}Cl$$

This produces chlorine-17, which has a magic number, i.e. 20 neutrons. Beta emission would give potassium-37, which is less stable.

23.46 The mass of the parent atom must be at least 0.001097 amu greater than the mass of the daughter atom. Assuming that the emitted positron undergoes an annihilation collision with an electron within the atom that loses the positron, this atom loses the mass of one positron plus the mass of one electron, since these two particles are transformed into electromagnetic radiation. The mass of the positron is the same as that of an electron, 0.0005486 amu, and the minimum loss in mass is thus twice the mass of an electron: 2×0.0005486 amu = 0.001097 amu.

23.47 Six half life periods correspond to the fraction 1/64 of the initial material. That is, one sixty-fourth of the initial material is left after

6 half lives:

3.00 mg × 1/64 = 0.0469 mg remaining

23.48 After four half life periods, one sixteenth of the original sample remains:

9×10^{-9} g × 1/16 = 5.6×10^{-10} g remaining

Transmutations

23.49 $^{53}_{24}Cr$

$$^{51}_{23}V + ^{2}_{1}D \rightarrow ^{53}_{24}Cr \rightarrow ^{1}_{1}H + ^{52}_{23}V$$

23.50 $^{19}_{9}F + ^{4}_{2}He \rightarrow ^{23}_{11}Na^* \rightarrow ^{22}_{11}Na + ^{1}_{0}n$

23.51 $^{80}_{35}Br$

23.52 $^{115}_{48}Cd + ^{1}_{0}n \rightarrow ^{116}_{48}Cd + \gamma$

23.53 $^{55}_{26}Fe$

$$^{55}_{25}Mn + ^{1}_{1}H \rightarrow ^{1}_{0}n + ^{55}_{26}Fe$$

23.54 $^{1}_{0}n$; $^{239}_{93}Np$

23.55 $^{1}_{0}n$; $^{256}_{100}Fm$

23.56

$$^{241}_{94}Pu\ (^{1}_{0}n,\ \gamma)\ ^{242}_{94}Pu\ (^{1}_{0}n,\ \gamma)\ ^{243}_{94}Pu \rightarrow ^{243}_{95}Am + ^{0}_{-1}e$$

$$^{0}_{-1}e + ^{244}_{96}Cm \leftarrow ^{244}_{95}Am \longleftarrow$$

Detecting and Measuring Radiations

23.57 The Geiger counter detects the ions that form in matter through which radiation is passing. This is illustrated in Figure 23.12 on page 975 of the text.

23.58 Radiation generates unstable and reactive ions within cells. These reactive ions can participate in reactions that can eventually lead to birth defects, tumors, mutations, and cancer.

23.59 The common unit of radioactivity is the curie (Ci), whereas the SI unit of radioactivity is the becquerel (Bq).

23.60 The unit of energy for radiation is the electron-volt, or its multiples.

23.61 This is the becquerel.

23.62 The common unit is the rad, and the SI unit is the gray.

23.63 This is one curie. It is also:

$$1.0 \text{ Ci} \times 3.7 \times 10^{10} \text{ Bq/Ci} = 3.7 \times 10^{10} \text{ Bq}$$

23.64 A dose in rems is calculated by multiplying the dose in rads by fractions that take into account the kind of radiation (amount of energy) that is involved. It also takes into account the different biological effects that various radiations have, even when their energies are the same.

Applications of Radionuclides

23.65 A short half life is desirable in medical work in order to minimize the exposure that a patient receives from radioactive materials. This reduces radiation-caused damage to the body. If the half life is too short, the radionuclide will not be present in the body long enough to permit useful measurements.

23.66 Alpha particles cannot reach a detector outside the body, because they cannot penetrate tissue. Furthermore, as the body captures alpha particles, they are damaged.

23.67 The element for which an analysis is sought is made radioactive by neutron bombardment. Other elements that may be present may also be made radioactive by the bombardment, and each such newly generated radionuclide gives off its radiation at frequencies and intensities that are unique. The identification of the elements in the sample is possible because each emission is characteristic of one particular element.

23.68 Cosmic rays produce carbon-14 in the atmosphere by the following reaction:

$$^{1}_{0}n + ^{14}_{7}N \rightarrow ^{15}_{7}N \rightarrow ^{14}_{6}C + ^{1}_{1}p$$

23.69 The ratio is constant because life processes continue to take place. In other words, in a living organism, the rate at which carbon dioxide (and other nutrients for that matter) is ingested is the same as the rate at which carbon dioxide is lost. The amount of carbon-14 in this carbon dioxide thus remains constant.

23.70 The assumption is that the lead in a mixture being analyzed has all come from the disintegration of uranium only.

23.71 Some of the forms of radiation that everybody experiences are cosmic rays, diagnostic X rays, radioactive pollutants, fallout from atmospheric tests of nuclear devices, and, in some cases, radiation used in medicine.

23.72 For a first order process, the rate constant is related to the half life by the equation:

$k = 0.693/t_{1/2}$ Hence, $k = 0.693/28.1\ yr = 2.47 \times 10^{-2}\ yr^{-1}$

Also, for a first order process, the concentration varies with time according to the equation:

$$\ln \frac{[A]_0}{[A]_t} = kt$$

which allows us to solve for the time, t, for the activity to decrease by the specified amount:

$$t = \frac{1}{k} \times \ln \frac{[A]_0}{[A]_t} = \frac{1}{(2.47 \times 10^{-2}\ yr^{-1})} \times \ln \frac{(0.245\ Ci/g)}{(1.00 \times 10^{-6}\ Ci/g)} = 502\ yr$$

23.73 This calculation makes use of the first order rate equation, where knowing $[A]_t$, we need to calculate $[A]_0$:

$$\ln \frac{[A]_0}{[A]_t} = kt$$

$k = 0.693/t_{1/2} = 0.693/8.07\ d = 8.59 \times 10^{-2}\ d^{-1}$

$$\ln \frac{[A]_0}{(25.6 \times 10^{-5}\ Ci/g)} = (8.59 \times 10^{-2}\ d^{-1})(28.0\ d) = 2.41$$

Taking the antiln of both sides of the above equation gives:

$$\frac{[A]_0}{(25.6 \times 10^{-5}\ Ci/g)} = \text{antiln } (2.41) = 11.1$$

Solving for the value of $[A]_0$ gives:

$$[A]_0 = 2.84 \times 10^{-3}\ Ci/g$$

23.74 The rate constant for the first order process is first determined:

$$k = 0.693/t_{1/2} = 0.693/6.02\ hr = 0.115\ hr^{-1}$$

Also, we know that:

$$\ln \frac{[A]_0}{[A]_t} = kt$$

$$\ln \frac{(4.52 \times 10^{-6}\ Ci)}{[A]_t} = (0.115\ hr^{-1})(8.00\ hr) = 0.920$$

Taking the antiln of both sides of this equation gives:

$$\frac{4.52 \times 10^{-6}\ Ci}{[A]_t} = \text{antiln } (0.920) = 2.51$$

$[A] = 1.80 \times 10^{-6}$ Ci, the activity after 8 hours.

23.75 The half life is determined as follows:

$$t_{1/2} = 0.693/k = 0.693/7.37 \times 10^{3}\ yr = 9.40 \times 10^{-5}\ yr^{-1}$$

A period equal to 20 half lives is thus:

$$(20)(7.37 \times 10^{3}\ yr) = 1.47 \times 10^{5}\ yr$$

$$\ln \frac{120\ Bq/g}{[A]_t} = (9.40 \times 10^{-5}\ yr^{-1})(1.47 \times 10^{5}\ yr)$$

$$[A]_t = 1 \times 10^{-4}\ Bq/g$$

23.76 In order to solve this problem, it must be assumed that all of the argon-40 that is found in the rock must have come from the potassium-40, i.e. that the rock contains no other source of argon-40.

If the above assumption is valid, then any argon-40 that is found in the rock represents an equivalent amount of potassium-40, since the stoichiometry is 1:1. Since equal amounts of potassium-40 and argon-40 have been found, this indicates that the amount of potassium-40 that remains is exactly half the amount that was present originally. In other words, the potassium-40 has undergone one half life of decay by the time of the analysis. The rock is thus seen to be 1.3×10^9 years old.

23.77 In a fashion similar to that outlined in the answer to Review Exercise 23.76, we conclude that, since the rock is one half life old (1.3×10^9 yr), there must be equal amounts of the two isotopes, one having been formed by decay of the other. The answer is thus 1.16×10^{-7} mol of potassium-40.

23.78 This calculation makes use of the Inverse Square Law:

$$\frac{I_1}{I_2} = \frac{d_2^{\,2}}{d_1^{\,2}}$$

$$\frac{8.4 \text{ rem}}{0.50 \text{ rem}} = \frac{d_2^{\,2}}{(1.60 \text{ meter})^2}$$

$$d_2 = 6.6 \text{ meters}$$

23.79 This calculation makes use of the Inverse Square Law:

$$\frac{I_1}{I_2} = \frac{d_2^{\,2}}{d_1^{\,2}}$$

$$\frac{50 \text{ mrem}}{I_2} = \frac{(0.50 \text{ m})^2}{(4.0 \text{ m})^2}$$

$$I_2 = 32 \times 10^2 \text{ mrem}$$

23.80 We shall employ equation 23.4, and we shall express the activities S_0 and S in units Ci/mL, i.e. in terms of the activities found per mL of solution.

$$S_0 = \frac{1.58 \text{ mg}}{500 \text{ mL}} \times \frac{9.67 \text{ Ci}}{1 \text{ mg}} = 0.0306 \text{ Ci/mL}$$

$$S = \frac{0.385 \text{ Ci}}{14.6 \text{ mL}} = 0.0264 \text{ Ci/mL}$$

$$n = 1.58 \text{ mg} \left[\frac{(0.0306 \text{ Ci/mL})}{(0.0264 \text{ Ci/mL})} - 1 \right] = 0.251 \text{ mg}$$

$0.251 \text{ mg} \times 1000\ \mu\text{g/mg} = 251\ \mu\text{g}$

$251\ \mu\text{g}/50$ capsules $= 5\ \mu\text{g}$ per capsule

23.81 We shall employ equation 23.4, and we shall express the activities in units Ci/mL, i.e. in terms of the activities found per mL of solution.

$$S_0 = \frac{1.12 \text{ mg}}{25.0 \text{ mL}} \times \frac{3.08 \text{ Ci}}{1 \text{ mg}} = 0.138 \text{ Ci/mL}$$

$$S = \frac{1.20 \text{ Ci}}{10.5 \text{ mL}} = 0.114 \text{ Ci/mL}$$

$$n = 1.12 \text{ mg} \times \left[\frac{0.138 \text{ Ci/mL}}{0.114 \text{ Ci/mL}} - 1 \right]$$

$n = 0.236$ mg of non-radioactive vitamin B_{12}

If we express this in terms of μg of non-radioactive vitamin per mL of broth, we have:

$0.236 \text{ mg} \times 1000\ \mu\text{g/mg} = 2.36 \times 10^2\ \mu\text{g}$ of vitamin, and:

$2.36 \times 10^2\ \mu\text{g} \div 25.0 \text{ mL} = 9.44\ \mu\text{g/mL}$ of non-radioactive vitamin B_{12}

23.82 We use equation 23.6, in which s is defined to be the specific activity in the sample, after t years since death have elapsed:

$$t = 8.26 \times 10^3 \text{ yr} \times \ln (918/s)$$

The value of s is, assuming the sample that was recovered to be pure carbon, determined as follows:

$$s = \frac{61.2 \text{ Bq}}{152 \text{ mg C}} \times \frac{1000 \text{ mg C}}{1 \text{ g C}} = 403 \text{ Bq/g C}$$

$t = 8.26 \times 10^3$ yr $\times$ ln (918/403) = 6.8×10^3 yr

23.83 We use equation 23.6, in which s is defined to be the specific activity in the sample, after t years since death have elapsed:

$t = 8.26 \times 10^3$ yr $\times$ ln (918/s)

The value of s is, assuming the sample that was recovered to be pure carbon, determined as follows:

$$s = \frac{271\ \text{Bq}}{458\ \text{mg}} \times \frac{1000\ \text{mg C}}{1\ \text{g C}} = 592\ \text{Bq/g C}$$

$t = 8.26 \times 10^3$ yr $\times$ ln (918/592) = 3.6×10^3 yr

Nuclear Fusion

23.84 Fusion, rather than being a spontaneous process, requires enormous collisional velocities in order to cause two nuclei to approach each other closely enough to allow the strong force to operate.

23.85 Plasmas are high energy fluids, often either gases or liquids, but always mobile phases with high energies. They typically consist of an overall neutral mixture of nuclei and detached electrons.

23.86 The Lawson criterion is the expectation that thermonuclear fusion will yield a net energy only if the product n × t has the minimum value:

$n \times t = 3 \times 10^{14}$ s/cm^3

where n is the particle density in units particles per cm^3, and t is the confinement time in seconds.

23.87 The lasers are used to cause a sudden and large energy input to the outer regions of the pellet. The resulting loss of material (explosion) from the resulting surface plasma causes a corresponding implosion of the remaining pellet material. This implosion compresses the pellet material, and increases its temperature, causing fusion to become possible.

23.88 $^{3}_{1}H + ^{2}_{1}H \rightarrow ^{4}_{2}He + ^{1}_{0}n$

23.89 The liquid lithium metal serves as a neutron capture medium:

$$^{6}_{3}Li + ^{1}_{0}n \rightarrow ^{4}_{2}He + ^{3}_{1}H$$

The energy released by this process is dissipated into heat exchangers.

23.90 Self ignition is the condition that the thermal energy of the alpha particles is sufficient to heat the unchanged fuel and to help compress the fuel prior to its continued fusion. Without self ignition, fusion using lasers is not pratical.

23.91 A tokamak, as illustrated in Figure 23.21, is a chamber for confining plasma by the use of magnetic fields.

Nuclear Fission

23.92 A nucleus can more easily capture a neutron than a proton because the neutron is neutral. The proton, having a positive charge, is repelled by a nucleus as it approaches.

23.93 A thermal neutron is one whose kinetic energy is governed by the temperature of its surroundings, even at room temperature.

23.94 Nuclear fission is the breakup of a heavy nucleus, normally caused by absorption of a neutron, to give two or more lighter nuclei, plus two or more high energy neutrons (i.e. not simply thermal neutrons).

23.95 A fissile isotope is one that is capable of undergoing fission following neutron capture.

23.96 The only naturally-occurring fissile isotope is uranium-235.

23.97 The common fissile by-products of a ^{235}U reactor are uranium-233 and plutonium-239.

23.98 The fission of each uranium-235 produces more than two new neutrons, which can react further to sustain a chain reaction. In other words, more neutrons are generated by each fission event than are needed to cause another such event.

23.99 The isotopes (e.g. krypton-94 and barium-139) that are produced by the fission of uranium-235 are neutron-rich materials. In other words, their neutron-to-proton ratios are too high, and they spontaneously emit the excess neutrons.

23.100 $^{235}_{92}U + ^{1}_{0}n \rightarrow ^{94}_{38}Sr + ^{139}_{54}Xe + 2^{1}_{0}n$

Chapter 23

23.101 Both will be beta emitters. Both lie above the band of stability, and they can mover closer to it by emitting beta particles.

23.102 These neutrons can either escape to the surroundings or be captured by unfissioned nuclei, initiating more fission.

23.103 The sub-critical mass is incapable of self-capturing all of the emitted neutrons, which are necessary for sustaining a chain reaction.

23.104 A moderator in a nuclear reactor is used to slow down the fast neutrons that are produced by fission.

23.105 The multiplication factor (k) is the ratio of neutrons at the end of a "cycle" to those at the start of the "cycle" of nuclear events in a reactor. It is therefore an indication of the rate of acceleration (when $k > 1$) or deceleration (when $k < 1$) of the chain reaction process.

23.106 Both reactors can have the same multiplication factor.

23.107 A critical reactor has a value of k equal to 1.

23.108 A power plant is kept from going super-critical by inserting and adjusting the control rods, which are designed to capture neutrons, thereby lowering the multiplication factor.

23.109 This can cause a steam expansion explosion.

23.110 Hydrogen gas might be formed through reduction of water by the hot cladding or the hot fuel rods.

23.111 The cladding tubes are also supposed to confine gaseous products that are generated by the fuel, as well as to protect the fuel from the primary coolant.

23.112 No critical mass can form because the atoms of the fissionable isotopes are greatly diluted by those of the non-fissile isotopes.

23.113 The fission reactor produces a greater number and variety of radioactive wastes because heavier nuclides are involved from the outset.

23.114 Molten sodium provides a safer and more convenient primary coolant because liquid sodium can be employed at higher temperatures than water, and because the reactor can be immersed in a pool of liquid sodium, which then is able to serve as an emergency core coolant. Also, see the answer to Review Exercise 23.115.

23.115 The use of liquid sodium coolant, coupled with a metallic form of the fuel, offers two principal advantages. The excess heat from a sudden increase in reactor temperature is readily conducted to the coolant, which has a high heat capacity. Secondly, the metallic fuel assemblies apparently expand enough at high temperatures to slow the chain process by separating individual fissile atoms from one another.

23.116 The environmental issue is the major one; safely storing the reactor by-products is technologically difficult, and the danger of a melt-down is worrisome, since a large quantity of radioactive material could be released. The proponents respond by predicting that safe storage will soon be technologically possible, and by citing the various safeguards that exist against melt-down.

Breeder Reactor

23.117 Uranium-238 can be transformed into a fissile isotope, namely uranium-239.

23.118 $^{1}_{0}n + ^{238}_{92}U \rightarrow ^{239}_{92}U \rightarrow ^{239}_{94}Pu + 2\,^{0}_{-1}e$

23.119 The plutonium-239 is not consumed by the net process; rather it is regenerated from uranium-238, as shown in Figure 23.24.

23.120 Although such a reactor is not currently feasible in economic terms, when it does become economically feasible, there is the potential to use the present stockpiles of uranium-238. It is estimated that these resources would last for centuries of future energy needs.

CHAPTER TWENTY-FOUR

PRACTICE EXERCISES

1. (a) 3-methylhexane
   (b) 4-ethyl-2,3-dimethylheptane
   (c) 5-ethyl-2,4,6-trimethyloctane

REVIEW EXERCISES

Classifications

24.1 These compounds as a class are generally important to the chemistries of living organisms.

24.2 Biochemistry constitutes the study of the chemistry associated with a living organism.

24.3 Molecular biology constitutes the study of nucleic acids and proteins.

24.4 The functional group is the unique and characteristic portion of a molecule that sets it into one of a number of families of similar substances, and that is the focus of the reactivity of the substance.

24.5 The inorganic compounds of carbon are generally taken to be the oxides and cyanides of carbon, as well as carbonates and bicarbonates of the metals. Thus the inorganic compounds in this list are (b), (d), and (e), although (c) - CCl_4 - might arguably be listed also.

Structural Formulas

24.6 This is a straight-chain hydrocarbon, since no group is attached to the non-terminal carbon atoms by a C—C bond.

24.7 (a) This is impossible, since there should be three hydrogen atoms attached to the first carbon atom.

(b) This is impossible, since there should be only two hydrogen atoms attached to the first carbon atom.

(c) This is impossible since there should be only one hydrogen atom attached to the third carbon atom.

24.8 (a)

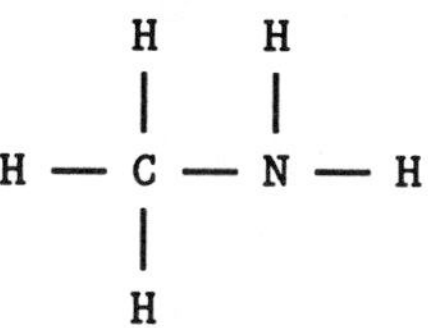

(b)

```
        H
        |
  Br —  C — Br
        |
        H
```

(c)

```
        Cl
        |
   H —  C — Cl
        |
        Cl
```

(d)

```
       H    H
       |    |
  H —  C —  C — H
       |    |
       H    H
```

(e)

```
       O
       ||
  H —  C — O — H
```

(f)

```
       O
       ||
  H —  C — H
```

(g)

```
       H
       |
  H —  N — O — H
```

(h)

```
  H — C ≡ C — H
```

(i)

```
       H    H
       |    |
  H —  N —  N — H
```

(j)

```
  H — C ≡ N
```

(k)

```
       H
       |
  H —  C — C ≡ N
       |
       H
```

(l)

```
       H
       |
  H —  C — O — H
       |
       H
```

24.9 (a) $CH_3CH_2CH_2CH_3$

(b)

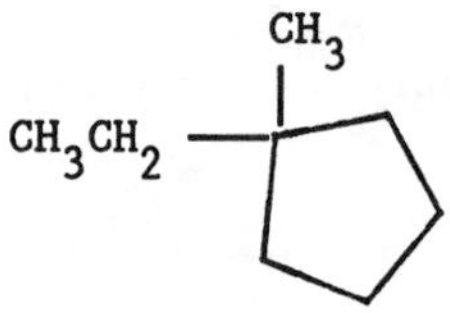

CH_3

CH_3CH_2

(c)

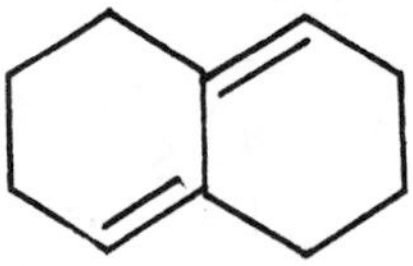

24.10 The functional groups typically impart polarity to what would be an otherwise nonpolar hydrocarbon, and they therefore give the molecule more of an opportunity for reaction than a simple hydrocarbon would have. In other words, since the functional group is the site of polarity, it is the likely site of attack by polar or ionic reactants. Amines, for instance, are Lewis and Brønsted bases, and they characteristically can be protonated as illustrated on pages 1013 and 1014 of the text.

24.11 (a) alkene
(b) alcohol
(c) ester
(d) carboxylic acid
(e) amine
(f) alcohol
(g) alkyne
(h) aldehyde
(i) ketone
(j) ether

24.12 The saturated compounds are b, e, f, and j.

Isomers

24.13 (a) These are identical, being oriented differently only.
(b) These are identical, being drawn differently only.
(c) These are unrelated, being alcohols with different numbers of carbon atoms.
(d) These are isomers, since they have the same empirical formula, but different structures.
(e) These are identical, being oriented differently only.
(f) These are identical, being drawn differently only.

(g) These are isomers, since they have the same empirical formula, but different structures.
(h) These are identical, being oriented differently only.
(i) These are isomers, since they have the same empirical formula, but different structures.
(j) These are isomers, since they have the same empirical formula, but different structures.
(k) These are identical, being oriented differently only.
(l) These are identical, being drawn differently only.
(m) These are isomers, since they have the same empirical formula, but different structures.
(n) These are unrelated.

Names of Hydrocarbons

24.14 (a) pentane
(b) 2-methylpentane
(c) 2,4-dimethylhexane
(d) 2,4-dimethylhexane
(e) 3-hexene
(f) 4-methyl-2-pentene

24.15 (a)

CH_3 (above the second carbon)

$CH_3CCH_2CH_2CH_2CH_2CH_2CH_3$

CH_3 (below the second carbon)

(c) CH_3CH_2 CH_2CH_3 (both on one carbon of a cyclohexane ring)

(b) CH_3 (on a cyclopentane ring), CH_3

(d)

CH_3CHCH_3 (on the first CH) CH_3 (on the last CH)

$CH_2=CHCH_2CH_2CHCH — CHCH_3$

CH_2CH_3 (on the first CH)

(e)

H H

C = C

CH_3 CH_2CH_3

Chapter 24

Properties of Unsaturated Hydrocarbons

24.16 Free rotation is prevented by the π-nature of the second of two bonds in the carbon-carbon double bond. This π bond is illustrated in Figure 8.14 on page 284 of the text.

24.17 Isomerism is possible only for (e) and (f).

(e) 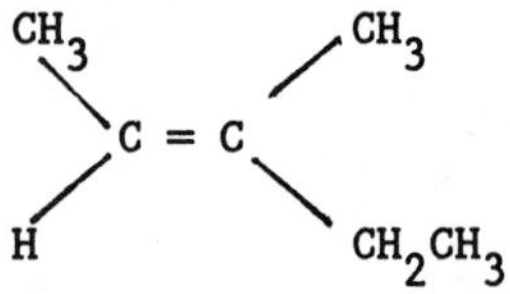 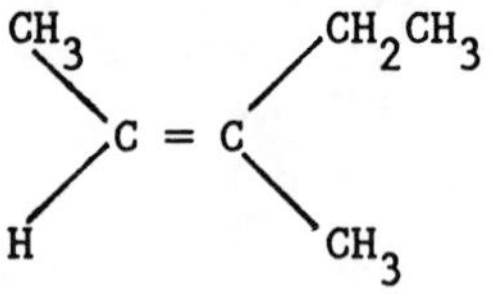 $(CH_3)(H)C=C(CH_3)(CH_2CH_3)$ and $(CH_3)(H)C=C(CH_2CH_3)(CH_3)$

(f) 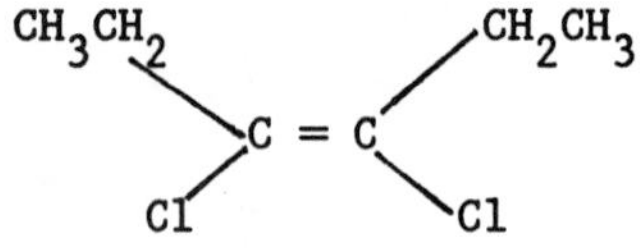 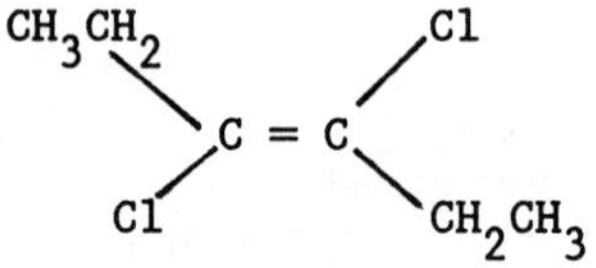 $(CH_3CH_2)(Cl)C=C(CH_2CH_3)(Cl)$ and $(CH_3CH_2)(Cl)C=C(Cl)(CH_2CH_3)$

24.18 $CH_3CH{=}CH_2 + H^+ \rightarrow CH_3CH{-}CH_3{}^+$

24.19 (a) CH_3CH_3

(b) $ClCH_2CH_2Cl$

(c) $BrCH_2CH_2Br$

(d) CH_3CH_2Cl

(e) CH_3CH_2Br

(f) CH_3CH_2OH

(g) $CH_3CH_2{-}O{-}SO_3H$

24.20 (a) $CH_3CH_2CH_2CH_3$

(b) 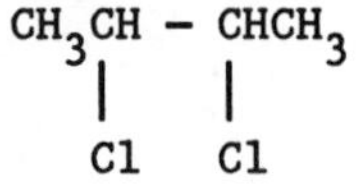 $CH_3CH(Cl) - CH(Cl)CH_3$

(c) 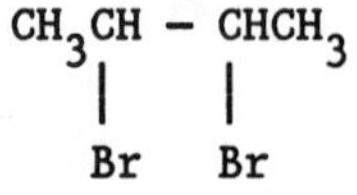 $CH_3CH(Br) - CH(Br)CH_3$

(d) $CH_3CH - CHCH_3$

H Cl

(e) $CH_3CH - CHCH_3$

H Br

(f) $CH_3CH - CHCH_3$

H OH

(g) $CH_3CH - CHCH_3$

H OSO_3H

24.21 (a)

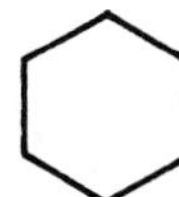

(b)

Cl

Cl

(c)

Br

Br

(d)

H

Cl

(e)

H

Br

(f) [cyclohexane ring bearing H and OH on adjacent carbons]

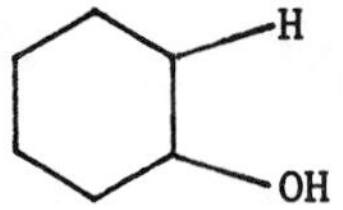

(g) [cyclohexane ring bearing H and OSO_3H on adjacent carbons]

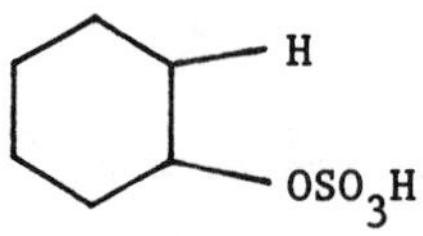

24.22

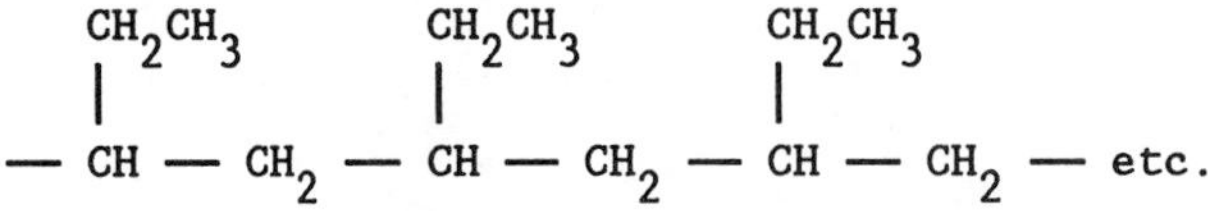

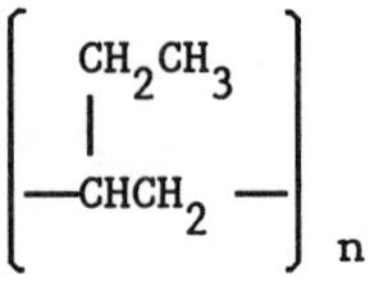

$$
\begin{array}{cccccccc}
 & CH_2CH_3 & & CH_2CH_3 & & CH_2CH_3 & & \\
 & | & & | & & | & & \\
— & CH & — CH_2 — & CH & — CH_2 — & CH & — CH_2 — & \text{etc.}
\end{array}
$$

$$
\left[\begin{array}{c} CH_2CH_3 \\ | \\ —CHCH_2— \end{array}\right]_n
$$

24.23

$$
\begin{array}{cccccccc}
 & CH_3 & & CH_3 & & CH_3 & & \\
 & | & & | & & | & & \\
— & CH & — CH_2 — & CH & — CH_2 — & CH & — CH_2 — & \text{etc.} \\
 & | & & | & & | & & \\
 & CH_3 & & CH_3 & & CH_3 & &
\end{array}
$$

$$
\left[\begin{array}{c} CH_3 \\ | \\ —CHCH_2— \\ | \\ CH_3 \end{array}\right]_n
$$

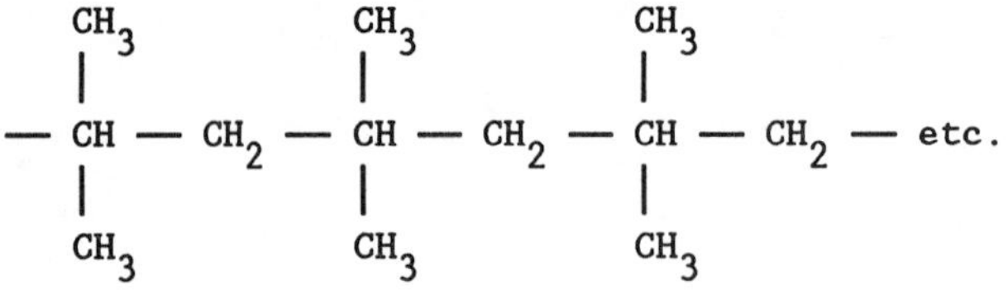

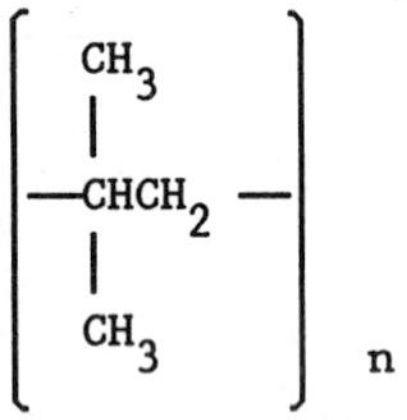

24.24 This sort of reaction would disrupt the π delocalization of the benzene ring. The subsequent loss of resonance energy would not be favorable.

24.25 Benzene does not "add" Cl_2. Rather, in the presence of a catalyst, it gives chlorobenzene:

$C_6H_6 + Cl_2 \rightarrow C_6H_5Cl + HCl$ ($FeCl_3$ catalyst)

24.26 It is the presence of one or more benzene rings that causes an organic substance to be classified as aromatic.

Organic Compounds of Oxygen

24.27 CH_3OH IUPAC name = methanol; common name = methyl alcohol

CH_3CH_2OH IUPAC name = ethanol; common name = ethyl alcohol

$CH_3CH_2CH_2OH$ IUPAC name = 1-propanol; common name = propyl alcohol

CH_3CHCH_3 IUPAC name = 2-propanol; common name = isopropyl alcohol
(with OH on the central C)

$$\begin{array}{l} CH_3\underset{\displaystyle OH}{\underset{|}{C}}HCH_3 \end{array}$$

24.28 (a) $CH_3CH_2CH_2CH_2OH$

(b) $$CH_3CH_2\underset{\displaystyle OH}{\underset{|}{C}}HCH_3$$

24.29 The proper IUPAC name is 2-butanol:

$$CH_3\underset{\displaystyle OH}{\underset{|}{C}}HCH_2CH_3$$

24.30 Ethyl alcohol can form a hydrogen bond in water, whereas ethane cannot.

24.31 The oxygen atom of an alcohol group acquires another proton when the alcohol is treated with sulfuric acid, and this weakens all of the bonds to oxygen, including the C — O bond:

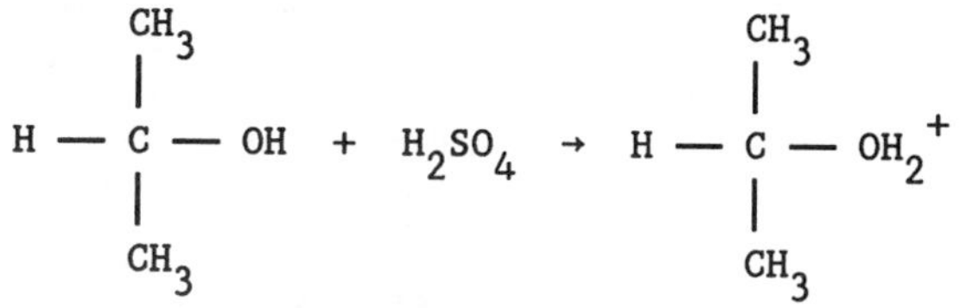

$$H - \overset{\displaystyle CH_3}{\underset{\displaystyle CH_3}{\overset{|}{\underset{|}{C}}}} - OH + H_2SO_4 \rightarrow H - \overset{\displaystyle CH_3}{\underset{\displaystyle CH_3}{\overset{|}{\underset{|}{C}}}} - OH_2^{\ +}$$

24.32 The elimination of water can result in a C=C double bond in two locations:

$CH_2{=}CHCH_2CH_3$ $\qquad$ $CH_3CH{=}CHCH_3$

24.33 The ratio should be 3:2, since there are three hydrogen atoms on the first carbon of 2-butanol, whereas there are only two hydrogen atoms on the third carbon atom of 2-butanol.

24.34 Statistical factors are not important in determining the distribution of products. Thermodynamic factors that must be considered are the relative strengths of the different C—H bonds that must be broken and the relative stabilities of the two different alkenes that are formed.

24.35 The aldehyde is more easily oxidized:

$$CH_3CH_2\overset{\overset{\displaystyle O}{\|}}{C}OH$$

24.36 The order is C < A < B. This is a result of the —OH group of the alcohol, which allows for hydrogen bonding, and a correspondingly high boiling point. Also, the trend is in keeping with the relative order of increasing molecular polarity, the ketone (A) having a higher boiling point than the alkane (B), which is nonpolar.

24.37 The increasing solubility in water is also C > A > B. This arises because the hydrocarbon (C) is nonpolar and unable to enter into hydrogen bonding with the solvent, whereas the alcohol (B) is both polar and capable of hydrogen bonding. The ketone (A) is polar, but not able to enter into hydrogen bonding with water solvent.

24.38 The order of increasing boiling points is C < A < B. The carboxylic acid (B) is highest boiling, because it has two polar groups that can engage in hydrogen bonding. The aldehyde (C) cannot enter into hydrogen bonding. The alcohol (A) is intermediate between the acid and the aldehyde, having less ability to form a hydrogen bond than the acid.

24.39 (a) A is easily oxidized, giving B.

(b) B is neutralized by sodium hydroxide:

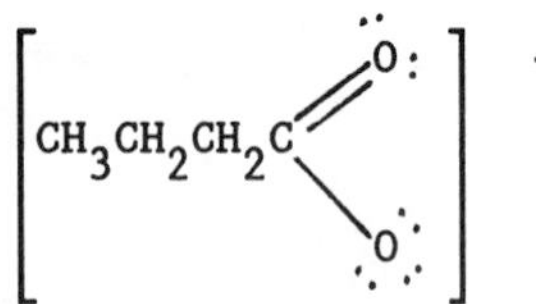

(c) B forms an ester with methyl alcohol:

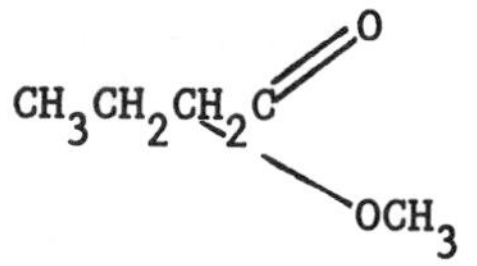

$$CH_3CH_2CH_2C(=O)OCH_3$$

24.40

$$-OCH_2CH_2O\overset{\displaystyle O}{\overset{\|}{C}}CH_2CH_2\overset{\displaystyle O}{\overset{\|}{C}}-$$

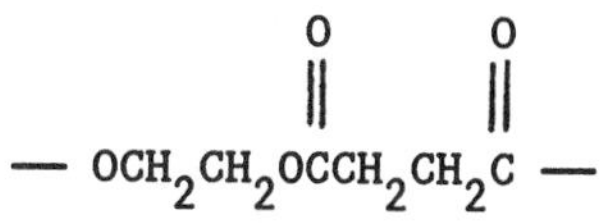

Carbohydrates

24.41 Carbohydrates are naturally occurring polyhydroxyaldehydes and polyhydroxyketones, or substances that give these on hydrolysis. They include the monosaccharides, the disaccharides and the polysaccharides. Examples are sucrose, starch and cellulose.

24.42 Each yields only glucose on complete hydrolysis.

24.43 The three forms differ from one another only in the manner in which the ring is either opened or closed, as illustrated in Figure 24.2 on page 1029 of the text. Furthermore, the open form (which gives the substance the characteristics of an aldehyde) is readily regenerated by the two dynamic and reversible equilibria that convert it into either the α or the β form.

24.44 The digestion of sucrose yields glucose and fructose.

24.45 The hydrolysis of lactose gives glucose and galactose.

Lipids

24.46 Lipids are substances that are found in living systems and that are soluble in nonpolar solvents such as ethers and benzene. The lipids include a diverse range of substances such as cholesterol, and hormones, as well as fats, oils, fatty acids, and other materials known as triacylglycerols.

24.47 It is soluble in nonpolar solvents, and it is a substance found in living things.

24.48 This particular type of unsaturation refers only to the presence of the C=C double bond.

24.49

$$\begin{array}{l} CH_2 - OH \\ | \\ CH - OH \\ | \\ CH_2 - OH \end{array} \qquad \begin{array}{l} HO - \overset{O}{\overset{\|}{C}} - (CH_2)_7CH = CH(CH_2)_7CH_3 \\ HO - \overset{O}{\overset{\|}{C}} - (CH_2)_{17}CH_3 \\ HO - \overset{O}{\overset{\|}{C}} - (CH_2)_7CH = CHCH_2CH = CH(CH_2)_4CH_3 \end{array}$$

Organic Compounds of Nitrogen

24.50 (a) amine
(b) amine
(c) amide
(d) amine plus a keto group
(e) amine
(f) amide

24.51 (a) The amine neutralizes HCl:

$$CH_3CH_2CH_2NH_2 + HCl \rightarrow CH_3CH_2CH_2NH_3^+ + Cl^-$$

(b) The amide is hydrolyzed:

$$CH_3CH_2\overset{O}{\overset{\|}{C}}NH_2 + H_2O \rightarrow CH_3CH_2\overset{O}{\overset{\|}{C}}OH + NH_3$$

(c) The alkylammonium cation neutralizes sodium hydroxide:

$$CH_3CH_2NH_3^+ + OH^- \rightarrow CH_3CH_2NH_2 + H_2O$$

24.52 The N–H bond is not as polar as the O–H bond, and the hydrogen bond of amines is therefore not as strong as that of alcohols.

24.53 The additional carbonyl group in an amide causes the boiling point of an amide to be much higher than that of a comparable amine.

24.54

$$-NHCH_2CH_2CH_2CH_2NHC(=O)-C_6H_4-C(=O)-$$

24.55 (a) The monomers of polypeptides are alike in having the nitrogen – carbon – carbonyl repeating backbone, one linked to another by the peptide bond.

(b) The monomers of polypeptides have different side groups, since each of the various amino acids could be employed at any point in the polymer.

24.56 The peptide bond is the amide bond, found in polypeptides.

24.57

$$^{+}NH_3CH_2C(=O)-NHCH_2CO(=O)^{-}$$

24.58

$$^{+}NH_3CH(CH_3)C(=O)-NHCH(CH_3)C(=O)-NHCH(CH_3)C(=O)-O^{-}$$

24.59

$$^{+}NH_3CH(CH_3)C(=O)-NHCH_2CO(=O)^{-} \qquad ^{+}NH_3CH_2C(=O)-NHCH(CH_3)CO(=O)^{-}$$

24.60 Two isomeric polypeptides can differ in the amino acid sequence.

24.61 Two dissimilar polypeptides can differ in (1) the number of amino acids, (2) the identities of the amino acids, and (3) the sequence of amino acids that comprise the chain.

24.62 Although some proteins are made up of a single polypeptide, others may contain two or more associated polypeptides, an organic group, as well as a metal atom.

Chapter 24

Nucleic Acids and Heredity

24.63 A particular amino acid sequence in a polypeptide is obtained through the following sort of "direction" from the gene that is responsible for a particular polypeptide synthesis. There is a one-to-one relationship between the series of side chain bases (taken in groups of three) in the DNA of a gene and the series of amino acids that are to be assembled in the polypeptide that is to be made under the "direction" of the gene. The gene directs the synthesis of ptRNA, which the cell possesses to make mRNA. It is mRNA that is the bearer of the genetic message from the nucleus of the cell to a site outside the nucleus, where the polypeptide is to be assembled. This is the process of transcription, and it is diagramed in Figure 24.10. Next, the mRNA accumulates at the ribosomes, awaiting the arrival of the correct amino acids, which are delivered by tRNA (diagramed in Figure 24.11). The process of polypeptides synthesis next involves translation of the information in mRNA into a specific amino acid sequence in the growing polypeptide (Figure 24.12).

24.64 The genetic messages are carried as a sequence of side chain bases on a DNA segment.

24.65 The two strands in DNA are held together by hydrogen bonds between specific base pairs, as shown in Figure 24.8.

24.66 This is shown in Figure 24.7 and 24.8. Only one particular base can be found opposite another in the double helix.

24.67 DNA and RNA differ in the identity of the monosaccharide. In RNA it is ribose. In DNA it is deoxyribose, as shown at the bottom of page 1039. Also, DNA uses the bases A, T, G, and C, whereas RNA uses the bases A, U, G, and C.

24.68 (a) In DNA, A pairs with T.
(b) In RNA, A pairs with U.
(c) C pairs with G.

24.69 A segmented gene is one in which the strands of DNA are divided or split, such that the gene is a conglomerate, exons being separated by introns.

24.70 Codons are found on mRNA.

24.71 Anticodons are found on tRNA.

24.72 (a) The ribosome is the site outside the nucleus where the polypeptide is made.

(b) mRNA carries the genetic message from the DNA in the nucleus to the ribosomes.

(c) tRNA brings the necessary amino acids for polypeptide synthesis to the ribosomes.

24.73 This is ptRNA, which is used in making mRNA.

24.74 Transcription begins with DNA and ends with the synthesis of mRNA.

24.75 Translation begins with mRNA, uses tRNA, and ends with a specific polypeptide.

24.76 Viruses consist of nucleic acids and proteins. They interrupt the normal genetic activity of a host cell, and use the cell functions to multiply rapidly.

24.77 Genetic engineering is able to cause the biological synthesis of important human polypeptides, enzymes etc, by tricking bacterial plasmids into synthesizing human proteins, through the introduction of recombinant DNA. This is diagramed in Figure 24.13.

24.78 The DNA that is present in a bacterial plasmid that has been altered (by the introduction of DNA that is foreign to the bacteria) is recombinant DNA.

CHAPTER TWENTY-FIVE

REVIEW EXERCISES

General Concepts and Procedures

25.1 (a) Cation Group 1 = Ag^+, Hg_2^{2+}, and Pb^{2+}

(b) Cation Group 2 = Bi^{3+}, Cu^{2+}, and Sn^{4+} (or Sn^{2+})

(c) Cation Group 3 = Al^{3+}, Ni^{2+}, Fe^{2+} (or Fe^{3+}), Mn^{2+}, and Zn^{2+}

(d) Cation Group 4 = Na^+, NH_4^+, Ca^{2+}, and Ba^{2+}

25.2 (a) Cation Group 1 is the silver or chloride group.
(b) Cation Group 2 is the hydrogen sulfide or the copper-arsenic group.
(c) Cation Group 3 is the ammonium sulfide or the aluminum-nickel group.
(d) Cation Group 4, in the shortened version presented here, has no name, although the term soluble group might apply to Na^+, K^+, and NH_4^+, whereas the term carbonate group might apply to Ca^{2+}, and Ba^{2+}.

25.3 Group reagents are used to divide the various cations into groups of ions having similar chemistries. The group separations are necessary in order to prevent the presence of one ion from interferring with the test for another ion.

25.4 (a) The group reagent for Cation Group 1 is 6 M HCl.
(b) The group reagent for Cation Group 2 is saturated (0.10 M) H_2S, and the test solution is first brought to a pH of 0.5. The H_2S is generated by the hydrolysis of thioacetamide.
(c) The reagent for Cation Group 3 is 0.1 M H_2S (from thioacetamide), and the solution is adjusted to a pH of 8 or 9, by the addition of 2M NH_4Cl followed by 6 M NH_3.

Chapter 25

25.5 The anions are organized into Group 1 (SO_4^{2-}, CO_3^{2-}, and PO_4^{3-}) and Group 2 (Cl^-, Br^-, I^-, and HS^-). Additionally, we have NO_3^-, which is in Group 3.

Procedures

25.6 The semimicro scale involves volumes of from one drop to 5-10 mL, and masses in the range 100-200 mg. Concentrations of ions are such as to give readily visible results on these amounts of sample.

25.7 Tap water might typically (and undesirably) contain Ca^{2+}, Mg^{2+}, Fe^{2+}, Fe^{3+}, Cl^-, HCO_3^-, and SO_4^{2-}.

25.8 Digestion makes the particles of a colloid become larger, facilitating their precipitation by causing them to conglomerate. This is done with heating, followed by slow cooling of the solution.

25.9 (a) The supernatant is the solution that is to be removed when recovering a precipitate.
(b) A centrifugate is the supernatant found above the precipitate after use of a centrifuge.
(c) Extraction is the process of removing one or more solids from a mixture of solids by dissolving certain of them in a solvent and separating the insoluble ones with a centrifuge.

25.10 Precipitates are washed either to extract certain solids or to remove unwanted residual centrifugate.

25.11 One should practice in order to become familiar with the various procedures and to see what the positive tests look like. Also this gives an opportunity to check the various reagents that are to be used.

Cation Group 1

25.12 See Figure 25.11.

25.13 $Ag^+(aq) + Cl^-(aq) \rightarrow AgCl(s)$

$Hg_2^{2+}(aq) + 2Cl^-(aq) \rightarrow Hg_2Cl_2(s)$

$Pb^{2+}(aq) + 2Cl^-(aq) \rightarrow PbCl_2(s)$

25.14 All of the Cation Group 1 chlorides are white.

25.15 A slight excess is used, first of all, in order to force the precipitation of a group 1 cation, by taking advantage of the common ion effect. Equations given in the answer to Review Exercise 25.13 apply. Secondly, one requires a slight excess of the hydrogen ion in order to shift the following equilibrium to the left, preventing hydrolysis of Bi^{3+}, which would cause an undesirable precipitation of BiOCl(s):

$Bi^{3+}(aq) + H_2O(\ell) + Cl^-(aq) \rightleftharpoons BiOCl(s) + 2H^+(aq)$

25.16 The solid chlorides might undesirably redissolve, owing to the formation of soluble chloro complexes:

$AgCl(s) + Cl^-(aq) \rightleftharpoons [AgCl_2]^-(aq)$

$PbCl_2(s) + 2Cl^-(aq) \rightleftharpoons [PbCl_4]^{2-}(aq)$

25.17 Lead is sometimes transferred into Cation Group 2 by heating, because lead chloride is more soluble than the other chlorides of Cation Group 2. Additionally, there is the possibility that lead might be carried into Cation Group 2 by the inadvertant use of too much chloride ion, which dissolves $PbCl_2(s)$ by the following equilibrium:

$PbCl_2(s) + 2Cl^-(aq) \rightleftharpoons [PbCl_4]^{2-}(aq)$

25.18 (a) $PbCl_2(s) \rightarrow Pb^{2+}(aq) + 2Cl^-(aq)$, on heating

$Pb^{2+}(aq) + CrO_4{}^{2-}(aq) \rightarrow PbCrO_4(s)$

(b) $Hg_2Cl_2(s) + 2NH_3(aq) \rightarrow Hg(\ell) + HgNH_2Cl(s) + NH_4{}^+(aq) + Cl^-(aq)$

(c) $AgCl(s) + 2NH_3(aq) \rightleftharpoons [Ag(NH_3)_2]^+(aq) + Cl^-(aq)$

$[Ag(NH_3)_2]^+(aq) + 2HCl(aq) \rightarrow AgCl(s) + 2NH_4{}^+(aq) + Cl^-(aq)$

25.19 See Figure 25.12.

25.20 $Pb^{2+}(aq) + S^{2-}(aq) \rightarrow PbS(s)$

$2Bi^{3+}(aq) + 3S^{2-}(aq) \rightarrow Bi_2S_3(s)$

$Cu^{2+}(aq) + S^{2-}(aq) \rightarrow CuS(s)$

$Sn^{4+}(aq) + 2S^{2-}(aq) \rightarrow SnS_2(s)$

25.21 Sulfides of Cation Group 2 are precipitated at a pH equal to 0.5.

25.22 The H_2S concentration is ca. 0.1 M.

22.23 This is done to provide an electrolyte that will prevent the formation of a colloidal precipitate.

22.24 PbS > CuS > SnS_2 > Bi_2S_3

22.25 (a) Sn^{2+} is always oxidized by oxygen in air to give Sn^{4+} anyway.

(b) SnS_2 is more of a crystalline solid than SnS, so oxidation to tin(IV) is advantageous, since it precludes the formation of a gelatinous SnS material.

(c) SnS does not subsequently form a soluble complex with S^{2-}, or at least SnS_2 does this more readily.

(d) It is desirable to have all of the tin in the form Sn^{4+}, so that the chloro complex is readily formed:

$$Sn^{4+}(aq) + 6Cl^-(aq) \rightarrow [SnCl_6]^{2-}(aq)$$

This complex does not oxidize H_2S to elemental sulfur, whereas the tin(II) form would.

25.26 (a) An oxidizing acid would interfere later, by converting S^{2-} into elemental sulfur, which is a bothersome yellow colloid.

(b) An acid is required to prevent the precipitation of Cation Group 3 sulfides, and to prevent the precipitation of hydroxides or hydrated oxides of Sn^{4+} or Bi^{3+}.

(c) The chloride ion is used to prevent the reduction of Sn^{4+} by H_2S.

This is accomplished by formation of a chloro complex of tin(IV):

$$Sn^{4+}(aq) + 6Cl^-(aq) \rightarrow [SnCl_6]^{2-}(aq)$$

25.27 The bismuth ion is hydrolyzed in aqueous acid:

$$Bi^{3+}(aq) + H_2O(\ell) \rightleftharpoons BiO^+(aq) + 2H^+(aq)$$

25.28 PbS and CuS are black. Bi_2S_3 is dark brown, and SnS_2 is yellow.

25.29 (a) $SnS_2(s) + H_2S(aq) + 2OH^-(aq) \rightleftharpoons [SnS_3]^{2-}(aq) + 2H_2O(\ell)$

(b) Tin(II) is not extracted by this procedure.

25.30 (a) This is mercury(II) chloride.

(b) The white precipitate is $Hg_2Cl_2(s)$:

$$[SnCl_4]^{2-}(aq) + 2HgCl_2(s) \rightarrow [SnCl_6]^{2-}(aq) + Hg_2Cl_2(s)$$

(c) The black product is $Hg(\ell)$:

$$[SnCl_4]^{2-}(aq) + Hg_2Cl_2(s) \rightarrow [SnCl_6]^{2-}(aq) + 2Hg(\ell)$$

(d) H_2S must be completely removed in order to prevent a false positive test resulting from the formation of black HgS, when Hg^{2+} is added.

(e) H_2S is removed by complete hydrolysis of thioacetamide using HCl, and thorough heating of the solution to drive off $H_2S(g)$.

(f) It is necessary to reduce Sn^{4+} (in $[SnCl_6]^{2-}$) to Sn^{2+} (in $[SnCl_4]^{2-}$) in order to generate a reducing agent.

(g) $[SnCl_6]^{2-}(aq) + Fe(s) \rightarrow [SnCl_4]^{2-}(aq) + Fe^{2+}(aq) + 2Cl^-(aq)$

25.31 $3PbS(s) + 8H^+(aq) + 2NO_3^-(aq) \rightarrow 3Pb^{2+}(aq) + 3S(s) + 2NO(g) + 4H_2O(\ell)$

$Bi_2S_3(s) + 8H^+(aq) + 2NO_3^-(aq) \rightarrow 2Bi^{3+}(aq) + 3S(s) + 2NO(g) + 4H_2O(\ell)$

$3CuS(s) + 8H^+(aq) + 2NO_3^-(aq) \rightarrow 3Cu^{2+}(aq) + 3S(s) + 2NO(g) + 4H_2O(\ell)$

25.32 A high concentration of H^+ tends to dissolve $PbSO_4(s)$ by shifting the following equilibrium to the right:

$$PbSO_4(s) + H^+(aq) \rightleftharpoons Pb^{2+}(aq) + HSO_4^-(aq)$$

The additional H^+ comes from HNO_3 as well as H_2SO_4. Heating decomposes nitric acid:

$$4HNO_3(aq) \rightarrow O_2(g) + 4NO_2(g) + 2H_2O(g)$$

as well as some of the sulfuric acid:

$$H_2SO_4(\ell) \rightarrow H_2O(g) + SO_3(g)$$

When water is added, the concentration of H^+ becomes too low to cause the $PbSO_4(s)$ to dissolve.

25.33 (a) This helps distinguish $PbSO_4$ from $(BiO)_2SO_4$.

(b) An acetate complex is formed:

$$PbSO_4(s) + 4C_2H_3O_2^-(aq) \rightleftharpoons [Pb(C_2H_3O_2)_4]^{2-}(aq) + SO_4^{2-}(aq)$$

(c) This separates $PbSO_4(s)$ from $(BiO)_2SO_4(s)$.

(d) One adds potassium chromate, and forms the bright yellow lead chromate:

$$[Pb(C_2H_3O_2)_4]^{2-}(aq) + CrO_4^{2-}(aq) \rightarrow PbCrO_4(s) + 4C_2H_3O_2^-(aq)$$

25.34 The confirmation calls for the addition of concentrated ammonia, and one sees a deep blue color due to the formation of a copper ammonia complex:

$$[Cu(H_2O)_4]^{2+}(aq) + 4NH_3(aq) \rightarrow [Cu(NH_3)_4]^{2+}(aq) + 4H_2O(\ell)$$

25.35 Ammonia molecules not only form the deep blue copper complex (Review Exercise 25.34), but also make the solution sufficiently basic to cause the precipitation of the insoluble hydroxide of bismuth:

$$BiO^+(aq) + OH^-(aq) \rightarrow BiO(OH)(s)$$

25.36 One adds freshly prepared stannite solution ($[Sn(OH)_3]^-(aq)$), and one sees the conversion of white BiO(OH)(s) into black Bi(s):

$$2BiO(OH)(s) + 3[Sn(OH)_3]^-(aq) + 3OH^-(aq) + 2H_2O(\ell) \rightarrow 2Bi(s) + 3[Sn(OH)_6]^{2-}(aq)$$

25.37 The stannite ion is both oxidized by O_2 from air and disproportionated when allowed to stand:

$$2[Sn(OH)_3]^-(aq) + O_2(aq) + 2OH^-(aq) + 2H_2O(\ell) \rightarrow 2[Sn(OH)_6]^{2-}(aq)$$

$$2[Sn(OH)_3]^-(aq) \rightarrow Sn(s) + [Sn(OH)_6]^{2-}(aq)$$

25.38 (a) The stannite ion is prepared using both NaOH(aq) and $SnCl_2(aq)$.

(b) $Sn^{2+}(aq) + 2OH^-(aq) \rightarrow Sn(OH)_2(s)$

$Sn(OH)_2(s) + OH^-(aq) \rightarrow [Sn(OH)_3]^-(aq)$

Cation Group 3

25.39 See Figure 25.13.

25.40 $Mn^{2+}(aq) + HS^-(aq) + NH_3(aq) \rightarrow MnS(s) + NH_4^+(aq)$

$Fe^{2+}(aq) + HS^-(aq) + NH_3(aq) \rightarrow FeS(s) + NH_4^+(aq)$

$Ni^{2+}(aq) + HS^-(aq) + NH_3(aq) \rightarrow NiS(s) + NH_4^+(aq)$

$Al^{3+}(aq) + 3OH^-(aq) \rightarrow Al(OH)_3(s)$

$Zn^{2+}(aq) + HS^-(aq) + NH_3(aq) \rightarrow ZnS(s) + NH_4^+(aq)$

25.41 The pH is adjusted using a NH_3/NH_4Cl buffer.

25.42 The hydroxide, $Al(OH)_3$, forms in place of the sulfide, because the hydroxide has a smaller value of K_{sp}.

25.43 MnS - peach
FeS - black
NiS - black
ZnS - white
$Al(OH)_3$ - white

25.44 Fe^{3+} is reduced to Fe^{2+} by H_2S:

$2Fe^{3+}(aq) + H_2S(aq) \rightarrow 2Fe^{2+}(aq) + S(s) + 2H^+(aq)$

25.45 A low value of pH suppresses the hydrolysis that gives the pale yellow $[Fe(H_2O)_5(OH)]^{2+}$.

25.46 Above pH 3, a colloidal suspension forms, and eventually gives a precipitate of red-brown hydrated iron(III) oxide.

25.47 $FeS(s) + 2H^+(aq) \rightarrow Fe^{2+}(aq) + H_2S(g)$, which is followed by oxidation:

$3Fe^{2+}(aq) + 4H^+(aq) + NO_3^-(aq) \rightarrow 3Fe^{3+}(aq) + NO(g) + 2H_2O(\ell)$

For the other sulfides, we have:

$MnS(s) + 2H^+(aq) \rightarrow Mn^{2+}(aq) + H_2S(g)$

$3NiS(s) + 8H^+(aq) + 2NO_3^-(aq) \rightarrow 3Ni^{2+}(aq) + 2NO(g) + 3S(s) + 4H_2O(\ell)$

$ZnS(s) + 2H^+(aq) \rightarrow Zn^{2+}(aq) + H_2S(g)$

$Al(OH)_3(s) + 3H^+(aq) \rightarrow Al^{3+}(aq) + 3H_2O(\ell)$

For the oxidation of H_2S, we write:

$3H_2S(aq) + 2H^+(aq) + 2NO_3^-(aq) \rightarrow 3S(s) + 2NO(g) + 4H_2O(aq)$

25.48 (a) $Cl^-(aq)$ interferes with the final test for manganese.

(b) $6Cl^-(aq) + 2NO_3^-(aq) + 8H^+(aq) \rightarrow 3Cl_2(g) + 2NO(g) + 4H_2O(\ell)$

25.49 An amphoteric substance is one that can react with both an acid and a base. This property is used to separate Al^{3+} and Zn^{2+} from the other members of Cation Group 3, once all have been redissolved using dilute HCl/HNO_3.

25.50 (a) These cations are separated into the insoluble hydroxides ($Mn(OH)_2$, $Fe(OH)_2$, and $Ni(OH)_2$) and the soluble hydroxides ($Al(OH)_4^-$ and $Zn(OH)_4^{2-}$).

(b) This is done using 6 M NaOH.

(c) $Al(OH)_3(s) + OH^-(aq) \rightarrow [Al(OH)_4]^-(aq)$

$Zn(OH)_2(s) + 2OH^-(aq) \rightarrow [Zn(OH)_4]^{2-}(aq)$

25.51 An oxidizing acid is needed to keep iron in the 3+ oxidation state. Although HCl would dissolve the precipitate, it would also furnish chloride ion, which would interfere with the subsequent test for manganese.

25.52 Sodium bismuthate is added, and one sees the formation of a purple color due to permanganate:

$$2Mn^{2+}(aq) + 5BiO_3^-(aq) + 14H^+(aq) \rightarrow 2MnO_4^-(aq) + 5Bi^{3+}(aq) + 7H_2O(\ell)$$

25.53 If the purple color fades, it is likely an indication that some Cl^- remained, when it should have been removed by Procedure 3B.

25.54 Neither Fe^{3+} nor Ni^{2+} can be oxidized by BiO_3^-, and neither forms a precipitate with this anion. Thus bismuthate is able to oxidize Mn^{2+} only.

25.55 Two soluble cations are formed instead: $[Ni(NH_3)_6]^{2+}$ and $[Mn(H_2O)_6]^{2+}$.

25.56 This causes the precipitation of the hydroxide:

$$Fe^{3+}(aq) + 3NH_3(aq) + 6H_2O(aq) \rightarrow Fe(OH)_3(H_2O)_3(s) + 3NH_4^+(aq)$$

25.57 (a) $Fe(OH)_3(s) + 3HCl(aq) + 3H_2O(\ell) \rightarrow [Fe(H_2O)_6]^{3+}(aq) + 3Cl^-(aq)$

$[Fe(H_2O)_6]^{3+}(aq) + SCN^-(aq) \rightarrow [Fe(H_2O)_5SCN]^{2+}(aq) + H_2O(\ell)$

(b) One sees the formation of an intense, blood-red color.

25.58 (a) Ammonia is added at Procedure 3E, in the separation of iron as the hydroxide. If Ni^{2+} is also present, the centrifugate is blue, owing to the formation of $[Ni(NH_3)_6]^{2+}(aq)$.

(b) The solution (centrifugate 3C5) is blue.

(c) $[Ni(NH_3)_6]^{2+}(aq)$

(d) $[Cu(NH_3)_4]^{2+}(aq)$

25.59 (a) One adds dimethylglyoxime, and one observes the formation of a brick-red color due to a chelate complex of nickel.

(b) $[Ni(NH_3)_6]^{2+}(aq) + 2H_2DMG \rightarrow Ni(HDMG)_2(s) + 4NH_3(aq) + 2NH_4^+(aq)$

(c) A bidentate ligand is one, like $HDMG^-$, that binds to a metal atom through two atoms of the ligand.

25.60 The reagent is aqueous ammonia. Zn^{2+} forms a soluble complex, but Al^{3+} precipitates as the hydroxide:

$[Zn(H_2O)_4]^{2+}(aq) + 4NH_3(aq) \rightarrow [Zn(NH_3)_4]^{2+}(aq) + 4H_2O(\ell)$

$[Al(H_2O)_6]^{3+}(aq) + 3NH_3(aq) \rightarrow Al(OH)_3(H_2O)_3(s) + 3NH_4^+(aq)$

25.61 After precipitation of $[Al(OH)_3(H_2O)_3](s)$ using aqueous ammonia, one adds aluminon reagent, followed by ammonia. The formation of a floculant red lake (solid) results if aluminum was present.

25.62 A floculant solid is one that floats on water, due to the low density of the solid. An example is the $Al(OH)_3$ - aluminon solid compound that confirms the presence of Al^{3+} by the formation of a red lake.

25.63 Lakes are colored, floculant solids that form in water.

25.64 (a) $[Zn(NH_3)_4]^{2+}(aq) + 4H^+(aq) \rightarrow Zn^{2+}(aq) + 4NH_4^+(aq)$

$2K^+(aq) + 3Zn^{2+}(aq) + 2[Fe(CN)_6]^{4-}(aq) \rightarrow K_2Zn_3[Fe(CN)_6]_2(s)$

(b) The presence of a pale, blue-green precipitate confirms the presence of Zn^{2+}.

Cation Group 4

25.65 See Figure 25.14.

25.66 NaOH is added to the original unknown sample, and the resulting gas (NH_3)

turns moist litmus paper blue.

25.67 Other test procedures call for the use of aqueous ammonia, making subsequent tests for ammonia pointless, since a positive result is guaranteed regardless of whether or not ammonium cation is present in the original unknown sample.

25.68 These cations are separated by use of CrO_4^{2-}; only barium forms a solid at this juncture. Calcium ion remains in solution.

$Ba^{2+}(aq) + CrO_4^{2-}(aq) \rightarrow BaCrO_4(s)$

25.69 Barium chromate is redissolved using HCl(aq), and it is then reprecipitated using sulfate ion:

$Ba^{2+}(aq) + SO_4^{2-}(aq) \rightarrow BaSO_4(s)$

Alternatively, a greenish yellow flame test may be obtained.

25.70 The white oxalate complex is formed:

$Ca^{2+}(aq) + C_2O_4^{2-}(aq) \rightarrow CaC_2O_4(s)$

25.71 The insoluble oxalate of calcium is redissolved in acid, and a red-orange flame test for this solution confirms the presence of calcium ion.

Anion Analysis

25.72 (a) This ion exists as either HPO_4^{2-} or $H_2PO_4^-$ at low pH.

(b) At low pH, carbonate exists in equilibrium with HCO_3^-, H_2CO_3, and $CO_2(g)$.

(c) At low pH, sulfide exists in equilibrium with HS^- and $H_2S(aq)$, some of which may be lost as $H_2S(g)$.

25.73 (a) The anions of Anion Group 1 are those that form insoluble barium salts:

SO_4^{2-}, CO_3^{2-}, and PO_4^{3-}

(b) $Ba^{2+}(aq) + SO_4^{2-}(aq) \rightarrow BaSO_4(s)$

$Ba^{2+}(aq) + CO_3^{2-} \rightarrow BaCO_3(s)$

$3Ba^{2+} + 2PO_4^{3-} \rightarrow Ba_3(PO_4)_2(s)$

25.74 The barium salts of these anions constitute the precipitate.

(a) If CO_3^{2-} is present as $BaCO_3(s)$, the addition of HNO_3 causes the evolution of $CO_2(g)$:

$$BaCO_3(s) + 2H^+(aq) \rightarrow CO_2(g) + H_2O(\ell) + Ba^{2+}(aq)$$

(b) If PO_4^{3-} is present as $Ba_3(PO_4)_2(s)$, the solid will dissolve, but unless $BaSO_4$ is absent, it will not be easy to detect this dissolving, since any $BaSO_4$ that is present will not dissolve.

$$Ba_3(PO_4)_2(s) + 4H^+(aq) \rightarrow 3Ba^{2+}(aq) + 2H_2PO_4^-(aq)$$

(c) If $BaSO_4(s)$ is present, it is identified and distinguished from the phosphate by the fact that it does not dissolve in acid.

25.75 After redissolving the barium phosphate solid in acid, the phosphate is confirmed by adding ammonium molybdate reagent and observing the formation of a yellow solid, ammonium phosphomolybdate:

$$HPO_4^{2-}(aq) + 12MoO_4^{2-}(aq) + 23H^+(aq) + 3NH_4^+(aq) \rightarrow (NH_4)_3PO_4(MoO_3)_{12}(s) + 12H_2O(\ell)$$

25.76 On adding acid to the solid $BaCO_3$, one observes the formation of a gas that causes a white precipitate to form in $Ca(OH)_2(aq)$ - lime water:

$$Ca^{2+}(aq) + CO_2(g) + H_2O(\ell) \rightarrow CaCO_3(s) + 2H^+(aq)$$

25.77 Since $BaSO_4$ remains insoluble when the carbonate and phosphate are redissolved in acid, the presence of an insoluble solid at this juncture (Procedure A-2) confirms the presence of sulfate.

25.78 See Figure 25.15, as well as Procedures A-5, A-6, and A-7. Chloride is removed as the silver salt, and the usual confirmation test is performed, i.e. dissolving AgCl in ammonia solution. Iodide and bromide are step-wise oxidized with OCl^-, and the colors of I_2 or Br_2 are "developed" in dichloromethane. The sulfide is liberated from its solution with acid, and the presence of $H_2S(g)$ is indicated by lead acetate paper.

25.79 (a) This concerns nitrate ion.

(b) $[FeNO]^{2+}$

(c) $Fe^{2+} + NO(aq) \rightarrow [FeNO]^{2+}(aq)$

(d) First, sulfide is removed:

$Pb^{2+}(aq) + S^{2-}(aq) \rightarrow PbS(s)$

and then halide ions (represented generally as X^-) are removed:

$Ag^+(aq) + X^-(aq) \rightarrow AgX(s)$

25.80 Since HSO_4^- is acidic, its solution ought to turn litmus paper blue.

25.81 Both CO_3^{2-} and HS^- are absent. Any carbonate ion should decompose to $CO_2(g)$ at such a pH. Also, if sulfide were present, the low pH should cause the solution to smell of $H_2S(g)$. Other anions may be present.

25.82 Only the nitrate ion (of those in the present analysis scheme) can be present, since all of the other possible anions are known to form insoluble salts with at least one of these cations.

Additional Questions

25.83 $[Mn(H_2O)_6]^{2+}$ pale pink

$[Ni(H_2O)_6]^{2+}$ green

$[Fe(H_2O)_6]^{2+}$ pale green

$[Fe(H_2O)_6]^{3+}$ pale violet

$[Cu(H_2O)_4]^{2+}$ light blue

$[Fe(H_2O)_5OH]^{2+}$ amber

25.84 (a) white (b) white (c) black (d) white (e) black (f) white (g) yellow (h) white (i) none (j) white (k) dark brown (l) black (m) blue (n) none (o) yellow (p) none (q) none (r) black (s) purple (t) black (u) black (v) white (w) none (x) black (y) yellow (z) yellow

25.85 (a) white (b) violet (c) blue (d) yellow (e) brown (f) brown (g) violet (h) blue-green (i) brick red (j) amber (k) peach (l) white (m) brown (n) white (o) none (p) blood red (q) yellow (r) brown (s) brown-orange (t) none (u) none

25.86 (a) greenish yellow
(b) yellow
(c) red-orange
(d) pale violet

25.87 (a) A flame test would quickly distinguish sodium (yellow) from potassium (violet).
(b) Addition of silver nitrate reagent would cause a precipitation if it was NaCl, whereas $NaNO_3$ would not give a precipitate.
(c) Addition of barium nitrate solution would give a precipitate if it was Na_2SO_4, otherwise no precipitate should form.
(d) Addition of aqueous ammonia should give a blue color if it is copper nitrate, whereas no color is expected if it is zinc nitrate.
(e) Only the nickel solution should give a brick red precipitate on the addition of H_2DMG solution.
(f) The addition of aqueous ammonia should cause a precipitate to form if it is aluminum nitrate, otherwise no precipitate is expected.
(g) On addition of aqueous ammonia, a precipitate should form if it is iron nitrate, whereas a blue solution should form if it is nickel nitrate.
(h) Silver nitrate solution should cause the precipitation of silver chloride from the chloride-containing solution, whereas barium nitrate solution should cause the precipitation of barium sulfate from the sulfate containing solution.
(i) Potassium oxalate solution should cause a precipitate to form from the calcium chloride solution, whereas it would not cause the formation of a precipitate from the sodium chloride solution. Also a flame test could distinguish sodium from calcium.
(j) On the addition of chlorine water along with dichloromethane, the iodide solution should give a purple color, whereas the bromide

solution should develop a brown color.
(k) Any chloride solution, such as HCl, should cause the silver nitrate solution to give a precipitate; otherwise, no precipitate is expected.
(l) On the addition of an acid, the carbonate should effervesce. Otherwise the solution contains phosphate.
(m) On the addition of an acid, the sulfide-containing solution should liberate a foul smelling gas. Otherwise it is the bicarbonate solution, which gives an odorless gas.
(n) The chromate-containing substance should be yellow.
(o) The addition of sodium hydroxide solution should produce the odor of ammonia if the material is the ammonium salt.

25.88 (a) $Pb^{2+}(aq) + CrO_4{}^{2-}(aq) \rightarrow PbCrO_4(s)$

(b) $Hg_2Cl_2(s) + 2NH_3(aq) \rightarrow Hg(\ell) + HgNH_2Cl(s) + NH_4{}^+(aq) + Cl^-(aq)$

(c) $2BiO^+(aq) + 3H_2S(aq) \rightarrow Bi_2S_3(s) + 2H_2O(\ell) + 2H^+(aq)$

(d) $Sn^{4+}(aq) + 2H_2S(aq) \rightarrow SnS_2(s) + 4H^+(aq)$

(e) $2[SnCl_4]^{2-}(aq) + 3HgCl_2(aq) \rightarrow Hg(\ell) + Hg_2Cl_2(s) + 2[SnCl_6]^{2-}(aq)$

(f) $3PbS(s) + 8H^+(aq) + 2NO_3{}^-(aq) \rightarrow 3Pb^{2+}(aq) + 3S(s) + 2NO(g) + 4H_2O(\ell)$

(g) $H_2SO_4(\ell) \rightarrow H_2O + SO_3$

(h) $Sn^{2+}(aq) + 3OH^-(aq) \rightarrow [Sn(OH)_3]^-(aq)$

(i) $AgCl(s) + 2NH_3(aq) \rightarrow [Ag(NH_3)_2]^+(aq) + Cl^-(aq)$

(j) $2Sn^{2+}(aq) + O_2(g) + 4H^+(aq) \rightarrow 2Sn^{4+}(aq) + 2H_2O(\ell)$

(k) $[SnCl_6]^{2-}(aq) + Fe(s) \rightarrow [SnCl_4]^{2-}(aq) + Fe^{2+}(aq) + 2Cl^-(aq)$

(l) $Bi_2S_3(s) + 8H^+(aq) + 2NO_3{}^-(aq) \rightarrow 2Bi^{3+}(aq) + 3S(s) + 2NO(g) + 4H_2O(\ell)$

(m) $PbSO_4(s) + 4C_2H_3O_2{}^-(aq) \rightarrow [Pb(C_2H_3O_2)_4]^{2-}(aq) + SO_4{}^{2-}(aq)$

(n) $BiO^+(aq) + OH^-(aq) \rightarrow BiO(OH)(s)$

(o) $Bi_2S_3(s) + 2H_2O(\ell) + 2H^+(aq) \rightarrow 2BiO^+(aq) + 3H_2S(aq)$

(p) $AgCl(s) + Cl^-(aq) \rightarrow [AgCl_2]^-(aq)$

(q) $PbS(s) + 2H^+(aq) \rightarrow Pb^{2+}(aq) + H_2S(g)$

(r) $Sn^{4+}(aq) + 2H_2S(aq) \rightarrow SnS_2(s) + 4H^+(aq)$

(s) $Bi^{3+}(aq) + H_2O(\ell) \rightarrow BiO^+(aq) + 2H^+(aq)$

25.89 (a) $[SnS_3]^{2-}(aq) + 6H^+(aq) + 6Cl^-(aq) \rightarrow [SnCl_6]^{2-}(aq) + 3H_2S(aq)$

(b) $[SnCl_4]^{2-}(aq) + Hg_2Cl_2(s) \rightarrow 2Hg(\ell) + [SnCl_6]^{2-}(aq)$

(c) $3CuS(s) + 8H^+(aq) + 2NO_3^-(aq) \rightarrow 3Cu^{2+}(aq) + 3S(s) + 2NO(g) + 4H_2O(\ell)$

(d) $[Pb(C_2H_3O_2)_4]^{2-}(aq) + CrO_4^{2-}(aq) \rightarrow PbCrO_4(s) + 4C_2H_3O_2^-(aq)$

(e) $PbCl_2(s) + 2Cl^-(aq) \rightarrow PbCl_4^{2-}(aq)$

(f) $Cu^{2+}(aq) + H_2S(aq) \rightarrow CuS(s) + 2H^+(aq)$

(g) $SnS_2(s) + H_2S(aq) + 2OH^-(aq) \rightarrow [SnS_3]^{2-}(aq) + 2H_2O(\ell)$

(h) $[SnCl_4]^{2-}(aq) + 2HgCl_2(aq) \rightarrow Hg_2Cl_2(s) + [SnCl_6]^{2-}(aq)$

(i) $2BiO(OH)(s) + 3[Sn(OH)_3]^-(aq) + 3OH^-(aq) + 2H_2O \rightarrow$

$2Bi(s) + 3[Sn(OH)_6]^{2-}(aq)$

(j) $Al^{3+}(aq) + 3OH^-(aq) \rightarrow Al(OH)_3(s)$

(k) no reaction

(l) $Zn^{2+}(aq) + 4OH^-(aq) \rightarrow [Zn(OH)_4]^{2-}(aq)$

(m) $Fe^{3+}(aq) + 3NH_3(aq) + 6H_2O(\ell) \rightarrow Fe(OH)_3(H_2O)_3(s) + 3NH_4^+(aq)$

(n) $NH_4^+(aq) + OH^-(aq) \rightarrow NH_3(g) + H_2O(\ell)$

(o) no reaction

(p) no reaction

(q) $Mn^{2+}(aq) + HS^-(aq) + NH_3(aq) \rightarrow MnS(s) + NH_4^+(aq)$

(r) $FeS(s) + 2H^+(aq) \rightarrow Fe^{2+}(aq) + H_2S(aq)$

(s) $Zn^{2+}(aq) + HS^-(aq) + NH_3(aq) \rightarrow ZnS(s) + NH_4^+(aq)$

25.90 (a) $Al^{3+}(aq) + 4OH^-(aq) \rightarrow [Al(OH)_4]^-(aq)$

(b) no reaction

(c) $Ba^{2+}(aq) + CrO_4^{2-}(aq) \rightarrow BaCrO_4(s)$

(d) $HOCl(aq) + Cl^-(aq) + H^+(aq) \rightarrow Cl_2(aq) + H_2O(\ell)$

(e) $Cl_2(aq) + 2I^-(aq) \rightarrow I_2(aq) + 2Cl^-(aq)$

(f) $Fe^{2+}(aq) + HS^-(aq) + NH_3(aq) \rightarrow FeS(s) + NH_4^+(aq)$

(g) $2Fe^{3+}(aq) + 3HS^-(aq) + 3NH_3(aq) \rightarrow 2FeS(s) + S(s) + 3NH_4^+(aq)$

(h) $Al(OH)_3(s) + 3H^+(aq) \rightarrow Al^{3+}(aq) + 3H_2O(\ell)$

(i) $2Mn^{2+}(aq) + 5BiO_3^-(aq) + 14H^+(aq) \rightarrow 2MnO_4^-(aq) + 5Bi^{3+}(aq) + 7H_2O(\ell)$

(j) $Ca^{2+}(aq) + C_2O_4^{2-}(aq) \rightarrow CaC_2O_4(s)$

(k) $HCO_3^-(aq) + H^+(aq) \rightarrow CO_2(g) + H_2O(\ell)$

(l) $Cl_2(aq) + 2Br^-(aq) \rightarrow Br_2(aq) + 2Cl^-(aq)$

(m) $Ni^{2+}(aq) + HS^-(aq) + NH_3(aq) \rightarrow NiS(s) + NH_4^+(aq)$

(n) $2Fe^{3+}(aq) + H_2S(aq) \rightarrow 2Fe^{2+}(aq) + S(s) + 2H^+(aq)$

(o) $MnS(s) + 2H^+(aq) \rightarrow Mn^{2+}(aq) + H_2S(aq)$

(p) $6Cl^-(aq) + 2NO_3^-(aq) + 8H^+(aq) \rightarrow 3Cl_2(g) + 2NO(g) + 4H_2O(\ell)$

(q) $BaCO_3(s) + 2H^+(aq) \rightarrow Ba^{2+}(aq) + CO_2(g) + H_2O(\ell)$

(r) no reaction

(s) no reaction

1 - GYPSUM

1.1 Gypsum is composed of the following proportions of elements:

1 Ca, 1 S, 6 O, 4H

1.2 $2CaSO_4 \cdot 2H_2O \rightarrow (CaSO_4)_2 \cdot H_2O + 3H_2O$

Alternatively we can write:

$CaSO_4 \cdot 2H_2O \rightarrow CaSO_4 \cdot \frac{1}{2}H_2O + \frac{3}{2}H_2O$

1.3 This is the opposite of the one written for question 2 above:

$(CaSO_4)_2 \cdot H_2O + 3H_2O \rightarrow (CaSO_4)_2 \cdot 4H_2O$

1.4 The addition of water to partially hydrated calcium sulfate (plaster of Paris) gives gypsum, which has an expanded structure. In this expanded structure, the crystals have become interlocked so as to make the material hard and rigid.

1.5 Keen's cement is a tough plaster finish that is made from fully dehydrated calcium sulfate.

1.6 This is alabaster.

1.7 Rain water that seeps through a gypsum deposit dissolves some of the calcium sulfate, making the water "hard". The result is that calcium ions form an insoluble compound with most soaps, rendering them less effective.

2 - SODIUM CHLORIDE

2.1 Sodium chloride is obtained mainly from the ocean and from rock salt mines.

2.2 Salt was important as a nutrient, a flavoring agent, and as a means of preserving meats.

2.3 The average amount of NaCl in the ocean is 3.3 % by weight.

2.4 These are chlorine (Cl_2), hydrochloric acid (HCl), and sodium hydroxide (NaOH).

3 - SODIUM HYDROXIDE AND HYDROCHLORIC ACID

3.1 (a) $NaOH + HCl \rightarrow NaCl + H_2O$

(b) $NaHCO_3 + HCl \rightarrow NaCl + CO_2 + H_2O$

(c) $Na_2CO_3 + 2HCl \rightarrow 2NaCl + CO_2 + H_2O$

3.2 $KOH + HCl \rightarrow KCl + H_2O$

3.3 Sodium hydroxide pellets are deliquescent, meaning that they readily absorb atmospheric moisture. Also, pellets and solutions of NaOH alike absorb carbon dioxide from the atmosphere to form $NaHCO_3$.

3.4 Hydrochloric acid is used in pickling iron and other metals, and as a means of passivating lead (by coating it with a film of $PbCl_2$).

Sodium hydroxide is used in the manufacture of glass, soap and other laundry supplies, as well as in the production of pharmaceuticals, and in the baking industry.

4 - SODIUM CARBONATE

4.1 Soda ash is sodium carbonate - Na_2CO_3.

Potash is potassium carbonate - K_2CO_3.

4.2 Sodium carbonate is chiefly used in the manufacture of glass, soaps, detergents, and other cleansers.

4.3 It was found to be useful in making soap.

4.4 $NaCl + H_2SO_4 \rightarrow NaHSO_4 + HCl(g)$

$NaHSO_4 + NaCl \rightarrow Na_2SO_4 + HCl$

$Na_2SO_4 + CaCO_3 + 2C \rightarrow Na_2CO_3 + CaS + 2CO_2$

4.5 Hydrochloric acid is harmful to the environment, contributing to acid rain, etc.

4.6 $NaCl + NH_3 + CO_2 + H_2O \rightarrow NaHCO_3 + NH_4Cl$

$2NaHCO_3 \rightarrow Na_2CO_3 + CO_2 + H_2O$

4.7 (a) $Na_2CO_3{\cdot}H_2O + 2HNO_3 \rightarrow 2NaNO_3 + CO_2 + 2H_2O$

(b)

$$20{,}000 \text{ gal} \times \frac{3.786 \text{ L}}{1 \text{ gal}} \times \frac{1000 \text{ mL}}{1 \text{ L}} \times \frac{101 \text{ g}}{100 \text{ mL}} = 7.65 \times 10^{7} \text{ g } HNO_3$$

$$7.65 \times 10^{7} \text{ g} \times \frac{1 \text{ mol } HNO_3}{63.0 \text{ g}} \times \frac{1 \text{ mol } Na_2CO_3 \cdot H_2O}{2 \text{ mol } HNO_3} = 6.07 \times 10^{5} \text{ mol } Na_2CO_3 \cdot H_2O$$

$$6.07 \times 10^{5} \text{ mol} \times \frac{124 \text{ g}}{\text{mol}} = 7.53 \times 10^{7} \text{ g } Na_2CO_3 \cdot H_2O$$

$$7.53 \times 10^{7} \text{ g} \times \frac{1 \text{ ton}}{9.07 \times 10^{5} \text{ g}} = 83.0 \text{ ton}$$

5 - SILICON

5.1 $SiO_2(s) + 2C(s) \rightarrow Si(s) + 2CO(g)$

5.2 $Si + 2Cl_2 \rightarrow SiCl_4$

$SiCl_4 + 2H_2 \rightarrow Si + 4HCl$

5.3 This is the valence band.

5.4 This is the conduction band.

5.5 The large energy gap between an insulator's filled valence band and its empty conduction band prevents the promotion of electrons into the conduction band, at least at normal temperatures.

5.6 Semiconductors have a relatively small energy gap between their filled valence band and their empty conduction band. Thermal energy is sufficient to cause some population of the conduction bands in these materials.

6 - SILICATE MINERALS AND SOIL

6.1 The bonding in pure silica is of the network-covalent type. In minerals, however, ionic interactions become important.

6.2 The oxygen atoms form a tetrahedron around each silicon atom, in keeping with VSEPR concepts.

6.3 Other ions and molecules such as water are trapped or located in the various interstices of a silicate mineral, as diagramed for instance, in Figure 6c.

6.4 The various ions, molecules, and groups are rearranged and dispersed in numerous ways into igneous rock by weathering processes. One clay that results, for example, is diagramed in Figure 6c. The water is located in particular layers between sheets of silicate and ions.

6.5 The water is located in layers between ion-bearing silicate sheets.

7 - AIR AS A NATURAL RESOURCE

7.1

$$78 \text{ g } N_2 \times \frac{1 \text{ mol}}{28.0 \text{ g}} = 2.8 \text{ mol } N_2$$

$$21 \text{ g } O_2 \times \frac{1 \text{ mol}}{32.0 \text{ g}} = 0.66 \text{ mol } O_2$$

The result is approximately a mol ratio of 2.8/0.66 = 4.2 to one.

7.2 This process is the fixation of nitrogen, which is the source of ammonia for the nitrogen cycle, diagramed in Figure 7b.

7.3 $N_2 + 3H_2 \rightarrow 2NH_3$

$NH_3 + 2O_2 \rightarrow HNO_3 + H_2O$

The nitrogen-containing fertilizers are urea, ammonium nitrate, and ammonia.

7.4 The use of oxygen by the biosphere is in respiration by animals, combustion, and decay.

7.5 Liquid oxygen is a pure form of O_2, whereas only a portion of liquid air is made up of O_2.

8 - GLASS

8.1 A glass is an inorganic material, made by thermal fusion of solids, and cooled, without crystallization, until solid.

8.2 Pure quartz has a very high melting point, making it difficult to melt and to work into a shape.

8.3 A flux is an agent that aids in melting by reducing the melting point and increasing the temperature range of the useful molten state.

8.4 Calcium carbonate serves as a flux in the manufacture of glass (soda-lime-silica glass), and it gives a glass that is less soluble in water than "water glass," which is made using only sodium carbonate.

8.5 These are calcium oxide, CaO, and sodium oxide, Na_2O.

8.6 Borosilicate glass has a lower relative thermal expandability. This makes it less likely to shatter on cooling. It can also be worked, fused, and annealed more effectively.

8.7 This is lead oxide, PbO, which causes the cut product to sparkle nicely.

9 - LIMESTONE, CHEMICAL EROSION, AND LIMESTONE CAVERNS

9.1 Limestone is principally calcium carbonate, $CaCO_3$. Dolomite is $MgCa(CO_3)_2$, in which roughly half of the calcium ions of limestone are replaced by magnesium ions.

9.2 (a) $CO_2 + H_2O \rightleftharpoons H_2CO_3$

(b) $H_2CO_3 \rightleftharpoons H^+ + HCO_3^-$

(c) $CaCO_3 + H^+ \rightleftharpoons Ca^{2+} + HCO_3^-$

(d) $CO_2(g) \rightleftharpoons CO_2(aq)$

9.3 $CaCO_3 + 2H^+ \rightarrow Ca^{2+} + CO_2 + H_2O$

9.4 Water that drips from ceiling cracks loses some of its dissolved carbon dioxide. This shifts equilibrium (2) to the left in order to restore this carbon dioxide to the water:

$H_2CO_3 \rightarrow CO_2 + H_2O$

Next, equilibrium (3) shifts to the left to replenish carbonic acid:

$HCO_3^- + H^+ \rightarrow H_2CO_3$

Finally, equilibrium (4) shifts to the left to replenish the H^+ in the drop that emerges from the ceiling crack:

$HCO_3^- + Ca^{2+} \rightarrow H^+ + CaCO_3(s)$

The result is the formation of solid calcium carbonate, i.e. a stalactite.

10 - PHOTOGRAPHY

10.1 These are silver chloride (AgCl), silver bromide (AgBr), and silver iodide (AgI).

10.2 $Ag^+ + Cl^- \rightarrow AgCl$

10.3 The developer is the reducing agent that transforms Ag^+ into Ag(s) in those regions of the film that have accumulated Ag from the exposure process, i.e. where the latent image is found.

10.4 $AgBr(s) + 2S_2O_3^{2-}(aq) \rightarrow Ag(S_2O_3)_2^{3-}(aq) + Br^-(aq)$

10.5 The three colors are red, green, and blue.

11 - AMMONIA

11.1 Natural gas, CH_4, is reacted with steam, over a nickel catalyst, to generate hydrogen:

$CH_4 + H_2O \rightleftharpoons CO + 3H_2$

$CH_4 + 2H_2O \rightleftharpoons CO_2 + 4H_2$

The mixture of CO_2, CO, and H_2 serves, then, as the source of the hydrogen for the Haber-Bosch process.

11.2 This is transformed into CO, and eventually CO_2, which is then precipitated as potassium bicarbonate:

$$CO_2 + H_2O + K_2CO_3 \rightarrow 2KHCO_3$$

11.3 Oxygen in air is reduced by hydrogen to give water.

11.4 A high temperature decreases the yield of ammonia because the equilibrium that produces ammonia from nitrogen and hydrogen is exothermic.

11.5 High pressure favors the right side of the equilibrium:

$$N_2 + 3H_2 \rightleftharpoons 2NH_3$$

11.6 Carbon monoxide acts as a poison to the final catalyst in the process.

12 - ION SELECTIVE ELECTRODES

12.1 This is the Nernst equation.

12.2 Glass electrodes may be used to measure H^+, Li^+, Na^+, K^+, Ag^+, and NH_4^+, i.e. singly charged cations.

12.3 This is shown in Figure 12b. It contains a membrane that separates an aqueous reference solution from an organic liquid.

12.4 A silicone membrane separates blood to be measured from a solution of $NaHCO_3$. The membrane passes only CO_2. The transfer of CO_2 from the blood, across the membrane and into the $NaHCO_3$ solution, causes the following equilibrium to shift to the right, making more HCO_3^-:

$$CO_2 + H_2O \rightleftharpoons HCO_3^- + H^+$$

This causes a measurable change in pH, which is detected by a standard glass membrane.

12.5 An enzyme is held in a rigid but permeable gel at the end of a normal glass membrane electrode. The substrate is able to permeate the gel, and hydrolysis results. This causes a change in pH that can be measured by the glass membrane electrode. Figure 12d shows such an electrode made sensitive to urea by packing the gel with the enzyme urease.

13 - SOME ATMOSPHERIC CHEMISTRY

13.1 $O_2 \rightarrow 2O$

13.2 Two chain processes are initiated because the net result is the formation of two oxygen atoms, each of which enters the cycle:

$O + O_2 + M \rightarrow O_3 + M$

$O_3 \rightarrow O_2 + O^*$

$O_2 + O^* + M \rightarrow O_3 + M$

13.3 Skin cancer is caused in humans, and photosynthesis in plants is reduced.

13.4 N_2O reacts with the excited oxygen atoms, giving NO:

$O^* + N_2O \rightarrow 2NO$

The NO then participates in the conversion of O_3 to O_2:

$NO + O_3 \rightarrow NO_2 + O_2$

$NO_2 + O \rightarrow NO + O_2$

$O_3 + O \rightarrow 2O_2$ net reaction

13.5 This catalyzes the conversion of ozone to oxygen, as shown in the answer to question 13.4 above.

13.6 These are chlorofluorocarbons, which are gases that are employed heavily in industry as solvents, gas propellants, and refrigerants.

13.7 First, chlorine atoms are obtained when CFC-11 absorbs UV energy:

$CCl_3F \rightarrow CCl_2F + Cl$

The chlorine atoms react directly with ozone:

$Cl + O_3 \rightarrow ClO + O_2$

$ClO + O \rightarrow Cl + O_2$

$O_3 + O \rightarrow 2O_2$ net reaction

14 - INORGANIC POLYMERS

14.1 Organic polymers suffer from the tendency to:
(a) become brittle at low temperature.
(b) deteriorate when hot.
(c) swell in organic solvents.
(d) burn readily, giving toxic gases in some cases.
(e) be of limited use in medicine.

14.2

```
      R           R           R
      |           |           |
 — Si — O — Si — O — Si — O   etc.
      |           |           |
      R           R           R
```

14.3 The $—Si(CH_3)_3$ group terminates a chain, as shown in Figure 14a. The chain would branch at a site where the CH_3SiCl_3 group was used.

14.4

```
          X         X         X
          |         |         |
 — N — P = N — P = N — P =   etc.
          |         |         |
          X         X         X
```

14.5 These crystals conduct electricity along the chain length, as if the material were metallic. Thus it is an electrically conducting polymer. They are also superconductors at low temperatures.

15 - IRON AND STEEL

15.1 (a) hematite - Fe_2O_3 (b) magnetite - Fe_3O_4 (c) limonite - $2Fe_2O_3 \cdot 3H_2O$

15.2 The reducing agent (CO) is generated from coke (carbon):

$$Fe_3O_4 + 4CO \rightarrow 3Fe + 4CO_2$$

15.3 The reaction is:

$$Fe_2O_3 + 3CO \rightarrow 2Fe + 3CO_2$$

Therefore 2/3 mol of Fe arise from each mol of C. Since 1 ton is equal to 2000 lb, the number of moles of C is:

$$2000 \text{ lb} \times \frac{453.6 \text{ g}}{1 \text{ lb}} \times \frac{1 \text{ mol}}{12.0 \text{ g}} = 7.56 \times 10^4 \text{ mol C}$$

The number of moles of iron is:

$$7.56 \times 10^4 \text{ mol C} \times \frac{2/3 \text{ mol Fe}}{1 \text{ mol C}} = 5.04 \times 10^4 \text{ mol Fe}$$

The mass of iron is:

$$5.04 \times 10^4 \text{ mol Fe} \times \frac{55.8 \text{ g}}{\text{mol}} \times \frac{1 \text{ lb}}{453.6 \text{ g}} \times \frac{1 \text{ ton}}{2000 \text{ lb}} = 3.10 \text{ ton}$$

15.4 The three classes of steel alloy are carbon steels, low-alloy steels, and high-alloy steels.

(a) Tungsten imparts strength at high temperatures.
(b) Manganese makes the steel impact-resistant.
(c) Nickel imparts resistance to pulling stresses.
(d) Chromium is employed in stainless steels.

Nickel as well as chromium is found in stainless steels of one variety or another.

15.5

$$65.0 \text{ g Fe} \times \frac{1 \text{ mol}}{55.8 \text{ g}} = 1.16 \text{ mol Fe}$$

$$1.16 \text{ mol Fe} \times \frac{1 \text{ mol } Fe_2O_3}{2 \text{ mol Fe}} = 0.58 \text{ mol } Fe_2O_3$$

$$0.58 \text{ mol } Fe_2O_3 \times \frac{160 \text{ g}}{\text{mol}} = 92.8 \text{ g } Fe_2O_3$$

In other words, such an ore is 92.8 % hematite.

16 - ETHYL ALCOHOL

16.1 These are ethyl alcohol (C_2H_5OH) and carbon dioxide (CO_2).

16.2 When the alcohol concentration reaches 14 %, the enzymes that drive the fermentation process are inactivated.

16.3 110-Proof alcohol is 50 % (v/v).

16.4 (a)

$$1.00 \times 10^3 \text{ g} \times 0.20 = 200 \text{ g glucose}$$

$$200 \text{ g } C_6H_{12}O_6 \times \frac{1 \text{ mol}}{180 \text{ g}} = 1.11 \text{ mol } C_6H_{12}O_6$$

$$1.11 \text{ mol } C_6H_{12}O_6 \times \frac{2 \text{ mol } C_2H_5OH}{1 \text{ mol } C_6H_{12}O_6} = 2.22 \text{ mol } C_2H_5OH$$

$$2.22 \text{ mol } C_2H_5OH \times \frac{46 \text{ g}}{\text{mol}} = 102 \text{ g } C_2H_5OH$$

(b) The total mass of the product solution is less than the starting mass, due to the loss of carbon dioxide:

final mass = 1000 - (200 - 102) = 902 g

$$\% \ C_2H_5OH = \frac{102 \text{ g}}{902 \text{ g}} \times 100 = 11.3 \% \text{ (w/w)}$$

(c)

$$102 \text{ g} \times \frac{1 \text{ mL}}{0.789 \text{ g}} = 129 \text{ mL}$$

(d) The volume of the solution is:

$$902 \text{ g} \times \frac{1 \text{ mL}}{0.982 \text{ g}} = 919 \text{ mL}$$

$$\% \ C_2H_5OH = \frac{129}{919} \times 100 = 14.0 \% \text{ (v/v)}$$

This is 2 × 14.0 = 28.0 proof.

66711

NOTES

NOTES

NOTES

NOTES

NOTES

NOTES

NOTES

NOTES

NOTES

NOTES

NOTES

NOTES

NOTES